This book deals with issues of fluid flow and solute transport in complex geologic environments under uncertainty. The resolution of such issues is important for the rational management of water resources, the preservation of subsurface water quality, the optimization of irrigation and drainage efficiency, the safe and economic extraction of subsurface mineral and energy resources, and the subsurface storage of energy and wastes. Over the last two decades, it has become common to describe the spatial variability of geologic medium flow and transport properties using methods of spatial (or geo-) statistics. According to the geostatistical philosophy, these properties constitute spatially correlated random fields. As medium properties are random, the equations that govern subsurface flow and transport are stochastic.

This volume describes the most recent advances in stochastic modeling. It takes stock of mathematical and computational solutions obtained for stochastic subsurface flow and transport equations, and their application to experimental field data, over the last two decades. The book also attempts to identify corresponding future research needs. This volume is based on the second Kovacs Colloquium organised by the International Hydrological Programme (UNESCO) and the International Association of Hydrological sciences. Fifteen leading scientists with international reputations review the latest developments in this area of hydrological research.

The book is a valuable reference work for graduate students, research workers and professionals in government and public institutions, interested in hydrology, environmental issues, soil physics, petroleum engineering, geological engineering and applied mathematics.

Subsurface Flow and Transport:
A Stochastic Approach

INTERNATIONAL HYDROLOGY SERIES

The **International Hydrological Programme** (IHP) was established by the United Nations Educational, Scientific and Cultural Organisation (UNESCO) in 1975 as the successor to the International Hydrological Decade. The long-term goal of the IHP is to advance our understanding of processes occurring in the water cycle and to integrate this knowledge into water resources management. The IHP is the only UN science and educational programme in the field of water resources, and one of its outputs has been a steady stream of technical and information documents aimed at water specialists and decision-makers.

The **International Hydrology Series** has been developed by the IHP in collaboration with Cambridge University Press as a major collection of research monographs, synthesis volumes and graduate texts on the subject of water. Authoritative and international in scope, the various books within the Series all contribute to the aims of the IHP in improving scientific and technical knowledge of fresh water processes, in providing research know-how and in stimulating the responsible management of water resources.

TITLES IN PRINT IN THE SERIES:

M. Bonell, M. M. Hufschmidt and J. S. Gladwell *Hydrology and Water Management in the Humid Tropics: Hydrological Research Issues and Strategies for Water Management*
Z. W. Kundzewicz *New Uncertainty Concepts in Hydrology*
R. A. Feddes *Space and Time Scale Variability and Interdependencies in the Various Hydrological Processes*
J. Gibert, J. Mathieu and F. Fournier *Groundwater and Surface Water Ecotones: Biological and Hydrological Interactions and Management Options*
G. Dagan and S. Neuman *Subsurface Flow and Transport: A Stochastic Approach*

INTERNATIONAL HYDROLOGY SERIES

Subsurface Flow and Transport: A Stochastic Approach

Edited by

Gedeon Dagan *Tel Aviv University*
Shlomo P. Neuman *University of Arizona*

UNESCO
IAHS AISH
CAMBRIDGE UNIVERSITY PRESS

CAMBRIDGE UNIVERSITY PRESS
Cambridge, New York, Melbourne, Madrid, Cape Town, Singapore, São Paulo

Cambridge University Press
The Edinburgh Building, Cambridge CB2 2RU, UK

Published in the United States of America by Cambridge University Press, New York

www.cambridge.org
Information on this title: www.cambridge.org/9780521572576

First published 1997
This digitally printed first paperback version 2005

A catalogue record for this publication is available from the British Library

Library of Congress Cataloguing in Publication data

Subsurface flow and transport : a stochastic approach / Gedeon Dagan,
Shlomo P. Neuman [editors].
p. cm. – (International hydrology series)
ISBN 0 521 57257 6 (hardbound)
1. Groundwater flow. I. Dagan, G. (Gedeon), 1932– .
II. Neuman, S. P. III. Series.
GB1197.7.S825 1997
551.49–dc21 96–37796 CIP

ISBN-13 978-0-521-57257-6 hardback
ISBN-10 0-521-57257-6 hardback

ISBN-13 978-0-521-02009-1 paperback
ISBN-10 0-521-02009-3 paperback

Contents

Contributors

PROF. MARY P. ANDERSON
Department of Geology and Geophysics, University of Wisconsin-Madison, 1215 West Drayton Street, Madison, WI 53706, USA

CARL AXNESS
Sandia National Laboratories, Albuquerque, New Mexico 87185-1328, USA

PROF. JESÚS CARRERA
Departmento de Ingeniería del Terreno y Cartográfica, Escuela Técnica Superior de Ingenieros de Caminos, Canales y Puertos, Universitat Politècnica de Catalunya, 08034 Barcelona, Spain

PROF. JOHN H. CUSHMAN
Center for Applied Mathematics, Math Sciences Building, Purdue University, West Lafayette, IN 47907, USA

PROF. VLADIMIR CVETKOVIC
Department of Water Resources Engineering, Royal Institute of Technology, S-10044 Stockholm, Sweden

PROF. GEDEON DAGAN
Faculty of Engineering, Department of Fluid Mechanics and Heat Transfer, Tel Aviv University, Ramat Aviv, Tel Aviv, 69978 Israel

AKHIL DATTA-GUPTA
Lawrence Berkeley National Laboratory, 1 Cyclotron Road, Berkeley, CA 94720, USA *and* Department of Petroleum Engineering, Texas A&M University, College Station, TX 77843, USA

FEI-WEN DENG
Center for Applied Mathematics, Math Sciences Building, Purdue University, West Lafayette, IN 47907, USA

CHRISTINE DOUGHTY
Lawrence Berkeley National Laboratory, 1 Cyclotron Road, Berkeley, CA 94720, USA

PROF. RICHARD E. EWING
Institute for Scientific Computation, Texas A&M University, 236 Teague Research Center, College Station, TX 77843-3404, USA

PROF. LYNN W. GELHAR
Room 48-237, Department of Civil Engineering, Massachusetts Institute of Technology, Cambridge, MA 02139, USA

PROF. STEVEN M. GORELICK
Department of Geological and Environmental Sciences, Stanford University, Stanford, CA 94305-2115, USA

KEVIN HESTIR
Lawrence Berkeley National Laboratory, 1 Cyclotron Road, Berkeley, CA 94720, USA *and* Department of Mathematics, Utah State University

BILL X. HU
Center for Applied Mathematics, Math Sciences Building, Purdue University, West Lafayette, IN 47907, USA

PROF. PETER K. KITANIDIS
Civil Engineering Department, Stanford University, Stanford, CA 94305-4020, USA

DR JANE C. S. LONG
Lawrence Berkeley National Laboratory, 1 Cyclotron Road, Berkeley, CA 94720, USA

AGUSTÍN MEDINA
Deparamento de Ingeniería del Terreno y Cartográfica, Escuela Técnica Superior de Ingenieros de Caminos, Canales y Puertos, Universitat Politècnica de Catalunya, 08034 Barcelona, Spain

PROF. SHLOMO P. NEUMAN
Department of Hydrology and Water Resources, The University of Arizona, Tucson, AZ 85721, USA

PROF. JACK C. PARKER
Environmental Systems & Technologies, Inc., Blacksburg, VA 24070-6326, USA

PROF. YORAM RUBIN
Department of Civil Engineering, 435 Davis Hall, University of California, Berkeley, CA 94720, USA

DR DAVID RUSSO
Department of Soil Physics, Institute of Soils and Water, Agricultural Research Organization, The Volcani Center, PO Box 6, P.A., Bet Dagan 50-250, Israel

PROF. F. JAVIER SAMPER CALVETE
Escuela Técnica Superior de Ingenieros de Caminos, Canales y Puettos, Universidad de La Coruña, Campus de Elviña, 15192 La Coruña, Spain

DON VASCO
Lawrence Berkeley National Laboratory, 1 Cyclotron Road, Berkeley, CA 94720, USA

D. A. ZIMMERMAN
Gram, Inc., 8500 Menoul Boulevard, Albuquerque, New Mexico, USA

Preface

This book contains the refereed and edited proceedings of the Second IHP/IAHS George Kovacs Colloquium on Subsurface Flow and Transport: The Stochastic Approach, held in Paris, France, during January 26–30, 1995. The Colloquium was convened by Professors Gedeon Dagan and Shlomo P. Neuman under the auspices of UNESCO's Division of Water Sciences as part of its International Hydrological Programme (IHP), and the International Association of Hydrological Sciences (IAHS).

The book is devoted to issues of fluid flow and solute transport in complex geologic environments under uncertainty. The resolution of such issues is important for the rational management of water resources, the preservation of subsurface quality, the optimization of irrigation and drainage efficiency, the safe and economic extraction of subsurface mineral and energy resources, and the subsurface storage of energy and wastes. Over the last two decades, it has become common to describe the spatial variability of geologic medium flow and transport properties using methods of statistical continuum theory (or geostatistics). According to the geostatistical philosophy, these properties constitute spatially correlated random fields. As medium properties are random, the equations that govern subsurface flow and transport are stochastic. This book takes stock of mathematical and computational solutions obtained for stochastic subsurface flow and transport equations, and their application to experimental field data over the last two decades. The book also attempts to identify corresponding future research needs.

The book contains invited articles on selected topics by 15 leading experts in the emerging field of stochastic subsurface hydrology. All 15 authors have made seminal contributions to this field during its early formative years. The book opens with a broad retrospective on stochastic modeling of subsurface fluid flow and solute transport by G. Dagan. It then proceeds with three papers devoted to the characterization and estimation of subsurface medium properties that control flow and transport. The paper by M.P. Anderson emphasizes geological considerations in the characterization of subsurface heterogeneity that by J. Samper describes methods of geostatistical inference while J. Carrera addresses practical and theoretical aspects of parameter estimation by inversion (the so-called inverse problem). Flow modeling and aquifer management are discussed in three articles by P K. Kitanidis, R.E. Ewing, and S.M. Gorelick. The first of these three articles concerns computer modeling of flow in randomly heterogeneous porous media; the second surveys and assesses the state of the art in numerical simulation of multiphase flows in such media; and the third shows how to incorporate uncertainty into computer models of aquifer management. Four articles are devoted to solute transport in randomly hetrogeneous porous media. Y. Rubin presents an overview of purely adjective transport; V. Cvetkovic extends the treatment to reactive solutes; J.H. Cushman highlights nonlocal effects on transport; and L.W. Gelhar explains how stochastic transport theories have been used in the interpretation of field-scale tracer tests. The difficult topic of flow and transport in fractured rocks is tackled in a specialty paper by J.C. S. Long. It is followed by two papers on multiphase phenomena: one by D. Russo on stochastic analysis of transport in partially saturated heterogeneous soils, and the other by J.C. Parker on field-scale modeling of multiphase flow and transport. The book closes with a view to the future by S.P. Neuman.

Gedeon Dagan, Faculty of Engineering, Tel Aviv University, Ramat Aviv, Tel Aviv, Israel.
Shlomo P. Neuman, Department of Hydrology and Water Resources, The University of Arizona, Tucson, Arizona 85721, USA.

Acknowledgments

The editors want to thank all those who have contributed to the success of the Second IHP/IAHS George Kovacs Colloquium and this book. We thank the sponsoring organizations and their dedicated officers, especially Dr Andrasz Szollossi-Nagy, Director of UNESCO's Division of Water Sciences, and Dr Uri Shamir, President of IAHS, whose support and active help were instrumental in bringing about the Colloquium and publishing this book. We are grateful to UNESCO's Division of Water Sciences staff, and particularly to Dr Alicia Aureli and M. Bonnell, who ensured the success of the organization of the meeting and of publishing the book. We are most grateful to the authors for accepting our invitation to share their expertise and erudition with the participants of the Colloquium and the readers of this book. The person who worked hardest on the final editing of this book, and deserves kudos for its professional appearance, is Ms Bette Lewis; we acknowledge with gratitude her dedication to the task.

I

Introduction

1 Stochastic modeling of flow and transport: the broad perspective

GEDEON DAGAN
Tel Aviv University

1 INTRODUCTION

Stochastic modeling of subsurface (unsaturated and saturated zones) flow and transport has become a subject of wide interest and intensive research in the last two decades. In principle, this approach recognizes that hydrological variables are affected by uncertainty and regards them as random. This randomness leads to defining models of flow and transport in a stochastic context, and predictions are made in terms of probabilities rather than in the traditional deterministic framework.

This approach is not new and was adopted by many disciplines in physics and engineering a long time ago. The closest field is of course that of surface hydrology, which relies traditionally on time series analysis in predicting floods and other extreme events. However, subsurface modeling deals mainly with spatial variability, the uncertainty of which is of a more complex nature. Besides, the physics of the phenomena is accounted for through the differential equations of flow and transport and various constitutive equations. These equations, regarded as stochastic, are intended to provide a general theoretical framework, rather than particular, empirical, statistical procedures. In this respect the subject is closer in outlook and methodology to the advanced statistical theories of continuum mechanics and of solid state physics.

It is beyond the scope of this presentation to attempt to review the various applications of stochastic modeling or even to try to cover the specific area of subsurface flow that was discussed in depth in the Colloquium. Instead, we shall try to discuss a few issues of principle, which are probably analyzed to a lesser extent in the following chapters.

Rather than a general discussion, the development will start with a representative example of flow and transport and will try to touch a few issues faced by the modeler, from the formulation of the problem to its numerical solution. Since I do not attempt to carry out a systematical review, there will be only a few references to the works directly quoted in the exposition. However, the presentation draws from the large body of knowledge accumulated during the years, and I apologize for not giving full credit here to all those who have contributed to the development of the subject. I am confident these contributions will be amply discussed in the following sections.

The plan of the presentation is as follows. In Section 2 we start with an example of a problem of flow and transport in a natural formation and devote the rest of this section to the selection of a conceptual model and a stochastic framework to solve the problem by a numerical procedure. In Section 3 we discuss the solution of the flow problem, which leads to deriving the velocity field, and in Section 4 we examine the solution of the transport problem, which relies on the previous steps. We devote special attention to two fundamental concepts, namely that of effective hydraulic properties and of macrodispersion, in Sections 3 and 4, respectively.

2 SELECTION OF THE MODEL

2.1 Representative examples of subsurface flow and transport problems

A common problem encountered these days all over the world is that of groundwater pollution. For instance, a contaminant source, on or near the ground surface, creates a plume that reaches groundwater which spreads further due to natural or forced water flow. Modeling of the phenomenon is needed in order to predict the development of the plume, though there are cases in which tracing back the history of the plume is the salient question. The first step toward solving such a problem is to formulate a hydrogeological conceptual model, involving boundaries of the formation and its significant geological features. The first stage would generally be carried out by incorporating all

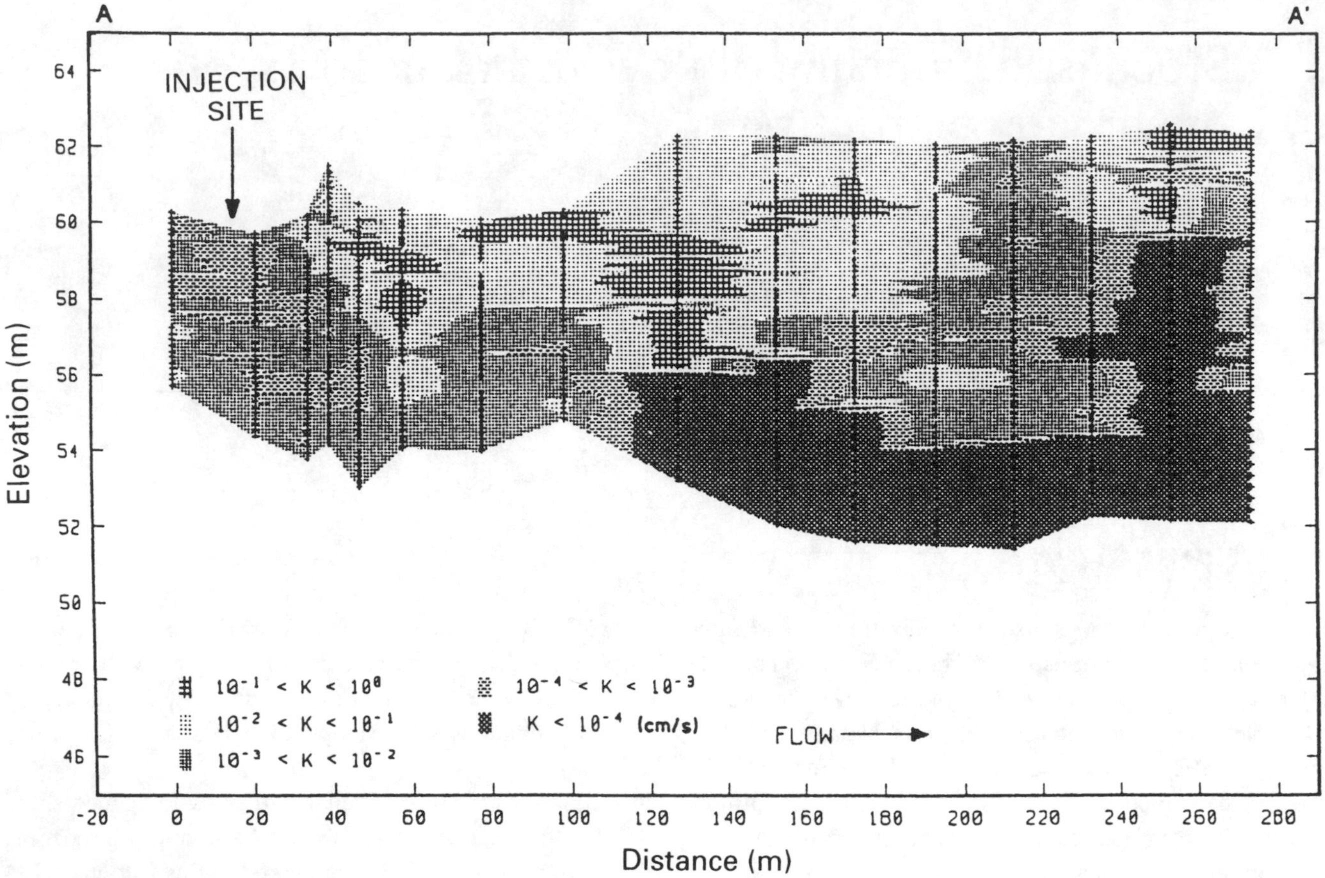

Fig. 1 Hydraulic conductivity spatial distribution in a vertical cross-section at the transport field site at Columbus Air Force Base (from Adams & Gelhar, 1992).

the information available from geological, geophysical, geochemical and hydrological field investigations. The striking feature of these findings is that natural formations are generally heterogeneous, with hydraulic properties varying widely and in an irregular manner over various scales. To illustrate this point, we reproduce in Figs. 1 and 2, taken from the literature, cross-sections through two aquifers.

Fig. 1 is a vertical cross-section of the aquifer in which a recent tracer test has been conducted at the Columbus Air Force Base in the USA (Adams & Gelhar, 1992). The figure represents the distribution of the hydraulic conductivity by interpolation of the dense measurement network of wells. Fig. 2 provides two cross-sections through an alluvial deposit in the Netherlands (Bierkens, 1994), with different lithologic units that were identified from geological data.

Both figures illustrate the complex subsurface structures one usually encounters in applications. To fix our ideas, let us assume that for the formation of Fig. 2, a contaminant source originating from a pond is present on the surface in a certain area (see Fig. 2), and that the problem is to predict the long range change of the contaminant concentration in groundwater.

Since in the case of Fig. 1 an actual field experiment, though of a relatively short duration, has been conducted, we reproduce in Fig. 3 cross-sections of the plume of an inert solute after a few time intervals from the injection of the tracer. This figure illustrates the complex character of the transport phenomenon related to the spatially variable structure. It suggests that a detailed, deterministic prediction of the solute concentration distribution in space and time is impossible.

Returning to the hypothetical case of Fig. 2, the objective of a modeling effort may be two-fold: a qualitative, scientific one and a quantitative, predictive mode. In the first case, the objective is to grasp the main mechanisms involved in the flow and transport processes toward their better understanding. In the second mode, the model has to provide estimates of the solute concentration that can be used for engineering and management purposes. While proceeding toward answering the second task, we shall raise a few topics of principle along the road.

At this stage it is already of importance to define the problem and to specify the aim of the modeling effort. Thus, it is relevant to know whether it is the *maximum* or the *average* local concentration that is sought, in pumping wells

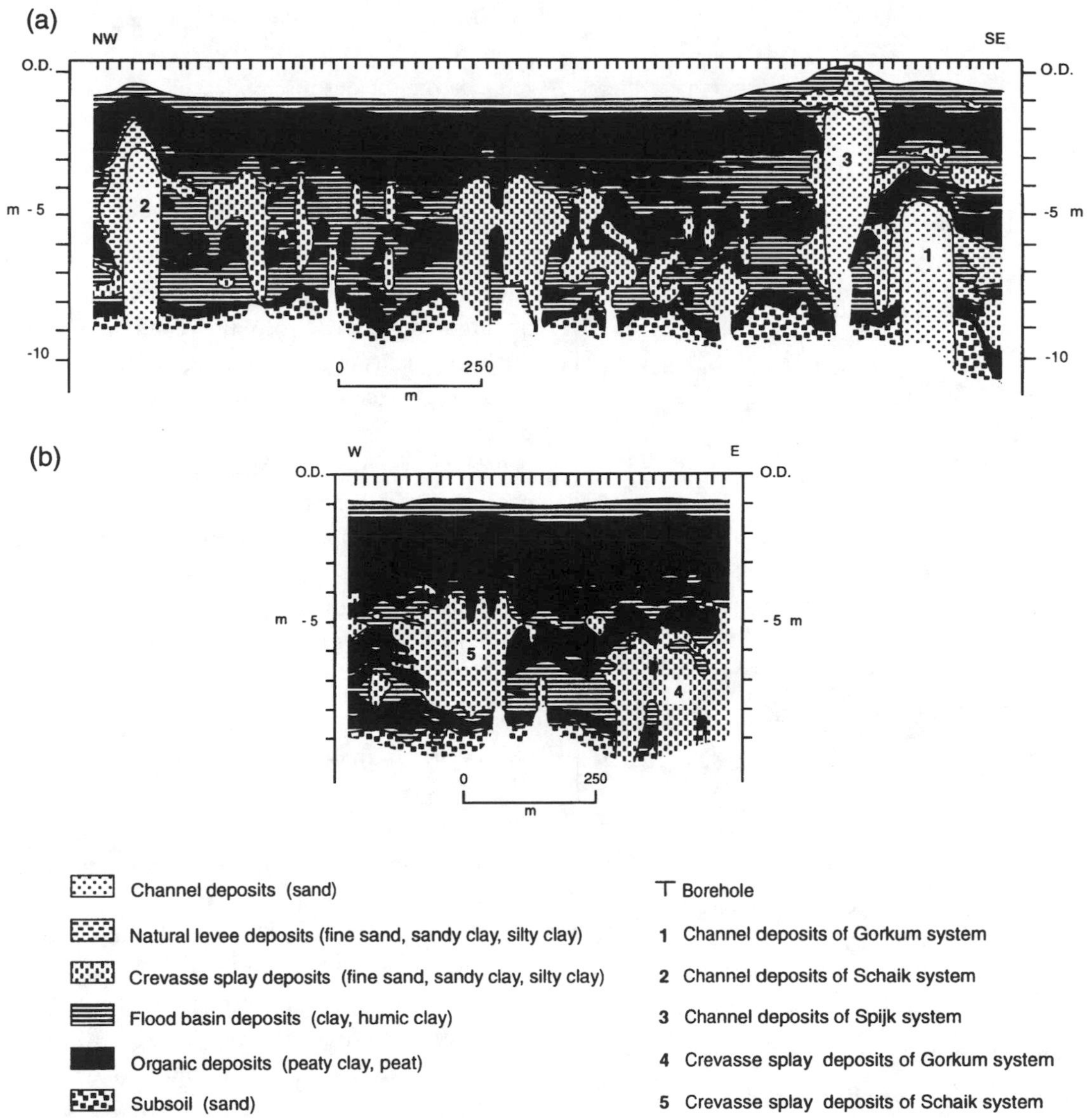

Fig. 2 Two cross-sections through a conductive fluvial deposit in the west-central district of the Netherlands (from Bierkens, 1994).

or in a control-plane boundary of the aquifer. Is it the long range plume behavior, say at a distance of a few kilometers from the source, which is of interest, or the local behavior close to the injection site? As we shall see, each of these questions has a profound impact upon modeling and upon the accuracy of predictions. This remark anticipates similar observations that indicate that our models are problem oriented and very much related to specific applications which dictate the choice of tools we are going to employ.

2.2 Selection of the geohydrological setup

The first step toward modeling flow and transport consists in defining a geohydrological setup of the formation. It is assumed that most of the contributions to the Colloquium will proceed by assuming that this stage is already accomplished, and I feel, therefore, that it is worthwhile to discuss it briefly.

The selection of the geohydrological setup consists of making a few decisions about, e.g.,

(i) the boundaries of the formation domain to be modeled;
(ii) the geohydrological distinct units within the domain;
(iii) the areas of recharge from the surface and rates of infiltration; and
(iv) the additional relevant boundary conditions, such as given heads or fluxes, location of outlets of the formation, seasonal variations and their influence, long range time trends, pumping wells, existing or to be developed, and their discharges.

Such a selection has preceded, for instance, the definition of the structure shown in Fig. 2. Generally, the process leading to the selection of the geohydrological setup involves incorporation of information from various sources, e.g. geological, hydrological, meteorological, geochemical, agricultural,

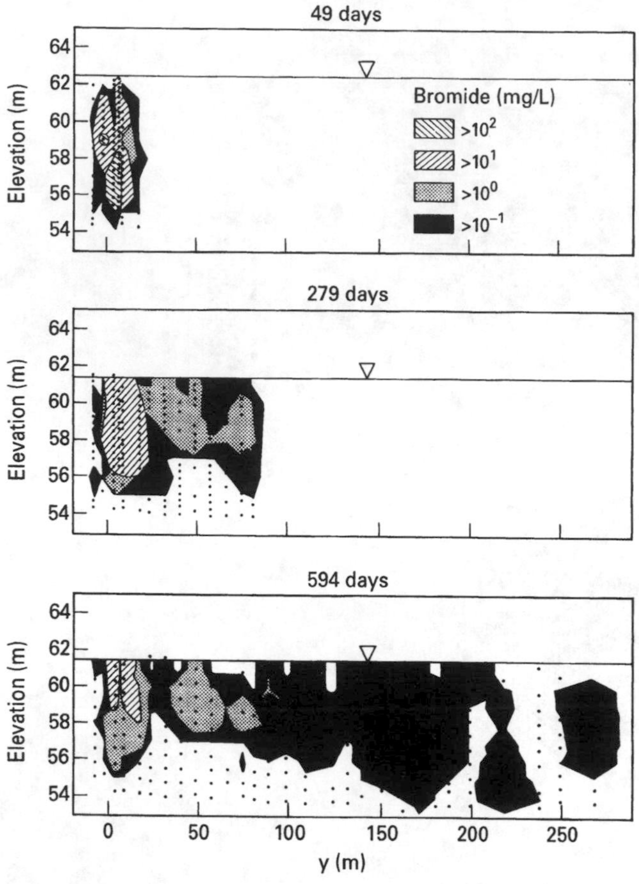

Fig. 3 Vertical cross-sections along the longitudinal axis of the bromide plume in the field study at Columbus Air Force Base (from Boggs *et al.*, 1992).

water resources authorities, etc. This process is highly dependent on the aim of the modeling effort. Thus, if one is interested in local pollution, over a relatively short distance from the source, say of the order of tens of aquifer depths, the local structure of the formation is of interest. In that case, the far boundaries are less relevant if enough information is available about the local flow conditions (heads, conductivities, soil characteristics, etc). In contrast, the large scale features and all the aforementioned aspects play a role in cases of long range pollution over an extended period of time and over large distances from the source.

Additional assumptions have to be made about the nature of the contaminant source, which is not defined in a clear-cut manner in many applications. Thus, if the pollutant escapes through the walls of a container and no precise measurements are available, we have to specify the rate of leakage as well as the effluent concentration. We may also have to make assumptions about the chemical changes of the solute over time due to reactions taking place at the source itself, before it enters the subsurface structure.

Many of these initial choices about the geohydrological and environmental setup are based on qualitative information and 'soft data', which are difficult if not impossible to quantify, and they require understanding and reliance on various disciplines, as well as experience, intuition and sound judgment.

2.3 Selection of the conceptual model

After having decided upon one or a few geohydrological and environmental scenarios, our next step is to formulate a quantitative model to solve the flow and transport problem. Such a model is usually expressed by a set of balance partial differential equations and constitutive equations that depend on parameters which are generally space and time dependent. However, we have various options at this stage, and our choices have a considerable impact upon the tools we may use, upon the computational effort and upon the accuracy of solutions.

To illustrate the point, listed below are a few alternative conceptual models we have to contemplate when solving the flow problem:

(i) Are we going to model the heterogeneous structure (permeability, porosity) of the aquifer as a three-dimensional one or rather as a planar, depth averaged, two-dimensional one? This choice is dictated mainly by whether contamination is local or long-range (over tens to hundreds of depths).

(ii) Similarly, can the flow in the unsaturated zone be approximated by a vertical one, or do we have to model it as fully three-dimensional? This choice depends to a large extent upon the ratio between the source size and the thickness of the unsaturated zone.

(iii) One of our main concerns is the effect of heterogeneity, as illustrated in Figs. 1 and 2, upon transport. Here we have to decide on how to separate the formation in a number of distinct subunits, each of them having a narrower range of variability, as compared with the extreme case of modeling the saturated and unsaturated zones as one unit each, and how to incorporate the variability of the subunits in a broad heterogeneous distribution.

(iv) Is flow regarded as unsteady, or can we approximate it as steady due to the small seasonal variations of the natural gradient? Is a similar approximation valid for pumping wells?

(v) If the aquifer is phreatic, can we linearize the free-surface motion around a constant saturated thickness, or do we need to account for the change of thickness?

(iv) If the formation is constituted from fractured rock, are we going to model the fractures as discrete or should we use an equivalent continuum approach?

Similarly, when selecting the appropriate framework to treat transport we face further choices, such as:

(i) Can we model the solute as conservative, or do we have to account for reactive properties?
(ii) Can we regard the solute as inert, or do we need to consider its influence upon the density and viscosity of the solution?
(iii) Are flow and transport immiscible ?
(iv) Do we have to account for pore-scale dispersion, or are we going to be satisfied with modeling the large scale mixing associated with heterogeneity?
(v) Are we going to seek local concentrations, and on what scales, or are we going to be satisfied with some averages over the depth or control planes?

Although we could expand this list, the variety of choices we have enumerated so far illustrates the point we want to make: before applying any quantitative model, we have to make a series of decisions. We shall discuss later, in Section 2.6, the significance of this stage to the modeling process.

2.4 Stochastic modeling

After the completion of the previous steps, one, or a few combinations of a, well defined geometry and the types of boundary conditions and processes are selected. The objective of the model, i.e. the solution for the concentration, is also defined. The next step comprises the selection of the values of different parameters characterizing the system. Such parameters can be conveniently separated into two sets :

(i) The distributed ones in space, e.g. the hydraulic and transport properties such as permeability, storativity, porosity, retention curves, pore-scale dispersivity and sorption coefficients. These are functions of the coordinate $\mathbf{x}$ (possibly of time) and they have to be specified in a manner appropriate to the analytical or numerical models we are going to use.
(ii) Discrete parameters, such as constant values of head and recharge on the boundaries, well discharge, initial concentration, etc. A few of these may be time dependent.

As a rule, these parameters are not known accurately and their values are affected by uncertainty. The stochastic approach is precisely addressing this uncertainty in a rational, quantitative framework by using probability theory. However, the uncertainty associated with the aforementioned two types of parameters is handled in a different manner from a mathematical point of view and it has different physical interpretations.

PARAMETRIC UNCERTAINTY

The parametric uncertainty, i.e. one of the discrete parameters given in (ii) above, can be treated along the same lines as classical statistics. Denoting by θ the parameter's vector values, the solution of the flow and transport problem, whether analytical or numerical, is a function of θ. If the θ values are regarded as random variables characterized by their joint p.d.f. (probability density function), $f(\theta)$, the solution is also random. For example, the contaminant concentration C conditioned on θ can be written as $C(\mathbf{x},t|\theta)$, and its statistical moments can be derived by repeated integration over $f(\theta)$. Conceptually, the procedure is simple, and when commonly referring to stochastic modeling we do not consider this type of uncertainty. Still, it may have a serious impact upon predictions, and I believe it should be given more attention in applications.

To account for parametric uncertainty in the manner indicated above, one has to know the p.d.f., $f(\theta)$, and this knowledge is seldom available. If uncertainty is associated with measurement errors, it is common to regard θ as normal and to represent $f(\theta)$ in terms of the mean and variance–covariance matrix of θ. Alternatively, to account for equal probability in a range of values, rectangular distributions may be used, with the same number of statistical moments used to characterize θ. Even if such prior information could be achieved, the computational burden of calculating the moments of C can be quite heavy. A common simplification is to assume that the θ values vary little around their mean $\langle\theta\rangle$, i.e. if $\theta=\langle\theta\rangle+\theta'$ we assume that $CV(\theta)=\sigma_\theta/\langle\theta\rangle$ is much smaller than unity. Then, an expansion of the function of interest, e.g. C, yields at first-order

$$C(\mathbf{x},t|\theta)\simeq C(\mathbf{x},t|\langle\theta\rangle)+\theta'_i\,\frac{\partial C((,t\times\theta)}{\partial\langle\theta_i\rangle}\ldots$$

i.e.

$$\langle C(\mathbf{x},t)\rangle\simeq C(\mathbf{x},t|\langle\theta\rangle);\ \sigma_C^2(\mathbf{x},t)\simeq\sigma_{\theta_i}\sigma_{\theta_j}\frac{\partial C((,t|\theta)}{\partial\langle\theta_i\rangle}\frac{\partial C((,t|\theta)}{\partial\langle\theta_j\rangle} \qquad (1)$$

The coefficients $\partial C(\mathbf{x},t|\theta)/(\partial\langle\theta_i\rangle)$ are known as sensitivity coefficients and they play an important role in assessing the impact of the uncertainty of θ_i upon that of C or other similar functions.

An important point of principle is that reduction of uncertainty can be achieved only by a better characterization of the parameters.

If parameters are time dependent (e.g. recharge from precipitation), the approach is similar, though of increasing computational difficulty.

We shall not dwell further upon parametric uncertainty, and we now concentrate the discussion on spatial variability.

SPATIAL VARIABILITY

Spatial variability of properties seems to be a ubiquitous feature of natural formations, as illustrated convincingly by Figs. 1 and 2, and the topic will be discussed at length in most contributions to this book.

We consider for illustration the hydraulic conductivity $K(\mathbf{x})$, a property which varies over a few orders of magnitude (Fig. 1) and has a large impact on transport. The conductivity is defined as that of a well core at $\mathbf{x}$, over a support which is large compared with the pore scale, but much smaller than any other scale of the problem. Measurements show clearly that the spatial variation of K is irregular and cannot be captured by interpolation among a few measured values. This uncertainty is modeled by regarding K as RSF (random space function) of $\mathbf{x}$, or a regionalized variable in the geostatistical terminology. Its randomness is carried over into the flow and transport variables that depend on K.

The statistical characterization of an RSF is very complex, since it represents an infinite set of random variables. For most practical purposes, it is enough to know the joint p.d.f. of K at a few arbitrary points. Thus, $f_K(\mathbf{x})$ is known as the univariate distribution, whereas $f_K(\mathbf{x},\mathbf{y})$ is the two-point, bivariate distribution, etc.

Assuming that the moments of various orders are known, the solution of the flow and transport problem by stochastic modeling consists in deriving the statistical moments of the dependent variables of interest in terms of those of the spatially variable parameters. In simple words, we generate an ensemble of realizations of the formation and seek the statistics of concentration, solute flux, etc., by solving the flow and transport problem for this set of realizations.

The first problem of principle we face in this process is that only one formation exists, the actual one, and that in fact the ensemble is fictitious, being a tool to assess uncertainty. Thus, we have to identify, from measurements in the only existing realization, the statistics of the ensemble in order to derive results which apply to the same unique realization!

The way to break this apparent deadlock is to invoke stationarity, which in principle makes possible identification of the ensemble statistics from one record. This in itself creates a vicious circle, but I shall assume that the statistical methodology makes this possible. Furthermore, by using geostatistical methods we may rely on some type of generalized stationarity, a simple example being that of stationary increments. In the latter case we filter out the mean and identify a variogram of a process that may have an ever increasing scale. More involved methods filter out trends, but then the identification becomes data intensive. Unfortunately, in most hydrological applications data are not that abundant to permit one to use other than simple models. Furthermore, the same limitation precludes determining more than the univariate distribution and the two-point covariance or variogram. Thus, because of lack of information, freedom is left about selecting the higher moments and multi-point distributions. It is customary, for instance, to assume multivariate normal distributions for $Y=\ln K$, the logarithm of the hydraulic conductivity. Such an assumption simplifies the simulations considerably since it reduces the representation of the entire stationary structure of the permeability in terms of a constant mean $\langle Y \rangle$ and a two-point covariance $C_Y(\mathbf{x},\mathbf{y})$, or variogram, that depends only on $\mathbf{r}=\mathbf{x}-\mathbf{y}$, the distance between the points. Furthermore, by selecting some analytical form for C_Y, we may reduce the representation of the entire structure to three parameters: the mean $\langle Y \rangle$, the variance σ_Y^2 and the integral scale I_Y. Additional features may be incorporated by assuming simple trends or by dividing the domain into subdomains that, although significantly different, are each modeled as stationary. It is emphasized that this simplicity is achieved on the basis of a few significant, hard to validate assumptions.

To summarize, identifying the statistical structure of distributed variables from a set of scarce measurements of a single realization has many degrees of freedom. Under usual constraints, it is not a unique procedure, and it requires a good dose of judgment and experience.

2.5 Scales of the problem and selection of the numerical model

At the completion of the former stages of the setting up of the conceptual model (or a few alternative models), we can define a few length (and associated or additional time) scales characterizing the formation and the processes of interest. Such typical scales are:

(i) The formation horizontal and vertical extent, L_{fh} and L_{fv}, respectively. These scales belong to the entire formation as a hydrogeological unit, and generally $L_{fh} \gg L_{fv}$ (e.g. in Fig. 2, $L_{fh} \simeq 1600$ m, $L_{fv} \simeq 8$ m). The unsaturated zone is characterized by a vertical length scale L_{unv}, which is generally of the order of or smaller than L_{fv}.

(ii) The scales related to the transport problem we wish to solve. First, L_{trh} is the distance traveled by the solute plume to reach the accessible environment or the target area; conversely, the travel time T_{trh} could be considered. A second scale is that characterizing the contaminant source extent L_{trin}. Finally, l_{trav} is defined as the scale over which the concentration is averaged at the accessible environment. It may be very large, of the order of L_{fh} if we are interested, for instance, in the mass of contaminant reaching a river or a lake in which the aquifer discharges, or extremely small if we seek local dosages. An intermediate case of considerable interest in practice is that of pumping wells, for which $l_{trav} \simeq L_{fv}$.

(iii) Scales characterizing spatial variability of distributed parameters which impact flow and transport, e.g. the hydraulic conductivity K. Here we may discriminate between the local scale and the regional one.

The correlation scales of $Y=\ln K$, within domains of the order of the formation thickness, were coined as a local scale

(Dagan, 1986). Such scales are generally determined by analyzing permeability measurements of cores taken from a few wells or by multilever samplers. In most unconsolidated sedimentary formations it is found that the structure of three-dimensional heterogeneity is anisotropic, with I_{Yv}, the vertical correlation scale, smaller than I_{Yh}, the horizontal correlation scale, while both are smaller than L_{fv}, the formation thickness. In a few recent field investigations of sedimentary aquifers, I_{Yh} was found to be of the order of meters, whereas I_{Yv} was an order of magnitude smaller.

When considering flow and transport over L_{fh}, the formation horizontal scale, we encounter spatial variability characterized by correlation scales much larger than the thickness. These may be the scales of the geological subunits into which we have divided the formation (Fig. 2), or may result from statistical analysis of properties averaged over the formation thickness. Thus, in practice, common measurements are those of transmissivity T and storativity S. These are obtained by pumping tests, which in principle provide a kind of average over blocks of vertical size of order L_{fv} and similar horizontal scale. T may be viewed as the effective conductivity of such blocks, incorporating the effect of the local heterogeneity, whereas S is a volume average. It is clear that T and S are two-dimensional spatially variable properties, in the horizontal plane. Their statistical structure analysis leads to horizontal correlation scales I_Y, where $Y=\ln T$ is now the logtransmissivity, which are much larger than L_{fv}. Such scales were found to be of the order of thousands of meters (Delhomme, 1979; Hoeksema & Kitanidis, 1985) or even evolving over the formation, with a cutoff of the order L_{fh} (Desbarats & Bachu, 1994).

By definition, the local and the regional scales are widely separated. Of course, this separation is possible if we accept the results of the so-far limited number of field investigations in which it was found that the local I_{Yh} is of the order of meters only.

(iv) Scales related to the nonuniformity and temporal variations of the flow variables. Thus, flows driven by natural gradients, e.g. by a drop of constant heads between the boundaries, can be regarded as being close to uniform in the mean. As a contrasting example, flows caused by a recharge area of a horizontal scale L_{rh} which is much smaller than L_{fh} are nonuniform in the mean.

If we model the entire zone of interest as a stationary heterogeneous unit, we may use some analytical approaches to solve the problem in the first instance. However, the complex structure revealed in Fig. 2 and flow spatial and temporal nonuniformities generally call for a numerical solution. The various relative magnitudes of the aforementioned scales characterize different types of problems. Thus, a large ratio L_{trin}/L_{fv} between the contaminant source input zone and formation thickness, or a large L_{trin}/I_Y, leads to what is known as a non-point-source problem, while the opposite case is that of a point source. The case, $L_{trh}/L_{fh}=\mathbf{O}(1)$, i.e. transport distance of the order of the formation extent, calls for a regional modeling. In contrast, $L_{trh}/L_{fv}=\mathbf{O}(1)$, i.e. transport over scales of the order of the formation thickness, requires modeling of the three-dimensional local structure.

At this stage, we have to take the next step in our modeling campaign, namely to design a numerical model to solve the flow and transport problem in a stochastic context. It is beyond the scope of this chapter to discuss the host of available numerical tools; these will be considered in detail in other chapters. A point of principle is that most numerical schemes involve a discretization in space, with elements of scale l. This scale, and more precisely its relative magnitude with respect to other scales, has a very significant impact on the nature of the solution. Ideally, a very fine discretization, with l much smaller than the local heterogeneity scale I_{Yv}, would provide a solution of the flow and transport problem at any level. However, modeling a formation at the regional scale and in three dimensions at such a level of detail leads to a huge number of elements and to requirements of computer memory and times that are beyond the capability of present and near future machines. This difficulty is compounded by the need to carry out a repetitive numerical solution for a large number of realizations, e.g. when considering a few alternative conceptual models and in a Monte Carlo framework. Fortunately, usually the scale of the solution is large enough and it does not require the level of detail implied by the aforementioned partition of the domain.

Thus, in applications we have to compromise in the selection of a discretization scale l commensurate with available computing resources, taking into account the requirements of the problem and our ability to correct for discretization effects. Such a choice involves a good understanding of the scales of the problem, of the nature of the solution and of numerical techniques.

For illustration, the formation of Fig. 2 has been discretized (Bierkens, 1994) in the manner shown in Fig. 4, with blocks of horizontal scale $l_h \simeq 5$ m and $l_v \simeq 0.5$ m. We shall discuss in Sections 3 and 4 the relationship between the computational scale and the nature of the solution.

2.6 A few summarizing remarks

In this section, we have tried to describe the process leading from the general formulation of a subsurface flow and contamination problem to the setting of models to provide quantitative solutions, say by numerical methods.

Two main points are worthy of mention at this stage. First, in the case of natural formations, the selection of the model is always affected by uncertainty. This uncertainty manifests

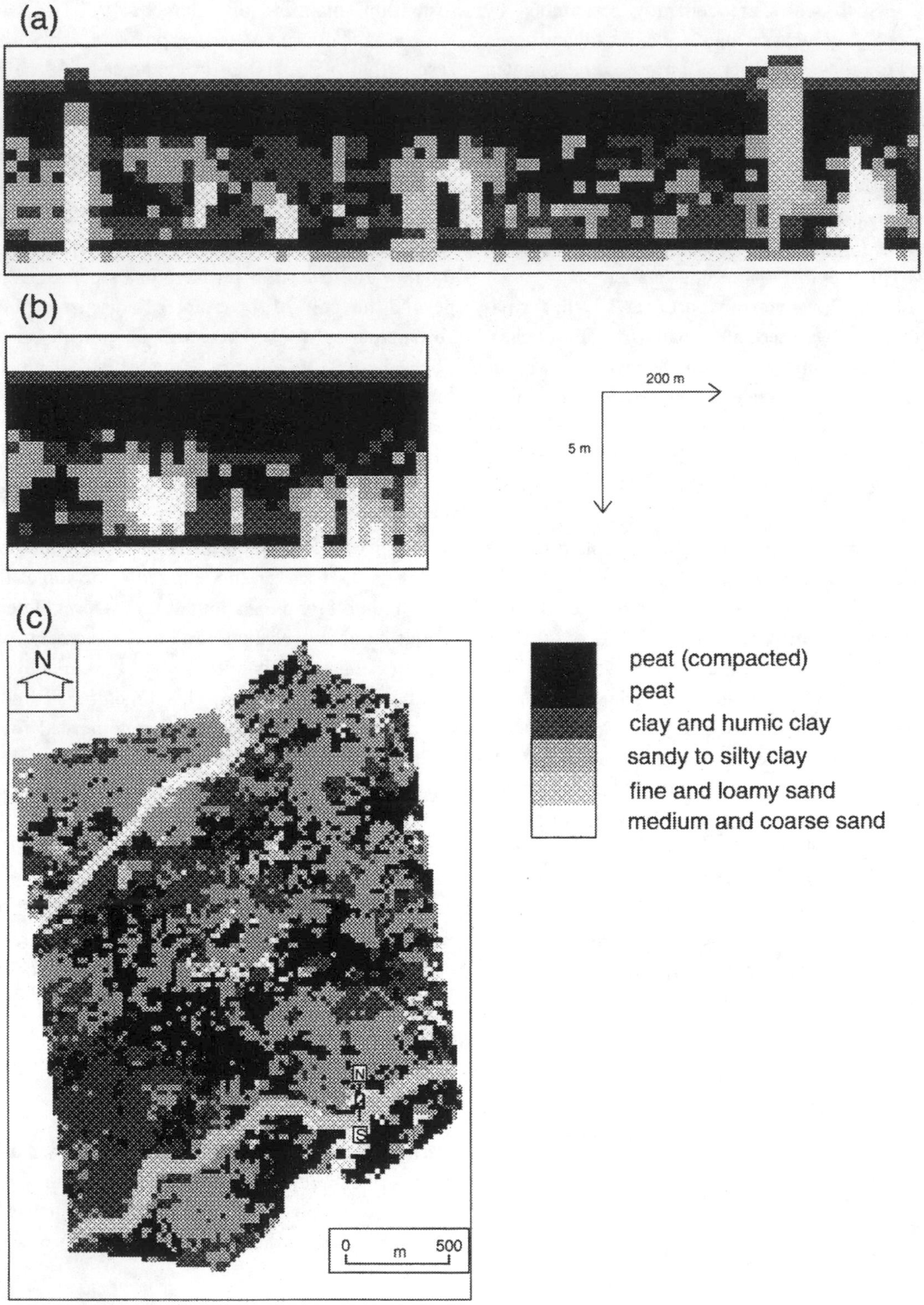

Fig. 4 Illustration of partition in numerical blocks of the formation of Fig. 2 (from Bierkens, 1994).

first of all at the conceptual level, and it leads to a few possible scenarios. Unfortunately, assigning probabilities and incorporating conceptual models in a formal, quantitative framework has not been given sufficient attention in the literature, and generally only one such model is chosen by modelers. Uncertainty affecting parameters or spatial variability, subjects which were set in quantitative terms in the last two decades, will be discussed extensively in the following.

Secondly, I hope I have amply demonstrated that in setting a conceptual and computational model we have to make many choices of a qualitative nature. These choices require an understanding of the physical, chemical and biological processes taking place in the formation, concerning the geology, hydrology and geochemistry, of the engineering aspects of the problem and of available theoretical and modeling tools. Furthermore, these choices may have a considerable effect upon the validity and accuracy of the solution. In the following chapters we will read about modern and sophisticated techniques of modeling and solving problems by the stochastic approach. These developments may obscure the fact that the process of setting and selecting the conceptual model is based on experience, intuition and good hydrological and engineering judgment. Thus, our field is not just an area of applied mathematics and physics, but a combination of art and science in which understanding of the nature of the problem, of its engineering and management aspects, is as important if not more so, than that of the theoretical concepts and methods.

3 SOLUTION OF THE FLOW PROBLEM

3.1 Introduction

At the completion of the previous stages of development of the model, we are in possession of a discrete representation of the formation (Fig. 4) and a set of partial differential equations for the pressure head and other flow variables (e.g. the moisture content in the unsaturated zone or the saturation of the wetting phase in immiscible flow). These equations comprise mass conservation, Darcy's law and constitutive equations. The hydraulic properties appearing as coefficients in these equations are regarded as random space functions of given statistical structure, in the simplest approach by a parametrization of the moments of stationary distributions. The boundary and initial conditions are also given, though the parameters appearing there, as well as those of the statistical distributions, may be regarded as random variables of given p.d.f.

The solution of the flow problem consists in determining the head and velocity fields, which are random space functions. Ideally, one would like to characterize them completely by their various multi-point joint p.d.f., but in practice we may be satisfied with knowing a few statistical moments. It is emphasized that in the problem selected as a prototype here, our aim is to solve the transport problem. In this case, the solution of the flow problem is an intermediate step intended to provide the velocity field statistical structure, which is a prerequisite to solving transport. However, in many applications the solution of the flow problem may be the ultimate objective.

To simplify the discussion, we shall concentrate on the simple case in which the only random property is the hydraulic conductivity (under saturated flow conditions) K and disregard parametric uncertainty (see Section 2.4). The flow problem may be classified under a few different criteria as follows: direct versus inverse, unconditional versus conditional and three-dimensional versus two-dimensional.

The direct problem is the one stated above, whereas in the inverse problem the formation properties are also partly unknown and have to be determined with the aid of measurements of the head. We shall concentrate here on the direct problem, while the opposite problem will be discussed in another chapter.

In the unconditional mode, the realizations of the permeability field underlying the stochastic flow model are derived from the given p.d.fs. Measurements were used in order to infer these p.d.fs., either by statistical analysis or by solving the inverse problem. In the conditional mode, the authorized realizations belong to a subset in which measured values of permeability are honored. A common procedure followed to carry out conditioning on measurements is by *kriging* or by the closely related Gaussian conditional probability distribution. Conditioning is at the heart of geostatistics, and its use in stochastic modeling of flow and transport is one of the most powerful and distinctive tools of this discipline. We shall discuss these alternative modes in the following.

As for the distinction between three- and two-dimensional flows, we have discussed the issue in Section 2.5. In the case of the formation of Figs. 1 and 2, the modeling is at the local scale and of a three-dimensional nature. However, for larger horizontal distances a regional model is necessary. We shall discuss separately the two types of models, since this choice is of definite significance.

3.2 Three-dimensional flow (the local scale)

The heterogeneous formation is now represented in a discretized form, say by a division in numerical blocks of dimensions $l_h \times l_h \times l_v$, as shown in Fig. 4. The vertical block size l_v is sufficiently small compared with the formation depth L_{fv}, so we can capture accurately the variability at that scale. Generally, such a representation precludes modeling the entire formation, and the flow boundary conditions are either known or are found by solution of the problem at the regional scale, preceding the three-dimensional one. However, the blocks are not that small as to capture accurately the local scale heterogeneity, i.e. the requirements $l_v \ll I_{Yv}$ and $l_v \ll I_{Yh}$ are generally not met. Thus, the numerical blocks are usually of a size comparable with the local heterogeneity scale. At this point, in order to solve the flow problem by Monte Carlo simulations., various replicates of the conductivity of the blocks, $K_b(\mathbf{x}_i)=\exp[Y_b(\mathbf{x}_i)]$, have to be generated, $\mathbf{x}_i$ being the centroid of the ith block. We recall that the logconductivity is modeled as an RSF which has a mean appropriate to the subunit to which the block belongs and a stationary residual of known two-point covariance structure, which in turn is expressed in terms of a few parameters, e.g. σ_Y^2, I_{Yh} and I_{Yv}. Thus, in general, the conductivity of the numerical blocks is also random, but the statistics of $Y_b(\mathbf{x})$ is different from that of $Y(\mathbf{x})$ because of the finite size of the block. The process of transferring the pointwise statistics to that of numerical block values is known as *upscaling*, and it is an essential component of the numerical solution. It is quite astonishing that this topic has been discussed only recently in the literature; in fact, it is still a matter of debate. It is emphasized that upscaling is not identical to space averaging, as will be shown below in the extreme case of large blocks. Indeed, if the numerical size of the blocks, l, is much larger, say tens to hundreds of heterogeneity scales, the upscaled conductivity tends to a deterministic value known as K_{ef}, the effective conductivity.

There are few topics which have been addressed more extensively in the physics and engineering literature than that of effective properties of heterogeneous media, and we do not intend to review it here. We limit the discussion to recalling the definition of effective conductivity and the various assumptions which underlie its derivation. Thus the usual definition and values apply to:

(i) the media of stationary heterogeneous structures characterized by finite correlation scales I_Y;
(ii) the unbounded domain Ω, or any domains of dimensions much larger than the correlation scales, while the flow variables are considered in the central part of the domain, far from the boundaries;
(iii) a uniform, time independent, gradient $-\mathbf{J}$ applied on the boundary;
(iv) Darcy's law for the mean flux $\langle \mathbf{q} \rangle$ and mean gradient $\mathbf{J}$, i.e. $\langle \mathbf{q} \rangle = K_{ef}\,\mathbf{J}$, which serves as a definition of the effective conductivity.
(v) Under the conditions of (i)–(iii), ergodicity prevails, i.e. $\langle \mathbf{q} \rangle$ and $\mathbf{J}$ can be exchanged with their space averages in any realization.

The considerable interest in effective properties stems from the fact that the ratio between the different 'macroscopic' scales (domain extent, scale of mean field nonuniformity) on the one hand, and the heterogeneity correlation scales on the other, are extremely large in many applications of physics and engineering. In such cases, (v) applies and the heterogeneous material can be replaced by a homogeneous one of effective properties.

However, the usefulness of the concept is more limited for flow in the subsurface, precisely because there is no such a disparity between scales. Thus, neither the condition of remoteness from boundaries nor that of average uniform flow is always satisfied. An extreme, but important, application case is that of flow toward a pumping or injecting well, which has not yet been solved in a comprehensive manner.

Considerable progress has been made in evaluating K_{ef} for formations of lognormal multivariate distributions. Simple expressions were found by small perturbation expansions in the variance σ_Y^2, and they are quite robust, especially for isotropic covariances. An interesting point is that the first-order approximation in σ_Y^2 does not imply normality and only at higher-order are assumptions about the structure needed.

Recently, attempts were made to relax a few of the constraints enumerated above, in particular those of flow steadiness and uniformity. It is not always possible to relate $\langle \mathbf{q} \rangle$ to $\langle \mathbf{J} \rangle$ with the aid of coefficients that depend only on the structure, on $\mathbf{x}$ and time. A few approaches led to nonlocal relationships in which the mean flux at $\mathbf{x}$ depends, through a convolution, on the head gradient in a volume surrounding the point. The kernel of the convolution, depending only on the structure and coordinates, generalizes K_{ef}. However, such a result raises the question of applicability of the concept to a single realization. Unless the flow nonuniformity or time dependence is such that one can identify surfaces over which the kernel is localized, no ergodic arguments can be invoked and the mean values of the flux and gradient have a statistical meaning only, not exchangeable with any kind of space averages. These issues have been investigated only recently, and they depart from the traditional literature on effective properties.

Even if an effective property is adopted, our interest resides not only in the mean flux and gradient, but also in their fluctuations, which are essential to modeling transport. Again, this topic is not in the mainstream of the physics and engineering literature on heterogeneous media. Summarizing this topic, setting the values of the properties of numerical blocks by upscaling replaces the actual random medium by a fictitious one. The upscaled properties are not just space averages and their derivation is a complex and unresolved issue. Furthermore, even if an accepted procedure is adopted, 'high frequency' fluctuations of the random functions, at the subgrid scale, are lost. This subgrid variability has to be recovered in some way for transport modeling.

There are two relatively simple conceptual approaches which can be employed to avoid the upscaling complications: (i) to model heterogeneous structures by numerical blocks that are much smaller than the correlation scales, which is prohibitive for three-dimensional structures. Furthermore, the level of detail achieved is redundant if we are interested in some space averages of the flow variables; (ii) to model formations as fictitious, homogeneous, ones of effective properties. The latter route has the limitations mentioned above and in any case provides the mean head and flux fields only.

Hence, to characterize statistically the fluctuations, one has to go a step further.

Everything so far has been in the framework of unconditional probability, i.e. the input information was the unconditional p.d.f. of distributed parameters. However, in the presence of actual measurements, we may carry out all the previous steps (upscaling, derivation of mean values) by using the p.d.f. of Y^c, the logconductivity conditioned on measurements of $Y(\mathbf{x}_i)$, where $\mathbf{x}_i$ are the coordinates of the measurement points. These p.d.fs can be written in a general form by using Bayes' theorem, but they have simple expressions for multivariate normal variables. In the latter case, the mean and conditional covariances are identical to those obtained by cokriging. Once we move into the conditional mode, stationarity is lost and the concept of effective properties as medium dependent parameters becomes meaningless. However, if we are interested in some space averages of the flow variables and not their point value mean and fluctuations, conditioning loses its impact. This point can be underscored by thinking about the extreme case in which measurements are available on a very dense grid, so that Y^c is practically deterministic. However, under conditions of ergodicity, the space averaged flux is related to the space averaged head gradient by the effective property, derived as a statistical mean, and there is no advantage in conditioning. Conditioning may affect upscaling; this is a central topic in petroleum reservoir 'architecture', a subject which is outside our scope. We shall return to the concept later in relation to two-dimensional flow and transport.

In spite of the progress achieved in the last two decades in stochastic subsurface modeling, many important issues have barely been touched, as mentioned above. For instance, much has to be done in order to apply the concepts we have discussed in problems of two-phase immiscible and unsaturated flows or to fractured media.

3.3 Two-dimensional flow (regional scale)

The main difference between regional and local modeling, besides the reduction in dimension, is the disparity in scales mentioned in Section 2.5. Thus, the integral scale of the log-transmissivity Y may be of the order of kilometers to tens of kilometers. For instance, for a transport travel distance of the order of kilometers, the formation may be viewed as one of ever-increasing scale, characterized by a variogram without a sill.

The numerical modeling is much less demanding in two dimensions, though the problem of upscaling of properties to the numerical block scale is still present.

The concepts of effective transmissivity and of unconditional simulation are of limited value in regional modeling. Indeed, the ergodic conditions that make effective properties useful may be met only for global entities such as the whole formation water budget. However, modeling transport or predicting the potential of wells in areas lacking measurements are affected by uncertainty, which cannot be reduced significantly unless we operate in the conditional mode.

An important aspect of the solution of flow equations by using conditional probability distributions of Y is that, as a rule, we have at our disposal not only transmissivity measurements but also measurements of heads (made with the aid of piezometers) and sometimes concentration measurements of existing plumes. Identification of the statistical moments of Y^c, conditioned on head measurements, calls for the so-called inverse problem, a subject of considerable interest which will be discussed in a later chapter. Here we only mention that, in the stochastic context, the aim of the solution of the inverse problem is to identify the statistical structure of Y^c not its deterministic values over the numerical blocks. In this respect, the problems of uniqueness and identifiability have a different meaning than those pertaining to a deterministic spatial distribution.

Another point of interest is how to incorporate the effect of the local, three-dimensional, variability upon the fluctuations of the head and velocity at regional scale. We shall return to this point in Section 4.3.

Summarizing, although it deals with solving partial differential equations of flow, stochastic modeling at the regional scale departs from the traditional approach of statistical continuum theories and is closer to geostatistical concepts. In my view, it is in the area bridging these two different theoretical approaches that stochastic modeling of subsurface flow has made its distinct imprint.

4 THE SOLUTION OF THE TRANSPORT PROBLEM

4.1 Discussion of field findings

We shall limit the discussion at present to transport of inert solutes (tracers). The concentration $C(\mathbf{x},t)$ is defined as mass of solute per volume of fluid, and it is, like the hydraulic conductivity, defined as an average over a volume surrounding $\mathbf{x}$ that is large compared with the pore scale.

Many field measurements carried out in the past comprised measurements of concentration as a function of time in a few isolated wells. These measurements were subsequently interpreted by assuming that C is the solution of an advection–dispersion equation with a constant dispersivity coefficient, as observed in the laboratory. Such a solution implies that the solute plume has a smooth Gaussian-bell-like shape and that surfaces of constant concentration are, e.g. for a small initial plume, confocal ellipsoids. Apparently, data showed that the interpreted dispersion coefficients were much larger than those observed in the laboratory and that they changed with the travel distance.

With the advent of the stochastic approach, it has become clear that the basic premises of measurement interpretation and of prediction were not founded. Since then, a few elaborate field experiments have been carried out, and C was measured by extracting numerous cores, by multilever samplers or by measuring radiation, at a large number of points and at different times, such as to obtain detailed snapshots of the solute plume.

A few such snapshots, for the field test at Columbus Air Force Base (Boggs *et al.*, 1992), are reproduced in Fig. 3. Without going into details, the main features revealed by these and other measurements are:

(i) The solute plume has an irregular shape, far from the Gaussian one. Plots of C display an erratic behavior, and C does not lend itself to representation by smooth functions. Describing or predicting the concentration distribution in deterministic terms does not seem to be feasible.
(ii) The plume longitudinal spread, characterized by any quantitative measure, is orders of magnitude larger than the one related to pore-scale dispersion.
(iii) The vertical and transverse spreads are much smaller than the longitudinal one, but still larger than those observed in the laboratory. These findings have a clear interpretation in view of the discussion of Section 2: transport and solute spreading are related to the large scale spatial variability of hydraulic properties, and particularly of the permeability. Indeed, one of the main motivations for the development of the stochastic approach was the need to understand and predict transport at field scale.

In the following sections we discuss a few issues of principle; the subject itself is covered by several other chapters.

4.2 Two basic approaches to model transport

Rather than giving a general discussion, we shall develop the subject along the example of Section 2 (see Figs. 2 and 4). Thus, at the completion of the solution of the flow problem, we have arrived at a statistical characterization of the fluid velocity field, say in terms of the mean Eulerian velocity $\langle \mathbf{V}(\mathbf{x},t)\rangle=\langle \mathbf{q}\rangle/n$ (where n is the effective porosity) and the two-point covariance $u_{ij}(\mathbf{x},\mathbf{x}',t,t')=\langle u_i(\mathbf{x},t)u_j(\mathbf{x}',t')\rangle (i,j=1,2,3)$. Here, $\mathbf{u}=\mathbf{V}-\langle\mathbf{V}\rangle$ stands for the velocity fluctuation and furthermore, for a spatially stationary field, u_{ij} is a function of $\mathbf{r}=\mathbf{x}-\mathbf{x}'$. It is emphasized that if $\mathbf{V}$ results from numerical flow solutions, say Monte Carlo simulations, we may arrive at the statistical structure of $\mathbf{V}$ in a comprehensive manner. On the other hand, the information at a scale smaller than l, the size of numerical blocks (Fig. 4), is lost, and besides we get only a finite number of replicates of $\mathbf{V}$.

Now, at this point, we have to solve the transport problem and determine C subjected to initial and boundary conditions, e.g. a given solute body of concentration C_0 in a finite volume V_0, or a continuous injection source of $C=C_0$.

Since $\mathbf{V}$ is modeled as an RSF, so is $C(\mathbf{x},t)$, and its prediction can be made only in terms of p.d.fs or of statistical moments. This goal may be achieved by two basic approaches: the Eulerian and the Lagrangian.

THE EULERIAN FRAMEWORK

The starting point is the advection–dispersion equation

$$\frac{\partial C}{\partial t}+\mathbf{V}\cdot\nabla C=\nabla(\mathbf{D}_d\nabla C)-\frac{\partial N}{\partial t} \tag{2}$$

where $\mathbf{D}_d$ represents the pore-scale dispersion tensor or any other local, diffusive, mechanism, whereas $\partial N/\partial t$ is a source term, representing decay or transfer (at present we consider conservative solutes with $N\equiv 0$).

In the Eulerian approach, the solution $C(\mathbf{x},t)$ is achieved by solving (2), with $\mathbf{V}$ and $\mathbf{D}_d$ random. Thus, if Monte Carlo simulations are carried out in order to solve the flow problem and to determine $\mathbf{V}$, in each of them (2) may be solved toward obtaining replicates of C.

The first problem we encounter in such a procedure is that the velocity field obtained from a numerical solution is upscaled and smoothed out, with consequent loss of information about the fine scale spreading mechanism of the solute. If we want to avoid this loss of information by solving for both $\mathbf{V}$ and C on a fine grid at the heterogeneity scale, we end up with an insurmountable numerical problem. Furthermore, if (2) is solved by discretization over numerical blocks, the solution C is no longer a local value, but represents a space average over these blocks as well, i.e. a numerical dilution is present. We shall return to these aspects in the following sections.

The advantages of the Eulerian approach are: the existence of a large body of literature on this type of equations, the ability to obtain directly the concentration C and a relatively easy incorporation of $\partial N/\partial t$ as a function of C.

The drawbacks of this methodology are: severe numerical problems for velocity fields that vary abruptly in space, leading to poor mass balances, and problems of numerical dispersion, resulting in smearing of fronts. These problems are particularly acute for large σ_Y^2, for which the coefficients of variation of $\mathbf{V}$ are large. It seems that deriving methods which can tackle these difficulties is still a matter of research and debate.

THE LAGRANGIAN FRAMEWORK

The plume is regarded as a collection of particles represented by material points and the mass density of these particles renders C. The number or density of such particles selected in order to represent the plume is dictated by the entity we wish to simulate with their aid. Thus, if the aim is C averaged over blocks of scale l, we need to ensure that a sufficiently large number of particles is present in each such block within V_0. However, this requirement can be relaxed if we seek only $\langle C(\mathbf{x},t)\rangle$ and, say, σ_C^2. We may use even a smaller number of particles if we seek the spatial moments of the plume

$$M=\int nC(\mathbf{x},t)d\mathbf{x},\mathbf{R}=\frac{1}{M}\int n\mathbf{x}C(\mathbf{x},t)d\mathbf{x} \qquad (3)$$

$$S_{ij}=\frac{1}{M}\int n(x_i-R_i)(x_j-R_j)C(\mathbf{x},t)d\mathbf{x}$$

where n is the effective porosity and M, $\mathbf{R}$ and S_{ij} $(i,j=1,2,3)$ are the total mass, centroid and second spatial moments of the plume, respectively. A similar global representation that can be achieved accurately with a relatively small number of particles is the breakthrough curve (total mass of solute per unit time crossing a fixed control plane).

The basic entity in the Lagrangian representation is the particle trajectory, $\mathbf{x}=\mathbf{X}(t;t_0,a)$, which satisfies

$$d\mathbf{X}=\mathbf{V}[\mathbf{X}(t),t]dt+d\mathbf{X}_d;\ t=t_0,\ \mathbf{X}=\mathbf{a} \qquad (4)$$

where $d\mathbf{X}_d$ stands for a Wiener process to represent the diffusive process of coefficients $\mathbf{D}_d$ (2). Obviously, $\mathbf{X}$ is random because of the randomness of $\mathbf{V}$ and $d\mathbf{X}_d$ and it is characterized by its moments

$$\langle\mathbf{X}\rangle,X_{ij}=\langle X_i'(t;t_0,\mathbf{a})X_j'(t';t_0,\mathbf{a}')\rangle\ldots$$

where $\mathbf{X}'=\mathbf{X}-\langle\mathbf{X}\rangle$. The spatial moments (3) of C can be easily expressed in terms of the $\mathbf{X}$ statistical moments through the basic relation $dC(\mathbf{x},t)=C_0(n_0/n)\delta(\mathbf{x}-\mathbf{X})d\mathbf{a}$, where δ is the Dirac operator or its numerical equivalent.

The advantages of the Lagrangian approach are: mass balance is obeyed exactly, high velocity gradients are handled easier by refining the time step along the trajectory, numerical dispersion can be reduced in a similar manner and the codes are very simple. The main drawbacks are: the method does not lend itself to a simple calculation of the local C and accounting for reactive behavior or multiphase flow is more complicated than in the Eulerian framework. It seems, however, that the advantages of the Lagrangian approach make it the preferred methodology both on numerical and theoretical grounds, and we shall adhere to it in most of the following discussion.

4.3 Three-dimensional (local scale) transport

As in the flow problem (Section 3), we refer here to travel distances of the order of the formation thickness. The mechanism of the plume spreading is dominated in this case by heterogeneity of a three-dimensional structure. It is emphasized that all recent field tests were conducted at this scale for obvious practical reasons: the time required to cover much larger travel distances may be prohibitively large. Thus, the development of the plume of Fig. 3 and prediction of transport for the structure of Fig. 2 close to the source are influenced mainly by local heterogeneity.

Returning to our example of Figs. 2 and 4, we assume that after solving the flow problem the statistics of the velocity field is known, and our next step is to solve numerically the transport problem, e.g. by using the discretization displayed in Fig. 4. Whether we use the Eulerian or the Lagrangian methodology, we face the same problem of *upscaling*, as already discussed in Section 3.2. To be more specific, the spatial variability of $\mathbf{V}$ at scales smaller than l is filtered out and the dispersive effect of these variations is lost. As in the flow problem (Section 3) we may try to evade this dilemma by either operating with a very fine numerical grid, which is prohibitive, or by using large elements with $l \gg I_u$, where I_u is the velocity correlation scale. In the latter case, the field $\mathbf{V}$ becomes very smooth, it is well approximated by $\langle\mathbf{V}\rangle$ and the mechanism of plume spreading is lost, except for the pore-scale effect of $\mathbf{D}_d$. The natural extension of the approach in the flow problem would be to use an *effective* or macrodispersion tensor to represent the combined effects of velocity fluctuations and pore-scale dispersion. We shall discuss this problem at some length and examine a few alternative modes to achieve this goal.

THE MEAN CONCENTRATION AND THE MACRODISPERSION CONCEPT

First, a naive approach, which was the traditional one before the advent of stochastic modeling, was to assume that C is deterministic and satisfies the same eq. (2), but with $\mathbf{D}_d$ replaced by $\mathbf{D}$, the effective or macrodispersion tensor, assumed to be constant. This is not correct, since the resulting concentration field is smooth and Gaussian, contradicting the essential features of observed plumes (Fig. 3). Furthermore, the actual extent of the plume grows quicker than the one predicted with the aid of a constant $\mathbf{D}$ tensor. Unfortunately, this approach was used in the past to interpret breakthrough curves measured in isolated wells. We shall show later that this may lead to erroneous results.

A more rational approach, in line with that leading to the concept of K_{ef}, is to introduce the effective dispersion coefficients in an advection–dispersion equation for $\langle C(\mathbf{x},t)\rangle$, the ensemble averaged concentration, i.e.

$$\frac{\partial\langle C\rangle}{\partial t}+\langle\mathbf{V}\rangle\cdot\nabla\langle C\rangle=\mathbf{D}\nabla^2\langle C\rangle \qquad (5)$$

Such a representation poses two problems of principle:

(i) How does one validate (5) or use it for prediction in a single given realization in which C, and not $\langle C\rangle$, is the available variable? In the flow problem, we could overcome this problem by using the ergodic argument and exchanging ensemble and space averages. This is not possible for C because, in any conceivable case of interest, $\langle C\rangle$ varies in space and time and C is not ergodic. (A hypothetical experiment leading to ergodic C would be a steady state distribution created by application of two constant concentrations on two parallel and distant planes, and defining C as the space average over planes parallel to the boundary.) Thus, $\langle C\rangle$ has a statistical meaning at best and it is not a measurable entity.

(ii) Even under the last restriction, it is not obvious at all that $\langle C\rangle$, as determined for instance with the aid of computer simulations, satisfies (5) with constant coefficients.

The Lagrangian theory leads to the following conditions needed in order to ensure the validity of (5), by using the Focker–Planck equation for the p.d.f. of trajectories:

(i) the velocity field is stationary, in particular $\langle \mathbf{V}\rangle$ is constant and u_{ij} are functions of $\mathbf{r}$ and have *finite* integral scales;

(ii) the travel distance of the plume from the input zone is much larger than the correlation scale. By arguments relying on the central limit theorem, $\mathbf{X}$ becomes normal at this limit and the effective dispersion coefficients tend to

$$D_{ij}=\frac{1}{2}\frac{d\mathbf{X}_{ij}}{dt}=\int_0^{\infty} u_{ij}[\mathbf{X}(t)-\mathbf{X}(t')]dt' \quad (i,j=1,2,3) \tag{6}$$

The last restriction of (ii) can be relaxed if $\mathbf{X}$ is normal at any time, and then

$$D_{ij}(t;t_0)=\frac{1}{2}\frac{dX_{ij}}{dt}=\int_0^{t} u_{ij}[\mathbf{X}(t,t_0),\mathbf{X}(t',t_0)]dt' \quad (i,j=1,2,3) \tag{7}$$

It is emphasized that D_{ij} (Eqs. (6) and (7)) is a *nonlocal* variable since it depends on the travel time t. For example, for two solute plumes injected in the formation at different times, one cannot define a macrodispersion coefficient for the concentration to satisfy (5).

If $\mathbf{X}$ is not normal, it has been suggested in the recent literature on the subject that $\langle C\rangle$ satisfies a nonlocal, more general, equation than (5), namely

$$\frac{\partial\langle C\rangle}{\partial t}+\langle\mathbf{V}\rangle\cdot\nabla\langle C\rangle=\int_0^t\int D_{ij}(\mathbf{x},\mathbf{x}',t,t')\frac{\partial^2\langle C(\mathbf{x}',t')\rangle}{\partial x'_i\partial x'_j}d\mathbf{x}'dt' \tag{8}$$

The meaning of the generalized equation (8) is a matter of debate, and this topic will be covered elsewhere in the book. In any case the time t in (8) is also a Lagrangian time, and the same observation as the one following (7) applies also to (8).

As we have already mentioned above, the usefulness of the ensemble mean concentration $\langle C\rangle$ is questionable, as is that of the macrodispersion coefficients, because of nonergodicity of C.

THE CONCENTRATION FLUCTUATIONS AND A LAGRANGIAN DEFINITION OF EFFECTIVE MACRODISPERSION COEFFICIENTS

A simple inspection of the concentration distribution in field tests, e.g. in Fig. 3, and its comparison with the smooth solutions of (5) leads to the conclusion that C has large fluctuations. This can be put in a more formal framework by deriving the concentration variance σ_C^2 and the coefficient of variation $\sigma_C/\langle C\rangle$. The calculation of σ_C^2 is not a simple matter and will be discussed elsewhere. However, if pore-scale diffusive effect is neglected and if the initial concentration C_0 is constant, a simple upper bound for σ_C^2 is given by

$$\sigma_C^2=\langle C\rangle(C_0-\langle C\rangle) \tag{9}$$

showing indeed that $\sigma_C/\langle C\rangle$ becomes very large due to the drop of $\langle C\rangle$ with time. Pore-scale dispersion reduces σ_C^2 (9), but, due to its small coefficients, $\sigma_C/\langle C\rangle$ is still very large.

Under these circumstances, the usefulness of predicting $\langle C\rangle$ and σ_C^2 is doubtful and we are not able to capture the actual C distribution of a given realization, like that of Fig. 3, or even to delimit it within reasonable bounds.

The theory can perform much better in characterizing plumes by a few global measures, and we shall focus the discussion here on the spatial moments (3). These moments, except M, are generally *random variables of time* and are therefore characterized in terms of their statistical moments $\mathbf{R}$, R_{ij}, S_{ij}, $VAR(S_{ij})$, . . .

The actual dispersion coefficient of a plume D_{ij} are usually defined as

$$D_{ij}=\frac{1}{2}\frac{dS_{ij}}{dt} \tag{10}$$

i.e. half the rate of change of the second spatial moment (3). The centroid trajectory $\langle\mathbf{R}\rangle$, as well as D_{ij} (10), are measurable entities, and indeed analysis of field tests like that of Fig. 3 has been carried out with their aid.

It can be shown that the spatial moments (3) and the associated $d\mathbf{R}/dt$ and D_{ij} (10) become ergodic if the plume transverse initial scale is much larger than the velocity transverse correlation scale. Since this condition is generally fulfilled for three-dimensional transport at the local scale, one can write

$$\frac{d\mathbf{R}}{dt}\simeq\frac{d\langle\mathbf{R}\rangle}{dt}=\langle\mathbf{V}\rangle;\ S_{ij}(t)\simeq\langle S_{ij}\rangle$$

$$D_{ij}(t)\simeq\langle D_{ij}\rangle=\frac{1}{2}\frac{d\langle S_{ij}\rangle}{dt} \tag{11}$$

We suggest that defining D_{ij} (11) as *effective dispersion coefficients* is both useful and theoretically sound. Furthermore,

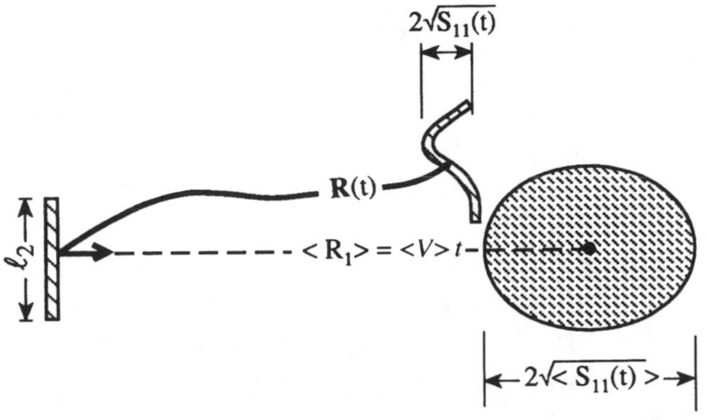

Fig. 5 Schematic representation of the motion of a solute body by a two-dimensional velocity field (Quinodoz & Valocchi, 1990). Full line: transport in a realization of the velocity field; dashed line, the ensemble averaged concentration.

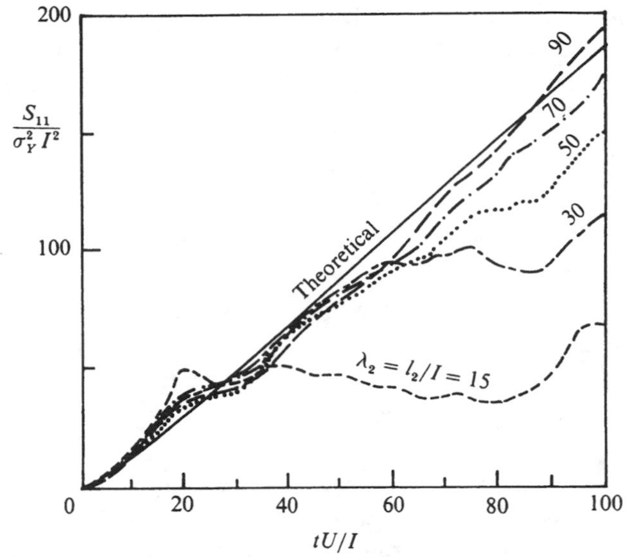

Fig. 6 The dependence of the longitudinal spatial moment S_{11} upon time in a realization of the two-dimensional velocity field and for solute bodies of different transverse initial size (advection only). The flow is of average uniform velocity U and the two-dimensional formation is of normal logtransmissivity of exponential covariance of integral scale I and variance σ_Y^2 (from Quinodoz & Valocchi, 1990).

D_{ij} tend to their usual definition in (5) if $\langle C\rangle$ satisfies such an equation.

The usefulness of D_{ij}, and even more so of S_{ij}, is that, in conjecture with $\langle \mathbf{R}\rangle$, they provide the primary information about the location of the plume and its spatial extent. They do not lead to estimates of $\langle C\rangle$ and σ_C^2 unless we adopt additional assumptions of the type mentioned above.

Even this type of information may not be accessible for the case of nonergodic plumes, which is discussed next.

4.4 Two-dimensional (regional) transport

We consider now transport of a plume over distances that are compared with the formation thickness. We also assume that the plume is large enough to ensure ergodicity of its spatial moments or other global measures resulting from the random local heterogeneity effect. At these large travel distances, the regional scale heterogeneity may start to affect the transport process. Unfortunately, systematic field experiments in which plumes have been monitored extensively are not available for transport at this scale for the practical reason that the travel times may be of the order of tens to hundreds of years. Although such long term controlled experiments do not seem feasible, we are often required to make predictions over such periods. While, in principle, modeling and simulating two-dimensional flow and transport is simpler than for the three-dimensional case, regional transport poses different and additional problems.

The main difference between local and regional heterogeneity is in the correlation scales: as we have mentioned already in Section 3, the logtransmissivity integral scale may be of the order of hundreds to thousands of meters. For travel distances smaller than or of the order of this scale, heterogeneity may appear to an observer that moves with the plume as one of an evolving, ever increasing scale.

Another difference of principle is that the plume size is generally smaller than, or of, the order of the logtransmissivity correlation scale, and the ergodic hypothesis cannot be invoked for the plume spatial moments. This is illustrated in Fig. 5, which shows in a schematic manner the plume motion. Thus, the centroid no longer moves with the mean velocity $\langle \mathbf{V}\rangle$ in any realization, and $\mathbf{R}$, its coordinate, is a random variable of time. While $d\langle \mathbf{R}\rangle/dt=\langle \mathbf{V}\rangle$, the variance $R_{ij}(t)$ characterizes the uncertainty of the centroid trajectory. In a similar vein, S_{ij}, the second spatial moment, is random and $D_{ij}=(1/2)d\langle S_{ij}\rangle/dt$ *depends on the plume size l*, decreasing with it. This point is illustrated convincingly in Fig. 6, which depicts the evolution of S_{11}, the longitudinal moment of a plume of initial size l_2, transverse to the mean flow direction, as a function of time. Each curve corresponds to a different computer generated realization of an isotropic logtransmissivity field of variance $\sigma_Y^2=0.5$ and of integral scale I, the flow being driven by a constant head drop applied on the boundary. Two main features revealed by Fig. 6 are worth discussing:

(i) First, as the initial transverse plume size l_2, relative to the integral scale I, decreases, its longitudinal spread diminishes. The simple kinematical interpretation of this result is that dispersion is caused by velocity fluctuations over scales smaller than l, whereas larger scales cause the wandering of the plume as a whole and manifest in its centroid random motion.

(ii) Secondly, it is seen that in a given realization, S_{22} is fluctuating and may even temporarily decrease. Thus, the actual dispersion coefficient defined above as $D_{11}=(1/2)dS_{11}/dt$ may become negative!

It still makes sense to define an effective dispersion coefficient as in eq. (11), i.e. $D_{ij}=\langle D\rangle_{ij}=(1/2)d\langle S_{ij}\rangle/dt$, which may be identified in each realization by a time averaging (Fig. 6).

Hence, under nonergodic conditions, the best we can do, based on the knowledge of the *unconditional* statistics of the transmissivity and velocity fields, is to predict the mean and intervals of confidence of the centroid trajectory and of higher spatial moments.

Assuming ergodicity under these circumstances has dramatic consequences upon prediction: the centroid moves with the mean velocity and the plume spreads considerably, resulting in large, unrealistic dilution. Furthermore, if the travel distance is smaller than the logtransmissivity integral scale, the *ergodic* dispersion coefficient seems to grow indefinitely with travel time, a phenomenon known as superdiffusion. It can be shown that the ergodic dispersion coefficient and the associated spatial moments provide envelopes of plume configurations over many realizations. This topic has only recently been discussed in the literature, and it will be addressed elsewhere in this book.

A final point is that conditioning, unlike the case of local transport, may have a large effect in reducing uncertainty of spatial moments, and this is the most promising avenue to take to improve prediction in regional transport. Once again, incorporating measurements in a systematic manner in stochastic modeling is one of the main novel facets of this theory.

4.5 Reactive solutes

Many contaminants and tracers are not conservative and they undergo various physical and chemical transformations, resulting in their transfer from the moving water phase to other phases. This subject, as well as that of immiscible, multiphase flow will be discussed later in this book.

A point of principle, however, is that kinetics, characterizing either sorption or transfer into zones of stagnant or very slowly moving fluid, may result in long tailing of breakthrough curves and a large increase of second spatial moments. Ignoring reactive behavior and attributing the phenomenon to heterogeneity related macrodispersion may lead to misinterpretation and to inference of unwarranted large scales of spatial variability. Another important point is that such reactive behavior may be a field scale process which is not detected by simple laboratory batch or column experiments.

5 CONCLUDING REMARKS

The aim of this chapter was to put the developments of the last two decades of stochastic modeling of subsurface flow and transport into a broad perspective. Rather than trying to cover systematically most of the advancements of the field, which is the task of the following chapters, we have analyzed a few points of principle. This has been done by going through a hypothetical field case and examining a few issues which surface in the process. We would like to emphasize here a few conclusions which seem to be particularly relevant.

First, it was shown that the process of analyzing and predicting water flow and contaminant transport in natural formations is based on a series of decisions of a conceptual and qualitative nature. The choices require applying sound judgment based on a multidisciplinary approach, intuition, experience and engineering considerations. Thus, although stochastic modeling has evolved as a branch of science, its application to real life problems is a combination of art and science.

Secondly, it is important to remember that uncertainty, whose quantitative characterization is at the heart of the stochastic approach, stems from a few sources: conceptual, parametric and spatial variability. Incorporating all these factors in prediction is important and should be kept in mind.

Thirdly, it was shown that, in spite of the tremendous progress achieved in this active field, many problems of interest, e.g. upscaling, have only been touched upon and much is left to be done. These unfinished or even untouched topics will be discussed in the following chapters.

Finally, it is fair to single out as one of the main contributions of stochastic subsurface modeling to the scientific knowledge its rather unique approach to incorporating field data by conditioning into the solution. The transition from largest uncertainty associated with the unconditional probability distributions of the properties of interest, to a practically deterministic characterization of flow and transport in presence of many measurements, is a concept of considerable theoretical and practical value.

REFERENCES

Adams, E. E. & Gelhar, L. W. (1992). Field study of dispersion in a heterogeneous aquifer 2. Spatial moments analysis. *Water Resources Research*, 28, 3293–3307.

Bierkens, M. F. P. (1994). *Complex Confining Layers: A Stochastic Analysis of Hydraulic Properties at Various Scales.* Utrecht: Netherlands Geographical Studies, p. 262.

Boggs, J. M., Young, S.C. & Beard, L. M. (1992). Field study of dispersion in a heterogeneous aquifer 1. Overview and site description. *Water Resources Research*, 28, 3281–3291.

Dagan, G. (1986). Statistical theory of groundwater flow and transport: pore to laboratory, laboratory to formation and formation to regional scale. *Water Resources Research,* 22, 120–135.

Delhomme, J. P. (1979). Spatial variability and uncertainty in groundwater flow parameters: a geostatistical approach. *Water Resources Research,* 15, 269–280.

Desbarats, A. J. & Bachu, S. (1994). Geostatistical analysis of aquifer heterogeneity from the core scale to the basin scale: A case study. *Water Resources Research,* 30, 673–684.

Hoeksema, R. J. &. Kitanidis, P. K. (1985). Analysis of spatial structure of properties of selected aquifers. *Water Resources Research,* 21, 563–572.

Quinodoz, H. A. M. & Valocchi, A. J. (1990). Macrodispersion in heterogeneous aquifers: numerical experiments. In *Proceedings of the Conference on Transport and Mass Exchange Processes in Sand and Gravel Aquifers: Field and Modeling Studies,* vol. 1, ed. G. Moltyaner. Atomic Energy Canada, pp. 465–468.

II

Subsurface characterization and parameter estimation

1 Characterization of geological heterogeneity

MARY P. ANDERSON
University of Wisconsin-Madison

Longitudinal distortion attributable to macroscopic structure can be determined on a scale consistent with the variance of hydraulic conductivity. (Kovacs, 1983)

1 INTRODUCTION

Heterogeneity of porous media has been a troublesome topic from the very beginning of groundwater hydrology as a quantitative science. Darcy (1856) recognized the necessity to quantify flow through porous media using a macroscopic, rather than a microscopic, viewpoint; he defined a flux based on an average linear flow path through a representative volume of porous media. Meinzer (1932) called heterogeneity the 'most formidable difficulty' in quantifying aquifer parameters. Shortly after this, Theis (1935) offered a solution to the heterogeneity problem by developing a way of calculating effective aquifer parameters. He demonstrated that by measuring the drawdown of water levels in response to pumping, it is possible to use an analytical solution to calculate effective aquifer parameters for average transmission and storage characteristics. Theis' method in essence replaces the heterogeneous aquifer with an equivalent homogeneous porous medium.

Theis' technique for dealing with heterogeneity allowed groundwater hydrologists to ignore geological heterogeneity for approximately 40 years. Then, Freeze (1975) called attention to the effect of uncertainty in hydraulic conductivity on the head distribution computed using groundwater models. About the same time, researchers were beginning to attempt to use the advection–dispersion equation to describe the transport of contaminant plumes in groundwater (Bredehoeft & Pinder, 1973; Pinder, 1973), and they were confronted with the problem of quantifying the dispersion coefficient. Slichter (1905) had earlier recognized that the spreading he observed in tracer experiments could not be due solely to the effects of molecular diffusion. He attributed dispersion to 'the repeated branching and subdivision of the capillary pores around the grains of the sand and gravel'. We now know that dispersion is largely caused by the presence of macroscopic heterogeneities in the subsurface.

Theis himself recognized that the approach he had pioneered based on the concept of the equivalent homogeneous porous medium would not work for problems involving solute transport. Results of a laboratory experiment demonstrating the effects of heterogeneities on the spread of tracers (Skibitzke & Robinson, 1963) may have led Theis (1967) to observe that:

> Studies of actual aquifers have shown, however, that the applicable dispersion parameters are larger by several orders of magnitude than those of the laboratory and that dispersion in actual aquifers is a phenomenon of first importance in all transport studies.

He also observed that:

> It seems obvious that mixing processes are involved in real aquifers that are not reproduced in dispersion experiments in the laboratory. It also seems obvious that the heterogeneous character of clastic sediments and other porous rocks must be involved in these mixing processes. The characteristics of the heterogeneity of sediments must probably be investigated before we can hope to understand the movement of ground water through them and to understand the phenomena of transport of particular dissolved substances, wanted or unwanted through them.

Theis (1967) also commented on the inadequacies of the concept of the equivalent homogeneous porous medium:

> The type of aquifer study in which our homogeneous model of ground water flow is most grossly inadequate is that dealing with transport phenomena. ...the simple and useful model for problems of well field development will mislead us if we apply it to problems of transport, in which we are concerned with the actual detailed movement of the water.

Theis also recognized the importance of the geological setting:

> I consider it certain that we need a new conceptual model, containing the known heterogeneities of the natural aquifer, to explain the phenomenon of transport in ground water

and he anticipated the need for field tracer studies to quantify field scale dispersion in a diversity of geological settings:

> ...the quantitative expression of the mixing techniques and the possible correlation of these with types of sedimentation and other phases of the geologic history of a formation must await many more detailed tracer studies in the field.

These were startling statements in 1967, but the implications of his remarks were not recognized at the time.

Quantifying aquifer heterogeneity is of interest, of course, for purely academic reasons, but it also has practical significance. The presence of heterogeneities in the subsurface is critical to understanding the movement of contaminants and the possibility for their removal by various remediation technologies. For example, in a recent analysis of alternatives for groundwater cleanup (National Research Council, 1994), the following research questions were identified (p. 73):

> How can the variability of aquifer properties be characterized over a sufficient range of scales – millimeters to kilometers – to understand the effect of variability on contaminant transport and cleanup system performance?
>
> Can the variability of aquifer properties be assessed adequately with statistical information developed from common geologic environments?
>
> What new ideas and techniques can contribute to development of a reliable three-dimensional map of subsurface geology and ground water flow patterns at a site?

Essaid, Herkelrath & Hess (1993) simulated movement of a crude oil plume at a site in Minnesota, USA, and showed that representation of heterogeneity was critical in reproducing the large-scale features of the plume. The small-scale details of the observed oil distribution, however, were not reproduced in the simulation owing to uncertainties about the spatial variability of hydraulic properties that actually exist in the field, even though the sediments at this site are relatively homogeneous. Theoretical investigations by other researchers (e.g. Frind, Sudicky & Schellenberg, 1989; Tompson & Gelhar, 1990) showed that heterogeneity dominates the movement of the plume particularly at early times and that the initial configuration of the plume influences its evolution in the long term.

The search for a way to incorporate heterogeneity into hydrogeological analyses was given direction by Freeze (1975), who launched a new era of stochastic analysis of groundwater systems when he called attention to the importance of uncertainty by exploring the consequences of characterizing heterogeneity stochastically. Variability was initially assumed to be spatially periodic and hydraulic conductivity was assumed to be a random variable. The statistical moments (e.g. the mean, the variance and the covariance) of the hydraulic conductivity field can be used to compute effective parameters (e.g. effective hydraulic conductivity and effective dispersivities). Also, the governing equations for flow and transport can be solved for the mean and the variance in the head and concentration distributions, thereby incorporating uncertainty into the analysis.

In this paper, the term stochastic is understood to apply to any method that requires estimates of uncertainty in parameter values and yields ranges in head and/or concentration values. This is in contrast to a deterministic approach, which assumes that parameter values are known with certainty. A deterministic model yields only one solution for heads and/or concentrations.

Much of the early literature on stochastic analysis of groundwater systems assumes that geological media are spatially periodic. But this assumption has been questioned, and investigators have proposed other conceptual models for describing heterogeneity, including fractal models and geological facies models. Analysis of spatially periodic media typically invokes Gaussian statistical models. Recently, indicator geostatistics was introduced for use with other types of conceptual models. Specifically, stochastic analysis can be applied to three classes of conceptual models: continuous, discrete, or mixed. In a continuous model, variability is spatially periodic (i.e. hydraulic conductivity is a random variable) or evolving (i.e. hydraulic conductivity is fractal). In a discrete model of heterogeneity, geological structures are included directly and special attention is focused on the continuity of high conductivity units. Discrete models may be generated using models that directly simulate geological deposition of sediments or models that rely on analysis of geological information using geostatistical approaches, such as indicator geostatistics. In a mixed model, variability is considered to be random within discrete structures or facies.

This chapter focuses on describing heterogeneity in nonfractured sedimentary deposits. Characterization of fractured media is more complex (see review papers by Wang, 1991, and Long *et al.*, this volume, Part V, Chapter 1) and fewer field studies have been conducted in this type of material. Some of the same approaches discussed below, e.g. stochastic simulation, have been applied to fractured rocks (e.g. Schwartz & Smith, 1988; Tsang & Tsang, 1989; Follin, 1992). Also, Neuman (1990, 1994, 1995) applies his universal scaling method to both fractured and non-fractured media. But a detailed review of the application of these methods to the characterization of heterogeneity in fractured rocks is beyond the scope of this chapter.

Since 1975, numerous theoretical papers have been published based on a stochastic treatment of aquifer heterogeneity (e.g. see review papers by Neuman, 1982; Sudicky & Huyakorn, 1991; Yeh, 1992), but the central question of whether the stochastic method is applicable to real aquifers under field conditions has not been definitively answered. Gelhar (1993) has observed that the state of the art in stochastic characterization of field heterogeneity can be described as 'confused' and that the limitation of obtaining field data to quantify the covariance structure of hydraulic conductivity is often 'the weakest link in the process of applying stochastic analysis' (p. 283). In view of this, we might ask whether it is appropriate to use stochastic techniques at all to describe geological materials. Can these methods capture enough about the geological structure to describe the dispersion of solutes in the subsurface?

2 CONTINUOUS HETEROGENEITY

Most commonly, a quantitative description of continuous heterogeneity invokes a Gaussian model, in which hydraulic conductivity is represented as a log normal distribution with a single correlation scale. The theory of Gaussian models is well developed, and stochastic differential equations that utilize this model can be solved analytically. The Gaussian model views heterogeneity as spatially periodic. Continuous heterogeneity that is characterized by nested scales of heterogeneity is called evolving heterogeneity and can be represented by a fractal model.

There are two main approaches in stochastic analysis that utilize continuous models. Using an analytical approach, stochastic differential equations are used to derive explicit expressions for effective (mean) hydraulic conductivity and dispersivity. The concept of the effective parameter is an extension of the paradigm of the equivalent homogeneous porous medium in deterministic modeling. In stochastic modeling, however, there is an attempt to account for uncertainty by specifying both the mean and the standard deviation of parameters. In a numerical approach, spatial statistical methods, such as random field generators, are used to produce hydraulic conductivity fields. This representation of heterogeneity is then input to numerical flow and transport models.

2.1 Gaussian models – effective parameters

Researchers have attempted to verify the validity of expressions for effective parameters derived from stochastic theory by comparing theoretical results with results from field tracer studies and with results from numerical experiments. Two types of parameters are of interest: effective hydraulic conductivity (or transmissivity) and effective dispersivities. Calculation of these parameters from stochastic theory requires information about the geostatistics of the hydraulic conductivity field, which is typically represented as log normal. Thus, the random variable of interest is ln *K*, where *K* is the hydraulic conductivity.

Geologists have been studying variability in the subsurface for more than 100 years. However, the approach has been largely descriptive. Thus, it is difficult to extract information from the sedimentological literature that is directly useful to geostatistical characterization of subsurface heterogeneity. Shortly after Freeze's (1975) seminal paper was published, hydrogeologists began to try to quantify geological heterogeneity in the field. Gelhar (1993) recently summarized most of these studies in tabular form. His table is reproduced here as Table 1. At most of the sites, the standard deviation of the natural logarithm of hydraulic conductivity (ln *K*) is approximately equal to or less than one, indicating a relatively homogeneous porous medium. Exceptions are the regional limestone aquifers studied by Delhomme (1979, and personal communication) and the alluvial aquifer at the Columbus site (Rehfeldt *et al.*, 1989).

HYDRAULIC CONDUCTIVITY

Theoretical expressions for effective hydraulic conductivity were recently summarized by Gelhar (1993) and Fenton & Griffiths (1993). Expressions for effective hydraulic conductivity derived from stochastic theory assume uniform flow. The existence of an effective hydraulic conductivity has been assumed ever since scientists began to apply Darcy's law. That is, we assume that we can find a value for hydraulic conductivity that reproduces the average flow behavior of the system. But stochastic analysis allows a quantitative evaluation of the effects of variability and thereby provides a way of addressing uncertainty in the resultant heads and flows caused by uncertainty in the hydraulic conductivity field.

Field measurements of effective hydraulic conductivity are limited, and comparisons with theory are somewhat tenuous (Gelhar, 1993, p. 150). However, a number of researchers have used numerical models to calculate an effective hydraulic conductivity (or transmissivity) and then to compare the result with the value expected from theory. The approach is to generate an hydraulic conductivity field and then calculate flow rates through the system under an imposed head gradient. The effective hydraulic conductivity is calculated as the average flow rate divided by the applied head gradient. Most of these experiments use a hydraulic conductivity field generated by a random-field generator using a Gaussian model. Multiple versions of the hydraulic conductivity field may be produced using Monte Carlo simulations, or a single realization may be used.

Table 1. *Standard deviation (S.D.) and correlation scale of the natural logarithm of hydraulic conductivity or transmissivity* (Modified from Gelhar, 1993.)

				Correlation scale (m)		Overall scale (m)	
Source	Medium	Type	S.D.	Horizontal	Vertical	Horizontal	Vertical
Aboufirassi and Marino (1984)	alluvial-basin aquifer	T	1.22	4000		30 000	
Bakr (1976)	sandstone aquifer	A	1.5–2.2		0.3–1.0		100
Binsariti (1980)	alluvial-basin aquifer	T	1.0	800		20 000	
Byers and Stephens (1983)	fluvial sand	A	0.9	>3	0.1	14	5
Delhomme (1979)	limestone aquifer	T	2.3	6300		30 000	
Delhomme (pers. commun.)							
Aquitane	sandstone aquifer	T	1.4	17 000		50 000	
Durance	alluvial aquifer	T	0.6	150		5000	
Kairouan	alluvial aquifer	T	0.4	1800		25 000	
Normandy	limestone aquifer	T	2.3	3500		40 000	
Nord	chalk	T	1.7	7500		80 000	
Devary and Doctor (1982)	alluvial aquifer	T	0.8	820		5000	
Gelhar *et al.* (1983, Fig. 6.15)	fluvial soil	S	1.0	7.6		760	
Goggin *et al.* (1988)	eolian sandstone outcrop	A	0.4	8	3	30	60
[1]Hess (1989)	glacial outwash sand	A	0.5	5	0.26	20	5
Hoeksema and Kitanidis (1985)	sandstone aquifer	T	0.6	4.5×10^4		5×10^5	
Hufschmied (1986)	sand and gravel aquifer	A	1.9	20	0.5	100	20
Loague and Gander (1990)	prairie soil	S	0.6	8		100	
Luxmoore *et al.* (1981)	weathered shale subsoil	S	0.8	<2		14	
[2]Rehfeldt *et al.* (1989)	fluvial sand and gravel aquifer	A	2.1	13	1.5	90	7
Russo (1984)	gravelly loamy sand soil	S	0.7	500		1600	
Russo and Bressler (1981)	Homra red mediterranean soil	S	0.4–1.1	14–39		100	
Sisson and Wierenga (1981)	alluvial silty-clay loam soil	S	0.6	0.1		6	
Smith (1978); Smith (1981)	glacial outwash sand and gravel outcrop	A	0.8	5	0.4	30	30
[3]Sudicky (1986)	glacial lacustrine sand aquifer	A	0.6	3	0.12	20	2
Viera *et al.* (1981)	alluvial soil (Yolo)	S	0.9	15		100	

T, transmissivity; S, soils; A, three-dimensional aquifer
[1]Cape Cod site
[2]Columbus site
[3]Borden site

Some researchers have reported success in finding an effective hydraulic conductivity that matches theory. For example, Ababou *et al.* (1989) used supercomputer simulations involving on the order of one million nodes and found good agreement with theory for statistically isotropic media. Desbarats & Srivastava (1991) found moderately good agreement between the transmissivities predicted using Gelhar & Axness's (1983) theory for an infinite, two-dimensional, statistically anisotropic field, but found excellent agreement with the theoretical value of effective transmissivity predicted for an infinite, two-dimensional, statistically isotropic field, which is the geometric mean. This good comparison is surprising inasmuch as Desbarats & Srivastava's (1991) single realization was for a simulated hydraulic conductivity field that was two-dimensional, bounded, anisotropic, and nonstationary. The good match may have been fortuitous.

The existence of a unique value for effective hydraulic conductivity has been questioned by a number of researchers, including Smith & Freeze (1979), El-Kadi & Brutsaert

(1985), Gomez-Hernandez & Gorelick (1989), and Neuman & Orr (1993), all of whom based their conclusions on the results of Monte Carlo simulations of hypothetical aquifers. Williams (1988) questioned the concept on geological grounds. It appears that it is not possible to define a unique value for effective hydraulic conductivity for a heterogeneous aquifer with irregular boundaries in which a number of wells are pumped (Gomez-Hernandez & Gorelick, 1989). If either the pumping rates or the well placement changes, the value of the effective hydraulic conductivity also changes.

DISPERSIVITIES

While field measurements of hydraulic conductivity are always affected by uncertainty, it is at least possible to obtain these estimates using relatively straightforward laboratory and field methods. A pumping test, for example, can be completed within several days and requires only a pumping well and one or more observation wells. However, it is much more difficult to obtain field values of dispersivity. These measurements require tracer tests performed over hundreds of days and involve installation of a three-dimensional network of monitoring wells. Hence, the possibility of calculating dispersivities from analytical expressions that use parameters based on the geostatistics of the hydraulic conductivity field is appealing, although Hess, Wolf & Celia (1992), who used over 1500 measurements of hydraulic conductivity to describe glacial outwash sand at the Cape Cod site, observed that the effort needed to collect enough data to quantify the variability in ln K for this relatively homogeneous site was considerable.

There has been much effort expended in testing expressions for effective dispersivities derived from stochastic theory. Theoretically calculated dispersivities are compared with dispersivities calculated from field data and/or with dispersivities computed from numerical simulations of transport in statistically generated porous media.

Comparisons with tracer studies

There has been some success in using theoretically derived expressions to calculate dispersivities for aquifers that approximate a statistically homogeneous (stationary) medium with respect to ln K. For example, Hess *et al.* (1992), using a sample population of 1500 hydraulic conductivity measurements, calculated the statistics of the sand and gravel aquifer at the Cape Cod site in Massachusetts, USA (see Table 1). Using the stochastic theories of Dagan (1988) and Gelhar & Axness (1983), they computed values for longitudinal dispersivity that agreed well with those obtained from a tracer experiment. Values of transverse dispersivity estimated from theory, however, were too low compared with field measurements. This discrepancy was attributed to transients in the flow field. Several researchers (e.g. Kinzelbach & Ackerer, 1986; Rehfeldt, 1988; Naff, Yeh & Kemblowski, 1989; Goode & Konikow, 1990; Rehfeldt & Gelhar, 1992) have demonstrated that temporal variations in hydraulic gradient can enhance transverse dispersion. When corrected for transient effects using the theory developed by Rehfeldt (1988) (see also Rehfeldt & Gelhar, 1992) better agreement with field-measured values was found.

Farrell *et al.* (1994) summarized some of the attempts to calculate dispersivities for the glacial-lacustrine sand aquifer at the Borden site. Most attempts to date have produced less than satisfactory results, particularly in the transverse direction when compared with dispersivities computed by Freyberg (1986) from tracer test data using moments analysis. Freyberg (1986) calculated longitudinal dispersivity equal to 0.36 m and horizontal transverse dispersivity equal to 0.039 m. Sudicky (1986) used stochastic theory to calculate an asymptotic longitudinal dispersivity of 0.61 m, which is at least of the same order of magnitude. However, Sudicky's (1986) theoretical estimate for asymptotic horizontal transverse dispersivity was zero. Rubin (1990) developed a more computationally intensive quasi-analytical theoretical approach to calculate dispersion coefficients. He reported relatively good agreement between results from his theory and Freyberg's field-based results for both longitudinal and transverse spreading of the tracer plume at the Borden site.

The failure of standard stochastic theory (Gelhar & Axness, 1983; Dagan, 1988) to predict transverse dispersivity for relatively homogeneous aquifers like the Borden and Cape Cod sites is discouraging. However, it appears that there is a fair amount of transience in the flow field at the Borden site. Farrell *et al.* (1994) noted that while temporal variations in hydraulic gradient are small at the Borden site, they account for almost all of the horizontal transverse spreading. Farrell *et al.* (1994) applied geostatistical analysis to 1989–1991 water level data in order to test whether the transient theories of Rehfeldt (1988) (see also Rehfeldt & Gelhar, 1992) and Naff *et al.* (1989) could be used to correct for transience and produce a better comparison between theoretical and field-measured transverse dispersivities at the Borden site. These theories assume that it is possible to model the effects of transience in the flow field by using a contaminant transport model under a steady-state flow field but with enhanced dispersion parameters derived from theory. Farrell *et al.* (1994) found that Rehfeldt's (1988) theory adequately approximated the asymptotic horizontal transverse dispersivity but that Naff *et al.*'s (1989) model did not reproduce the early time behavior of the transverse spreading of the plume. Farrell *et al.* (1994) concluded that there is no theory that adequately describes transient effects at early time and that '...a more general stochastic framework for plume evolution and dispersion in a heterogeneous media [*sic*] with an unsteady mean flow field is required.'

While transience in the flow field is likely to be a complicating factor at many sites, at geologically complex sites like the Columbus site (Boggs *et al.*, 1992) there are the additional problems of relatively large variance in ln K (Table 1) and the presence of deterministic trends. The aquifer at the Columbus site consists of poorly sorted to well-sorted alluvial terrace deposits of sand and gravel with minor amounts of silt and clay. Large-scale trends in hydraulic conductivity are evident. Calculation of dispersivity values is complicated by the presence of a non-uniform flow field caused by the trend. Adams & Gelhar (1992) used spatial moments analysis of tracer test results to calculate a range in longitudinal dispersivity between 5 and 10 m. Owing to complications in the field data, it was not possible to calculate transverse dispersivities. Rehfeldt, Boggs & Gelhar (1992) calculated longitudinal dispersivities in the range 1.5–1.8 m and a horizontal transverse dispersivity between 0.3 and 0.6 m. The theory does at least give values that are of the correct order of magnitude.

Comparisons with numerical experiments

Several investigators used random-field generators to produce hydraulic conductivity fields and then simulated tracer movement through the random hydraulic conductivity fields. In this way, dispersivities can be computed by analyzing the simulated spread of the tracer.

For example, Frind, Sudicky & Schellenberg (1989) used a turning bands random-field generator to produce three realizations of a two-dimensional hydraulic conductivity field based on the statistics of the Borden aquifer. They used grids of up to one million nodal points and ran the simulations on a supercomputer. The asymptotic value of the longitudinal dispersivity (macrodispersivity) was calculated to be 1.15 m, which was reached after the plume had traveled about 50 correlation lengths. This can be compared with the range of 0.8–1.11 m computed from stochastic theories and with the value of 0.36 m calculated from field data by Freyberg (1986). The value of horizontal transverse dispersivity calculated from the numerical results was essentially zero, in agreement with stochastic theory but at odds with Freyberg's (1986) value of 0.039 m. (Rajaram & Gelhar (1991) calculated a similar value (0.05 m) for the horizontal transverse dispersivity from an independent analysis of the field data.) Frind *et al.* (1989) also found that each realization of the hydraulic conductivity field produced a somewhat different approach to asymptotic conditions and may yield a different value for the macrodispersivity. This observation led Frind *et al.* to conclude that macrodispersivity should be calculated as an average over a number of realizations.

MacQuarrie & Sudicky (1990) simulated a tracer experiment at the Borden site involving a solution of benzene, toluene, and xylenes. They used the same procedure as Frind *et al.* (1989) to generate a random field to represent the variation in hydraulic conductivity. The organic tracer plume was affected by biodegradation, which caused the plume to diminish in size along its travel path. They found that the longitudinal dispersion coefficient calculated from the numerical experiment was much less than that predicted from standard stochastic theory, which is based on the transport of chemically conservative plumes. This finding led Frind *et al.* to conclude that the use of stochastic theory in this case is inappropriate and furthermore that it may not be possible to define a macrodispersivity for solutes undergoing biodegradation. Also see Sudicky *et al.* (1990).

Burr, Sudicky & Naff (1994) presented yet another attempt to calculate the dispersivities for the Borden site. They used numerical stochastic simulations to calculate dispersivity, following similar attempts by other investigators. They also reanalyzed the field data to compute a longitudinal dispersivity of 0.62 m. Comparisons of their numerical experiments with various theories led them to conclude that:

> The rather consistent difference between the theory and the centroid-based numerical results presented here raises some concern about the ability of traditional first-order stochastic-analytic theory to predict the actual spreading tendencies of these solutes in a real aquifer. This concern is deepened when one considers the small variance of the ln $[K(x)]$ process used in the simulations.

Rajaram & Gelhar (1993) attributed the tendency of conventional stochastic theory to overestimate the longitudinal dispersivity to the use of the second spatial moment of the ensemble mean concentration field. They found a significant reduction in the longitudinal dispersivity when the theory is based on the ensemble average of the second spatial moment of the concentration distribution. When applied to the Borden tracer experiment, their results yielded a reduction factor of 0.79 for the longitudinal dispersivity over conventional stochastic theory, bringing the theoretical results into closer agreement with field data.

Zheng, Jiao & Neville (1994) simulated the tracer experiment at the Columbus site using a three-dimensional transport model and a supercomputer. They interpolated hydraulic conductivity values from point field data and found that the results were very sensitive to the interpolation scheme. By trial-and-error calibration of the transport model to concentration data collected during the tracer experiment, they found that longitudinal dispersivities ranging from 0.2 to 5 m best represented the tracer plume. This result can be compared with a longitudinal dispersivity of 1.5 m computed from stochastic theory (Rehfeldt *et al.*, 1992) and with a range of 5 to 10 m computed from spatial moments analysis of field data (Adams & Gelhar, 1992).

A number of investigators have attempted to test stochastic theory against results from numerical experiments of solute movement in hypothetical porous media. Smith & Schwartz (1980) examined Monte Carlo simulations of random porous media using a nearest-neighbor model. Tompson & Gelhar (1990) used single realizations of an exponentially correlated three-dimensional isotropic random field generated using the turning bands method. They used different source locations and three different values for the standard deviation of ln *K*. Good agreement with theory was found when the standard deviation of ln *K* was equal to unity. With larger standard deviations, there were larger discrepancies with theory.

Jussel, Stauffer & Dracos (1994a,b) generated ten realizations of a synthetic porous medium based on field data from a gravel pit in Switzerland. They collected statistics on the geometry and density of gravel lenses in a matrix of finer material and then generated random hydraulic conductivity fields that incorporated these structural elements (Jussel *et al.*, 1994a). Jussel *et al.* (1994b) then simulated tracer movement through the hydraulic conductivity field and calculated dispersivities based on the spread of the tracer cloud after a travel distance of 100 m. Given that the generated hydraulic conductivity field does not meet the assumptions of standard stochastic theory, it is probably not surprising that there is not good agreement with theory. The value of longitudinal dispersivity calculated from the numerical experiment was 10 m, compared with a value of 0.90 m using the theory of Gelhar & Axness (1983) and 16.2 m using the theory of Dagan (1988). Transverse dispersivities calculated from theory were much smaller than those from the numerical experiment.

Desbarats & Srivastava (1991) examined transport through a deterministically generated nonstationary hydraulic conductivity field for a hypothetical porous medium. The hydraulic conductivity field showed mild heterogeneity (variance of ln *K* approximately equal to unity) but with strong trends characteristic of a fluvial environment. Dispersivity increased with distance traveled and was reasonably well predicted by theory. Flow was dominated by channeling of contaminants through high permeability material; dispersion was non-Fickian. The maximum dispersivity was lower than the macrodispersivity predicted from theory. Desbarats & Srivastava concluded that the simulated field length was insufficient to allow the development of asymptotic conditions.

In short, comparisons of theory with field data and results from numerical experiments show mixed results, with some successes and some failures.

2.2 Fractal models and nonstationarity

Gaussian models require that the ln *K* process is statistically homogeneous or stationary. Simply stated, this implies that variability is independent of location in space. See Myers (1989), for example, for a detailed discussion of this concept. While it is likely that geological media are statistically homogeneous at some scales, stationarity certainly will not apply at all scales. Observations at even relatively homogeneous sites, such as the Borden site, suggest that stationarity is not appropriate at the scale typical of most hydrogeological investigations.

While most of the literature on stochastic modeling deals with the use of Gaussian models to describe the ln *K* process, other models are possible and are becoming more common in the literature. For example, some investigators have advocated fractal models (e.g. Neuman, 1990, 1994, 1995; Neuman, Zhang & Levin, 1990; Desbarats & Bachu, 1994; see also the overview comments by Dagan (1994) and the review paper by Wheatcraft & Cushman (1991)) to represent so-called evolving heterogeneity. According to Neuman *et al.* (1990) geological media consist of discrete units that exhibit statistical homogeneity (stationarity) on a hierarchy of scales. In such multiscale media there is no one set of uniquely defined effective parameter values, and dispersion is non-Fickian. Furthermore, different effective parameters apply at different scales, and dispersion is Fickian only intermittently when a contaminant plume is confined to a discrete statistically homogeneous unit for sufficiently long transit times so that dispersion becomes Fickian. Theory consistent with this viewpoint was developed by Neuman & Zhang (1990) and Zhang & Neuman (1990), reviewed by Neuman *et al.* (1990), enhanced by Neuman (1993a), and applied by Zhang & Neuman (1995).

After examining field data summarized by Gelhar, Welty & Rehfeldt (1992), Neuman (1990) suggested a universal scaling law for longitudinal dispersivity (a_L) such that

$$a_L = 0.017 s^{1.5} \qquad (1)$$

where $s = s(t)$ is the mean travel distance of the plume and $s \leq 3500$ m. Neuman (1994, 1995), using field data summarized by Gelhar (1993), presented evidence to support his contention that the log-permeability field can be represented by a variogram of the form

$$\gamma(s) = cs^{2w} \qquad (2)$$

where c is a constant and w is a Hurst coefficient approximately equal to 0.25. The log-permeability field then is fractal. The geological basis for fractal models was discussed by Anderson (1991) and Neuman (1991, 1993b).

Neuman *et al.* (1990) reported on the application of their theory to the Borden tracer test. Their analysis shows that longitudinal spreading is non-Fickian for 2.8 years of transit time when the asymptotic value of dispersivity of 0.51 m is reached. Transverse spreading, however, does not become Fickian until 82 years of transit, when an asymptotic value of transverse dispersivity of 0.0095 m is achieved.

While fractal models do allow consideration of nested scales of heterogeneity and of time varying dispersivities, both fractal and Gaussian models are inappropriate for aquifers having strong deterministic trends. A few researchers have treated the problem of nonstationarity for continuous variation of heterogeneity (e.g. Li & McLaughlin, 1991, 1995; Loaiciga *et al.*, 1993), by applying a nonstationary spectral method to analyze flow through a continuous porous medium with a systematic trend in ln K.

Even identifying whether nonstationarity exists is not straightforward. Russo & Jury (1987), for example, observed that nonstationarity cannot always be recognized from variograms. The scale of analysis is critical when addressing the problem of stationarity. Stochastic representation of geological heterogeneity by a model that assumes stationarity hinges on whether geological systems are stationary at the scale of a hydrogeological investigation. At a large enough regional scale, geological systems are certainly nonstationary. But hydrogeological site characterization requires quantitative description of geological heterogeneities at a scale smaller than the contaminant plume, which is typically smaller than a regional scale. At relatively homogeneous sites, such as the Borden site (Sudicky, 1986) and the Cape Cod site (Hess *et al.*, 1992), an assumption of stationarity may be appropriate. But even then, heterogeneities can affect transport. For example, the tracer plume at the Borden site split into two segments during the early portion of the test (Freyberg, 1986).

At geologically complex sites where there are definite trends and geological structures (e.g. at the Columbus site (Boggs *et al.*, 1992)), the assumption of stationarity is not appropriate. Rehfeldt *et al.* (1992) argued that the trend in hydraulic conductivity at the Columbus site could be removed by third order polynomial detrending, which reduced the variance in ln K from 4.5 (Table 1) to 2.7. Similarly, the horizontal and vertical correlation lengths were reduced from 12.8 m and 1.6 m to 4.8 m and 0.8 m, respectively. Young, Herweijer & Benton (1990), however, noted that the structure of polynomial expressions used in detrending is different from the structure of geological deposits that control heterogeneity, leading them to question whether the trend could in fact be removed with polynomial detrending applied regionally across the site.

If nonstationarity exists, we are led to ask whether stochastic models can adequately represent geological nonstationarity. Stochastic models of nonstationary fields generally assume that the deterministic drift component of the random function is represented by a linear model, although other assumptions are possible (Russo & Jury, 1987). However, when performing a formal statistical removal of a deterministic trend it is necessary to express the trend mathematically. Geological studies, while generally descriptive in nature, do give some insight into the nature of trends in heterogeneity in geological systems. In order to incorporate trends into a continuous model of heterogeneity, it is necessary to quantify the trends and then remove them from the hydraulic conductivity field.

3 DISCRETE HETEROGENEITY

The Gaussian model assumes a continuous distribution of hydraulic conductivity and uses mean values for the correlation lengths, and the mean and variance of the entire hydraulic conductivity field. At some scale we know that geological media are not continuous but exhibit facies transitions and/or geological structures that affect the hydraulic conductivity field. Facies are units of similar characteristics. Geological structures include large-scale features such as faults and finer-scale features such as bedding planes and fractures. Bryant & Flint (1993, p. 15) observed that 'permeability variation is random or structured depending upon the scale of the observation'. While continuous models of heterogeneity may be appropriate for some aquifers at scales of interest in hydrogeological investigations, for most hydrogeological studies a discrete model of heterogeneity is likely to be more appropriate. The realization that continuous models of heterogeneity are inadequate to represent the geological complexities typical of most field sites has motivated attempts to apply stochastic methods to discrete models of heterogeneity.

Discrete models utilize geological information to help define units of similar hydraulic conductivity (hydrofacies). Petroleum geologists use detailed site-specific geological information together with qualitative facies models to describe geological heterogeneity for reservoir characterization (e.g. Hearn, Hobson & Fowler, 1986). Efforts to quantify discrete geological information include the use of facies models combined with a recently introduced geostatistical method known as indicator geostatistics (Journel, 1983). Depositional simulation models are also used to analyze discrete heterogeneity.

3.1 Facies models

Geological information is typically reported at two scales: a regional scale and a local scale the size of an outcrop. Additionally, petroleum geologists analyze borehole data for variations in permeability and porosity. Geologists use conceptual models to illustrate the shapes and relative relations of deposits in a particular geological environment, such as a fluvial, lacustrine, eolian, or shallow marine setting. These conceptual models usually take the form of block diagrams (Fig. 1) and/or stratigraphic columns that show the spatial

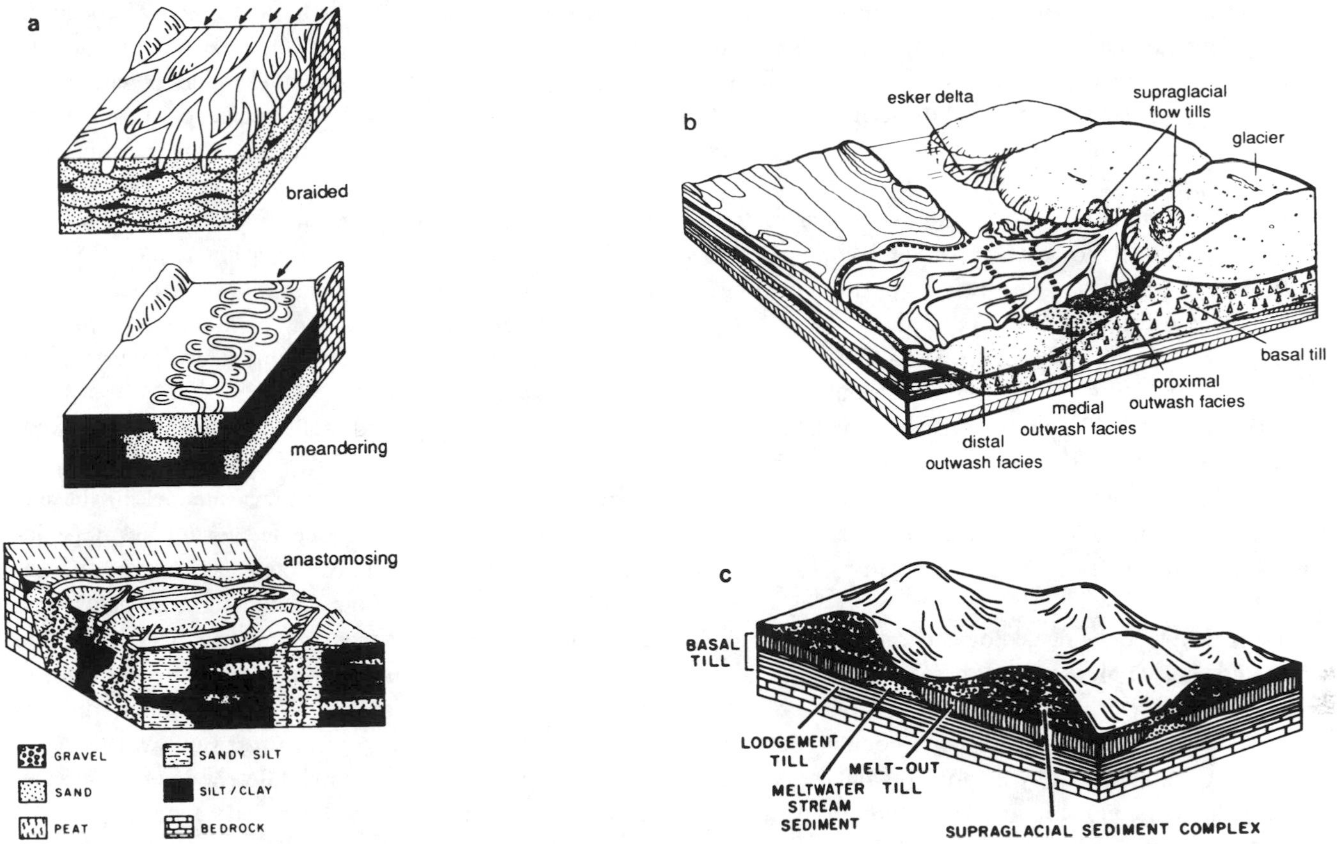

Fig. 1 Conceptual facies models. (a) Three types of river systems; (b) outwash deposits, including proximal, medial, and distal outwash facies, proceeding outward from the ice margin; (c) lodgement and basal till overlain by a supraglacial sediment complex. (From Anderson & Woessner, 1992.)

relations of facies. Hence, these conceptual models are called facies models. They may be constructed at a regional scale or a local scale.

The fundamental assumption inherent in a facies model is that facies transitions occur more commonly than would be expected if the process of deposition were random. Facies transition probabilities are computed using Markov chains (Carr, 1982) or embedded Markov analysis (Powers & Easterling, 1982; Le Roux, 1994). The facies model captures these trends, and is thought to be the most representative example of a particular sedimentary environment. Smith (1985) succinctly summarized the primary difficulty in applying facies models. He observed that the facies assemblage depicted in the facies model

> ...was in fact not actually seen in outcrop and could be thought of as the most representative depositional sequence that would have resulted had nature cooperated to its fullest. It never does, and therein lies a large part of the problem in recognizing cyclic deposits in complex environments.

Geologists compare the detailed site-specific information, which forms a site-specific facies model, with the trends predicted by a generic facies model representative of the relevant environment. In other words, geologists use a generic facies model, such as those in Fig. 1, as a guide in interpreting a site-specific depositional environment and to identify and/or verify the processes that formed the deposit.

Miall's (1985) system of architectural element analysis is one of the few geological conceptual models that addresses heterogeneity at a scale appropriate for hydrogeological analysis, which requires a scale smaller than regional but larger than an outcrop. Miall (1985) defined eight geometrically and lithologically distinct fluvial deposits, or elements, each of which consists of a suite of facies (a facies assemblage) and is associated with a specific depositional process in the fluvial environment, such as channel deposits, overbank deposits, and gravel bars. Miall's system is essentially a technique for organizing and categorizing deposits within a site-specific fluvial setting.

Anderson (1989) and Phillips, Wilson & Davis (1989) proposed the use of hydrogeological facies models, based on Miall's concept of architectural element analysis, to help quantify heterogeneity in hydrogeological site investigations. Anderson (1989) defined a hydrogeological facies to be 'a homogeneous but anisotropic unit that is hydrogeologically meaningful for purposes of field experiments and modeling'.

A hydrogeological facies is expected to have a horizontal correlation length that is finite and in most cases significantly greater than the vertical correlation length. In a similar vein, Poeter & Gaylord (1990) introduced the term hydrofacies for units of relatively homogeneous hydraulic properties that define interconnectedness of material, which controls channeling of contaminants. Just as geologists construct site-specific facies models for each area they study, hydrofacies ideally should be mapped routinely as part of a hydrogeological site investigation.

Phillips *et al.* (1989) advocated constructing a model of the 'permeability architecture' of the site that would incorporate the spatial statistics of the characteristic permeabilities of the component lithofacies, the spatial dimensions of the architectural elements, the joint probability densities for geometrical relations of the elements, the nature of the permeability transitions between elements, and the relative frequencies of the elements.

These attempts to define geological units that are useful in hydrogeological analysis hark back to Maxey's (1964) concept of the hydrostratigraphic unit, re-examined by Seaber (1988), who defined it to be 'a body of rock distinguished and characterized by its porosity and permeability'. Hydrostratigraphic units, however, are typically defined at a regional scale, whereas hydrofacies are defined at the scale of a hydrogeological site investigation.

Several investigators have attempted to apply the concept of facies to hydrogeological investigations in a qualitative way. For example, Shepherd & Owens (1981) used a fluvial facies model to site a water-supply well, and Groenewold, Rhem & Cherry (1981) used a fluvial depositional model to predict the potential for degradation of groundwater quality as a result of coal mining. Facies information can also be used to zone transmissivities for groundwater modeling studies (e.g. Davis, 1987; Danskin, 1988). Quantitative use of facies models in geostatistical characterization of heterogeneity will be much more difficult. Young *et al.* (1990) used a facies model to demonstrate the existence of nonstationarity at the Columbus site and identified the presence of 'irregular lenses of poorly-sorted to well-sorted sandy-gravel and gravelly-sand', but large uncertainties remained over the location of both the boundaries and the three-dimensional geometry of the facies. If the locations of facies boundaries could be identified, it might be possible to consider the ln K process as stationary within each facies and thereby effectively remove the nonstationarity. But it is clear that massive amounts of data are needed to develop a quantitative facies model, particularly for a site with complex geology.

GEOSTATISTICS OF FACIES DISTRIBUTIONS

Several investigators have calculated experimental variograms for geological facies. In an ideal world, we would like to discover that each facies type has a distinctive variogram. In the real world, such a conclusion is not yet supportable. Johnson (1995) summarized several examples of variograms computed for different types of sedimentary deposits. These analyses encompassed variograms based on bed thickness, permeability, and other indicators of lithologic type. The correlation distances ranged over three orders of magnitude, from 1 m to over 1000 m, for most of the reported studies. Analyses based on aquifer (pumping) tests yielded correlation ranges on the order of tens of kilometers, which is similar to the range reported in Table 1 for those estimates based on aquifer test results.

Davis *et al.* (1993) applied Miall's concept of architectural element analysis to an alluvial deposit in New Mexico in an attempt to develop a quantitative facies model for the site. They mapped architectural elements over a 0.16 km^2 peninsular section of outcrop and collected air permeability measurements at 33 points. Four architectural elements were delineated: a channel element dominated by gravelly and coarse-sand facies; a channel element dominated by sand; an overbank element consisting of fine-grained sediments deposited during flood stage; and a paleosol element.

Each facies had a significantly different mean log-permeability and a distinctive external geometry. Davis *et al.* (1993) suggested that the external geometries of the elements may exhibit a significant control on the spatial correlation structures of the log-permeability. Horizontal variograms constructed for the entire data set exhibited statistical anisotropy, with an orientation that seemed to correspond to the configuration of the paleo-drainage system. Vertical variograms of each element appeared to be controlled by the average thickness of the element. Weerts & Bierkens (1993) also found that thickness was a significant facies indicator. They demonstrated that variograms of the thickness of two different types of overbank deposits in a fluvial system in the Netherlands were significantly different and could be used to distinguish between the deposits. Variograms that showed small nugget values and a clear trend with no sill represented lithologically uniform deposits formed in a meandering river environment, while variograms with large nuggets and large sill values but a small range were representative of lithologically complex deposits formed in an anastomosing river environment (see Fig. 1).

Johnson (1995) developed variograms based on indicator geostatistics for proximal, medial, and distal alluvial fan facies assemblages for an aquifer in California, USA. He found distinct differences in the form of the variograms. The variogram for the proximal facies assemblage, which is formed near the source of the sediment, had a relatively small nugget effect and a large correlation range of about 500 m, while the distal assemblage had a large nugget effect and a smaller correlation range of 80 m. The medial assemblage had a large correlation range like the proximal assemblage, but a large nugget effect.

PREFERENTIAL FLOW PATHS

Facies models (Fig. 1) indicate that there are deterministic trends in geological environments and that interconnectedness of units is a prominent feature in many settings. Several researchers identify interconnectedness as the key to quantifying heterogeneity for purposes of hydrogeological investigations. Williams (1988) summarized the problem as follows:

> ...hydrogeologists have not demonstrated that the ensemble average (or any other average) of any hydraulic property is necessarily the critical element of interest in the theory of groundwater flow and transport.

Furthermore,

> ...identification of the so-called fastest path is critical. It is unreasonable to bury the identification of the fastest path in statistical parameters of the total rock population.

In a similar vein, Fogg (1986) noted that:

> One or two well-connected sands among a system of otherwise disconnected sands can completely alter a velocity field.

Preferential flow that causes channeling of contaminants was documented in numerical studies by several researchers, including Fogg (1986), Silliman & Wright (1988), Anderson (1990), Desbarats (1990), Desbarats & Srivastava (1991), Moreno & Tsang (1994), and Poeter & Townsend (1994). Berkowitz & Scher (1995) investigated the problem from a theoretical point of view.

Fogg (1990) reported the results of conditional stochastic simulations of sand-body distributions in a fluvial aquifer in Texas, USA. He presented interconnection probability graphs that show the probability with which sands interconnect across an entire flow region. These plots show that when the mean sand length exceeds 20% of the flow region length, the interconnection probability becomes nonzero and reaches a threshold. He suggested that interconnection probability is a useful indicator of early breakthrough of a contaminant.

Poeter & Gaylord (1990) constructed a site-specific facies model at the Hanford site, Washington, USA. Locations of preferential flow paths predicted using stratigraphic cross-sections and lithofacies maps were checked by studying the spread of a tritium plume over a ten-year period of record. While the lithofacies maps were useful in defining large-scale tritium pathways on the order of miles, they were not sufficiently detailed to serve as unambiguous predictors of preferential flow paths. Poeter & Gaylord concluded that delineation of hydrofacies, rather than lithofacies, where hydrofacies are defined to incorporate geological parameters more closely related to hydraulic properties such as degree of cementation, sorting, and sedimentary structures, would allow better prediction of preferential flow paths.

Complete characterization of the subsurface would ideally yield a site-specific three-dimensional facies model with delineation of the interconnectedness of high and low conductivity zones. Unfortunately, while generic facies models (Fig. 1) give some indication of the types of facies that may be present and the likelihood of preferential flow paths at a given site, there are no reliable techniques for accurately locating these structures in the subsurface.

Surface geophysical surveys provide some insight into subsurface heterogeneity (Olhoeft, 1994) and can yield information on overall aquifer geometry and on trends in hydraulic conductivity, but geophysical data require interpretation and results may be ambiguous. Geophysical data are used regularly for reservoir characterization in the petroleum industry. Relatively recent advances in geophysical techniques hold promise for enhanced use in both reservoir and aquifer characterization. Burns (1995), for example, has described how three-dimensional seismic surveys were used to help construct a geological model for an oil field in Saudi Arabia. Liner & Liner (1995) have discussed the use of ground-penetrating radar to map facies changes at a site in Arkansas. However, Bryant & Flint (1993) have observed that, despite these technological advances, geophysical measurements are not sufficient to describe heterogeneity in an unambiguous manner.

The work by Poeter & Gaylord (1990) and Davis *et al.* (1993) are promising first steps in the use of facies models for geostatistical characterization of subsurface heterogeneity. But it is evident that the detailed field work required to define locations and geometries of facies and their geostatistics, as well as delineation of interconnectedness, will not be practical in most situations. Furthermore, effective integration of all types of subsurface information, including lithological, geophysical, hydrological, and tracer data, are clearly helpful in characterizing subsurface heterogeneity. The problem, of course, is that rarely are multiple sets of data available for any one site. Perhaps detailed studies of many different site-specific settings will eventually lead to general principles that can be applied in the absence of detailed site data.

3.2 Indicator geostatistics

Several researchers have turned to indicator geostatistics, introduced by Journel (1983), in an attempt to represent discrete heterogeneity. For example, the studies by Davis *et al.* (1993) and Johnson (1995), discussed above, used indicator geostatistics to construct variograms for the facies assemblages they studied. In applying indicator geostatistics, the hydraulic conductivity field represented by the stochastic process, ln K, is transformed to a step function defined by cutoff values. For example, with a binary function, the data are grouped into two classes; the indicator function equals either 1 or 0. The variance and correlation length are measures of the connectivity of high conductivity units.

Desbarats (1987, 1990) used indicator geostatistics to simulate sand–shale sequences. Desbarats (1987) applied the method to a sandstone with small-scale shale features. He modeled hydraulic conductivity as a three-dimensional random function and estimated an effective permeability from simulations of flow through the dual-permeability field.

Desbarats (1990), following the approach of Desbarats (1987), used indicator geostatistics to generate a random hydraulic conductivity field, and then used Monte Carlo simulations with particle tracking to simulate tracer movement through the simulated media. The contrast in sand to shale conductivity was 10^4. He found that a Fickian model did not adequately describe longitudinal spreading in sand–shale sequences over finite fields owing to channeling of the tracer through high conductivity sands. The residence time distribution curves exhibited three segments. Over 95% of the particles arrived relatively quickly, having traveled along preferential flow paths; then followed a transition period in which very few particles arrived, followed by the gradual arrival of the rest of the particles that traveled at least partly through shale.

Johnson & Dreiss (1989) used binary indicator geostatistics to analyze more than 150 borehole logs from a complex alluvial deposit in the Santa Clara Valley, California. Their objective was to describe the boundaries of lithologic units larger than those considered by Desbarats (1987) but smaller than those traditionally mapped by sedimentologists. The resulting data set had a mean indicator value of 0.541 and a variance of 0.248. Variograms reflected observed stratigraphic features but also revealed details that could be used to infer changes in depositional environments. They also used indicator kriging to estimate probabilities of inferred high-permeability zones.

CONDITIONAL SIMULATION

In a purely stochastic approach, a random-field generator is used to create hydraulic conductivity distributions, making it possible to generate situations that are geologically unreasonable for a particular site. Such an unconditioned set of realizations may be appealing from a statistical viewpoint, but from a geological perspective it is logical to eliminate the unreasonable realizations from the data set. Another alternative is to constrain the simulation so that unreasonable realizations are not generated. Such is the rationale behind conditional simulation. Stochastic conditional simulation is also used in studying petroleum reservoirs (e.g. Ravenne & Beucher, 1988; Haldorsen and Damsleth, 1990).

A conditional simulation is constrained to maintain field-measured values of the parameter at appropriate points in the field. In a sense, the geostatistical model stochastically interpolates data so as to preserve the site-specific measured spatial variability. As explained by Fogg (1990), conditionally simulated values (Y') are calculated from

$$Y' = Y + e' \tag{3}$$

where Y represents the kriged value and e' is a stochastic term that is zero at data locations and varies quasi-randomly elsewhere, following the selected variogram structure and data distribution. When the simulated values (Y') form a normal distribution, the process is referred to as Gaussian conditioning. Multiple realizations are created by repeatedly generating e' with different random seeds using a random-field generator such as the turning bands method. Examples of studies that involved Gaussian conditioning include Wagner & Gorelick (1989) and Graham & McLaughlin (1990). An alternative to Gaussian conditioning is to use indicator geostatistics, as discussed above. Indicator geostatistics are used to create variograms for use in indicator kriging. As discussed below, several investigators advocate the use of indicator geostatistics with conditional simulation.

Teutsch, Hofmann & Ptak (1990) reported the results of conditional indicator simulations using a modified Monte Carlo approach with replacement transfer functions (Teutsch, 1992), which identify the realizations that give the fastest, slowest, and mean arrival of contaminants. They simulated three-dimensional porous media conditioned on hydraulic conductivity measurements from the Horkheimer Insel site in Germany (Schad & Teutsch, 1990). Particle tracking with replacement transfer functions was used to simulate a forced gradient tracer test, known to be affected by preferential flow.

Ritzi *et al.* (1994) used conditional indicator simulation to produce realizations of an outwash deposit in a buried-valley system in Ohio, USA. They assumed that the aquifer could be represented by two facies types: a low-permeability facies, consisting of till or lacustrine clay, and a high-permeability facies, consisting of sand and gravel outwash deposits. They used the computer code ISIM3D developed by Gomez-Hernandez & Srivastava (1990) to generate multiple realizations of the aquifer, two of which are shown in Figs. 2(a) and (b) and compared to a facies map produced using point indicator kriging (Fig. 2c). Bierkens & Weerts (1994b) used the computer code ISIM3D to generate maps of texture classes (e.g. peat, clay, and sand) for a confining layer in the Netherlands. Their results were conditioned on data obtained from drill holes.

Wen & Kung (1993) compared unconditioned Monte Carlo simulations that used a Gaussian random-field generator based on the turning bands technique, with conditional

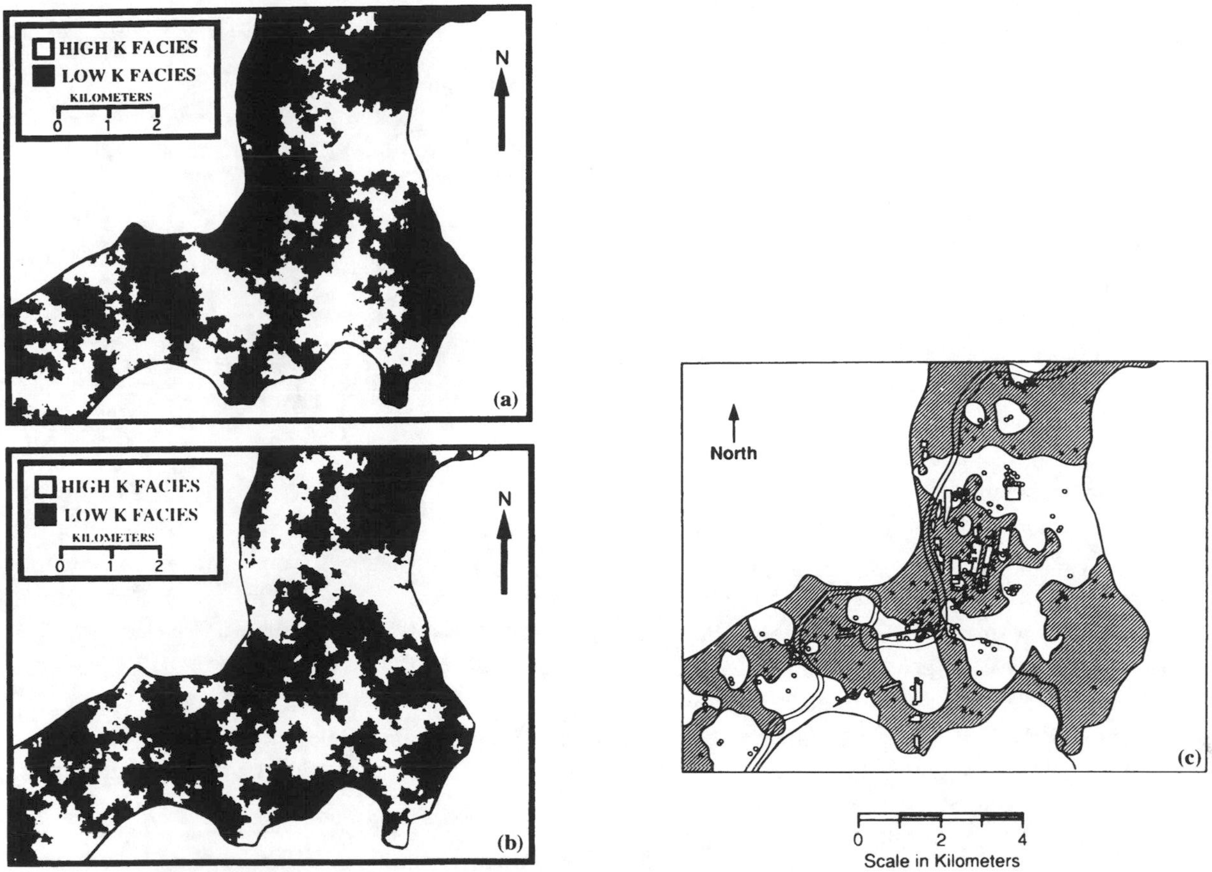

Fig. 2 Two realizations (a), (b) of an outwash aquifer with high- and low-permeability facies generated using conditional indicator simulation. (c) Distribution of high- and low- (cross-hatched areas) permeability facies based on indicator point kriging and a 0.65 probability cutoff. (From Ritzi *et al.*, 1994.)

and unconditional simulations based on indicator geostatistics. They used both methods to generate synthetic porous media and then performed particle tracking to simulate contaminant transport. They found that, in the absence of conditioning, both methods produced comparable results, but that conditional simulations predicted solute arrival time and position better than unconditional Gaussian simulations.

USE OF SOFT AND FUZZY DATA

Soft data are those data having significant associated uncertainty. Of course, hard data also have associated uncertainty, but the errors in these data are less than for soft data. For example, most geophysical data are considered to be soft, since there is significant uncertainty associated with their interpretation. Another category of data is fuzzy data, which are non-numeric data such as expert opinion.

McKenna & Poeter (1993) and Wingle & Poeter (1993) advocated the use of soft and fuzzy data to select geologically reasonable variograms. Because experimental variograms are always affected by uncertainty, Wingle & Poeter suggested the use of jackknifing to provide estimates of the error involved in constructing experimental variograms. In jackknifing, the experimental variogram is calculated with one or more data point(s) removed from the data set. The procedure is repeated for every point in the data set, yielding a series of variograms equal to the number of points in the data set. This allows the calculation of error bars and allows a likely range of reasonable variograms to be identified. From this range, a random set may be selected for stochastic simulation using Monte Carlo techniques.

Wingle & Poeter (1993) used latin-hypercube sampling combined with expert opinion to sample the variograms in order to avoid selection of variograms that are geologically unreasonable. They defended this approach as follows:

> There is little reason to evaluate solutions that are mathematically possible, but geologically improbable. Discarding geologically improbable solutions adds 'bias' to the results that may have to be defended later. However, omission of the bias means that we do not use all of the information available to us. When expert opinion is used wisely, the bias is likely appropriate, and will speed the site evaluation, thus limiting exploration and analysis costs.

Brannan & Haselow (1993) described a method for incorporating soft geologic information into stochastic simulation by using multiple indicator functions with associated cutoff levels to define hydraulic conductivity of geological materials. The distribution of ln *K* is assumed to be Gaussian within each geological unit defined by an indicator. The indicators can be defined at several scales from the microscale, to the scale of facies and facies assemblages, up to the regional scale of the hydrostratigraphic unit. The method generates lenses of geological units either randomly or constrained (conditioned) by the use of soft data. The constraints may include information on the geometry and orientation of the lenses for each indicator category.

Phillips & Wilson (1989) also advocated the use of soft geological information in stochastic modeling. They suggested that hydraulic conductivity can be correlated with soft geological information such as grain size or geophysical data. Maps of these kinds of data can then be used to estimate the geostatistics of the hydraulic conductivity field. Several other investigators have advocated combining geophysical data with hydrological data to help quantify subsurface heterogeneity (e.g. Rubin, Mavko & Harris, 1992; Copty, Rubin & Mavko, 1993; Lahm, Bair & Schwartz, 1995; Moline & Bahr, 1995). For example, it is well known that hydraulic conductivity and seismic velocity are related, and several empirical expressions have been suggested to quantify the relationship. Methods that use seismic data exploit this relation. Hyndman, Harris & Gorelick (1994) developed a parameter estimation code that uses cross-well seismic travel times and solute tracer concentrations to estimate the location and shape of large-scale heterogeneities, as well as the effective hydraulic conductivity and seismic velocity of each zone, and the effective small-scale dispersivity. James & Freeze (1993) described a Bayesian decision framework for assessing the worth of both hard and soft data.

While geophysical techniques potentially can delineate large-scale features such as buried stream channels and even finer-scale facies transitions (Liner & Liner, 1995), Teutsch *et al.* (1990) concluded that the geophysical methods available in 1990 did not have the high spatial resolution required to detect preferential flow paths. Instead they advocated the use of tracer tests and pumping tests (Ptak & Teutsch, 1990).

Wingle, Poeter & McKenna (1994) developed a computer code (UNCERT) for geostatistical analysis of uncertainty in groundwater flow and contaminant transport modeling. Software modules associated with this computer package allow the modeler to input field data, analyze the data using classical statistics, evaluate trends, and evaluate the data using geostatistics including variograms, kriging, and stochastic simulation. The philosophy used to build UNCERT stresses the importance of integrating all available data, including hard, soft, and fuzzy data, in order to constrain the results as much as possible while still allowing for reasonable variation in the interpretation (Fig. 3).

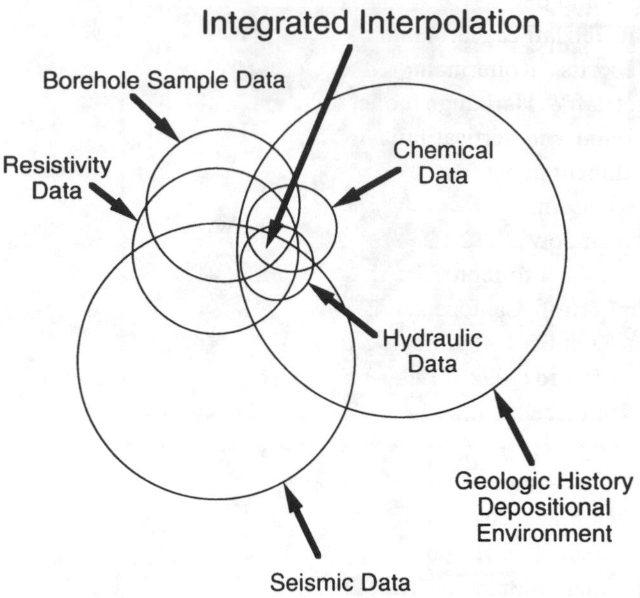

Fig. 3 Use of multiple data sets to constrain interpretations of geological heterogeneity at a site (from Wingle *et al.*, 1994).

3.3 Depositional simulation

The rationale behind depositional simulation is that heterogeneity is best approximated by simulating geological processes or structures directly. These types of models have been discussed in the geological literature over a number of years (see e.g., Harbaugh & Bonham-Carter, 1970; Bridge & Leeder, 1979; Bitzer & Harbaugh, 1987; Ross, 1989; Tetzlaff & Harbaugh, 1989; Howard, 1992; Cao & Lerche, 1994). An early paper by Price (1974), published in the hydrological literature, anticipated the interest hydrogeologists would later show in depositional simulation, but to date there have been few applications of these types of models to groundwater problems.

Depositional models that are theoretically rigorous use physically based deterministic governing equations of processes such as channel flow and transport to generate facies patterns. Other models use less rigorous approaches, such as a random walk model, to simulate channel flow. Koltermann & Gorelick (1996) distinguished between the two approaches by calling the theoretically rigorous approach one of process imitation and the less rigorous approach one of structure imitation using sedimentation patterns.

PROCESS IMITATION

Tetzlaff & Harbaugh (1989) developed a depositional process simulator capable of generating three-dimensional distributions of facies and used the model to simulate alluvial fans and deltas. Webb & Anderson (1990) used a version of the Tetzlaff & Harbaugh model to simulate braided stream deposits. Koltermann & Gorelick (1992) modified the Tetzlaff & Harbaugh model to include sea level change, horizontal and vertical fault motion, subsidence, compaction, sediment porosity, paleoclimate-driven fluctuations in floods and sediment loads, and stochastic generation of daily streamflow time series over geological time scales. They used the model to reproduce the geological evolution of an alluvial fan in California, USA, simulated on a regional scale through 600 000 years of geological time.

Howard (1992) developed a model that included equations of flow, bed topography, and sediment transport in meandering streams based on equations developed by Johannesson & Parker (1989), as well as equations for bank erosion and channel migration adapted from Howard and Knutson (1984) and a simple representation of floodplain sedimentation. Howard (1992), however, did not attempt to model individual sedimentary facies of the floodplain sediments, such as avulsion (splay) deposits, which develop as a result of a break in the levee deposits.

Juergens & Small (1990) extended the concepts presented by Allen (1978) and Bridge & Leeder (1979) in two dimensions to develop a code for simulating meandering stream deposition in three dimensions and analyzed dispersion characteristics through simulated porous media with particle tracking. Juergens (1994) added several refinements to the modeling procedure, including the use of equations formulated by Johannesson & Parker (1989). She used the code to simulate the development of four facies (channel fill, levee, splay, and overbank floodplain deposits) observed in a deposit in Texas, USA.

STRUCTURE IMITATION

Many of the models published in the geological literature contain elements of both process-imitating models and structure-imitating models. For example, Juergen's (1994) model used physically based equations to simulate channel migration but simulated avulsion using probabilistic rules controlling the timing of the avulsion and the location of a new channel. In a pioneering paper, Price (1974) used a random walk model to simulate deposition of sediments on an alluvial fan and suggested that depositional models could be used to predict hydraulic conductivity patterns for groundwater modeling. Webb (1994) used a random walk model to describe the formation of a braided channel network. He then used topographic information and relations for hydraulic geometry to construct a series of vertically stacked geomorphological surfaces (Fig. 4). Discharge through the channel network is considered, and sediment units are defined on the basis of the Froude number. For example, laminar bedded sand and trough cross-bedded fine gravel (Fig. 4(b)) could each constitute a facies defined by different Froude numbers.

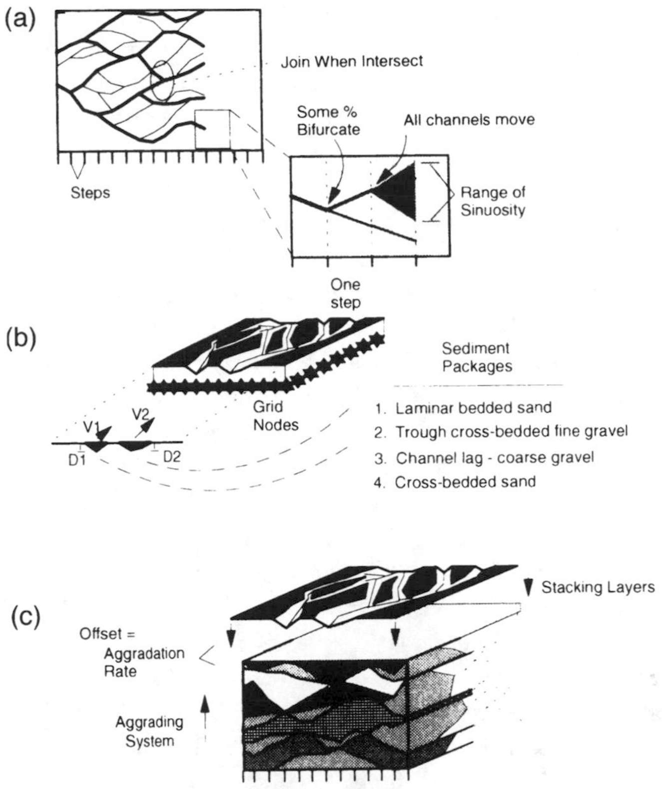

Fig. 4. Schematic representation of a structure-imitating depositional model for braided stream sediments. (a) Development of the topological channel pattern using a random walk model; (b) assignment of channel characteristics and sediment units to channel segments; (c) stacking of multiple topographic surfaces to produce a three-dimensional architecture. (From Webb, 1994.)

Webb & Anderson (1996) used Webb's model to produce a realization of a 2.8 m thick cross-bedded unit, 400 m long and 400 m wide. Hydraulic conductivities were assigned to each facies unit and a particle tracking code was used to trace out contaminant pathlines. Under four different assumptions about the relative permeability contrasts between facies and the number of facies considered, the results consistently showed that particles are channeled along preferential flow paths of high conductivity.

MODEL VALIDATION

A major challenge in testing depositional models is to demonstrate that the simulated patterns of facies are realistic. Geological data sets are generally inadequate for quantitative calibration of models of this type. Webb (1994) pointed out that appropriate calibration targets have not yet been defined for depositional models, and Howard (1992) observed that 'despite many years of observations, relevant data for model validation, calibration and revision remains fragmentary and inconclusive'.

Webb (1994) calibrated his simulated sequence of braided stream deposits to a composite set of measurements from two studies for systems with similar physical characteristics: the Ohau River in New Zealand and the Squamish River in British Columbia, Canada. Juergens (1994) calibrated her model to maps of facies distributions developed for a natural gas field. Howard (1992) found that his simulated pattern of channel migration produced spatial and temporal meander patterns that are visually similar to many meandering streams of high sinuosity observed in the field. However, statistical analysis of the simulated patterns revealed discrepancies with statistical measures from field observations, leading Howard to conclude that there are deficiencies in the theory upon which meandering stream models are based. Koltermann & Gorelick (1992) calibrated their model to 150 000 years of geological record and then simulated an additional 450 000 years of deposition. They compared the simulated deposits visually with geological cross-sections prepared from field data.

Howard (1992) observed that theoretical model development has outpaced our ability to validate these models with even empirical measures. Moreover, a method for conditioning depositional models with hard data has not yet been developed (Koltermann & Gorelick, 1996). And finally, the task of collecting the type of data needed to calibrate and validate these kinds of models is a formidable undertaking.

4 MIXED HETEROGENEITY

Continuous models of heterogeneity either eliminate the pattern of discrete structures found in geological materials by representing variability as a Gaussian random field, or assume that geological variability is fractal. Discrete models simplify the patterns produced by geological structures by reducing the variability to one that can be represented by indicator geostatistics or depositional simulation. A third alternative is to combine aspects of both types of conceptual models.

In a model of mixed heterogeneity, random variations in hydraulic conductivity are assumed to occur within discrete structures. Rubin & Journel (1991) used such a mixed model to study bimodal stratified formations by means of indicator geostatistics. Brannan & Haselow (1993) used a mixed model with indicator geostatistics to quantify discrete structures; the distribution of ln K was assumed to be Gaussian within each geological unit defined by an indicator. Similarly, Webb & Anderson (1996) assumed a Gaussian distribution of ln K within geological units produced by depositional simulation. Implicit in this approach is the assumption that variability within individual facies is stationary.

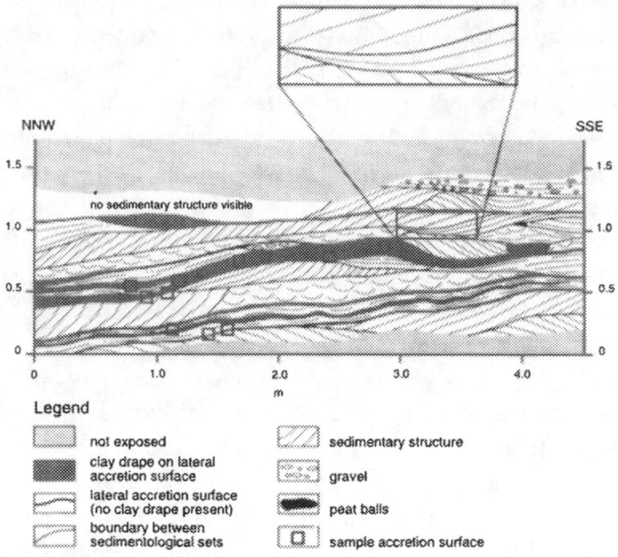

Fig. 5 Sketch of an outcrop of a point bar deposit of the Waal River in the Netherlands. The enlargement shows cross-bedding. (From Bierkens & Weerts, 1994a.)

4.1 Sedimentary structures

In order to assume that stationarity applies within individual facies, it is necessary to demonstrate that small-scale sedimentary structures do not give rise to nonstationarity. A sedimentary structure is a feature that is larger than the individual grains or mineral crystals that form the texture of the deposit. Examples of structures include bedding planes (e.g. Fig. 5), laminations, imbreccation (overlapping of pebbles), and joints (fractures). The effects of such structures have been studied relative to simulation of flow in petroleum reservoirs (e.g. Weber, 1982), where it is well known that the small-scale heterogeneities caused by sedimentary structures are important in hydrocarbon recovery (Weber, 1986).

In the hydrological literature, Bierkens & Weerts (1994a) used stochastic simulation to investigate the effect of cross-bedding on the components of the hydraulic conductivity tensor in grid blocks of a point bar deposit of a meandering river (Fig. 5). They found that when the slope angle of the accretion surfaces was larger than 6° and the core scale anisotropy ratio was larger than unity, the off-diagonal elements of the block conductivity tensor were as large as the main vertical conductivity of the block. The outcrop upon which they based their simulation (Fig. 5) has a core scale anisotropy ratio of 1.23 and a slope angle of 8–9°. Thus, Bierkens & Weert concluded that it may not be possible to ignore off-diagonal elements in numerical simulations of groundwater flow when cross-bedding is present. Similar conclusions were reached by Pickup *et al.* (1994).

Scheibe & Freyberg (1990) also developed a model of a point bar deposit as a composite of three-dimensional geometric forms. Scheibe & Cole (1994) used non-Gaussian particle tracking to simulate contaminant movement through this geologically generated field. Non-Gaussian behavior caused by the presence of the sedimentary structures was represented by modeling the correlation between components of the random dispersive vector applied at each time step in the random walk particle-tracking algorithm. They found that this treatment improved predictions of contaminant arrival times as compared with a standard Gaussian model.

The relatively small body of work completed to date on the significance of sedimentary structures in the transport of contaminants in groundwater suggests that these structures are important. However, the significance of these structures is undoubtedly dependent on the scale of the problem being considered.

5 CONCLUSIONS

The work reviewed above leads to the conclusion that at most scales likely to be of interest for hydrogeological investigations, porous media are statistically heterogeneous (nonstationary). Even in media with relatively low variance in ln *K*, such as at the Borden site, heterogeneities affect transport and cause non-Fickian dispersion. Furthermore, it seems that most plumes are likely to be affected by temporal variations in the head gradient.

Analysis of dispersion is further complicated when chemical reactions such as biodegradation or retardation are important. Analytical stochastic theories, which are based on a conceptual model of continuous heterogeneity (Brusseau, 1994), may be appropriate in a few cases where the scale of the plume is such that it falls within a single facies and is not affected by temporal variations in the flow field or chemical effects.

Depositional models are geologically appealing as a way of directly simulating facies patterns. However, they are cumbersome to use and calibration may be impossible. The most promising approach for simulating geological heterogeneity may be the use of indicator geostatistics with conditional stochastic simulation. This approach allows for statistical heterogeneity and offers a way to include facies in the simulated porous material. It can quantify the important property of connectivity of units and thereby capture preferential flow paths. Furthermore, the approach allows for conditioning with hard, soft and fuzzy data.

ACKNOWLEDGMENTS

This paper benefited from review comments by Shlomo Neuman, Donald Myers, Peter Riemersma, and Daniel Feinstein. The author benefited from conversations at the Kovacs Symposium with Georg Teutsch, Marc Bierkens, and Jaime Gomez-Hernandez.

REFERENCES

Ababou, R., McLaughlin, D., Gelhar, L. W. & Tompson, A. F. B. (1989). Numerical simulation of three-dimensional saturated flow in randomly heterogeneous porous media. *Transport in Porous Media*, 4, 549–565.

Aboufirassi, M. & Marino, M. A. (1984). Cokriging of aquifer transmissivities from field measurements of transmissivity and specific capacity. *Mathematical Geology*, 16(1), 19–35.

Adams, E. E. & Gelhar, L. W. (1992). Field study of dispersion in a heterogeneous aquifer, 2, spatial moments analysis. *Water Resources Research*, 28(12), 3293–3307.

Allen, J. R. L. (1978). Studies in fluviatile sedimentation: An exploratory quantitative model for the architecture of avulsion-controlled alluvial suites. *Sedimentary Geology*, 21, 129–147.

Anderson, M. P. (1989). Hydrogeologic facies models to delineate large-scale spatial trends in glacial and glaciofluvial sediments. *Geological Society of America Bulletin*, 101, 501–511.

Anderson, M. P. (1990). Aquifer heterogeneity – A geological perspective. In *Proceedings of the Fifth Canadian/American Conference on Hydrogeology, Banff, Alberta, Canada*, ed. B. Hitchon. Dublin, Ohio: National Water Well Association, pp. 3–22

Anderson, M. P. (1991). Comment on 'Universal scaling of hydraulic conductivities and dispersivities in geologic media' by S. P. Neuman. *Water Resources Research*, 27(6), 1381–1382.

Anderson, M. P. & Woessner, W. W. (1992). *Applied Groundwater Modeling: Simulation of Flow and Advective Transport*. San Diego: Academic Press, Inc.

Bakr, A. A. (1976). Stochastic analysis of the effects of spatial variations of hydraulic conductivity on groundwater flow, Ph.D. dissertation. Socorro: New Mexico Institute of Mining and Technology.

Berkowitz, B. & Scher, H. (1995). On characterization of anomalous dispersion in porous and fractured media. *Water Resources Research*, 31(6), 1461–1466.

Bierkens, M. F. P. & Weerts, J. J. T. (1994a). Block hydraulic conductivity of cross-bedded fluvial sediments. *Water Resources Research*, 30(10), 2665–2678.

Bierkens, M. F. P. & Weerts, J. J. T. (1994b). Application of indicator simulation to modelling the lithological properties of a complex confining layer. *Geoderma*, 62, 265–284.

Binsariti, A. A. (1980). Statistical analysis and stochastic modeling of the Cartaro aquifer in southern Arizona, Ph.D. thesis. Tucson: Department of Hydrology and Water Resources, University of Arizona.

Bitzer, K. & Harbaugh, J. (1987). DEPOSIM: A Macintosh computer model for two dimensional simulation of transport, deposition, erosion, and compaction of clastic sediments. *Computers and Geosciences*, 13, 611–637.

Boggs, J. M., Young, S. C., Beard, L. M., Gelhar, L. W., Rehfeldt, K. R. & Adams, E. E. (1992). Field study of dispersion in a heterogeneous aquifer, 1, overview and site description. *Water Resources Research*, 28(12), 3281–3292.

Brannan, J. R. & Haselow, J. S. (1993). Compound random field models of multiple scale hydraulic conductivity. *Water Resources Research*, 29(2), 365–372.

Bredehoeft, J. D. & Pinder, G. F. (1973). Mass transport in flowing groundwater, *Water Resources Research*, 9(1), 194–210.

Bridge, J. S. & Leeder, M. R. (1979). A simulation model of alluvial stratigraphy. *Sedimentology*, 26, 617–644.

Brusseau, M. L. (1994). Transport of reactive contaminants in heterogeneous porous media. *Reviews of Geophysics*, 32(3), 285–313.

Bryant, I. D. & Flint, S. S. (1993). Quantitative clastic reservoir geological modelling: problems and perspectives, In *Geological Modelling of Hydrocarbon Reservoirs and Outcrop Analogues*, eds. S. S. Flint & I. D. Bryant. Oxford: Blackwell Scientific Publications, pp. 3–20.

Burns, C. S. (1995). Seismic porosity: prediction and its impact on reservoir simulation. *Journal of Technology*, Saudi Aramco, Fall/Winter 1994–95, pp. 54–61.

Burr, D. T., Sudicky, E. A. & Naff, R. L. (1994). Nonreactive and reactive solute transport in three-dimensional heterogeneous porous media: mean displacement, plume spreading, and uncertainty. *Water Resources Research*, 30(3), 791–815.

Byers, E. & Stephens, D. B. (1983). Statistical and stochastic analyses of hydraulic conductivity and particle-size in a fluvial sand. *Soil Science Society American Journal*, 47, 1072–1081.

Cao, S. & Lerche, I. (1994). A quantitative model of dynamical sediment deposition and erosion in three dimensions. *Computers and Geosciences*, 20(4), 635–663.

Carr, T. R. (1982). Log-linear models, Markov chains, and cyclic sedimentation. *Journal of Sedimentary Petrology*, 52(3), 905–912.

Copty, N., Rubin, Y. & Mavko, G. (1993). Geophysical-hydrological identification of field permeabilities through Bayesian updating. *Water Resources Research*, 29(8), 2813–2825.

Dagan, G. (1988). Time-dependent macrodispersion for solute transport in anisotropic heterogeneous aquifers. *Water Resources Research*, 24(9), 1491–1500.

Dagan, G. (1994). The significance of heterogeneity of evolving scales to transport in porous formations. *Water Resources Research*, 30(12), 3327–3336.

Danskin, W. R. (1988). *Preliminary evaluation of the hydrogeologic system in Owens Valley, California*. Water-Resources Investigations Report 88–4003. U.S. Geological Survey.

Darcy, H. P. G. (1856). *Le fontaines publiques de la Ville de Dijon*. Paris: Victor Dalmont.

Davis, A. D. (1987). Determination of mean transmissivity values in the modeling of groundwater flow. In *Solving Problems With Ground Water Models*. Dublin, Ohio: National Water Well Association pp. 1162–1173.

Davis, J. M., Lohmann, R. C., Phillips, F. M., Wilson, J. L. & Love, D. W. (1993). Architecture of the Sierra Ladrones Formation, central New Mexico: depositional controls on the permeability correlation structure. *Geological Society of America Bulletin*, 105, 998–1007.

Delhomme, J. P. (1979). Spatial variability and uncertainty in groundwater flow parameters: A geostatistical approach. *Water Resources Research*, 15(2), 269–280.

Desbarats, A. J. (1987). Numerical estimation of effective permeability in sand-shale formations. *Water Resources Research*, 23(2), 273–286.

Desbarats, A. J. (1990). Macrodispersion in sand-shale sequences. *Water Resources Research*, 26(1), 153–163.

Desbarats, A. J. & Bachu, S. (1994). Geostatistical analysis of aquifer heterogeneity from the core scale to the basin scale: A case study. *Water Resources Research*, 30(3), 673–684.

Desbarats, A. J. & Srivastava, R. M. (1991). Geostatistical characterization of groundwater flow parameters in a simulated aquifer. *Water Resources Research*, 27(5), 687–698.

Devary, J. L. & Doctor, P. B. (1982). Pore velocity estimation uncertainties. *Water Resources Research*, 18(4), 1157–1164.

El-Kadi, A. I. & Brutsaert, W. (1985). Applicability of effective parameters for unsteady flow in nonuniform aquifers. *Water Resources Research*, 21(2), 183–198.

Essaid, H. I., Herkelrath, W. N. & Hess, K. M. (1993). Simulation of fluid distributions observed at a crude oil spill site incorporating hysteresis, oil entrapment, and spatial variability of hydraulic properties. *Water Resources Research*, 29(6), 1753–1770.

Farrell, D. A., Woodbury, A. D., Sudicky, E. A. & Rivett, M. (1994). Stochastic and deterministic analysis of dispersion in unsteady flow at the Borden tracer-test site. *Journal of Contaminant Hydrology*, 15(3), 159–185.

Fenton, G. A. & Griffiths, D. V. (1993). Statistics of block conductivity through a simple bounded stochastic medium. *Water Resources Research*, 29(6), 1825–1830.

Fogg, G. E. (1986). Groundwater flow and sand body interconnectedness in a thick, multiple-aquifer system. *Water Resources Research*, 22(5), 679–694.

Fogg, G. E. (1990). Architecture of low-permeability geologic media and its influence on pathways for fluid flow. In *Hydrogeology of Low Permeability Environments*, eds. S. P. Neuman & I. Neretnieks. Hanover: Verlag H. Heise, pp. 19–40.

Follin, S. (1992). Numerical calculations on heterogeneity of groundwater flow, Ph.D. dissertation. Stockholm: Royal Institute of Technology.

Freeze, R. A. (1975). A stochastic conceptual analysis of one-dimensional groundwater flow in non-uniform homogeneous media. *Water Resources Research*, 11(5), 725–741.

Freyberg, D. L. (1986). A natural gradient experiment on solute transport in a sand aquifer, 2, spatial moments and the advection and dispersion of nonreactive tracers. *Water Resources Research*, 22(13), 2031–2046.

Frind, E. O., Sudicky E. A. & Schellenberg, S. L. (1989). Micro-scale modelling in the study of plume evolution in heterogeneous media. In *Groundwater Flow and Quality Modelling*, eds. E. Custodio *et al.* Dordrecht: D. Reidel Publishing Co., pp. 439–461.

Gelhar, L. W. (1993). *Stochastic Subsurface Hydrology*, Englewood Cliffs, New Jersey: Prentice Hall.

Gelhar, L. W. & Axness, C. L. (1983). Three-dimensional stochastic analysis of macrodispersion in aquifers. *Water Resources Research*, 19(1), 161–180.

Gelhar, L. W., Welty, C. & Rehfeldt, K. R. (1992). A critical review of data on field-scale dispersion in aquifers. *Water Resources Research*, 28(7), 1955–1974.

Gelhar, L. W., Wierenga, P. J., Rehfeldt, K. R., Duffy, C. J., Simonett, M. J., Yeh, T-C. & Strong, W. R. (1983). *Irrigation return flow water quality monitoring, modeling and variability in the middle Rio Grande Valley, New Mexico*. U.S. Environmental Protection Agency Project Report, EPA-600/S2-83-072.

Goggin, D. J., Chandler, M. A., Kacurek, G. & Lake, L. W. (1988). Patterns of permeability in eolian deposits: Page sandstone (Jurassic), Northeastern Arizona. *SPE Formation Evaluation*, June, pp. 297–306.

Gomez-Hernandez, J. J. & Gorelick, S. M. (1989). Effective groundwater model parameter values: influence of spatial variability of hydraulic conductivity, leakance, and recharge. *Water Resources Research*, 25(3), 405–419.

Gomez-Hernandez, J. J. & Srivastava, R. M. (1990). ISIM3D: An ANSI-C three dimensional multiple indicator conditional simulation program. *Computers and Geosciences*, 16(4), 395–440.

Goode, D. J. & Konikow, L. F. (1990). Apparent dispersion in transient groundwater flow. *Water Resources Research*, 26(10), 2339–2351.

Graham, W. D. & McLaughlin, D. (1990). Stochastic modeling in groundwater: Application to a field tracer test. In *Calibration and Reliability in Groundwater Modeling*, ed. K. Kovar. IAHS Publ. no. 195, Wallingford: IAHS, pp. 501–510.

Groenewold, G. H., Rhem, B. W. & Cherry, J. A. (1981). Depositional setting and groundwater quality in coal-bearing sediments and spoils in western North Dakota. In *Recent and Ancient Non-Marine Depositional Environments: Models for Exploration*, eds. F. G. Ethridge and R. M. Flores. Special Publication 31. Tulsa: Society of Economic Paleontologists and Mineralogists, pp. 157–167.

Haldorsen, H. H. & Damsleth, E. (1990). *Stochastic Modeling*. Richardson, Texas: Society of Petroleum Engineers, pp. 404–412.

Harbaugh, J. W. & Bonham-Carter, G. (1970). *Computer Simulation In Geology*. New York: J. Wiley & Sons, Inc.

Hearn, C. L., Hobson, J. P. & Fowler, M. L. (1986). Reservoir characterization for simulation, Hartzog Draw Field, Wyoming. In *Reservoir Characterization*, eds. L. W. Lake & H. B. Carroll, Jr. Orlando, Florida: Academic Press, Inc., pp. 341–372.

Hess, K. M. (1989). Use of a borehole flow meter to determine spatial heterogeneity of hydraulic conductivity and macrodispersion in a sand and gravel aquifer, Cape Cod, Massachusetts. In *Proceedings of the Conference on New Field Techniques for Quantifying the Physical and Chemical Properties of Heterogeneous Aquifers*, eds. F. J. Molz *et al.* Dublin, Ohio: National Water Well Association, pp. 497–508.

Hess, K. M., Wolf, S. H. & Celia, M. A. (1992). Large-scale natural gradient tracer test in sand and gravel, Cape Cod, Massachusetts, 3, hydraulic conductivity variability and calculated macrodispersivities. *Water Resources Research*, 28(8), 2011–2027.

Hoeksema, R. J. & Kitanidis, P. K. (1985). Analysis of the spatial structure of properties of selected aquifers. *Water Resources Research*, 21(4), 563–572.

Howard, A. D. (1992). Modeling channel migration and floodplain sedimentation in meandering streams. In *Lowland Floodplain Rivers: Geomorphological Perspectives*, eds. P. A. Carling & G. E. Petts. Chichester: John Wiley & Sons, pp. 1–41.

Howard, A. D. & Knutson, T. R. (1984). Sufficient conditions for river meandering: a simulation approach. *Water Resources Research*, 20(11), 1659–1667.

Hufschmied, P. (1986). Estimation of three-dimensional, statistically anisotropic hydraulic conductivity field by means of single well pumping tests combined with flow meter measurements. *Hydrogeologie*, 2, 163–174.

Hyndman, D. W., Harris, J. M. & Gorelick, S. M. (1994). Coupled seismic and tracer test inversion for aquifer property characterization. *Water Resources Research*, 30(7), 1965–1977.

James, B. R. & Freeze, R. A. (1993). The worth of data in predicting aquitard continuity in hydrogeological design. *Water Resources Research*, 29(7), 2049–2065.

Johannesson, H. & Parker, G. (1989). Velocity redistribution in meandering rivers. *Journal of Hydraulic Engineering*, 115(8), 1019–1039.

Johnson, N. M. (1995). Characterization of alluvial hydrostratigraphy with indicator semivariograms. *Water Resources Research*, 31(12), 3217–3227.

Johnson, N. M. & Dreiss, S. J. (1989). Hydrostratigraphic interpretation using indicator geostatistics. *Water Resources Research*, 25(12), 2501–2510.

Journel, A. G. (1983). Nonparametric estimation of spatial distributions. *Mathematical Geology*, 15(3), 445–468.

Juergens, L. J. (1994). *Geologic process modeling and spatial analysis of fluvial/floodplain facies for subsurface characterization*. Ph.D. thesis, Carnegie-Mellon University.

Juergens, L. J. & Small, M. (1990). A model of channel deposition and resulting interconnectedness and dispersion. *EOS, American Geophysical Union*, 71(17), 511.

Jussel, P., Stauffer, F. & Dracos, T. (1994a). Transport modeling in heterogeneous aquifers: 1. Statistical description and numerical generation of gravel deposits. *Water Resources Research*, 30(6), 1803–1817.

Jussel, P., Stauffer, F. & Dracos, T. (1994b). Transport modeling in heterogeneous aquifers: 2. Three-dimensional transport model and stochastic numerical tracer experiments. *Water Resources Research*, 30(6), 1819–1831.

Kinzelbach W. & Ackerer, P. (1986). Modelisation de la propagation d'un contaminant dans un champ d'ecoulement transitoire. *Hydrogeologie*, 2, 197–205.

Koltermann, C. E. & Gorelick, S. M. (1992). Paleoclimatic signature in terrestrial flood deposits. *Science*, 256, 1775–1782.

Koltermann, C. E. & Gorelick, S. M. (1996). Heterogeneity in sedimentary deposits: A review of structure-imitating, process-imitating and descriptive approaches. *Water Resources Research*, 32(9), 2617–2658.

Kovacs, G. (1983). The influence of micro- and macro-structure of aquifers on the spreading of pollutants. In *Relation of Groundwater Quantity and Quality*, IAHS Publ. no. 146. Wallingford: IAHS, pp. 115–121.

Lahm, T. D., Bair, E. S. & Schwartz, F. W. (1995). The use of stochastic simulations and geophysical logs to characterize spatial heterogeneity in hydrogeologic parameters. *Mathematical Geology*, 27(2), 259–278.

Le Roux, J. P. (1994). Spreadsheet procedure for modified first-order embedded Markov analysis of cyclicity in sediments. *Computers and Geosciences*, 20(1), 17–22.

Li, S. & McLaughlin, D. (1991). A non-stationary spectral method for solving stochastic groundwater problems: Unconditional analysis. *Water Resources Research*, 27(7), 1589–1605.

Li, S. & McLaughlin, D. (1995). Using the nonstationary spectral method to analyze flow through heterogeneous trending media. *Water Resources Research*, 31(3), 541–552.

Liner, C. L. & Liner, J. L. (1995). Ground-penetrating radar: A near-surface experience from Washington County, Arkansas. *The Leading Edge*, pp. 17–21.

Loague, K. & Gander, G. A. (1990). R-5 revisited, 1, Spatial variability of infiltration on a small rangeland catchment. *Water Resources Research*, 26(5), 957–971.

Loaiciga, H. A., Leipnik, R. B., Marino, M. A. & Hudak, P. F. (1993). Stochastic groundwater flow analysis in the presence of trends in heterogeneous hydraulic conductivity fields. *Mathematical Geology*, 25(2), 161–176.

Luxmoore, R. J., Spaulding, B. P. & Munro, I. M. (1981). Areal variation and chemical modification of weathered shale infiltration characteristics. *Soil Science Society of America Journal*, 45, 687–691.

McKenna, S. A. & Poeter, E. P. (1993). Conditioning of hydrofacies simulations with data. In *Proceedings Ground Water Modeling Conference*. Golden, Colorado: International Ground Water Modeling Center, pp. 4-25–4-34

MacQuarrie, K. T. B. & Sudicky, E. A. (1990). Simulation of biodegradable organic contaminants in groundwater, 2, Plume behavior in uniform and random flow fields. *Water Resources Research*, 26(2), 223–239.

Maxey, G. B. (1964). Hydrostratigraphic units. *Journal of Hydrology*, 2, 124–129.

Meinzer, O. E. (1932). *Outline of methods for estimating ground-water supplies*. Water Supply Paper 638c, U.S. Geological Survey.

Miall, A. (1985). Architectural-element analysis: A new method of facies analysis applied to fluvial deposits. *Earth-Science Reviews*, 22, 261–308.

Moline, G. R. & Bahr, J. M. (1995). Estimating spatial distributions of heterogeneous subsurface characteristics by regionalized classification of electrofacies. *Mathematical Geology*, 27(1), 3–22.

Moreno, L. & Tsang, C-F. (1994). Flow channeling in strongly heterogeneous porous media: a numerical study. *Water Resources Research* 30(5), 1421–1430.

Myers, D. E. (1989). To be or not to be...stationary? That is the question. *Mathematical Geology*, 21(3), 347–362.

Naff, R. L., Yeh, J. T-C. & Kemblowski, M. W. (1989). Reply to comment on 'A note on the recent natural gradient tracer test at the Borden site', by G. Dagan. *Water Resources Research*, 25(12), 2523–2525.

National Research Council (1994). *Alternatives for Ground Water Cleanup*. Washington, D.C.: National Academy Press.

Neuman, S. P. (1982). Statistical characterization of aquifer heterogeneities: an overview. *Geological Society of America*, Special Paper 189, pp. 81–102.

Neuman, S. P. (1990). Universal scaling of hydraulic conductivites and dispersivities in geologic media. *Water Resources Research*, 26(8), 1749–1758.

Neuman, S. P. (1991). Reply to Comment by M. P. Anderson. *Water Resources Research*, 27(6), 1749–1758.

Neuman, S. P. (1993a). Eulerian-Lagrangian theory of transport in space-time nonstationary velocity fields: exact nonlocal formalism by conditional moments and weak approximation. *Water Resources Research*, 29(3), 633–645.

Neuman, S. P. (1993b). Comment on 'A critical review of data on field-scale dispersion in aquifers' by L. W. Gelhar, C. Welty and K. R. Rehfeldt. *Water Resources Research*, 29(6), 1863–1865.

Neuman, S. P. (1994). Generalized scaling of permeabilities: validation and effect of support scale. *Geophysical Research Letters*, 21(5), 349–352.

Neuman, S. P. (1995). On advective transport in fractal permeability and velocity fields. *Water Resources Research*, 31(6), 1455–1460.

Neuman, S. P. & Orr, S. (1993). Prediction of steady state flow in nonuniform geologic media by conditional moments: exact nonlocal formalism, effective conductivities, and weak approximation. *Water Resources Research*, 29(2), 341–364.

Neuman, S. P. & Zhang, Y-K. (1990). A quasi-linear theory of non-Fickian and Fickian subsurface dispersion, 1, Theoretical analysis with application to isotropic media. *Water Resources Research*, 26(5), 887–902.

Neuman, S. P., Zhang, Y-K. & Levin, O. (1990). Quasilinear analysis, universal scaling, and Lagrangian simulation of dispersion in complex geological media. In *Dynamics of Fluids in Hierarchical Porous Media*, ed. J. H. Cushman. San Diego: Academic Press, pp. 349–391.

Olhoeft, G. R. (1994). Geophysical observations of geological, hydrological and geochemical heterogeneity. In *Proceedings of a Symposium on the Application of Geophysics to Engineering and Environmental Problems*. Boston, MA.

Phillips, F. M. & Wilson, J. L. (1989). An approach to estimating hydraulic conductivity spatial correlation scales using geological characteristics. *Water Resources Research*, 25(1), 141–143.

Phillips, F. M., Wilson, J. L. & Davis, J. M. (1989). Statistical analysis of hydraulic conductivity distributions: a quantitative geological approach. In *Proceedings of New Field Techniques for Quantifying the Physical and Chemical Properties of Heterogeneous Aquifers*. Columbus, Ohio: National Ground Water Association.

Pickup, G. E., Rigrose, P. S., Jensen, J. L. & Sorbie, K. S. (1994). Permeability tensors for sedimentary structures. *Mathematical Geology*, 26(2), 227–250.

Pinder, G. F. (1973). A Galerkin-finite element simulation of groundwater contamination on Long Island, New York. *Water Resources Research*, 9(6), 1657–1669.

Poeter, E. & Gaylord, D. R. (1990). Influence of aquifer heterogeneity on contaminant transport at the Hanford Site. *Ground Water*, 28(6), 900–909.

Poeter, E. & Townsend, P. (1994). Assessment of critical flow path for improved remediation management. *Ground Water*, 32(3), 439–447.

Powers, D. W. & Easterling, R. G. (1982). Improved methodology for using embedded Markov chains to describe cyclical sediments. *Journal of Sedimentary Petrology*, 52(3), 913–923.

Price, W. E. (1974). Simulation of alluvial fan deposition by a random walk model. *Water Resources Research*, 10(2), 263–274.

Ptak, T. & Teutsch, G. (1990). Some new hydraulic and tracer measurement techniques for heterogeneous aquifer formations. In *Proceedings of The International Conference and Workshop on Transport and Mass Exchange Processes in Sand and Gravel Aquifers: Field and Modelling Studies*. Ottawa, Canada: AECL.

Rajaram, H. & Gelhar, L. W. (1991). Three-dimensional spatial moments analysis of the Borden tracer test. *Water Resources Research*, 27(6), 1239–1251.

Rajaram, H. & Gelhar, L. W. (1993). Plume scale-dependent dispersion in heterogeneous aquifers, Parts 1–2. *Water Resources Research*, 29(9), 3249–3276.

Ravenne, C. & Beucher, H. (1988). Recent development in description of sedimentary bodies in a fluvio deltaic reservoir and their 3D conditional simulations. *Society of Petroleum Engineers*, SPE 18310, pp. 463–476.

Rehfeldt, K. R. (1988). *Prediction of macrodispersivity in heterogeneous aquifers*. Ph.D. thesis, Massachusetts Institute of Technology.

Rehfeldt, K. R. & Gelhar, L. W. (1992). Stochastic analysis of dispersion in unsteady flow in heterogeneous aquifers. *Water Resources Research*, 28(7), 2085–2099.

Rehfeldt, K. R., Boggs, J. M. & Gelhar, L. W. (1992). Field study of dispersion in a heterogeneous aquifer, 3. Geostatistical analysis of hydraulic conductivity. *Water Resources Research*, 28(12), 3309–3324.

Rehfeldt, K. R., Gelhar, L. W., Southard, J. B. & Dasinger, A. M. (1989). *Estimates of macrodispersivity based on analyses of hydraulic conductivity variability at the MADE site*, EPRI EN-6405, Project 2485–5. Palo Alto, CA: Electric Power Research Institute Report.

Ritzi, R. W., Jayne, D. F., Zahradnik, A. J. Jr., Field, A. A. & Fogg, G. E. (1994). Geostatistical modeling of heterogeneity in glaciofluvial, buried-valley aquifers. *Ground Water*, 32(4), 666–674.

Ross, W. C. (1989). Modeling base-level dynamics as a control on basin-fill geometries and facies distribution: A conceptual framework. In *Quantitative Dynamic Stratigraphy*, ed. T. A. Cross. Englewood Cliffs, NJ: Prentice Hall.

Rubin, Y. (1990). Stochastic modeling of macrodispersion in heterogeneous porous media. *Water Resources Research*, 26(1), 133–141.

Rubin, Y. & Journel, A. G. (1991). Simulation of non-Gaussian space random functions for modeling transport in groundwater. *Water Resources Research*, 27(7), 1711–1721.

Rubin, Y., Mavko, G. & Harris, J. (1992). Mapping permeability in heterogeneous aquifers using hydrologic and seismic data. *Water Resources Research*, 28(7), 1809–1816.

Russo, D. (1984). Geostatistical approach to solute transport in heterogeneous fields and its applications to salinity management. *Water Resources Research*, 20(9), 1260–1270.

Russo, D. & Bressler, E. (1981). Soil hydraulic properties as stochastic processes, I, An analysis of field spatial variability. *Soil Science Society of America Journal*, 45, 682–687.

Russo, D. & Jury, W. A. (1987). A theoretical study of the estimation of the correlation scale in spatially variable fields, 2, nonstationary fields. *Water Resources Research*, 23(7), 1269–1280.

Schad, H. & Teutsch, G. (1990). Statistical and geostatistical analysis of field and laboratory hydraulic measurements at the 'Horkheimer Insel' field site. In *Proceedings of the International Conference and Workshop on Transport and Mass Exchange Processes in Sand and Gravel Aquifers: Field and Modelling Studies*. Ottawa, Canada: AECL.

Scheibe, T. D. & Cole, C. R. (1994). Non-Gaussian particle tracking: Application to scaling of transport processes in heterogeneous porous media. *Water Resources Research*, 30(7), 2027–2039.

Scheibe, T. D. & Freyberg, D. L. (1990). Impacts of geological structure on transport: creating a data base. In *Proceedings of the Fifth Canadian/American Conference on Hydrogeology, Banff, Alberta, Canada*, ed. B. Hitchon. Dublin, Ohio: National Water Well Association, pp. 56–71.

Schwartz, F. W. & Smith, L. (1988). A continuum approach for modeling mass transport in fractured media. *Water Resources Research*, 24(8), 1360–1372.

Seaber, P. R. (1988). Hydrostratigraphic units. In *Hydrogeology, The Geology of North America*, Vol. O-2, eds. W. Back, J. S. Rosehshein & P. R. Seaber. Boulder, CO: Geological Society of America, pp. 9–14.

Shepherd, R. G. & Owens, W. G. (1981). Hydrogeologic significance of Ogallala fluvial environments, the gangplank. In *Recent And Ancient Non-Marine Depositional Environments: Models for Exploration*, Special Publication 31. Society of Economic Paleontologists and Mineralogists, pp. 89–94.

Silliman, S. E. & Wright, A. L. (1988). Stochastic analysis of paths of high hydraulic conductivity in porous media. *Water Resources Research*, 24(11), 1901–1910.

Sisson, J. B. & Wierenga, P. J. (1981). Spatial variability of steady-state infiltration rates as a stochastic process. *Soil Science Society of America Journal*, 45, 699–704.

Skibitzke, H. E. & Robinson, G. M. (1963). *Dispersion in groundwater flowing through heterogeneous materials*. U.S. Geological Survey Professional Paper 386-B.

Slichter, C. S. (1905). *Field measurements of the rate of movement of underground waters*. U.S. Geological Survey Water-Supply and Irrigation Paper No. 140.

Smith, L. (1978). A stochastic analysis of steady state groundwater flow in a bounded domain. Ph.D. thesis, University of British Columbia.

Smith, L. (1981). Spatial variability of flow parameters in a stratified sand. *Mathematical Geology*, 13(1), 1–21.

Smith, L. & Freeze, R. A. (1979). Stochastic analysis of steady state groundwater flow in a bounded domain, 2. Two-dimensional simulations. *Water Resources Research*, 15(6), 1543–1559.

Smith, L. & Schwartz, F. W. (1980). A stochastic analysis of macroscopic dispersion. *Water Resources Research*, 16(2), 303–313.

Smith, N. D. (1985). Proglacial fluvial environment. In *Glacial Sedimentary Environments*. Short Course No. 16, eds. G. M. Ashley, J. Shaw & N. D. Smith. Tulsa: Society of Paleontologists and Mineralogists, pp. 85–134

Sudicky, E. A. (1986). A natural gradient experiment on solute transport in a sand aquifer: spatial variability of hydraulic conductivity and its role in the dispersion process. *Water Resources Research*, 22(13), 2069–2082.

Sudicky, E. A. & Huyakorn, P. S. (1991). Contaminant migration in imperfectly known heterogeneous groundwater systems. *Reviews of Geophysics*, Supplement, April 1991, pp. 240–253.

Sudicky, E. A., Schellenberg, S. L. & MacQuarrie, K. T. B. (1990). Assessment of the behaviour of conservative and biodegradable solutes in heterogeneous porous media. In *Dynamics of Fluids in Hierarchical Porous Media*, ed. J. H. Cushman. San Diego: Academic Press, pp. 429–461.

Tetzlaff, D. M. & Harbaugh, J. W. (1989). *Simulating Clastic Sedimentation*. New York: Van Nostrand Reinhold.

Teutsch, G. (1992). Stochastic groundwater transport simulation using replacement transfer functions (RTFs). In *Geostatistical Methods: Recent Developments and Applications in Surface and Subsurface Hydrology*, ed. A. Bardossy. Paris: UNESCO, pp. 32–39.

Teutsch, G., Hofmann, B. & Ptak, T. (1990). Non-parametric stochastic simulation of groundwater transport processes in highly heterogeneous porous formations. In *Proceedings of the International Conference and Workshop on Transport and Mass Exchange Processes in Sand and Gravel Aquifers: Field and Modelling Studies*. Ottawa, Canada: AECL.

Theis, C. V. (1935). The relation between the lowering of the piezometric surface and the rate and duration of discharge of a well using groundwater storage. *Transactions of the American Geophysical Union*, 2, 519–524.

Theis, C. V. (1967). Aquifers and models. In *Proceedings of the Symposium on Ground-Water Hydrology*, ed. M. A. Marino. Minneapolis: American Water Resources Association, p. 138.

Tompson, A. F. B. & Gelhar, L. W. (1990). Numerical simulation of solute transport in three-dimensional, randomly heterogeneous porous media. *Water Resources Research*, 26(10), 2541–2562.

Tsang, Y. W. & Tsang, C. F. (1989). Flow channeling in a single fracture as a two-dimensional strongly heterogeneous permeable medium. *Water Resources Research*, 25(9), 2076–2080.

Viera, S. R., Nielsen, D. R. & Biggar, J. W. (1981). Spatial variability of field measured infiltration rate. *Soil Science Society of America Journal*, 45, 1040–1048.

Wagner, B. J. & Gorelick, S. M. (1989). Reliable aquifer remediation in the presence of spatially variable hydraulic conductivity: from data to design. *Water Resources Research*, 25(10), 2211–2225.

Wang, J. S. Y. (1991). Flow and transport in fractured rocks. In *Contributions in Hydrology, U.S. National Report 1987–1990*. Washington, D.C.: American Geophysical Union, pp. 254–262.

Webb, E. K. (1994). Simulating the three-dimensional distribution of sediment units in braided-stream deposits. *Journal of Sedimentary Research*, B64(2), 219–231.

Webb, E. K. & Anderson, M. P. (1990). Tracking contaminant pathways in sandy, heterogeneous, glaciofluvial sediment using a sedimentary depositional model. In *Transport and Mass Exchange in Sand and Gravel Aquifers: Field and Modeling Studies*, ed. G. Moltayaner, AECL 10 308, vol. 1 Chalk River, Ontario: Atomic Energy of Canada Limited, pp. 243–354.

Webb, E. K. & Anderson, M. P. (1996). Simulation of preferential flow in three-dimensional, heterogeneous conductivity fields with realistic internal architecture. *Water Resources Research*, 32(3), 533–545.

Weber, R. J. (1982). Influence of common sedimentary structures on fluid flow in reservoir models. *Journal of Petroleum Technology*, March 1982, pp. 666–672.

Weber, R. J. (1986). How heterogeneity affects oil recovery. In *Reservoir Characterization*, eds. L. W. Lake & H. B. Carroll, Jr. Orlando: Academic Press, pp. 487–544.

Weerts, H. J. T. & Bierkens, M. F. P. (1993). Geostatistical analysis of overbank deposits of anastomosing and meandering fluvial systems: Rhine-Meuse delta, the Netherlands. *Sedimentary Geology*, 85, 221–232.

Wen, X-H. & Kung, C-S. (1993). Stochastic simulation of solute transport in heterogeneous formations: A comparison of parametric and nonparametric geostatistical approaches. *Ground Water*, 31(6), 953–965.

Wheatcraft, S. W. & Cushman, J. H. (1991). Hierarchical approaches to transport in heterogeneous porous media. *Reviews of Geophysics*, Supplement, March 1991, pp. 263–269.

Williams, R. (1988). Comment on 'Statistical theory of groundwater flow and transport: pore to laboratory, laboratory to formation, and formation to regional scale' by G. Dagan. *Water Resources Research*, 24(7), 1197–1200.

Wingle, W. L. & Poeter, E. P. (1993). Uncertainty associated with semi-variograms used for site simulation. *Ground Water*, 31(5), 725–734.

Wingle, W. L., Poeter, E. P. & McKenna, S. A. (1994). *UNCERT User's Guide, Draft*. Developed for the U.S. Bureau of Reclamation, Golden, CO: Colorado School of Mines.

Yeh, T-C. Jim, (1992). Stochastic modelling of groundwater flow and solute transport in aquifers. *Hydrological Processes*, 6, 369–395.

Young, S. C., Herweijer, J. & Benton, D. J. (1990). Geostatistical evaluation of a three-dimensional hydraulic conductivity field in an alluvial terrace aquifer. In *Fifth Canadian/American Conference on Hydrogeology Proceedings*. Dublin, Ohio: National Water Well Association, pp. 116–137.

Zhang, D. & Neuman, S. P. (1995). Eulerian-Lagrangian analysis of transport conditions on hydraulic data, Parts 1–4. *Water Resources Research*, 31(1), 39–88.

Zhang, Y-K. & Neuman, S. P. (1990). A quasi-linear theory of non-Fickian and Fickian subsurface dispersion, 2, Application to anisotropic media and the Borden site. *Water Resources Research*, 26(5), 903–914.

Zheng, C., Jiao, J. J. & Neville, C. J. (1994). Numerical simulation of a large-scale tracer test in a strongly heterogeneous aquifer, abstract. *Eos supplement*, Nov. 1, 1994, p. 284.

2 Application of geostatistics in subsurface hydrology

F. JAVIER SAMPER CALVETE

Universidad de La Coruña

ABSTRACT Geostatistics is a theory that was developed in the 1960s to deal with the analysis and estimation of spatially distributed variables having a stochastic spatial structure. Initially it was applied to mining engineering, but later found interesting applications in many other fields such as subsurface hydrology. A brief description of the geostatistical theory and a review of the most commonly applied geostatistical methods is first presented. Most relevant properties of the spatial correlation structure of some selected hydrogeological variables, including permeability, transmissivity and hydraulic head are described. Early applications of geostatistics in subsurface hydrology dealt with estimating hydrogeological variables at unsampled locations by means of point kriging and obtaining the corresponding map of estimation errors. Block kriging has been generally used to estimate block transmissivities in numerical flow models. With the increasing recognition of the paramount effects of spatial variability, geostatistical simulation gained more and more relevance. The particular properties of hydrogeological data (scarcity, variable support, measurement errors) compelled hydrogeologists to develop improved methods for estimating both drift and spatial covariance parameters. Geostatistical methods have been applied recently also to analyze hydrochemical and isotopic data. Another group of geostatistical applications in subsurface hydrology is related to optimum monitoring and observation network design.

1 INTRODUCTION

Geostatistics, a term coined by the French statistician G. Matheron of the Ecole des Mines Superieur de Paris in France, is a theory dealing with the estimation of regionalized variables. A regionalized variable (ReV) is any function z(**x**) that depends on the spatial location **x** and that exhibits a stochastic spatial structure. Examples of ReVs in subsurface hydrology are the hydraulic head in an aquifer and the concentration of a chemical species in groundwater. The spatial variation of these variables frequently shows a random component associated with erratic fluctuations, and a slowly varying aspect that reflects an overall trend or 'drift' of the phenomenon under study. Consider for example the spatial variation of the concentration c(**x**) of sodium in a regional aquifer. Due to cation exchange processes it is commonly seen that sodium concentrations increase from recharge to discharge areas. This concentration trend would correspond to the drift of c(**x**). Fluctuations around this trend may occur due to unpredictable local variations in the lithology of the aquifer. While the trend in c(**x**) can be adequately described using deterministic methods, its erratic fluctuations require a probabilistic interpretation. The need to describe the high-frequency erratic variability of a ReV was the motivation for introducing statistical tools in the analysis of spatial variability.

In geostatistics a ReV z(**x**) is interpreted as a realization of a random function (RF) Z(**x**). Two immediate questions arise when this probabilistic approach is taken. The first question is whether it makes sense to consider a natural phenomenon which is known to be unique as a random process. If the values of the variable were known at every point in space, the answer to this question would be negative. However, the lack of perfect and complete measurement makes uncertain our knowledge of the variable, thus justifying the probabilistic approach. The second question refers to the statistical inference of the RF. Generally it is not possible to infer the distribution function of a RF Z(**x**) from a single realization. Nonetheless, this inference problem can be solved if additional assumptions about Z(**x**) are introduced. The effect of these assumptions is to reduce the number of parameters needed to describe the distribution of Z(**x**). For instance, the stationarity assumption can be seen as if the RF would 'repeat' itself in space.

Early applications of geostatistics in subsurface hydrology focused on the analysis of hydraulic conductivity K (most often its logarithm), transmissivity T, fracture density f and hydraulic head h with the purpose of preparing contour maps of estimated values and their corresponding estimation errors. Examples of such applications are abundant. Kriging techniques were also found appropriate to derive prior block transmissivities for groundwater flow numerical methods.

Classical geostatistical techniques were developed for applications in mining engineering. Data availability, sampling patterns and especially the underlying processes of mining variables are very different from those in subsurface hydrology. For this reason, hydrogeologists soon found existing geostatistical methods insufficient for analyzing hydrogeological data. The limitations of available techniques and the challenges posed by the particular nature of hydrogeological variables compelled hydrogeologists to develop new techniques with which to provide more adequate answers.

One of the first problems found by hydrogeologists was related to the scarcity of data. Data availability in hydrology is orders of magnitude smaller than in mining engineering. These data often have variable supports, ranging from core scale K measurements up to field test T values. Most often, hydraulic heads and water samples for chemical analyses represent average values over the whole saturated thickness of the well. Transmissivity data are obtained from the interpretation of pumping tests. In this case, interpretation errors add to measurement errors, causing the data to contain large errors. A proper estimation of the parameters of both the spatial covariance function and the drift (when present) is crucial since most stochastic models of groundwater flow and solute transport assume they are given. Partly to fulfil this need and to overcome the limitations of current semivariogram and drift estimation methods, hydrologists have concentrated on parametric estimation methods.

We start by presenting a brief description of the basic geostatistical theory and assumptions. The theory of the most popular geostatistical applications, including semivariogram estimation, linear estimation methods of both intrinsic and nonintrinsic variables, nonparametric methods and simulation techniques, is also reviewed.

We later present a section on the most relevant geostatistical applications to subsurface hydrology, starting with a description of the spatial correlation behavior of some selected hydrogeological variables. Geostatistical applications to analyze hydrochemical and isotopic data are also reviewed. Finally, we describe applications related to network design.

2 GEOSTATISTICAL THEORY AND METHODS

2.1 Basic theory

In linear geostatistics, only the first two moments of the RF $Z(\mathbf{x})$ are required. The first order moment of $Z(\mathbf{x})$, $m(\mathbf{x})=E[Z(\mathbf{x})]$, generally a function of the location $\mathbf{x}$, is the drift or trend. Second order moments of the RF $Z(\mathbf{x})$ include the variance, the autocovariance, and the semivariogram. The latter is most common in geostatistics and is defined as

$$\gamma(\mathbf{x},\mathbf{x}+\mathbf{h})=(1/2)E\{[Z(\mathbf{x}+\mathbf{h})-Z(\mathbf{x})]^2\} \quad (1)$$

A random function $Z(\mathbf{x})$ is said to be strictly stationary if for any finite set of n points $\mathbf{x}_1,\mathbf{x}_2,\ldots,\mathbf{x}_n$ and for any vector $\mathbf{h}$, the joint distribution function of the n random variables $Z(\mathbf{x}_j)$, $j=1,2,\ldots,n$ is the same as that of the $Z(\mathbf{x}_j+\mathbf{h})$, $j=1,2,\ldots,n$. Inasmuch as only the first and second moments of the RF are used in linear geostatistics, stationarity can be weakened to stationarity of the first two moments. Accordingly, a RF $Z(\mathbf{x})$ is said to be second order stationary or weakly stationary if

$$E[Z(\mathbf{x})]=m \quad \text{for all } \mathbf{x} \quad (2)$$

$$\mathrm{Cov}[Z(\mathbf{x}+\mathbf{h}),Z(\mathbf{x})]=E[Z(\mathbf{x}+\mathbf{h})Z(\mathbf{x})]-m^2=C(\mathbf{h}) \quad \text{for all } \mathbf{x} \quad (3)$$

i.e. the mean and the autocovariance exist but are not dependent on $\mathbf{x}$. The stationarity of the autocovariance implies that the variance is finite and independent of $\mathbf{x}$, $\mathrm{Var}[Z(\mathbf{x})]=\sigma^2=C(0)$. Similarly, the semivariogram is stationary, independent of $\mathbf{x}$, and related to the autocovariance through

$$\gamma(\mathbf{h})=\sigma^2-C(\mathbf{h}) \quad (4)$$

Clearly for a second order stationary RF, the autocovariance and the semivariogram are equivalent as there is a one-to-one correspondence between the two functions. Random functions with finite second moments not satisfying conditions (2) and (3) are nonstationary. Among the nonstationary RFs, there is class of functions for which these conditions hold locally. If the drift function is smooth it can be considered nearly constant over small regions. A RF that satisfies (2) and (3) locally is said to be quasi-second order stationary or locally second order stationary. Another class of RF includes those having stationary first order increments. These functions are known as intrinsic and satisfy the following conditions:

$$E[Z(\mathbf{x}+\mathbf{h})-Z(\mathbf{x})]=0 \quad (5)$$

$$\mathrm{Var}[Z(\mathbf{x}+\mathbf{h})-Z(\mathbf{x})]=2\gamma(\mathbf{h}) \quad (6)$$

Thus, the increments $Z(\mathbf{x}+\mathbf{h})-Z(\mathbf{x})$ of an intrinsic RF have zero mean and variance equal to twice the semivariogram. While a second order stationary RF is always intrinsic, the reciprocal is not necessarily true. This is one reason why in geostatistical applications the intrinsic hypothesis is often preferred over second order stationarity.

The semivariogram of a second order stationary RF reaches a constant value called the sill, which is equal to the variance of the RF. The distance at which the semivariogram attains its sill is known as the range; beyond it Z is no longer autocorrelated. By definition, $\gamma(\mathbf{0})=0$; however, the semivariogram sometimes shows a discontinuity near the origin known as the nugget effect.

When conditions (5) and (6) are satisfied only locally, the RF is said to be quasi-intrinsic or locally intrinsic. Random functions that are neither stationary nor intrinsic are called nonintrinsic.

2.2 Estimation of intrinsic random functions

One important aspect of geostatistics is to estimate a two- or three-dimensional field from a set of measurements $Z_i=Z(\mathbf{x}_i)$, $i=1,2,\ldots,N$. This estimation problem is known as 'kriging', and was named by G. Matheron in honor of the South African mining engineer Daniel G. Krige, who first introduced an early version of the method. When all the information available about $Z(\mathbf{x})$ comprises the N measured values $Z_1,Z_2,\ldots,Z_N$, the best possible estimator (i.e. the one having the smallest estimation variance) of Z at a location $\mathbf{x}_0$ where Z has not been measured is the conditional expectation $E[Z_0|Z_1,Z_2,\ldots,Z_N]$. This expectation, however, requires knowledge of the joint distribution of the N+1 variables $Z_0,Z_1,Z_2,\ldots,Z_N$. In practice, the inference of this joint distribution is not possible because the information about the RF $Z(\mathbf{x})$ is limited to a single realization. This problem can be overcome by resorting to the ergodic hypothesis. A random function is said to be ergodic if all its statistics can be derived from a single realization (Papoulis, 1965). In practice, ergodicity is only required for the first two moments. $Z(\mathbf{x})$ is ergodic on the mean if the ensemble mean (the mean over all possible realizations) can be derived from averaging the values of a single realization. When $Z(\mathbf{x})$ is multivariate Gaussian, the conditional expectation can easily be obtained. In this case, only the first two moments of $Z(\mathbf{x})$ are required, and the conditional expectation is identical to the best linear estimator (Journel & Huijbregts, 1978). Between the linear estimator and the conditional expectation is the disjunctive kriging estimator, proposed by Matheron (1976), which is more accurate than the linear estimator and does not require knowledge of the conditional distributions. Other proposed nonlinear estimators include indicator and probability kriging (Journel, 1984a). Though better, most nonlinear estimators require more computation time than linear estimators (Kim *et al.*, 1977).

The kriging estimator Z_0^* of Z_0 (the value of Z at point $\mathbf{x}_0$) for an intrinsic RF with unknown mean (usually referred to as ordinary kriging) is given by a linear combination of the N measured values:

$$Z_0^*=\sum_{i=1}^{N}\lambda_i Z_i \tag{7}$$

The kriging weights λ_i are obtained upon requiring that Z_0^* be a minimum-variance unbiased estimator of Z_0. The lack of bias condition yields

$$\sum_{i=1}^{N}\lambda_i=1 \tag{8}$$

The estimation variance $E[Z_0-Z_0^*]^2$ is minimum when

$$\sum_{j=1}^{N}\lambda_j\gamma_{ji}+\mu=\gamma_{0i} \qquad i = 1,2,\ldots,N \tag{9}$$

where $\gamma_{ij}=\gamma(\mathbf{x}_j-\mathbf{x}_i)$, $\gamma_{0i}=(\mathbf{x}_0-\mathbf{x}_i)$, and μ is a Lagrange multiplier corresponding to condition (8). Equations (8) and (9) define a system of N+1 linear equations with N+1 unknowns, $\lambda_1,\lambda_2,\ldots,\lambda_N$ and μ. The kriging variance σ_0^2 is given by

$$\sigma_0^2=\sum_{i=1}^{N}\lambda_i\gamma_{i0}+\mu \tag{10}$$

When estimates are calculated at a number of locations, one can compute the covariance matrix of the estimates.

The kriging estimator is sometimes referred to as BLUE, i.e. best linear unbiased estimator. However, kriging is best only among linear estimators. Only if data are jointly Gaussian is it also best among all unbiased nonlinear estimators.

The kriging estimator of a second order stationary RF with known mean and covariance, known as simple kriging, has an estimation variance smaller than that of the ordinary kriging estimator (Matheron, 1971). When both $C(\mathbf{h})$ and m are estimated from the data using minimum-variance unbiased estimates, both simple and ordinary kriging provide the same estimation variance. Inasmuch as neither the mean nor the covariance function are known in practice, and since the semivariogram exists for a wider class of RF than the autocovariance, ordinary kriging is usually preferred to simple kriging. Given the relationship between the covariance and the semivariogram functions (eq. (4)), one can derive the kriging equations for the case of stationary RF with unknown mean from those of an intrinsic RF just by replacing $\gamma(\mathbf{h})$ by the covariance (Journel & Huijbregts, 1978).

The kriging estimator in (7) is an exact interpolator, that is $Z_i^*=Z_i$, for all the N measured values. Notice that no assumption is made about the type of distribution of $Z(\mathbf{x})$. When $Z(\mathbf{x})$ is Gaussian, the kriging estimator coincides with the conditional expectation. The kriging weights λ_i depend on the semivariogram $\gamma(\mathbf{h})$, the relative location of the sample points, and the locations of the point of estimation. Points close to $\mathbf{x}_0$ have generally more influence on the estimated value than points farther away because their corresponding weights are larger. If all the sample points are located at distances away from $\mathbf{x}_0$ that exceed the range of the semivariogram, the estimator reduces to the arithmetic mean with all the weights equal to 1/N and the estimation variance is equal to the variance of the random function. Notice that the kriging variance is determined by the semivariogram and the sample pattern but not by the sample values. This property has been widely used in the context of optimal network design (Delhomme, 1978; Hughes & Lettenmaier, 1981; Sophocleous, 1983; Carrera *et al.*, 1984; Bogardi *et al.*, 1985). The estimator in (7) can be a global estimator when all data are used for estimating Z_0. However, when the number of data is large, the size of the kriging system of equations becomes prohibitive. In this case, a local estimator based on a subset of the N available points is used instead. The local estimator assumes that the drift may change from one point to another.

It is known from inference theory that the conditional expectation E(Y|X) has a variance smaller than that of the variable Y. This explains why the kriging method usually produces smooth surfaces that coincide with the measured values at the sample locations.

Kriging is considered to be robust in that the kriging estimates are not sensitive to small deviations of the semivariogram from its true value. However, studies of the consistency and convergence of the kriging method are not abundant in the statistical literature. Yakowitz and Szidarovszky (1985) showed that in the absence of a drift, when the true semivariogram is known, continuous at the origin and the measurements contain no errors, the kriging estimator is consistent. Furthermore, Z_0^* tends to Z_0 when N increases regardless of the semivariogram used. If the measurements contain noise, Z_0^* converges to Z_0 when the semivariogram is correct. This means that the kriging estimator is robust. Yakowitz & Szidarovszky (1985) also showed that the computed kriging variance converges to its true value as N goes to infinity when the semivariogram is correct, is continuous near the origin and there is no drift. Clearly, a correct estimation of the semivariogram is most important for a meaningful evaluation of the kriging results.

Kriging can also be used when the measurements contain errors (Delhomme, 1978). The corresponding kriging system is similar to (9) except that the diagonal terms decrease by the amount of the variance of the measurement errors.

In some applications the average value of the RF over some support V is desired:

$$Z_V(\mathbf{x})=(1/V)\int_{(\mathbf{x}-\mathbf{u})\in V} Z(x+u)du \tag{11}$$

Here $\mathbf{x}$ denotes the centroid of V. The estimation of the average $Z_V(\mathbf{x})$ is known as block kriging. The estimator $Z_V^*(\mathbf{x}_0)$ has a form similar to (7), except that the kriging weights λ_i now satisfy equation (8) with a modified version of equation (9), in which λ_{0i} is replaced by $\gamma(\mathbf{x}_j,V)$, the spatial average of $\gamma(\mathbf{x}_j-\mathbf{s})$ over all $\mathbf{s}\in\mathbf{V}$. The kriging variance is given by

$$\sigma^2_V=-\gamma(V,V)+\sum_{i=1}^{N}\gamma(\mathbf{x}_i,V)+\mu \tag{12}$$

where $\gamma(V,V)$ is the average of $\gamma(\mathbf{s},\mathbf{t})$ over all $\mathbf{s},\mathbf{t}\in V$.

The kriging method can be easily extended to the case of several spatially correlated variables. In such cases the estimator is known as cokriging (Myers, 1982).

2.3 Estimation of nonstationary random functions

Many variables of practical interest do not satisfy the intrinsic hypothesis. Hydrogeological examples of variables that may exhibit a spatial drift are hydraulic head, aquifer thickness, and the concentration of dissolved species and environmental tracers (Myers *et al.*, 1982; Gilbert & Simpson, 1985; Samper & Neuman, 1989c). Nonintrinsic random functions are usually represented as the sum of a deterministic drift m(**x**) and a random component, $\epsilon(\mathbf{x})$ (Matheron, 1971). The latter is usually assumed to be intrinsic with zero mean and a semivariogram $\gamma_\epsilon(\mathbf{h})$. This decomposition of Z(**x**) is then

$$Z(\mathbf{x})=m(\mathbf{x})=\epsilon(\mathbf{x}) \tag{13}$$

with E[Z(**x**)]=m(**x**). The semivariogram of Z, $\gamma_Z(\mathbf{h})$ is given by

$$\gamma_Z(\mathbf{x},\mathbf{h})=(1/2)[m(\mathbf{x}+\mathbf{h})-m(\mathbf{x})]^2+\gamma_\epsilon(\mathbf{h}) \tag{14}$$

Proposed methods for estimating nonintrinsic random functions include the following:

(a) It is assumed that the RF is locally intrinsic and using local ordinary kriging.

(b) It is assumed that the drift function m(**x**) has a known form such as a low order polynomial and that the semivariogram $\gamma_\epsilon(\mathbf{h})$ is known. The universal kriging method (Matheron, 1963) stems from these assumptions.

(c) It is assumed that the drift m(**x**) has a known form and can be estimated from available data. Once an estimate m*(**x**) of the drift has been obtained, one performs kriging on the residuals R(**x**)=Z(**x**)−m*(**x**). In an early version of this method due to Gambolati and Volpi (1979) the form of the drift is deduced from physical considerations and its coefficients are then estimated using ordinary least squares. Neuman & Jacobson (1984) found this method internally inconsistent. They proposed a modified version in which the drift is estimated in two phases. In the first phase, which serves to identify the order of the polynomial drift, the residuals are assumed uncorrelated. In the second phase, the correlation among residuals is taken into account by using generalized least squares for estimating the drift. Inasmuch as the estimation of the semivariogram of the residuals $\gamma_R(\mathbf{h})$ affects the estimation of the drift, and vice versa, one has to estimate both $\gamma_R(\mathbf{h})$ and $m(\mathbf{x})$ iteratively until the sample semivariogram of the residuals in two consecutive iterations remains unchanged.

(d) It is assumed that increments of order k of the random function are second order stationary. This method, originally presented by Matheron (1973) and later developed by Delfiner (1976), is based on a generalization of the concept of intrinsic RFs. If taking first order increments filters out a linear drift, one expects to filter higher order drifts by considering higher order increments of the RF. In this method the drift is locally represented by a polynomial of order k. The increments of order k of the RF have a covariance structure $K(\mathbf{h})$ referred to as generalized covariance. When the kriging equations are written in terms of the generalized covariance, they attain a form similar to the equations arising from universal kriging, which are written in terms of the semivariogram $\gamma_\in(\mathbf{h})$ (Journel, 1989). The estimation of a RF using the generalized covariance method requires knowledge of the order k of the RF and the parameters of $K(\mathbf{h})$.

The primary application of the theory of nonintrinsic random functions in subsurface hydrology has been the estimation of groundwater hydraulic heads. For that purpose, Delhomme (1978) and Aboufirassi & Mariño (1984) used universal kriging; Gambolati & Volpi (1979) applied direct residual kriging; and Binsariti (1980), Fennessy (1982), Neuman & Jacobson (1984), and Neuman *et al.* (1987) used iterative residual kriging. Applications of the method of generalized covariance can be found in Neuman and Jacobson (1984), Hernández (1986), and Rouhani (1986).

2.4 Estimation of semivariogram and covariance functions

Estimation of the semivariogram (or the autocovariance) is critical for geostatistics. Methods for estimating semivariograms can be classified into five categories: (1) method of moments, (2) least squares methods, (3) maximum likelihood, (4) cross-validation, and (5) methods for estimating generalized covariance functions.

The earliest semivariogram estimator was proposed by Matheron (1963), who interpreted $2\gamma(\mathbf{h})$ to be the mean of $[Z(\mathbf{x}+\mathbf{h})-Z(\mathbf{x})]^2$. An unbiased estimator of $\gamma(\mathbf{h})$ obtained from $N(\mathbf{h})$ data pairs $[Z(\mathbf{x}_i), Z(\mathbf{x}_i+\mathbf{h})]$ is given by

$$\gamma^*(\mathbf{h})=\frac{1}{2N(\mathbf{h})}\sum_i [Z(\mathbf{x}_i+\mathbf{h})-Z(\mathbf{x}_i)]^2 \quad (15)$$

where $\gamma^*(\mathbf{h})$ is the sample or experimental semivariogram. When data locations are irregularly spaced, the number $N(\mathbf{h})$ of data pairs with separation distance $\mathbf{h}$ is generally small and the sample semivariogram has a very large variance. In order to increase the number of data pairs, it is common to consider a series of intervals (h_k,h_{k+1}) of length $L_k=h_{k+1}-h_k$ along the direction $\mathbf{h}$. The value of the sample semivariogram at some intermediate distance h_k^*, $h_k \leq h_k^* \leq h_{k+1}$, is computed using all data pairs $[Z(\mathbf{x}_i),Z(\mathbf{x}_j)]$ having separation vector $(\mathbf{x}_i - \mathbf{x}_j)$ that fall within the interval (h_k, h_{k+1}). For two- or three-dimensional variables this may not be enough to yield a large value of $N(\mathbf{h})$ since data locations are rarely aligned. In this case, it is common to define a window tolerance around the vector $\mathbf{h}$. In selecting the size of the intervals L_k and the tolerance, one has to compromise between ensuring a sufficiently large value of $N(\mathbf{h})$ and not losing accuracy due to the averaging process. There are no general rules for the optimal choice of L_k, and in general fluctuations in the values of $\gamma^*(\mathbf{h})$ tend to decrease as L_k increases (Myers *et al.*, 1982). The statistical properties of the sample semivariogram are difficult to study because its estimation variance $E[\gamma(\mathbf{h})-\gamma^*(\mathbf{h})]^2$ involves moments of fourth order, which are difficult to estimate from a single realization of a random function. It is known from inference theory that $\gamma^*(\mathbf{h})$ in eq. (15) is an optimal estimator only when $[Z(\mathbf{x}_i+\mathbf{h}),Z(\mathbf{x}_i)]$ are binormally distributed and all the observations are uncorrelated (Omre, 1984). These stringent conditions are rarely satisfied in real applications, which explains why $\gamma^*(\mathbf{h})$ has a very poor statistical behavior. For example, it is very sensitive to the presence of abnormally high values. In addition, when the random function has a highly skewed distribution (such as log-normal), the sample semivariogram shows an erratic behavior with frequent fluctuations. Kridge & Magri (1982) also found that the logarithmic transformation leads to a relatively stable sample semivariogram for highly skewed variables. Another advantage of the logarithmic transformation is that the effect of extremely high values (outliers) is diminished (Kridge and Magri, 1982).

As pointed out by Matheron (1963), the sample semivariogram suffers from large variability at large distances. Starks & Fang (1982) demonstrated with artificial data how misleading the sample semivariogram can be when a drift is present. To overcome the problems associated with sample semivariograms, resistant and robust semivariogram estimators have been proposed by various researchers (Verly *et al.*, 1984). According to Armstrong & Delfiner (1980), possible sources of nonrobust behavior of the sample semivariogram include artifacts, typographical errors, unsuitable choice of distance classes, the presence of two or more different populations due to the effect of different geological formations (heterogeneity), skewed distributions such as those of concentrations, and the presence of extremely high data values such as those commonly found for chemical concentrations in polluted aquifers. A careful analysis of the data using ordinary cleaning and screening methods helps to eliminate most of these problems. One should also take into account the physical nature of the data in distinguishing among outliers, skewed distributions, and mixed populations.

Equation (15) provides an estimate of the semivariogram at a given distance h. For kriging, however, one needs a semivariogram function that yields positive variances for any linear combination. For that, the semivariogram function must be conditionally negative definite (Journel & Huijbregts, 1978). Another property of the semivariogram is that the limit of $\gamma(\mathbf{h})/\mathbf{h}^2$ as $\mathbf{h}$ goes to infinity must be equal to zero (Matheron, 1971). A valid semivariogram function therefore cannot increase with distance faster than $\mathbf{h}^2$ does. Any function satisfying the previous conditions is a valid semivariogram model. Examples of valid models are the quadratic, the exponential, the Gaussian, the power model, the hole effect, and the nugget effect which corresponds to a purely random process. Inasmuch as it is not simple to check whether the sample semivariogram $\gamma^*(\mathbf{h})$ satisfies the conditions for a valid semivariogram, one has to adopt a valid model from those listed above. In practice, one fits a semivariogram model to the sample semivariogram. The traditional approach of fitting a semivariogram 'by eye', as proposed by Clark (1979), has its advantages, but is subjective, limited to intrinsic random functions, and its quality remains unknown.

Well established practice has shown that least squares fitting of a semivariogram model to the sample semivariogram, as proposed by Bastin & Gevers (1985), Tough & Leyshon (1985) and others, is not recommended.

The cross-validation method has been extensively used (Delfiner, 1976; Davis & David, 1978; Delhomme, 1978; Candy & Mao, 1981; Hughes & Lettenmaier, 1981; Baafi *et al.*, 1982, Starks & Fang, 1982; and Aboufirassi & Mariño, 1984), and recognized as an optimal estimation method (Gambolati & Volpi, 1979; Dowd, 1984). Cross-validation is a method for evaluating the adequacy of a semivariogram model. It consists of deleting each measured value Z_i and estimating it using the rest of the measured values. The estimate Z_i^* of Z_i is obtained by ordinary kriging. Repeating the estimation process at M ($M \leq N$) locations, one can compute the cross-validation errors as $e_i = Z_i^* - Z_i$. In trial and error cross-validation, one tries different sets of covariance parameters until the cross-validation results (the errors e_i and variances σ_i^2) have some desired statistical properties such as: (a) mean error close to zero, (b) minimum mean square error (Delhomme, 1978), (c) dimensionless mean squared error close to unity, (d) minimum average kriging variance, (e) cross-validation errors e_i nearly uncorrelated with measured values, (f) correlation coefficient of kriged and measured values close to unity, and (g) the normalized errors should plot as a straight line on normal probability paper (Starks and Fang, 1982). Different authors emphasize different criteria.

To reduce the effect of extreme data values (outliers), some authors suggest removing from the cross-validation process those points that have large normalized errors. This screening of outliers provides a better correlation between observed and kriged values, but can lead to underestimation of the true variance of the variable. The question of what constitutes an outlier is not trivial, and the use of a pragmatic criterion to eliminate them may lead to problems.

Though simple, the trial and error approach can be frustrating and time consuming because adjustments of the parameters to improve some of the criteria may result in large departures from satisfying other criteria. The only systematic and comprehensive approach to cross-validation is that of Bastin & Gevers (1985) and Samper & Neuman (1989a), who used maximum likelihood (ML). These and other ML estimators are described in Section 3.4.

2.5 Nonparametric models for uncertainty

Most deterministic interpolation techniques provide the estimates at unsampled locations, but give no measure of their reliability. One of the main advantages of geostatistical interpolation techniques such as ordinary kriging is that an error variance is attached to each estimate. This is one of the strongest arguments by geostatisticians in favor of kriging methods. A parametric distribution of spatial kriging errors has to be assumed in order to provide confidence intervals. Most often, Gaussian-related distribution models are assumed for the errors because they are fully characterized by the error mean and error variance. This assumption, which is highly questionable, has been strongly criticized by various researchers (Philip & Watson, 1986), who have also pointed out other weak aspects of geostatistical estimation such as the fact that the kriging error variance is data independent. These problems have been partly resolved with the development of nonparametric geostatistical methods. Instead of assessing the uncertainty of a particular estimate, nonparametric geostatistics places the priority on modeling the uncertainty. The uncertainty model takes the form of a probability distribution of the unknown rather than that of the estimation error. Nonparametric geostatistics aims at modeling the probability that a variable at an unsampled location is greater than a given threshold. These models account for the proximity and quality of neighboring data and depend on the particular threshold considered.

The conditional probability that the variable Z at location **x** exceeds a threshold value z_0 given N data values $Z_0, Z_1, \ldots, Z_N$, $\text{Prob}[Z(x)>z_0|N]=1-F(z_0;x|N)$, is written as

$$F(z_0;x|N)=\sum_{j=1}^{N} \lambda_j I(z_0,x_j) \tag{16}$$

where $I(z_0,x_j)$ is an indicator function defined as

$$I(z_0,x_j)=\begin{cases}0 & \text{if } z_0<Z(x_j)\\ 1 & \text{if } z_0\geq Z(x_j)\end{cases}$$

and λ_j are N coefficients derived from the solution of a set of indicator kriging equations. These indicator kriging equations are similar to those of ordinary kriging, except that the autocovariance of Z, C(**h**), is replaced by the autocovariance of the indicator variable, $C_I(\mathbf{h})$. An enhanced estimate is obtained by probability kriging, which makes use of both indicator data, $I(z_0, \mathbf{x}_j)$, and raw data conveniently scaled so that instead of the data $Z(\mathbf{x}_j)$ their rank order transformation is used. The indicator formalism allows the joint consideration of hard data, inequality constraints, and soft information, usually in the form of a prior probability distribution. In practice, the interval of variation of variable Z is discretized into a small number K of threshold values z_K. By repeating indicator/probability kriging for all K threshold values, a probability column $F(z_K;x|N)$ valued between 0 and 1 is obtained which can be used as a model for the uncertainty about the unknown value Z(**x**). Notice that this model is nonparametric since it assumes no prior distributions. The probability distributions characterize the uncertainty about the unknowns by using both data configuration and data values but being independent of the particular estimate Z*(**x**) considered. These distributions can be used for mapping: (1) the probability of exceeding a threshold value; (2) the risk of misclassification; and (3) the areas where the probability of exceedance is equal to a given value. Minimization of a user-defined loss function measuring the impact of interpolation errors leads to the optimal estimate. The most adequate loss function depends on the particular application being considered. Most developments in nonparametric geostatistics come from the group led by A. G. Journel at Stanford University (Journel, 1989). Besides mining and oil reservoir applications, this group has applied these techniques to characterize hazardous waste disposal sites (Isaaks, 1984; Journel, 1984b).

2.6 Simulation

Geostatistical estimation techniques (the whole family of kriging techniques) provide too smooth an image of the underlying reality because they aim at minimizing local estimation errors. In fact, the variance of kriging estimates is, in most cases, smaller than that of raw data.

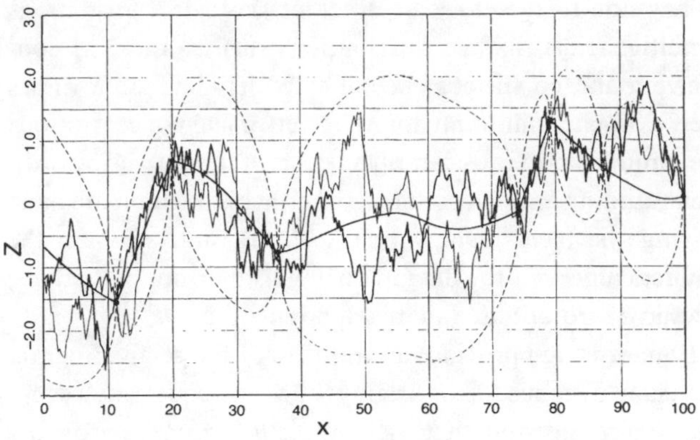

Fig. 1 Illustration of the difference between estimation (smooth kriged curve) and conditional simulations (fluctuating curves). Dots identify data values. Simulations correspond to a variable with a spherical variogram (range=20; sill=1).

Simulation allows the generation of alternative equiprobable realizations or images. These simulated images can be further generated so that they honor data values at sampled locations by means of conditional simulation techniques. Fig. 1 illustrates with a synthetic example taken from Samper & Carrera (1990) the difference between the real curve, the kriged curve, and a conditional simulation. Also shown in the figure is the Gaussian 95% confidence interval.

Conditional simulation values at a location **x**, $Z_{cs}(\mathbf{x})$, can be derived from unconditional simulation values $Z_s(\mathbf{x})$ by using (Journel and Huijbregts, 1978)

$$Z_{cs}(\mathbf{x})=Z^*(\mathbf{x})+[Z_s(\mathbf{x})-Z_s^*(\mathbf{x})] \tag{17}$$

where Z*(**x**) is the kriging estimate at **x** derived from the data and $Z_s^*(\mathbf{x})$ is the kriging estimate at **x** based on simulated values $Z_{cs}(\mathbf{x})$ at sample locations $Z_s(\mathbf{x}_j), j=1,2,\ldots,N$. In this way, conditionally simulated variables have the same mean and covariance as the original data. In addition, since kriging is an exact interpolator, simulated values coincide with measured values at sampled locations, i.e. $Z_{cs}(\mathbf{x}_j)=Z(\mathbf{x}_j)$.

Most simulation algorithms, such as the spectral method, the turning bands method, the LU-decomposition of the covariance matrix, and the fast Fourier transformation method, generate Gaussian fields. This property restricts their application to variables having a normal distribution. Highly non-Gaussian variables Z are usually transformed into a standard Gaussian variable, Y=f(Z). The transformation f(.) is based on the univariate histogram. Gaussian-model-related simulations suffer from introducing too much spatial disorder (Journel, 1989). These limitations are overcome by the sequential indicator conditional simulation (SICS) method (Journel, 1989; Gómez-Hernández and Srivastava, 1990). In addition, SICS allows both hard data and soft information to be honored. The SICS algorithm can be used to generate realization from either a multi-Gaussian or any non-multi-Gaussian random function as long as its conditional distribution is known. The key step consists of deriving the conditional probability distribution function (cpdf). In the multi-Gaussian case, the multivariate cpdf is the product of Gaussian univariate cpdf's, which have a mean and variance satisfying the solution of an indicator kriging system of equations. The SICS is extremely flexible, theoretically straightforward, is conditional by nature (requiring no additional calculations such as those of eq. (17)), can be applied on any grid regular or not, and is not limited to Gaussian functions. Gómez-Hernández & Journel (1993) have shown how SICS can be applied to the simulation of several correlated random functions.

3 APPLICATIONS

Geostatistics has found many other applications in subsurface hydrology, especially during the last few years. It is beyond the scope of this chapter to present a comprehensive review of all published applications. For our purposes, we have distinguished the following categories: (1) geostatistical estimation of hydrogeological variables; (2) spatial correlation structure of hydraulic conductivity and hydraulic head; (3) mapping and averaging; (4) improved methods for the estimation of drifts and covariance functions; (5) analysis of hydrochemical and isotopic data; (6) geostatistical inverse problem; and (7) geostatistical network design.

3.1 Direct estimation of hydrogeological variables

Geostatistical estimation of a variable follows two steps. The first one deals with what is known as the structural analysis, which includes determining the appropriate geostatistical model or hypothesis for the variable (stationary, intrinsic or not intrinsic) and quantifying the spatial correlation structure (semivariogram) and the trend whenever present. The second step includes the estimation process itself by which maps of both estimated values and estimation errors are obtained by means of kriging. This step is well defined and straightforward since it only requires the solving of the appropriate kriging equations (at present there are numerous software packages for this purpose). The methodology for the structural analysis, however, is not so well established because its determination is closely related to the physical meaning of the particular variable being analyzed. The conceptual significance of the variable together with the knowledge on other hydrogeological variables will help to determine whether the variable should have a trend or not, and how much correlation should be expected, etc. Available data must be carefully analyzed in order to detect possible data errors and aberrant values. Sample statistics and histograms are most useful as they provide information concerning the probability distribution of the variable, show the possible existence of several families, and help to decide whether to perform a transformation on the variable. As most geostatistical techniques are best suited for normally distributed variables, a transformation that renders the data Gaussian is always useful. The logarithm is an example of such a transformation which is often applied to transmissivity and concentrations of some dissolved species. Information about structure is contained in the semivariogram. A careful analysis of sample semivariograms will reveal important features of the variable such as a trend, the need to consider a transformation (it is usually observed for instance that sample semivariograms of log-concentrations behave much better than those of concentrations), and the possible directional dependency of spatial correlation. Structure analysis is therefore highly subjective. It implies carefully carrying out all steps discussed earlier. One should keep in mind, however, that in many cases the decision on what is the actual statistical structure of a variable can be arguable. To start with, there are no statistical tests to check for stationarity. In this respect, Journel (1989) states that stationarity and many other geostatistical hypotheses should not be considered as intrinsic properties of the process, but as properties of a geostatistical model. Fortunately, in many geostatistical applications the results are not too sensitive to the adopted hypothesis (or property of the model). In other applications (such as extrapolation), the identification of the appropriate geostatistical model is more critical.

3.2 Correlation structure of hydraulic conductivity and hydraulic head

Transmissivity and hydraulic conductivity were the first hydrogeological variables for which studies of spatial variability had been conducted due to the provoking conclusions of the pioneering works on stochastic subsurface hydrology (Freeze, 1975). The fact that permeability spatial variations play a dominant role in solute transport also favored the focus on studying the correlation structure of this variable.

There is a large body of experimental evidence supporting the fact that transmissivity data tend to follow a univariate log-normal distribution. This amounts to saying that log-transmissivities are Gaussian (Davis, 1969; Law, 1994). There are many studies concluding that log-T data pass statistical tests of normality (Hoeksema and Kitanidis, 1985). It has been observed also that the sample semivariogram of log-T behaves much better than that of T (Delhomme, 1978). Given the large degree of spatial variability, semivariograms of transmissivity often show a marked nugget effect. This is partly caused by interpretation errors associated with transmissivity determination from field tests. Often, log-transmissivities are estimated by linear regression against log-specific capacity data. Another part of the nugget effect is caused by the spatial variability at scales smaller than the smallest separation distance among all data pairs.

Log-transmissivities from field tests show correlation lengths which range from one to several tens of kilometers (Delhomme, 1978), depending on the geological environment and aquifer size. Reported correlation lengths tend to increase with aquifer size (Gelhar, 1984), which possibly reflects the superposition of nested scales of variability increasing in size. An extensive analysis of correlation patterns of transmissivities carried out by Hoeksema & Kitanidis (1985) on 31 aquifers covering a wide range of geological settings indicates that the total variance is similar in both sedimentary and fractured media, although the nugget effect is higher in the latter. This means that the structured variance is greater in sedimentary aquifers, reflecting the spatial continuity of depositional processes. There are a countless number of geostatistical analyses of transmissivity and hydraulic conductivity at various different scales, including core measurements, infiltration rates, pumping tests, large-scale experiments, and regional flow model calibration (in the order of kilometers). Schafmeister & Pekdeger (1993) present a detailed analysis of hydraulic conductivity measurements corresponding to nine sites where sample spacing ranges from 0.1 to 1000 m. Gelhar (1993) presents a detailed summary of published values of variance and correlation scale of hydraulic conductivity and transmissivity data. Both parameters are seen to increase with scale, an issue thoroughly discussed by Gelhar.

Hydraulic head is a continuous variable having continuous spatial derivatives except at interphases between different materials. For this reason, it usually shows a regular behavior. Its sample semivariograms are often continuous near the origin. Whenever there is groundwater flow, the hydraulic head shows a spatial trend and therefore is nonstationary. This trend explains why its semivariograms are not bounded and exhibit anisotropy. Usually the semivariograms increase with distance faster along the mean direction of groundwater flow. Theoretical analyses of hydraulic head residuals (obtained by subtracting the trend) under vertically averaged two-dimensional flow show that their semivariograms are anisotropic and have a parabolic behavior near the origin and a logarithmic shape at large distances (Chirlin & Dagan, 1980).

3.3 Mapping and averaging hydrogeological variables

Classical nonintrinsic geostatistical methods (universal kriging and generalized covariance methods), which do not explicitly compute the drift, may fail to provide a good representation of the drift. For this reason, some hydrogeologists have proposed estimating hydraulic heads using alternative methods such as residual kriging (Gambolati & Volpi, 1979; Neuman & Jacobson, 1984). This method was later applied to hydrochemical and isotopic data by Samper & Neuman (1989c). The need to better describe important variables such as K and T has led to techniques for data base augmentation. Specific capacity (SC), usually given at many more locations than T, has been used to improve T estimates. The most direct way to account for SC data is based on linear regression of log-T on log-SC (Delhomme, 1974, 1976). Values of log-T derived from this regression at unsampled locations have as prior estimation errors the ones obtained from regression. This procedure has been found useful in many cases (Delhomme, 1974, 1976; Clifton & Neuman, 1982; Fennessy, 1982; Ahmed & De Marsily, 1987). Multivariate geostatistical techniques such as cokriging have also been used for enlarging log-T data bases (Muñoz Pardo & García, 1989). Ahmed & De Marsily (1987) conducted a comparative study of linear regression, kriging with an external drift, and cokriging. All these methods seem to work equally well. Cokriging is the best method when the correlation coefficient of log-T and log-SC is high and the residuals of the regression are highly correlated. Transmissivity data have also been completed by resorting to geophysical data. Ahmed *et al.* (1988) present an example of cokriging of log-T, log-SC, and log-transverse resistance. Another interesting application of cokriging to enlarge hydrogeological data bases is that reported by Hoeksema *et al.* (1989), who improved the estimates of water table elevation by using data on ground surface elevation. In this particular case, cokriging not only improved the quality of groundwater table estimates, but also ensured that water elevations were always below the ground surface. Cokriging has also found applications for analyzing space-time processes such as hydraulic head under transient conditions (Rouhani & Myers, 1990; Rouhani & Wackernagel, 1990; Myers, 1992).

3.4 Improved methods for estimating drifts and spatial covariance functions

The particular features of hydrogeological data, together with the need to have good estimates of spatial covariance functions and drifts, have motivated several hydrogeologists to develop improved methods for estimating these functions. Semivariogram or autocovariance parameters can be estimated by maximizing the likelihood of (a) the observed data values Z_i, $i=1,2,\ldots,n$, or (b) the cross-validation errors $e_i=Z_i^*-Z_i$, Z_i^*, being the estimated value at location i. Hoeksema & Kitanidis (1985) developed a maximum likelihood estimator of type (a).

Kitanidis (1983, 1985) has proposed several maximum likelihood and minimum variance unbiased quadratic estimators for the parameters of generalized covariances. Feinerman *et al.* (1986) used multivariate normal conditional probability methods to derive estimates of both semivariogram parameters and their corresponding estimation errors.

Maximum likelihood cross-validation estimators have been derived by Bastin & Gevers (1985) and Samper (1986). Samper (1986) developed an adjoint state maximum likelihood cross-validation (ASMLCV) method, which uses adjoint state methods to maximize the likelihood of cross-validation errors. ASMLCV is based on maximum likelihood, and thus it is well suited for the problem of selecting the most appropriate semivariogram model among a set of alternative functional forms. The literature contains a number of model identification criteria developed in the context of maximum likelihood estimation by authors such as Akaike (1974, 1977), Hannan (1980) and Kashyap (1982). These criteria were described and tested by Carrera & Neuman (1986b) to distinguish between alternative parameterizations of transmissivities in dealing with the aquifer inverse problem. Hoeksema & Kitanidis (1985) used Akaike's criterion (1974) for the selection of covariance models. The results obtained by Samper & Neuman (1989b) in connection with ASMLCV indicate that Kashyap's criterion is the best among those mentioned because it not only always identifies the correct model but is also the most sensitive to the quality of the data.

ASMLCV can be combined with the stepwise iterative regression method of Neuman & Jacobson (1984) to estimate the global drift, m(x), and the semivariogram of the residuals of a nonintrinsic random function. The approach, described in detail by Samper (1986), differs from that of Neuman & Jacobson in the following ways: (a) instead of fitting a semivariogram model by eye to the sample semivariogram of the residuals at each generalized least squares iteration, the former is estimated by ASMLCV; (b) the generalized least squares iterative process is terminated not on the basis of a visual comparison between semivariograms computed at successive iterations but on the basis of the computed drift and semivariogram parameters. The iterative process ends when these parameters attain stable values. This modified ASMLCV version of the method by Neuman & Jacobson (1984) offers an important advantage over the original version in that it provides information about the quality of the estimated semivariogram. An application of the two versions to hydraulic heads from the Calera Basin in Mexico (Neuman *et al.*, 1987) has shown that the former yields higher estimates of sill and range than the latter.

Hydrologists often deal with data that represent spatial averages of a random function over variable supports. Traditional methods of estimating semivariograms are unsuitable to deal with such data; the only relevant literature is that concerning data regularized over vertical lines having a constant length. ASMLCV can easily cope with noisy data and variably regularized data merely by using kriging equations that have been suitably modified to account for the effect of such regularization. The manner in which this can be done has been described by Samper (1986) and Samper & Neuman (1989b). More recently, Samper *et al.* (1993) have presented a generalization of ASMLCV to the estimation of cross-covariances of several correlated variables.

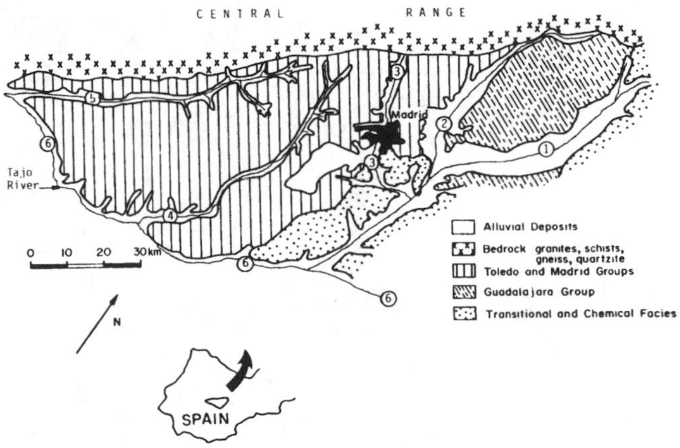

Fig. 2 Geological map of Madrid Basin showing the main rivers: (1) Henares, (2) Jarama, (3) Manzanares, (4) Guadarrama, (5) Alberche, and (6) Tajo.

3.5 Analysis of hydrochemical and isotopic data

There are only a few applications of geostatistics to hydrochemical and isotopic variables. Myers *et al.* (1982) used kriging to estimate the spatial variability of several chemical species. Gilbert & Simpson (1985) analyzed ^{241}Am in soil samples at the Nevada Test Site. Delgado García & Chica-Olmo (1993) applied disjunctive kriging to characterize the spatial extent of seawater intrusion in a coastal aquifer in Southern Spain. Goovaerts & Sonnet (1993) used factorial kriging to study the space-time behavior of calcium, chloride and nitrate contents in spring-waters in the Dyle aquifer in Belgium. Samper & Neuman (1989c) presented an exhaustive geostatistical analysis of natural chemical and isotopic species in a regional aquifer in Spain, which is discussed in detail in the following.

The study area is a large sedimentary basin surrounding Madrid in central Spain (Fig. 2). The basin is 160 km long, 30 to 60 km wide, and up to 3000 m deep. It is bounded by the Central Range on the north and by the Tajo and Henares rivers elsewhere. The surface varies in elevation from about 900 m above mean sea level in the northeast to 200 m in the southwest. It consists of gently rolling hills that protrude, at most, 200 m above valleys formed by the Tajo and its tributaries the Henares, Jarama, Manzanares, Guadarrama, and

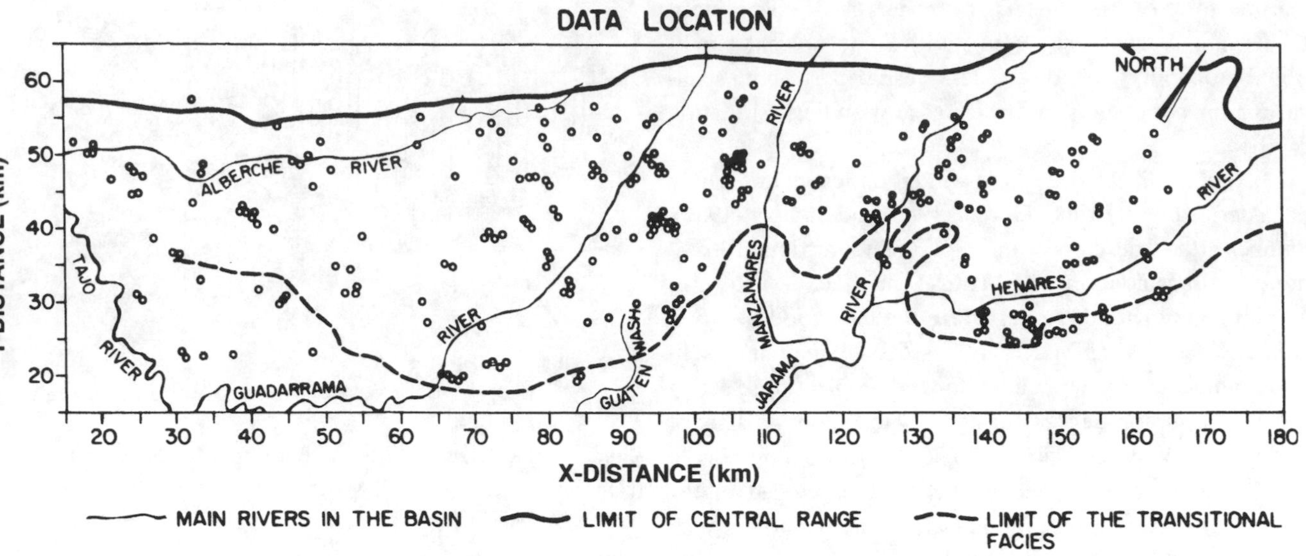

Fig. 3 Location of data and limits of the study area.

Alberche (see Fig. 2). The basin is a graben filled with arkosic sands containing variable amounts of silt and clay deposited in alluvial fans during the Miocene and Pleistocene. The main aquifer consists of marginal detritus which grades into a low-permeability chemical facies toward the center of the basin. The chemical facies in the central part is composed of gypsum, marl, and other evaporites. The transitional facies contain gradational deposits formed by the interfingering of the detritic with the chemical facies.

Mathematical groundwater flow models of the basin have revealed the presence of local shallow systems near the rivers with a deeper regional flow component from northeast to southwest. Recharge takes place by infiltration of precipitation in the interfluves. Discharge occurs along rivers and swampy lowlands and through springs. These features are qualitatively coherent with hydrochemical and isotopic studies (Fernández Uría, 1984; Rubio, 1984). Herráez (1983) found local averages of $\delta^{18}O$ in recharge areas ranging from -6 to -8 and decreasing with depth, whereas in discharge areas those values range from -8 to -9 (no correlation with depth was found, probably due to mixing of waters from various sources). Herráez (1983) also showed that $\delta^{18}O$ decreases with elevation at an average rate of 0.23 per 100 m. A joint interpretation of $\delta^{18}O$ and ^{14}C data made her conclude that older-waters must have been recharged under climatic conditions that were colder than current temperatures.

The data set used for geostatistical analyses is based on 287 groundwater samples from wells spread throughout the basin (Fig. 3). Well depths range from 50 m to more than 600 m. Most wells are screened selectively opposite sand lenses but are gravel packed so that mixing can take place during pumping over much of the saturated vertical interval they penetrate. Screen lengths range from 0 to 389 m with a mean length of 80 m. The variables analyzed include electric conductivity (EC), silica, pH, bicarbonate, calcium, nitrate, chloride, sulfate, sodium, magnesium, potassium, ^{18}O in SMOW per mil delta units, and ^{14}C in per cent modern carbon.

A base-10 log transformation applied to electric conductivity and all the major ions made their sample statistics correspond more closely to those of a Gaussian distribution than would otherwise be the case, brought their quantile plots closer to straight lines, and improved their sample semivariograms. Silica, pH, ^{18}O, and ^{14}C exhibit low variability with skewness and kurtosis close to 0 and 3, respectively, which explains why they were not transformed.

None of the sample semivariograms shows a discernible anisotropy and thus are averaged in all directions. The results of estimating theoretical semivariogram models by means of ASMLCV shows that only silica and ^{18}O exhibit a spatial drift. Spherical and exponential models with a nugget gave satisfactory results in all cases except ^{14}C. The choice between them was made on the basis of the sample semivariogram. A nugget is seen to affect all the variables, ranging from 9.7% of the sill for sodium, to 61.7% for silica residuals, to 100% for ^{14}C, with six values in excess of 50%. Three variables have an effective range of less than 10 km, the rest being between 16 and 53 km.

The nuggets associated with most variables appear to be much larger than could be attributed to noise stemming from sampling and laboratory procedures. A more important factor, however, is the two-dimensional nature of the data used in the geostatistical analysis. This is so because water samples have undergone mixing in boreholes which are in contact with the aquifer over randomly varying vertical intervals, while flow in the system is three dimensional. This

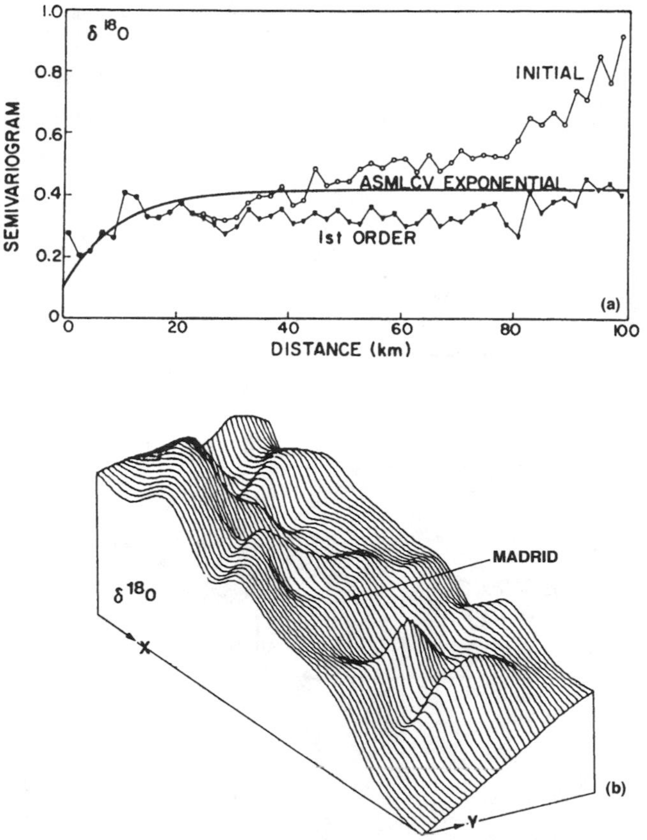

Fig. 4 Semivariograns (a) and three-dimensional view (b) of $\delta^{18}O$.

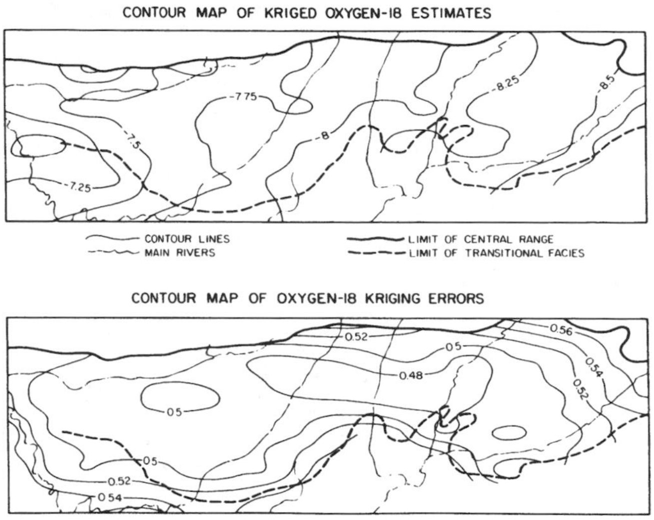

Fig. 5 Kriging estimates and errors for $\delta^{18}O$.

difficulty is compounded in regional discharge areas where groundwaters from various sources tend to mix prior to being intercepted by wells. Other factors contributing to the nugget are variations in groundwater quality at a scale smaller than the distance between neighboring sampling intervals due to nonuniformities in sediment composition, flow pattern, and temporal fluctuations. The fact that most variables do not exhibit a spatial drift does not come as a surprise; this, however, does not preclude several of them from showing regional variations in agreement with expected patterns.

A three-dimensional surface view of $\delta^{18}O$ data (Fig. 4) shows a near linear drift from the confluence of the Tajo and Alberche rivers on the southwest to the topographically elevated region on the northeast, with superimposed peaks over interfluves. The presence of such a drift is also indicated by the sample semivariogram in the same figure. Filtering out a first order polynomial drift by ordinary least squares (simple regression) is seen to stabilize the sill and thus confirms that a linear drift function is adequate to represent the data. Refinement by ASMLCV in conjunction with the iterative generalized least squares approach of Neuman & Jacobson (1984) was achieved after two iterations. The resulting model has a greater sill and range, and a smaller nugget, than that obtained by simple regression. Cross-validation is deemed successful as the dimensionless mean square error is virtually equal to unity; the correlation between kriged and computed residuals is reasonably high (0.6); and all the statistics of the normalized cross-validation errors lie within 95% confidence intervals about their appropriate values. Thc kriged map (Fig. 5) reflects the basin-wide drift but is too smooth to reproduce distinctly local peaks and lows. It yields a maximum drop in $\delta^{18}O$ from southwest to northeast slightly in excess of -1.75. An altitude effect at a gradient of $-0.23/100$ m, as determined by Herráez (1983), would result in a somewhat smaller drop of -1.20. Given that kriging estimation errors are relatively large (Fig. 4), due partly to a nugget that exceeds 24% of the sill, one cannot unambiguously ascribe the difference between these two numbers to real phenomena such as the proposed paleoclimatic change. The large nugget and kriging errors may help explain why some researchers have been reluctant to accept that ^{18}O in the Madrid Basin follows a regional drift. The existence of such a drift, however, is clearly indicated by the ASMLCV analysis which thus confirms, geostatistically, at least the altitude effect. It is seen that pH has a low variance. Its sample semivariograms along four equally spaced directions (Fig. 6) show a lack of anisotropy and no clear suggestion of a drift. The exponential model estimated by ASMLCV attests to a relatively low level of spatial correlation (effective range=3.29 km), small variance (sill=0.209), and sizeable nugget (36.4% of the sill).

The semivariograms of log-concentrations of some selected major ions are shown in Fig. 7. ASMLCV estimates lie somewhat above the sample semivariograms. Some influential points were eliminated during the cross-validation. These points were associated either with unusually high salinities or with extremely low concentrations, mostly below detection limits. Cross-validation was generally successful. Among the

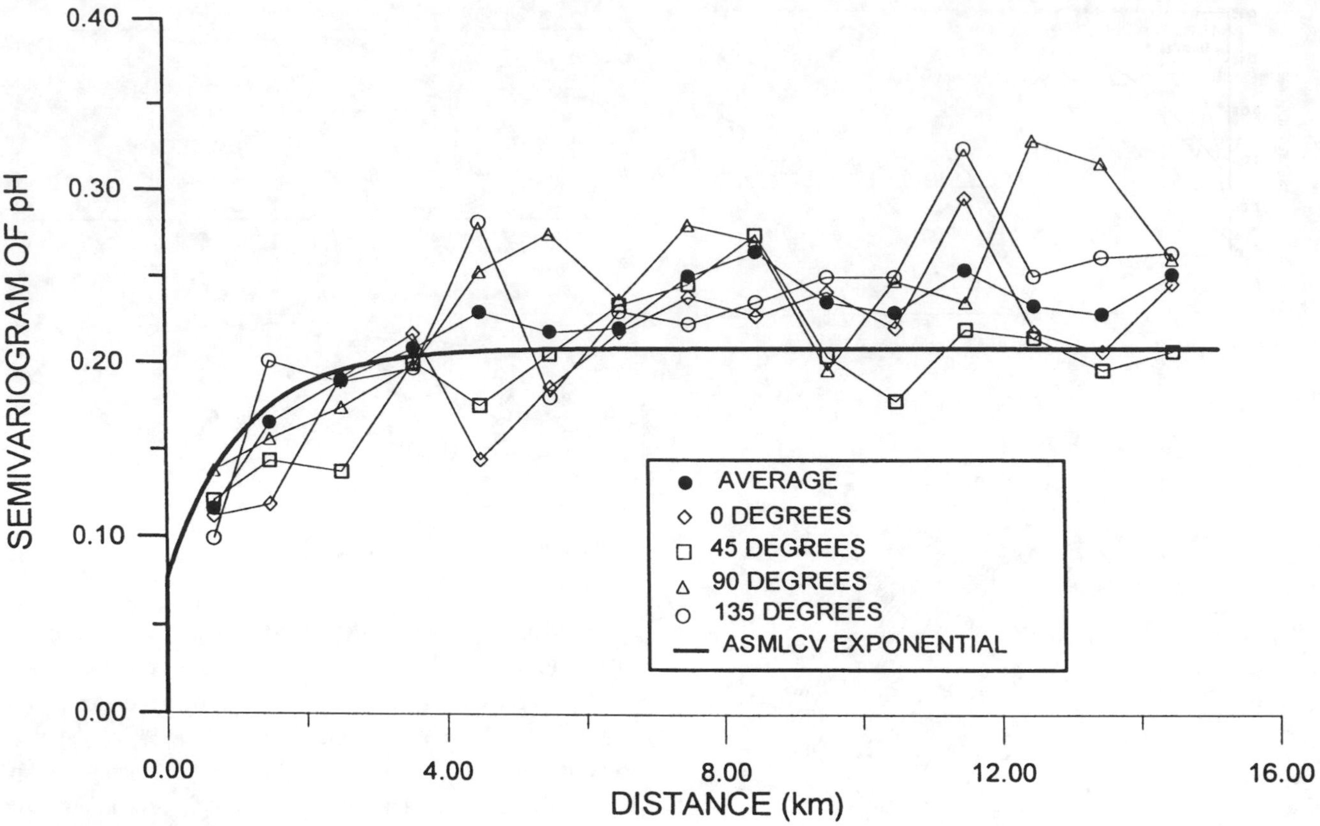

Fig. 6 Semivariograms of pH.

anions, bicarbonate possesses the strongest spatial correlation (effective range of 53 km) followed by sulfate (approximately 28 km), chloride (17 km), and nitrate (6 km). Bicarbonate also exhibits the smallest relative nugget (approximately 18%), followed by chloride (approximately 38%), nitrate (approximately 42%), and sulfate (approximately 54%). Among cations the degree of spatial correlation is less variable, the effective range being about 50 km for potassium and close to 30 km for calcium, sodium and magnesium. The relative nugget is smallest for sodium (approximately 10%), much larger for potassium (approximately 52%), and largest for calcium (approximately 62%). Kriging of log-concentration was performed using all data (including those not participating in cross-validation). The results are in agreement with hand drawn contours prepared by Fernández Uría (1984) and Rubio (1984). An advantage of the geostatistical approach is that it quantifies the associated errors, which, for most of the major ions in the Madrid Basin, are rather large.

Contrary to what was expected, the sample semivariogram of ^{14}C (Fig. 8) does not show a drift and strongly suggests that ^{14}C lacks spatial correlation on a regional scale. To verify this, ASMLCV was used to estimate the parameters of two spherical models and a pure nugget model. The spherical model with a higher sill and nugget corresponds to raw data; the other shows what happens when one filters out measurement errors. While filtering has little effect on the range and the nugget-to-sill ratio, it causes the sill to drop by about 137 ppm^2 and the nugget to drop to about 44 ppm^2. Both numbers are in line with the measurement variance. The pure nugget model is rated as the best by all the structure identification criteria considered. The criterion of Kashyap (1982), which most convincingly accounts for the noise, ranks the two spherical models as being only slightly inferior to the pure nugget model. Given the lack of spatial correlation implied by the preferred model, it is not surprising that the kriged and measured values exhibit a very low level of cross-correlation. Point kriging of ^{14}C with the selected model leads to the contours of Fig. 8. Due to the lack of spatial correlation, kriged estimates vary more smoothly and over a narrower range of values than the original data. However, as crude as the kriged map is, it does indicate distinct ^{14}C peaks over interfluves, and lows along the rivers. Samper & Neuman (1989c) noticed that, despite the large background noise, it was possible to ascribe a geostatistical significance to the increase in groundwater age from regional recharge zones to regional discharge areas (a fact inferred earlier deterministically by Llamas *et al.* (1982) and Herráez (1983)).

3.6 Geostatistical inverse problem

The construction of a hydrogeological model for an aquifer requires knowledge of the type and location of boundaries and the spatial distribution of the medium parameters. Once

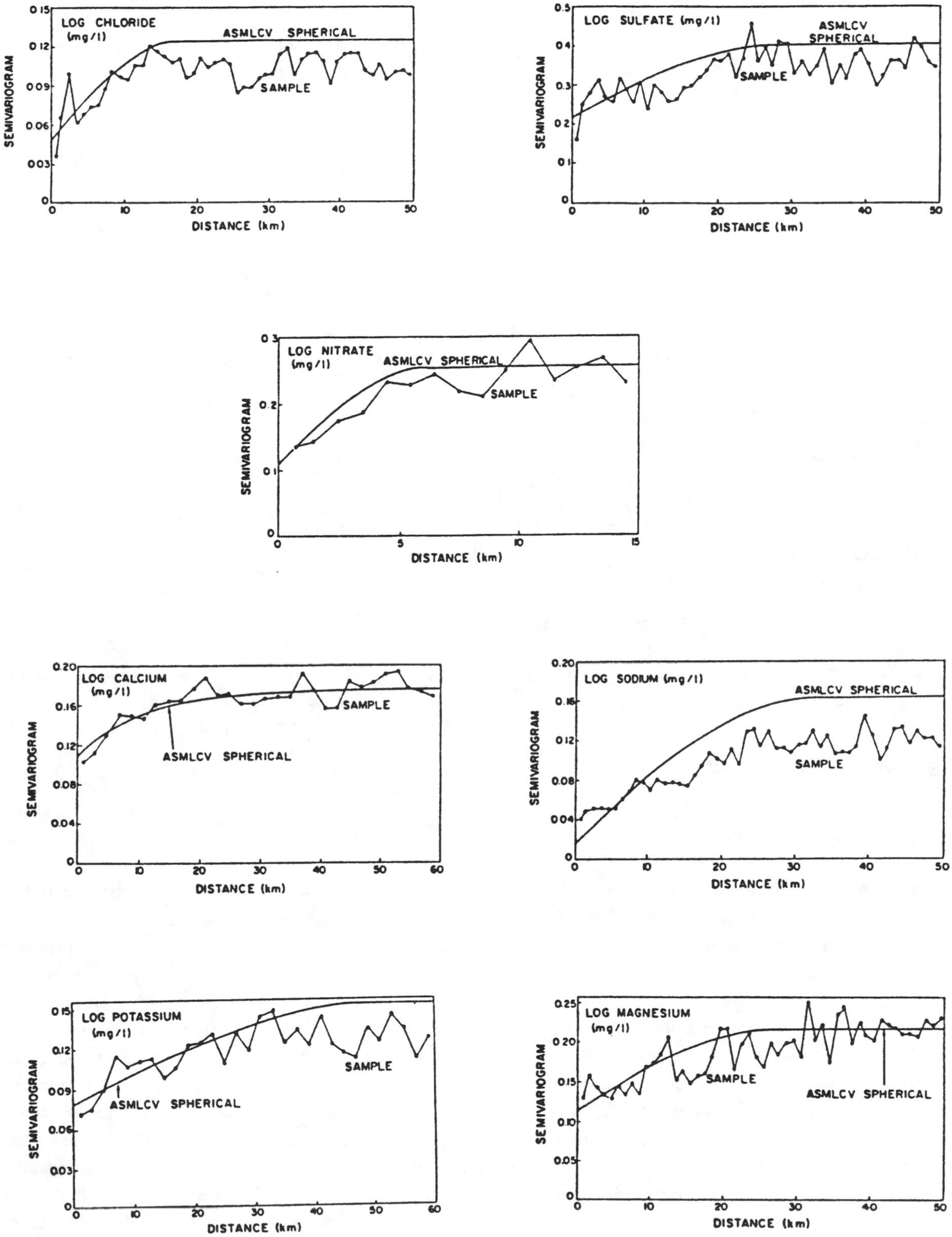

Fig. 7 Semivariograms of log-concentrations chloride, sulfate, nitrate, calcium, sodium, potassium, and magnesium.

these are known, the solution of the flow equation allows the computation of the dependent variable (hydraulic head). This is known as the direct problem. In practice, however, the information available on the parameters is insufficient to characterize the model. There are typically many more hydraulic head data. The inverse problem aims at improving the estimate of the parameters by using prior parameter information and available hydraulic head data.

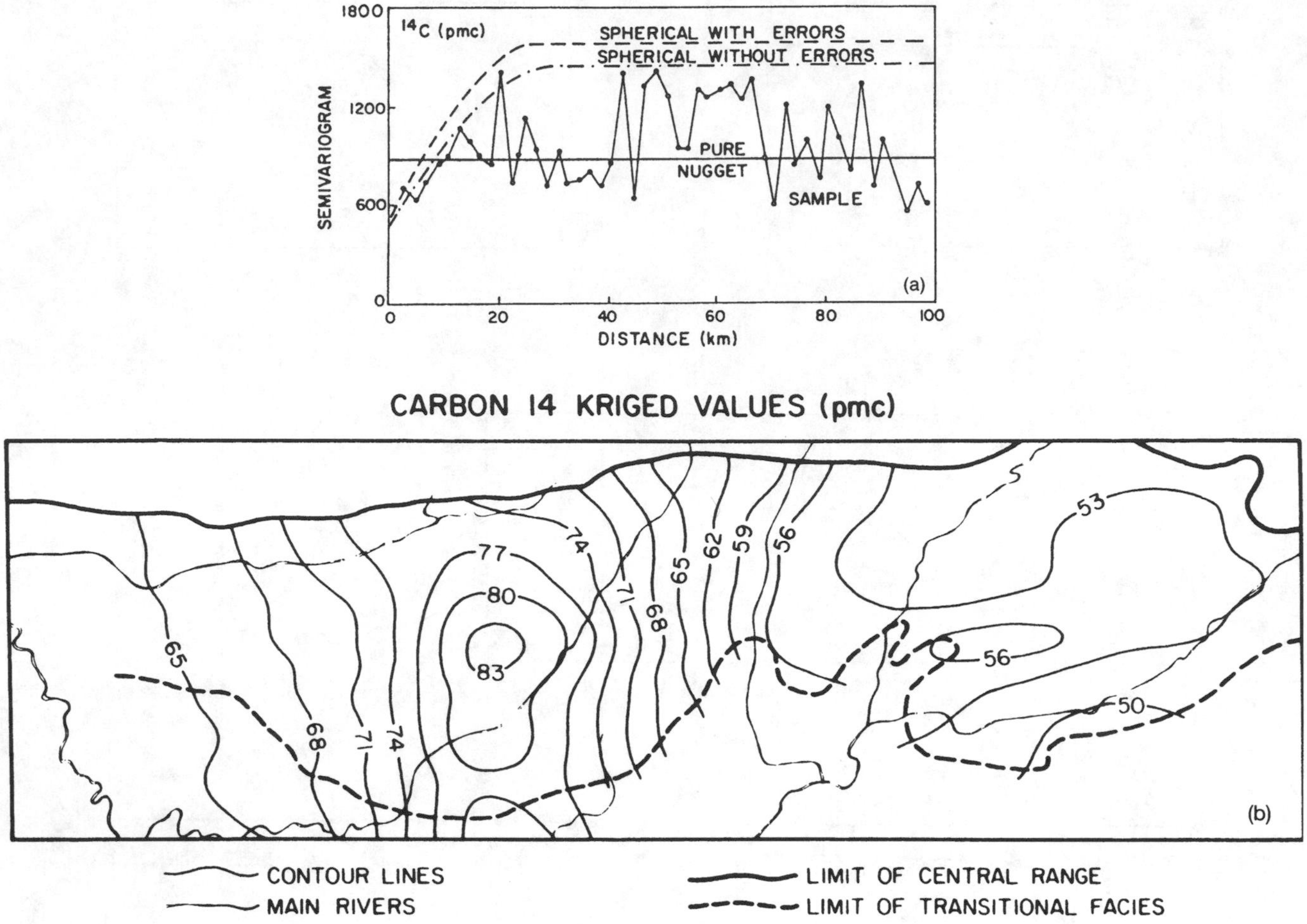

Fig. 8 Semivariograms (a) and kriging estimates (b) of ^{14}C.

The so-called deterministic inverse problem assumes that the parameters (log-T) are deterministic, although they are unknown at unsampled locations. Most of these inverse methods are based on a numerical discretization of the medium, so that parameters are assumed constant over some finite number of zones. The optimum model parameters are those minimizing a least square type criterion which measures the overall different between computed and measured heads. Formulated in this way, the inverse problem is prone to instability and nonuniqueness. To overcome these problems, Neuman (1980) proposed adding a plausibility criterion for the parameters which accounts for prior information on the parameters. Statistical formulations of the deterministic inverse problem (such as those of Neuman & Yakowitz (1979) and Carrera & Neuman (1986a)) still assume that the parameters are deterministic but that their prior estimates are uncertain and have a known covariance matrix. These authors initially used kriging, deriving both the prior estimates of the parameters and their associated covariance matrix. A review and a more detailed account of these methods can be found in Carrera & Neuman (1986a) and Yeh (1986). In fact the whole topic of inversion of random fields is addressed in depth by Carrera *et al.* in this volume, Part II, Chapter 3.

The so-called geostatistical inverse problem assumes from the beginning that both hydrogeological parameters and hydraulic heads are random fields or stochastic processes. This formulation, initially proposed by Kitanidis & Vomvoris (1983) and later expanded by Hoeksema & Kitanidis (1984), Dagan (1985), Rubin & Dagan (1987), and Dagan & Rubin (1988) emphasizes the need to characterize the correlation structure of log-T. There have been two approaches, which differ in the way in which the autocovariance of heads and the cross-covariance of heads and transmissivities are derived. These covariances can be computed analytically (Dagan, 1985; Rubin & Dagan, 1987; Dagan & Rubin, 1988) or numerically from the solution of a numerical groundwater flow model (Kitanidis & Vomvoris, 1983; Hoeksema & Kitanidis, 1984). The application of the inverse methodology using the numerical approach follows these steps: (1) proposal of a geostatistical model for log-T; (2) derivation of the correlation structure of hydraulic heads based on the covariance of log-T and the Jacobian (a matrix containing the sensitivity of hydraulic heads at measurement locations with respect to the log-T values at the blocks of the model); (3) maximum likelihood estimation of the statistical parameters defining the covariance of log-T based on both log-T and **h** data – this step requires specifying a joint Gaussian distribution; and (4) estimation of the log-T values by means of cokriging using log-T and hydraulic head data.

As shown by Carrera & Glorioso (1991), under certain conditions the geostatistical inverse method is a particular case of the maximum likelihood deterministic inverse method of Carrera & Neuman (1986a).

Geostatistical inverse methodologies have also been applied recently to the simultaneous solution of groundwater flow and solute transport by Carrera *et al.* (1993).

3.7 Network design

Data collection in hydrology is costly and time consuming. For this reason it is always desirable to reduce as much as possible the amount of data needed to characterize a given process. Geostatistics lends itself to provide answers to the problem of optimizing monitoring and observation networks. As already pointed out by Matheron (1963), kriging variances are not dependent on data values and thus can be computed in advance for a given data configuration. Locating additional sampling points can be addressed by minimizing an uncertainty measure which can be: (1) the estimation variance of the mean value of Z over a given domain, (2) the largest point kriging variance over the whole domain, and (3) a norm of the covariance matrix of kriging estimates. Optimization of network design can be formulated either (a) by minimizing the total cost while keeping the variance below a certain limit or (b) minimizing the variance for a given cost of sampling.

The simplest case for analysis corresponds to a situation where the location of an additional sampling point must be determined in order to minimize the variance of the mean value. This additional location is obtained by the fictitious point method, which consists of computing the relative reduction of variance $R(\mathbf{x})$ obtained when the fictitious point is located at a location $\mathbf{x}$. Knowing $R(\mathbf{x})$ at a large number of locations allows one to select the optimum location of the additional point. Delhomme & Delfiner (1973) applied this method to optimize raingauge networks. In some cases it is more convenient to locate the additional measurement in areas where the uncertainty is largest (Delhomme, 1978).

For data regularly spaced along a square grid, the variance can be related to grid size. The joint analysis of variance reduction and the cost increase associated to refining the measurement grid serves to find the optimum grid size. This approach was taken by Sophocleous (1983) for optimizing the groundwater management networks of Kansas Districts.

A more complex problem arises when the data base is not regularly spaced and has to be augmented by selecting among N existing wells the optimum subset of n ($n<N$) points. Efficient numerical search algorithms have to be used in order to explore all possible combinations. Carrera *et al.* (1984) give an account of these methods and present an application to the optimum selection of additional measurement locations for characterizing natural groundwater fluoride contents at the San Pedro River basin in Arizona. Network design has also been applied using multicriteria (Bogardi *et al.*, 1985) and in connection with numerical groundwater flow and solute transport models (Olivella & Carrera, 1987; Candela *et al.*, 1988).

4 SUMMARY AND CONCLUSIONS

For the last 20 years geostatistics has found interesting applications in subsurface hydrology. Early applications dealt with estimating hydrogeological variables at unsampled locations by means of point kriging and obtaining the corresponding map of estimation errors. Block kriging has been extensively used to estimate block transmissivities in the context of numerical flow models. Other groups of geostatistical applications in subsurface hydrology are related to optimum observation and monitoring network design. With the increasing recognition of the paramount effects of spatial variability of permeability, geostatistical simulation tools have gained more and more relevance. The particular properties of hydrogeological data (scarcity, variable support, measurement errors) compelled hydrogeologists to develop improved geostatistical methods for estimating both drift and spatial covariance parameters and use multivariate geostatistical methods for extracting the most information from all available data. More recently, geostatistical methods have been applied to analyze hydrochemical and isotopic data as well as properties of unsaturated media.

Active research areas which will enlarge the range of geostatistical applications in the near future include: (1) development of methods for combining hard and soft data and descriptive geological information; (2) bearing more hydrodynamics into the geostatistical estimation and simulation methods; (3) improving multivariate geostatistical methods; (4) development of improved theories for statistically heterogeneous media in order to capture more realistically the observed geological features; and (5) development of improved methods for estimating spatial covariance structures from limited data.

ACKNOWLEDGMENTS

Parts of this work were funded by CICYT Grant NAT91-1001, from the Spanish Department of Education and

Science, and Grant XUGA11803A94, from the Autonomous Government of Galicia.

REFERENCES

Aboufirassi, M. & Mariño, M. A. (1984). Cokriging of aquifer transmissivities from field measurements of transmissivity and specific capacity. *Journal of Mathematical Geology*, 16(1), 19–35.

Ahmed, S. & De Marsily G. (1987). Comparison of geostatistical methods for estimating transmissivity using data on transmissivity and specific capacity. *Water Resources Research*, 23(9), 1717–1737.

Ahmed, S., De Marsily G. & Talbot, A. (1988). Combined use of hydraulic and electrical properties of an aquifer in a geostatistical estimation of transmissivity. *Groundwater*, 26, 1, 78–86.

Akaike, H. (1974) A new look at statistical model identification. *IEEE Transactions on Automatic Control*, AC-19, 716–722.

Akaike, H. (1977). An entropy maximization principle. In *Applications of Statistics*, ed. P. R. Krishnaiah. Amsterdam: North-Holland.

Armstrong, M. & Delfiner, P. (1980). Towards a more robust variogram: A case study on coal. Note n-671, C.G.M.M., Fontainebleau, France.

Baafi, E. Y., Lenergan, J. E., Barua, S. L. & Kim, Y.C. (1982). *Condensed User's Manual for Basic Geostatistics Systems*. Department of Mining and Geological Engineering, University of Arizona.

Bastin, G. & Gevers, M. (1985). Identification and optimal estimation of random fields from scattered point-wise data. *Automatica*, 21(2), 139–155.

Binsariti, A.A. (1980). Statistical analysis and stochastic modeling of the Cortaro Aquifer in Southern Arizona. Ph. D. Dissertation. University of Arizona.

Bogardi, J., Bardossy, A. & Duckstein, (1985). Multicriterion network designing using geostatistics. *Water Resources Research*, 21(2), 199–208.

Candela, L., Olea, R. & Custodio, E. (1988). Lognormal kriging for the assessment of reliability in groundwater quality control observation networks. *Journal of Hydrology*, 103, 67–84.

Candy, J. V. & Mao, N. (1981). Nuclear waste repository characterization: A spatial estimation/identification approach. *Proceedings of the 8th Triennial World Congress IFAC Control Science and Technology*, Kyoto, pp. 629–636.

Carrera, J. & Glorioso, L. (1991). On geostatistical formulations of the groundwater flow inverse problem. *Advances in Water Resources*, 14(5), 273–283.

Carrera, J. & Neuman, S. P. (1986a). Estimation of aquifer parameters under transient and steady state conditions, 1, Maximum likelihood method incorporating prior information. *Water Resources Research*, 22(2), 199–210.

Carrera, J. & Neuman, S. P. (1986b). Estimation of aquifer parameters under transient and steady state conditions, 3, Application to synthetic and field data. *Water Resources Research*, 22(2), 228–242.

Carrera, J., Usunoff, E. & Szidarovsky, F. (1984). A method of optimal observation well designing for groundwater management. *Journal of Hydrology*, 73, 147–163.

Carerra, J., Medina, A. & Sánchez-Vila, X. (1993). Geostatistical formulations of groundwater coupled inverse problems. In *Geostatistics Troia'92*, vol. 2, ed. A. Soares. Dordrecht: Kluwer Academic Publishing, pp. 779–792.

Chirlin, G. R. & Dagan, G. (1980). Theoretical head variograms for steady-flow in statistically homogeneous aquifers. *Water Resources Research*, 16(6), 1001–1015.

Clark, I. (1979). *Practical Geostatistics.* Essex: Applied Science Publishers, p. 169.

Clifton, P. M. & Neuman, S. P. (1982). Effects of kriging and inverse modeling on conditional simulation of the Avra Valley Aquifer in Southern Arizona. *Water Resources Research*, 18(4), 1215–1234.

Dagan, G. (1985). Stochastic modeling of groundwater flow by unconditional and conditional probabilities: The inverse problem. *Water Resources Research*, 21(1), 65–72.

Dagan, G. & Rubin, Y. (1988). Stochastic identification of recharge, transmissivity and storativity in aquifer transient flow: A quasi-steady approach. *Water Resources Research,* 24(10), 1698–1710.

Davis, B. M. & David, M. (1978). Automatic kriging and contouring in the presence of trends. *Journal of Canadian Petroleum Technology*, 17(1), 1–10 & 90–8.

Davis, S. N. (1969). Porosity and permeability of natural materials. In *Flow Through Porous Media*, ed. J. M. De Wiest. New York: Academic Press.

Delfiner, P. (1976). Linear estimation of non-stationary spatial phenomena. In *Advanced Geostatistics in the Mining Industry*, eds. M. Guarascio, M. David & C. Huijbregts. Dordrecht: D. Reidel Publishing Company, pp. 49–68.

Delgado García, J. & Chica-Olmo, M. (1993). Nonlinear approach to the marine intrusion process modelling. In *Geostatistics Troia '92*, ed. A. Soares, vol. 2, pp. 685–694. Dordrecht: Kluwer Academic Publishers.

Delhomme, J. P. (1974). La cartographie d'une grandeur physique a partir de données du differentes qualites. *Memoires International Association of Hydrogeology*, 10(1), 185–194.

Delhomme, J. P. (1976). Applications de la théorie des variables regionalées dans les sciencies de l'eau. Doctoral Thesis, Ecole des Mines de Paris, Fontainebleau, France.

Delhomme, J. P. (1978). Kriging in the hydrosciences. *Advances in Water Resources*, I, 251–266.

Delhomme, J. P. & Delfiner, P. (1973). Application du krigeage a l'optimisation d'une campagne pluviométrique en zone aride. In *Proceedings on the Designing of Water Resources Projects with Inadequate Data*. Madrid: Unesco.

Dowd, P. A. (1984). The variogram and kriging: Robust and resistant estimators. In *Geostatistics for Natural Resources Characterization,* vol. 1, eds. G. Verly *et al.*, NATO ASI Series. Dordrecht: D. Reidel Publishing Company, pp. 91–106.

Feinerman, E. S., Dagan, G. & Bresler, E. (1986). Statistical inference of spatial random functions. *Water Resources Research*, 22(6), 935–942.

Fennessy, P. J. (1982). Geostatistical analysis and stochastic modelling of the Tajo Basin Aquifer, Spain. MSc. Thesis, University of Arizona.

Fernández Uría, A. (1984). Hidrogeoquímica de las aguas subterráneas en el Sector Oriental de la Cuenca de Madrid , Ph.D. Dissertation, Department of Geology, Autonomous University of Madrid.

Freeze, A. (1975). A stochastic-conceptual analysis of one-dimensional groundwater flow in nonuniform homogeneous media. *Water Resources Research*, 11(5), 725–741.

Gambolati, G. & Volpi, G. (1979). Groundwater contour mapping in Venice by stochastic interpolators, 1 Theory. *Water Resources Research*, 15(2), 281–290.

Gelhar, L. W. (1984). Stochastic analysis of flow in heterogeneous porous media. In *Fundamentals of Transport Phenomena in Porous Media*, eds. J. Bear & M.Y. Corapciouglu. Englewood Cliffs, NJ: Prentice Hall, pp. 673–720.

Gelhar, L. W. (1993). *Stochastic of Subsurface Hydrology*. Englewood Cliffs, NJ: Prentice Hall.

Gilbert, R. O. & Simpson, J.C. (1985). Kriging for estimating spatial pattern of contaminants: potential and problems. *Environmental Monitoring and Assessment*, 5, 113–135.

Gómez-Hernández, J. J. & Journel, A. G. (1993). Joint sequential simulation of multigaussian fields. In *Geostatistics Troia'92*, vol. 1, ed. A. Soares. Dordrecht: Kluwer Academic Publishing, pp. 85–94.

Gómez-Hernández, J.J. & Srivastava, R. M. (1990). ISIM3D: An ANSI-C three dimensional multiple indicator conditional simulation program. *Computer and Geosciences*, 16(4), 395– 40.

Goovaerts, P. & Sonnet, Ph. (1993). Study of spatial and temporal variations of hydrochemical variables using factorial kriging analysis. In: *Geostatistics Troia'92*, vol. 1, ed. A. Soares. Dordrecht: Kluwer Academic Publishing, pp. 745–756.

Hannan, E. S. (1980). The estimation of the order of an ARMA process. *Annals of Statistics*, 8, 1971–1981.

Hernández, M. (1986). Application of the algebraic technological function to the optimization of groundwater abstraction from an unconfined aquifer in Zacatecas, México. M. Sc. Thesis, University of Arizona.

Herráez, I. (1983). Análisis de las variaciones de los isótopos ambientales estables en el sistema acuífero terciario detrítico de Madrid. Ph. D. Dissertation, Department of Geology, Autonomous University of Madrid.

Hoeksema, R. J. & Kitanidis, P. K. (1984). An application of the geostatistical approach to the inverse problem in two-dimensional groundwater modeling. *Water Resources Research,* 20, 1003–1020.

Hoeksema, R. J. & Kitanidis, P. K. (1985). Analysis of the spatial structure of properties of selected aquifers. *Water Resources Research*, 21(4), 563–572.

Hoeksema, R. J., Clapp, R. B., Thomas, A. L., Hunley, A. E., Farrow, N. D. & Dearstone, K.C. (1989). Cokriging model for estimation of water table elevation. *Water Resources Research*, 25(3), 429–438.

Hughes, J. & Lettenmaier D. (1981). Data requirements for kriging estimation and network design. *Water Resources Research*, 17(6), 1641–1650.

Isaaks, E. H. (1984). Risk qualified mappings for hazardous wastes: a case study in non-parametric geostatistics. MSc. Thesis, Stanford University.

Journel, A. G. (1984a). The place of non-parametric geostatistics. In *Geostatistics for Natural Resources Characterization*, eds. G. Verly *et al.* NATO ASI Series Dordrecht: D. Reidel Publishing Co.

Journel, A. G. (1984b). Indicator approach to toxic chemical sites. EPA, EMSL-Las Vegas, report of Project CR-811235-02-0.

Journel, A. G. (1989). *Fundamentals of Geostatistics in Five Lessons*. Short Course in Geology, vol. 8. Washington DC: American Geophysical Union.

Journel, A. G. & Huijbregts, CH. J. (1978). *Mining Geostatistics*. New York: Academic Press.

Kashyap, R. L. (1982). Optimal choice of AR and MA parts in autoregressive moving average models. *IEEE Transactions on Pattern Analysis Machine Intelligence*, PAMI-4(2), 99–104.

Kim,Y. C., Myers, D. E. & Kundsen, H. P. (1977). Advances in geostatistics in ore reserve estimation and mini planning practitioner's guide. *Information for the U.S. Energy Research and Development Administration*, N° 76-003-E, Phase II.

Kitanidis, P. K. (1983). Statistical estimation of polynomial generalized covariance functions and hydrologic applications. *Water Resources Research*, 19(4), 909–921.

Kitanidis, P. K. (1985). Minimum-variance unbiased quadratic estimation of covariance of regionalized variables. *Journal of Mathematical Geology*, 17(2), 195–208.

Kitanidis, P. K. & Vomvoris, E. G. (1983). A geostatistical approach to the inverse problem in groundwater modelling (steady state) and one-dimensional simulations. *Water Resources Research*, 19(3), 677–690.

Kridge, D. G. & Magri, E. J. (1982). Studies of the effect of outliers and data transformation on variogram estimates for a base metal and a global ore body. *Journal of Mathematical Geology*, 14(6), 557–564.

Law, J. (1994). A statistical approach to the statistical heterogeneity of sand reservoirs. *Transactions of AIME*, 55, 202–222.

Llamas, M. R., Simpson, E. S. & Martinez Alfaro P. E. (1982). Groundwater age distribution in Madrid Basin, Spain. *Groundwater*, 20(6), 688–695.

Matheron, G. (1963). Principles of geostatistics. *Economic Geology*, 58, 1246–1266.

Matheron, G. (1971). The theory of regionalized variables and its applications. *Les Cahiers du CIMM*, Fasc. No. 5, ENSMP, Paris.

Matheron, G. (1973). The intrinsic random function and its applications. *Advances in Applied Probability*, 5, 438–468.

Matheron, G. (1976). A simple substitute for conditional expectation: disjunctive kriging. In *Advanced Geostatistics in the Mining Industry*, eds. M. Guarascio *et al.* Dordrecht: D. Reidel, pp. 221–236.

Muñoz Pardo, J. & García, R. (1989). Estimation of the transmissivity of the aquifer of Santiago, Chile, using different geostatistical methods. In *Groundwater Management: Quantity and Quality*, eds. A. Sahuquillo, J. Andreu & T. O'Donell. IAHS Publ. no. 188, pp. 77–84. Wallingford: IAHS.

Myers, D.E. (1982). Matrix formulation of co-kriging. *Journal of Mathematical Geology*, 14(3), 249–257.

Myers, D.E. (1992). Spatial-temporal geostatistical modeling in hydrology. In *Geostatistical Methods*, ed. A. Bárdossy. UNESCO: International Hydrology Programme, pp. 62–71.

Myers, D. E., Begovich, C. L., Butz, T. R. & Kane, V. E. (1982). Variogram models for regional groundwater geochemical data. *Journal of Mathematical Geology*, 14(6), 629–644.

Neuman, S. P. (1980). A statistical approach to the inverse problem of aquifer hydrology, 3. Improved solution method and added perspective. *Water Resources Research*, 16(2), 331–346.

Neuman, S. P. & Jacobson, E. A. (1984). Analysis of non intrinsic spatial variability by residual kriging with application to regional groundwater levels. *Journal of Mathematical Geology*, 16(5), 499–521.

Neuman, S. P. & Yakowitz, S. (1979). A statistical approach to the inverse problem of aquifer hydrology, 1, Theory. *Water Resources Research*, 15(4), 845–860.

Neuman, S. P., Samper F. J. & Hernández, M. (1987). Reply to comments by G. Gambolati and G. Galaeti on analysis of nonintrinsic spatial variability by residual kriging with application to regional groundwater levels. *Journal of Mathematical Geology*, 19(3), 259–266.

Olivella, S. & Carrera, J. (1987). Un método geostadístico para el diseño óptimo de redes de observación. *Hidrogeología y Recursos Hidráulicos*, 12, 587–598.

Omre, H. (1984). The variogram and its estimation. In *Geostatistics for Natural Resources Characterization*, vol. 1, eds. G. Verly *et al.*, pp. 107–125. NATO ASI Series, Reidel, Hingham, Mass.

Papoulis, A. (1965). *Probability, Random Variables and Stochastic Processes*. New York: McGraw-Hill.

Philip, G. M. & Watson, D. F. (1986). Matheronian geostatistics. Quo Vadis? *Journal of Mathematical Geology*, 18(1), 93–117.

Rouhani, S. (1986). Comparative study of ground water mapping techniques. *Groundwater*, March–April 24(2), 207–216.

Rouhani, S. & Myers, D. E. (1990). Problems in kriging of spatial-temporal geohydrological data. *Journal of Mathematical Geology*, 22, 611–623.

Rouhani, S. & Wackernagel H. (1990). Multivariate geostatistical approach to space-time data analyses. *Water Resources Research*, 26, 585–591.

Rubin, Y. & Dagan G. (1987). Stochastic identification of transmissivity and effective recharge in steady groundwater flow: 1. Theory. *Water Resources Research*, 23(7), 1185–1192.

Rubio, P. L. (1984). Hidrogeoquímica de las aguas subterráneas del sector occidental de la Cuenca de Madrid. Ph.D. Dissertation, Department of Geology, Autonomous University of Madrid.

Samper, F. J. (1986). Statistical methods of analyzing hydrological, hydrochemical and isotopic data from aquifers. Ph.D. Dissertation, Department of Hydrology and Water Resources, University of Arizona.

Samper, F. J. & Carrera, J. (1990). *Geoestadística: Aplicaciones a la Hidrología Subterránea*. Barcelona: Centro Internacional de Métodos Numéricos en Ingeniería.

Samper, F. J. & Neuman S. P. (1989a). Estimation of spatial covariance structures by adjoint state maximum likelihood cross-validation: 1. Theory. *Water Resources Research*, 3(25), 351–362.

Samper, F. J. & Neuman S. P. (1989b). Estimation of spatial covariance structures by adjoint of state maximum likelihood cross-validation: 2. Synthetic experiments. *Water Resources Research*, 3(25), 363–372.

Samper, F. J. & Neuman S. P. (1989c). Estimation of spatial covariance structures by adjoint of state maximum likelihood cross-validation: 3. Application to hydrochemical and isotopic data. *Water Resources Research*, 3(25), 373–384.

Samper, F. J., Cuchi , J. C. & Poncela, R. (1993). Estimation of spatial cross-covariances by maximum likelihood cross-covalidation: application to hydraulic heads and transmissivities. In *Geostatistics Troia'92*, vol. 2, ed. A. Soares. Dordrecht: Kluwer Academic Publishing, pp. 721–732.

Schafmeister, M.Th. & Pekdeger, A. (1993). Spatial structure of hydraulic conductivity in various porous media: Problems and experiences. In *Geostatistics Troia'92*, vol. 2, ed. A. Soares. Dordrecht: Kluwer Academic Publishing, pp. 733–744.

Sophocleous, M. (1983). Groundwater observation network designing for the Kansas groundwater management districts. *Journal of Hydrology*, 61(4), 371–389.

Starks, T. H. & Fang, J. H. (1982). The effect of drift on the experimental semivariogram. *Journal of Mathematical Geology*, 14(4), 309–319.

Tough, J. G. & Leyshon, P. R. (1985). SPHINX - A program to fit spherical and exponential models to experimental semivariograms. *Computers and Geosciences*, 11(1), 95–99.

Verly, G., David, M., Journel, A. G. & Marechal, A. (1984). Geostatistics for natural resources characterization, *NATO Advanced Science Institute Series*. Dordrecht: D. Reidel Publishing Company.

Yakowitz, S. & Szidarovszky, F. (1985). A comparison of kriging with non-parametric regression methods. *Journal of Multivariate Analysis*, 16(1), 21–53.

Yeh, W. W. G. (1986). Review of parameter identification procedures in groundwater hydrology: the inverse problem. *Water Resources Research*, 22(2), 95–108.

3 Formulations and computational issues of the inversion of random fields

JESÚS CARRERA
Universitat Politècnica de Catalunya

AGUSTÍN MEDINA
Universitat Politècnica de Catalunya

CARL AXNESS
Sandia National Laboratories

TONY ZIMMERMAN
Gram, Inc.

ABSTRACT Three issues related to the inversion of random fields in groundwater flow problems are reviewed: problem formulation, computational methods and application methodology. Regarding the first, we show that most current methods can be viewed as particular cases of minimum variance estimation. While theory suggests that non-linear methods should lead to more accurate estimation and more realistic evaluation of uncertainty, differences obtained in intercomparison exercises appear to be much less significant than expected. This can be partly attributed to the fact that differences are more significant in the case of large variances, in the presence of sink/source terms, for large head data density and for small head measurement errors. Regarding computational aspects we present an improved version of the adjoint state method which should reduce significantly the computational burden of transient geostatistical inversion problems. Finally, application methodology is illustrated with a case comprising abundant steady-state head data and drawdowns from three pumping tests observed at a large set of points.

1 INTRODUCTION

1.1 Background

The groundwater flow inverse problem consists of estimating groundwater flow parameters on the basis of head measurements and, possibly, prior independent data about the parameters themselves. Aquifer modelers soon became interested in this problem after noticing that entering measured parameters directly into flow models led to poor results. That is, heads computed with such models did not agree with actual head measurements. Such behavior can be attributed to many factors. Spatial variability of these parameters, notably transmissivity and hydraulic conductivity, can be large and cannot be properly characterized by a few isolated measurements. Moreover these are prone to errors and may represent only local conditions, so that they might not be applicable to large-scale models.

In order to overcome some of these difficulties, one needs to adopt a proper representation of spatial variability. This is the motivation for treating transmissivity as a regionalized variable. Since geostatistics is the study of these variables, it is natural to term the resulting problem the 'geostatistical inverse problem'.

Early solutions of the inverse problem placed the main emphasis on providing a good fit between computed and measured heads. Posed in this way, the problem tends to be highly unstable. Many solutions can be found that yield virtually identical computed heads. At first, attempts were made to resolve instability by seeking either smoothness (Emsellem & de Marsily, 1971) or by imposing closeness to prior estimates (Neuman, 1973). However, it soon became apparent that these solutions did not address the basic problem, that is, the essential heterogeneity of natural media. The fact that the above authors worked together, and with Delhomme, at Fontainebleau in the late 1970s, probably motivated the eventual coupling of inverse problem techniques with geostatistics. This led to the pilot point method (de Marsily, 1979; de Marsily *et al.*, 1984; La Venue & Pickens, 1992) and to least squares with a term that penalizes departures of inverse solution from kriging estimates (Clifton & Neuman, 1982). The latter was generalized by Carrera & Neuman (1986a), who adopted Maximum Likelihood Theory to allow estimation of geostatistical parameters. Reviews of inverse problem literature are given by Yeh (1986), Carrera (1987) and, for unsaturated flow, Kool & Parker (1988).

In parallel with these developments, Kitanidis & Vomvoris (1983) proposed using cokriging, while Dagan (1985) proposed using conditional expectation. Both of these methods concentrate on formulating the problem by starting from the posterior joint distribution of heads and log-transmissivities. The latter author then estimates the log T as the expected value conditioned on point measurements of head and log T, while the former seek a linear estimate (linear combination of such measurements) leading to minimum variance. As we shall see later, the two approaches are identical when a Gaussian distribution is assumed for heads and log-transmissivities, as is usually done.

1.2 Scope

This chapter consists of three parts, which are presented in Sections 2, 3 and 4, respectively. In Section 2, we examine the two families of methods described above in order to conclude that they are not as different as they might look at first sight. The discussion is based on the discussion of Carrera, Medina & Galarza (1993).

Computer time requirements place limits on the development and practical application of geostatistical estimation methods. In fact, any method requires establishing the dependence between heads and log T. This may become a numerically intensive task, unless the dependence can be found analytically. Section 3 is devoted to a review of this topic.

An example, based on data from a recent exercise comparing various methods is presented in Section 4. This allows us to discuss in detail the application procedure, which should not differ significantly from method to method.

The chapter ends with a discussion of the relative merits of different methods and practical application issues.

2 A FRAMEWORK FOR GEOSTASTISTICAL FORMULATIONS OF THE INVERSE PROBLEM

We will restrict our discussion to indirect formulations of the inverse problem that are specifically aimed at geostatistical treatment of the variability of hydraulic conductivity. Still, much of what is described in this section can be extended to other formulations. We have made an effort to unify all formulations, possibly at the cost of losing some of the subtleties of each. Hence, we start by discussing the general framework for conditional estimation and examining its application by different authors.

2.1 Conditional expectation

COKRIGING

Let us consider a flow problem in which we have obtained head measurements at a discrete set of observation points. Let $\boldsymbol{h}_m$ be the n_h-dimensional vector of such measurements. For the purpose of this section, it is not relevant whether the measurements are obtained under steady-state or transient conditions. Let us further assume that we have obtained n_y point measurements of Y=log T. Actually, the theory does not require Y measurements to be point-wise. This assumption is made only for the purpose of simplifying subsequent derivations. We will call $\boldsymbol{z}_m=(\boldsymbol{h}_m,\boldsymbol{y}_m)$ the vector of all measurements. Let us denote by $\boldsymbol{y}$ the vector of entities (dimension m) we want to estimate. In many instances, $\boldsymbol{y}$ will be the vector of block log-transmissivities at all cells of a finite-element or finite-differences grid. However, $\boldsymbol{y}$ may represent point values of log T as well.

With these definitions, the best prediction of $\boldsymbol{y}$ (minimum mean squared error) is given by the conditional expectation of $\boldsymbol{y}$ given $\boldsymbol{z}_m$ (Graybill, 1976, p. 432). In general, obtaining this conditional expectation can be extremely difficult, because it requires knowledge of the joint probability density function of $\boldsymbol{y}$, $\boldsymbol{h}_m$ and $\boldsymbol{y}_m$. However, if we restrict ourselves to linear estimates, then the 'best' linear estimator and its expected square error are (Graybill, 1976, p. 435):

$$\hat{\boldsymbol{y}}=E(\boldsymbol{y})+\boldsymbol{Q}_{yz}\boldsymbol{Q}_{zz}^{-1}(\boldsymbol{z}_m-E(\boldsymbol{z}_m)) \tag{1}$$

$$E((\boldsymbol{y}-\hat{\boldsymbol{y}})(\boldsymbol{y}-\hat{\boldsymbol{y}})')=\boldsymbol{Q}_{yy}-\boldsymbol{Q}_{yz}\boldsymbol{Q}_{zz}^{-1}\boldsymbol{Q}_{zy} \tag{2}$$

where $E(\cdot)$ stands for expected value, $\boldsymbol{Q}_{yz}$ is the cross-covariance matrix between $\boldsymbol{y}$ and $\boldsymbol{z}_m$, i.e. $\boldsymbol{Q}_{yz}=E[(\boldsymbol{y}-E(\boldsymbol{y}))(\boldsymbol{z}_m-E(\boldsymbol{z}_m))']$, $\boldsymbol{Q}_{zz}$ is the covariance matrix of $\boldsymbol{z}_m$ and $\boldsymbol{Q}_{yy}$ is the covariance matrix of $\boldsymbol{y}$.

Equations (1) and (2) are the basis of linear estimation. As we shall see in Section 2.3, they can be extended to non-linear estimation. A graphical illustration is shown in Fig. 1, which displays a typical dependence between head at a point and

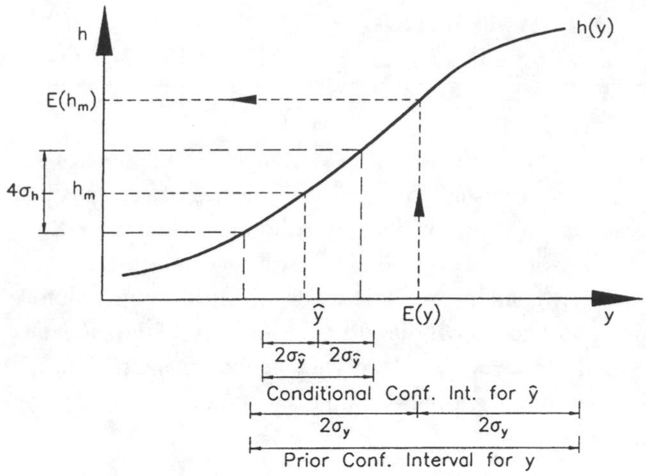

Fig. 1 Illustration of the concept of conditional expectation. Introducing head measurements, h_m, leads to displacing estimates of log T from their prior mean, $E(y)$, so that computed heads are consistent with measurements. Notice also the reduction in log T uncertainty, which depends on the size of head measurement errors σ_h

log T at a block, $h(y)$. This figure allows us to make two points. First, introducing head measurements displaces the estimate of log T, $\hat{y}$, from its prior value, $E(y)$, towards values which are consistent with such measurements. Consistency here should be understood in the sense that computed heads, $h(\hat{y})$, are close to h_m, compared with σ_h. Actually, the size of the displacement depends on both the prior uncertainty on y and the size of measurement error, as quantified by σ_y and σ_h in Fig. 1, respectively. This interplay is represented by Q_{zz} in eq. (1). The second point of Fig. 1 relates to eq. (2) and, more specifically, to the reduction in uncertainty of $\hat{y}$ being inversely proportional to the size of measurements errors. Equations (1) and (2) can be viewed as the basis of many well known techniques (Kriging, Kalman filtering, linear least squares) and deserve further comment.

Comment 1

Using eq. (1) does not require inversion of $\boldsymbol{Q}_{zz}$. In fact, all one needs are linear coefficients of $z_m - E(z_m)$. Separating the $\boldsymbol{y}_m$ and $\boldsymbol{h}_m$ components, eq. (1) is equivalent to:

$$\hat{\boldsymbol{y}} - E(\boldsymbol{y}) = \Lambda(\boldsymbol{y}_m - E(\boldsymbol{y}_m)) + \boldsymbol{M}(\boldsymbol{h}_m - E(\boldsymbol{h}_m)) \tag{3}$$

where Λ and $\boldsymbol{M}$ are matrices of dimensions $m \times n_y$ and $m \times n_h$, respectively. They are the solution of:

$$\begin{pmatrix} Q_{h_m h_m} & Q_{h_m y_m} \\ Q_{y_m h_m} & Q_{y_m y_m} \end{pmatrix} \begin{pmatrix} M \\ \Lambda \end{pmatrix} = \begin{pmatrix} Q_{h_m y} \\ Q_{y_m y} \end{pmatrix} \tag{4}$$

where $\boldsymbol{Q}_{h_m y_m}$, $\boldsymbol{Q}_{h_m y}$, $\boldsymbol{Q}_{y_m y}$ are cross-covariance matrices. The components of $\boldsymbol{M}$ and Λ are often referred to as estimation weights.

Comment 2

Equations (1) and (2) are valid regardless of the joint p.d.f. of $\boldsymbol{y}$ and z_m.

Comment 3

When the joint distribution of $\boldsymbol{y}$ and z_m is multi-Gaussian, then eq. (1) is the best (in the sense of minimum mean squared error) of all estimators, whether linear or non-linear. This is why the problem should be formulated in terms of log-transmissivities, which are known to follow a Gaussian distribution in many cases, rather than transmissivities. Moreover, in this case, $\hat{\boldsymbol{y}}$ is the conditional expectation of $\boldsymbol{y}$ given z. This is also why eq. (1) is often referred to as 'Gaussian conditional expectation'. While (2) is valid for any p.d.f., it only represents a covariance matrix when $\hat{\boldsymbol{y}} = E(\boldsymbol{y}/z)$, that is, when $\boldsymbol{y}$ and z_m are jointly multi-Gaussian.

Comment 4

$E(\boldsymbol{y})$ does not need to be an unconditional expectation. In fact, if any independent set of data (e.g. geophysical data) is available, it can be used for conditioning $\boldsymbol{y}$ or it can be appended to z. In the first option, both $E(\boldsymbol{y})$ and $\boldsymbol{Q}_{yz}$ should be conditioned on the additional data. In the second option, the additional data must be appended to z, so that $\boldsymbol{Q}_{yz}$ and $\boldsymbol{Q}_{zz}$ have to be extended to accommodate the covariances between these data and $\boldsymbol{y}$, $\boldsymbol{y}_m$ and z_m. In the multi-Gaussian case, these two options are identical (Carrera & Glorioso, 1991).

Comment 5: Simple kriging equations

When $z = \boldsymbol{y}_m$ (that is, when no head data are available), eq. (1) is known as the simple block-kriging equations. In this case, covariances in (1) are given by

$$Q_{y_m y_m ij} = cov(y_{mi}, y_{mj}) = c_y(x_i - x_j) + \delta_{ij}\sigma^2_{y_m i} \tag{5}$$

$$Q_{y y_m ij} = cov(y_i, y_{mj}) = \bar{c}_y(V_i, \boldsymbol{x}_j) \tag{6}$$

where y_{mi} is the Y (log T) measurement at location $\boldsymbol{x}_i$ (ith component of $\boldsymbol{y}_m$); y_i is the ith component of $\boldsymbol{y}$ (estimate of log T over block V_i); c_y is the autocovariance function of log T, which is assumed to be a sole function of $\boldsymbol{x}_i - \boldsymbol{x}_j$ rather than of $\boldsymbol{x}_i$ and $\boldsymbol{x}_j$; and $\bar{c}_y(V_i, \boldsymbol{x}_j)$ is the average value of c_y between $\boldsymbol{x}_j$ and the points of V_i. It is easy to check (see, e.g., Journel & Huijbregts, 1978; Samper & Carrera, 1990) that plugging (5) and (6) into (1) or, even better, into (3), leads to the simple kriging equations.

Comment 6: Ordinary kriging

One of the difficulties that might arise when trying to apply eq. (1) to real problems is that neither $E(\boldsymbol{y})$ nor $E(z)$ are initially known. The traditional approach taken to overcome this difficulty in geostatistics consists of filtering out these expectations from (1). This is achieved by writing $\hat{\boldsymbol{y}}$ as a linear function of $\boldsymbol{y}_m$, instead of $\boldsymbol{y}_m - E(\boldsymbol{y}_m)$. That is

$$y_i = \sum_j \lambda_i^j y_{mj} \tag{7}$$

where λ_i^j are called kriging weights. Taking expectations in (7) and recognizing that $E[y] = E[y_{mj}] = E[y(\boldsymbol{x})]$ leads to the well known unbiasedness condition of ordinary kriging:

$$\sum_j \lambda_i^j = 1 \text{ for all } i \tag{8}$$

Adding this constraint to eq. (4), while assuming that no head data are available, leads to ordinary kriging equations.

Additional constraints, similar to (8), may have to be imposed if $E(y(x))$ is not constant, but a function of $\boldsymbol{x}$. The equations resulting from including these additional constraints in (4) are known as universal kriging equations.

Comment 7: Cokriging

Equation (8) can be incorporated into eqs. (1) or (4), which allows us to eliminate $E(\boldsymbol{y})$ and $E(\boldsymbol{y}_m)$ from (3), resulting in:

$$\hat{\boldsymbol{y}} = \Lambda \boldsymbol{y}_m + \boldsymbol{M}[\boldsymbol{h}_m - E(\boldsymbol{h}_m)] \tag{9}$$

where Λ and $\boldsymbol{M}$ are now the solution of:

$$\begin{pmatrix} Q_{h_m h_m} & Q_{h_m y_m} & \omega_{n_h \times n_p} \\ Q_{y_m h_m} & Q_{y_m y_m} & F_m \\ \omega_{n_p \times n_h} & F_m^t & \omega_{n_p \times n_p} \end{pmatrix} \begin{pmatrix} M \\ \Lambda \\ L \end{pmatrix} = \begin{pmatrix} Q_{h_m y} \\ Q_{y_m y} \\ F \end{pmatrix} \tag{10}$$

where $\omega_{n \times m}$ is the zero matrix of dimensions $n \times m$; $\boldsymbol{L}$ is a matrix of Lagrange multipliers; $\boldsymbol{F}_m$ is the matrix of n_p drift functions evaluated at Y measurement points; and $\boldsymbol{F}$ is the matrix of drift functions evaluated at estimation points (blocks). In the case discussed in the previous comment (constant drift), $n_p = 1$ and $\boldsymbol{F}_m$ is a $n_y \times 1$ vector of 1's, while $\boldsymbol{F}$ is a $m \times 1$ vector of 1's. In general, columns may have to be added to $\boldsymbol{F}_m$ and $\boldsymbol{F}$ to account for more complex drifts. The reader should notice that because of constraints (8), or similar ones, on Λ, either (9) or (3) can be used to compute $\hat{\boldsymbol{y}}$. Both lead to identical results.

In what follows, we will call the coefficients and right-hand side matrices in (10) $\boldsymbol{Q}_{zze}$ and $\boldsymbol{Q}_{yze}$ (i.e. extended $\boldsymbol{Q}_{zz}$ and $\boldsymbol{Q}_{yz}$), respectively. This, together with the above paragraph, implies that eq. (1) is still valid when $\boldsymbol{Q}_{zz}$ and $\boldsymbol{Q}_{yz}$ are substituted by $\boldsymbol{Q}_{zze}$ and $\boldsymbol{Q}_{yze}$ and $\boldsymbol{z}_m$ is extended to $\boldsymbol{z}_{me}$ by adding n_p zeros after $\boldsymbol{h}_m$ and $\boldsymbol{y}_m$.

Application of (1) or any of its variations, (3) or (9), then requires evaluation of the expectations (at least, $E(\boldsymbol{h}_m)$) and covariance matrix. This is the subject of Sections 2.2 and 2.3.

2.2 Linear estimation

Conditional estimation has been the subject of much work. Essentially, two families of approaches can be considered. The approach of Dagan (1985) and Rubin & Dagan (1987) consists of estimating all covariance matrices analytically. The approach of Kitanidis & Vomvoris (1983) and Hoeksema & Kitanidis (1985) consists of evaluating these matrices numerically. Both methods must start with the computation of statistical parameters.

ESTIMATION OF STATISTICAL PARAMETERS

It is standard to decompose $Y = \log$ T into a drift function, $E(y(\boldsymbol{x}))$, plus a deviate, which is assumed stationary. The drift is usually expressed as a linear combination of drift elementary functions ($1, x, y, xy$, etc.) Autocovariance of deviates is normally expressed by means of the covariance function (recall eq. (5)). Therefore, statistical parameters that have to be estimated to complete the stochastic characterization of Y are the coefficients of the drift elementary functions and the parameters of c_y (at least variance and range). We will include all of them in a vector θ of statistical parameters. We want to stress that characterization of measurement errors (at least variances of head and log T measurement errors) may also be included in θ.

Inasmuch as heads can be derived from log T through the flow equation, spatial variability of heads can be related to that of log T. This can be achieved by perturbation techniques (Hoeksema & Kitanidis, 1984), by Monte Carlo simulations (Hoeksema & Clapp, 1990) or by standard sensitivity analysis, which is the approach we shall use. First order expansion of heads leads to

$$\boldsymbol{h}_m - \bar{\boldsymbol{h}}_m = (\boldsymbol{h}_m - \boldsymbol{h}_m(y) + \boldsymbol{h}_m(y) - \bar{\boldsymbol{h}}_m) \simeq \epsilon_h + \boldsymbol{J}(y - \bar{y}) \tag{11}$$

where ϵ_h represents measurement and model errors, $\bar{\boldsymbol{y}}$ is the vector of block-averaged mean log T at the m blocks comprising the solution domain and $\bar{\boldsymbol{h}}_m$ is the vector of heads computed at observation points using $\bar{\boldsymbol{y}}$. $\boldsymbol{J}$ is the Jacobian matrix $(\partial \boldsymbol{h}_m / \partial y)$, the computation of which is discussed later. In assessing the meaning of ϵ_h, the reader should notice that, in the absence of discretization and conceptual errors, $\boldsymbol{h}_m(\boldsymbol{y})$ would be the 'true' head values. Equation (11) allows us to express the covariance of head deviates $(\boldsymbol{h}_m - \bar{\boldsymbol{h}}_m)$ and the cross-covariance between these and log T deviates as:

$$\boldsymbol{Q}_{hh} = E[(\boldsymbol{h}_m - \bar{\boldsymbol{h}}_m)(\boldsymbol{h}_m - \bar{\boldsymbol{h}}_m)^t] = \boldsymbol{J}\boldsymbol{Q}_{yy}\boldsymbol{J}^t + \boldsymbol{C}_h \tag{12}$$

$$\boldsymbol{Q}_{hy} = E[(\boldsymbol{h}_m - \bar{\boldsymbol{h}}_m)(\boldsymbol{y}_m - \bar{\boldsymbol{y}}_m)^t] = \boldsymbol{J}\boldsymbol{Q}_{yy} \tag{13}$$

where $\boldsymbol{Q}_{yy}$ is the covariance matrix of block-averaged log T, which can be easily obtained by averaging c_y. A difference between the formulation presented here and that of Hoeksema & Kitanidis (1984) is the term $\boldsymbol{C}_h$, which we include to represent head measurement and model errors. These are assumed to be independent of log T deviates. The role of this term will be discussed later. Notice that $\boldsymbol{C}_h$ may be a function of θ.

The vector of statistical parameters θ is obtained by maximizing its likelihood given head and log-transmissivity measurements:

$$L(\theta/z_m)=\frac{1}{[(2\pi)^M|Q_{zz}|]^{1/2}}\exp\left[-\frac{1}{2}(z_m-\bar{z}_m)'Q_{zz}^{-1}(z_m-\bar{z}_m)\right] \quad (14)$$

where M is the dimension of z_m, $M=n_h+n_y$, and Q_{zz} is the covariance matrix of z_m, given by

$$Q_{zz}=\begin{bmatrix} Q_{h_m h_m} & Q_{hy_m} \\ Q_{y_m h} & Q_{y_m y_m} \end{bmatrix} \quad (15)$$

The cross-covariance between $\boldsymbol{h}_m$ and $\boldsymbol{y}_m$, Q_{hy_m}, is obtained as Q_{hy} in (13), which leads to

$$Q_{hy_m}=JQ_{yy_m} \quad (16)$$

where Q_{yy_m} can be easily obtained from c_y (recall eq. (6)). The log T measurement errors are often very large and should be taken into account. They enter in $Q_{y_m y_m}$, which is given by (5). Details on the maximization of (14) are given by Hoeksema & Kitanidis (1984, 1985).

ESTIMATION OF THE LOG T FIELD. NUMERICAL APPROACH

Once the statistical parameters have been computed, direct application of (1), (3) or (9) produces the log T field. The only issue we have not yet solved is the computation of the Jacobian matrix $\boldsymbol{J}$. This matrix can be obtained in several ways. Conceptually, the simplest one is to compute $\boldsymbol{J}$ by direct derivation of the flow equation. Using any discretization scheme (finite differences, finite elements, etc.), the solution to the steady-state flow equation can be written as

$$\boldsymbol{A}\boldsymbol{h}=\boldsymbol{b} \quad (17)$$

where A is an $n_n \times n_n$ matrix of coefficients that depends on transmissivity, discretization and boundary conditions; $\boldsymbol{h}$ is the vector of heads at the n_n nodes; and $\boldsymbol{b}$ is an n_n-dimensional vector of nodal sink and sources. Taking derivatives in (17), we obtain:

$$\boldsymbol{A}\frac{\partial \boldsymbol{h}}{\partial \boldsymbol{y}}=-\frac{\partial \boldsymbol{A}}{\partial \boldsymbol{y}}\boldsymbol{h}+\frac{\partial \boldsymbol{b}}{\partial \boldsymbol{y}} \quad (18)$$

where $\partial A/\partial \boldsymbol{y}$ is easy to compute because A is a linear function of transmissivity. It should be noticed that (18) represents m linear systems of equations, one for each transmissivity block. While solving all of them can be very cumbersome when m is large, they all have the same structure. Therefore, the Cholesky decomposition (or the inversion) performed for solving (17) can be used for solving the m systems in (18).

The works of Kitanidis & Vomvoris (1983) and Hoeksema & Kitanidis (1985) were originally motivated by steady-state flow. However, it is clear that the formulation can be equally applied to transient flow or other types of problems (Carrera & Glorioso, 1991; Sun & Yeh, 1992).

ANALYTICAL APPROACH

The analytical approach to estimating y is based on deriving analytically the covariance matrices that are required in Section 2.1. Clear descriptions of the methods are given by Dagan (1985), for steady-state flow with constant mean gradient, and by Rubin & Dagan (1987) for the case of uniform unknown recharge. In essence, the idea is to derive a first order approximation for $h'=h-E(h)$. In the absence of recharge, h' is given by (Gelhar & Axness, 1983; Dagan, 1985, etc.):

$$\nabla^2 h'=\nabla\cdot(Y'\boldsymbol{j}) \text{ in } \Omega \quad (19)$$

where $Y'=Y-E(Y)$ is the perturbation on log T, $\boldsymbol{j}$ is the mean gradient and Ω is the flow domain. Boundary conditions for (19) are the homogeneous equivalent of whatever boundary conditions were used for the original problem. Since $\boldsymbol{j}$ is assumed known, h' can be expressed in terms of Y' as:

$$h'(x)=-\int_\Omega \nabla(Y(\boldsymbol{x}')\boldsymbol{j})G(\boldsymbol{x},\boldsymbol{x}')d\boldsymbol{x}' \quad (20)$$

where G is Green's function for the Laplace equation with the appropriate boundary conditions. Substituting h' in the definition of the covariance function, one can easily obtain integral representations of the cross-covariance between head and log T and of the head's autocovariance function. Actually, since head variance is unbounded, one is forced to replace the latter by the variogram. Dagan (1985) and Rubin & Dagan (1987) were able to find closed-form expressions for these functions in the case of infinite domain and exponential covariance, without and with recharge, respectively.

Dagan & Rubin (1988) extended the method to transient flow by sequentially conditioning on each time step. Although, ideally, computed parameters would be constant, in practice they vary over time, a problem which is overcome by averaging them over time. Sun & Yeh (1992) showed, however, that superior results were obtained by conditioning simultaneously on all time varying data.

2.3 Non-linear estimation

Most traditional approaches to the inverse problem were formulated in an iterative manner. Especially relevant to this chapter is the work of Neuman (1980a) and Clifton & Neuman (1982). These authors formulated the inverse problem as that of minimizing the weighted sum of squared errors in heads with a regularizing term which kept the estimated $\boldsymbol{y}$ values close to their prior estimates $\boldsymbol{y}^*$. Clifton & Neuman (1982) proposed estimating these by kriging. Carrera & Neuman (1986a) showed that posing the inverse problem as the maximization of the likelihood function leads to the same type of objective function, and, moreover, that this allowed them to estimate statistical parameters. For this reason, we will first describe the formulation of Carrera &

Neuman (1986a) and later see how it relates to the conditional expectation formula.

ITERATIVE ALGORITHM

Our formulation of the inverse problem is based on considering a vector of parameters β comprising both model parameters $\boldsymbol{p}$ and statistical parameters θ. The objective is to estimate β on the basis of both head measurements $\boldsymbol{h}_m$ and prior estimates of model parameters $\boldsymbol{p}^*$. The maximum likelihood estimate of β, given data $\boldsymbol{z}^*=(\boldsymbol{h}_m,\boldsymbol{p}^*)$, is the value of β that maximizes the probability density of $\boldsymbol{z}^*$ given β. In what follows we shall assume that the only uncertain parameters are log T and $\boldsymbol{y}$. Here, we write the likelihood function under the following assumptions:

(1) Head errors $\boldsymbol{h}-\boldsymbol{h}_m$, including both measurement and model errors, are normally distributed with zero mean and covariance matrix $\boldsymbol{C}_h$.
(2) Prior estimation errors, $\boldsymbol{y}-\boldsymbol{y}^*$, are also normally distributed with zero mean and covariance matrix $\boldsymbol{C}_y$.
(3) Head and prior estimation errors are uncorrelated. It is worth emphasizing here that this does not imply that heads are uncorrelated with transmissivities. In fact, heads are a function of transmissivities through the flow equation, and their covariances were explicitly included in section 2.2.

Under these hypotheses, the likelihood of parameters β given $\boldsymbol{z}^*$ (the likelihood function) can be written as (Edwards, 1972):

$$L(\beta/\boldsymbol{z}^*)=f(\boldsymbol{z}^*/\beta)=\frac{1}{[(2\pi)^N|\boldsymbol{C}_z|]^{1/2}}\exp\left[-\frac{1}{2}(\hat{\boldsymbol{z}}-\boldsymbol{z}^*)\boldsymbol{C}_z^{-1}(\hat{\boldsymbol{z}}-\boldsymbol{z}^*)\right] \tag{21}$$

where $\boldsymbol{C}_z$ is a block diagonal matrix composed of two diagonal blocks $\boldsymbol{C}_h$ and $\boldsymbol{C}_y$, $(\boldsymbol{z}^*)^t=(\boldsymbol{h}_m,\boldsymbol{y}^*)^t$ and N is the dimension of $\boldsymbol{z}$. L depends on model parameters through $\hat{\boldsymbol{z}}$, its first n_h components being heads computed with such parameters at the n_h observation points and the remaining m components being the computed parameters themselves. On the other hand, the vector θ of statistical parameters may be needed to specify the values of $\boldsymbol{C}_h$, $\boldsymbol{C}_y$ and $\boldsymbol{y}^*$. In fact, considering that prior estimation of variances is often difficult, covariance matrices are written as

$$\boldsymbol{C}_h=\sigma_h^2\boldsymbol{V}_h \tag{22}$$
$$\boldsymbol{C}_y=\sigma_y^2\boldsymbol{V}_y \tag{23}$$

where σ_h^2 and σ_y^2 are unknown scaling parameters and $\boldsymbol{V}_h$ and $\boldsymbol{V}_y$ are positive-definite matrices representing the initial guess of the actual covariance matrices. The ratio $\lambda=\sigma_h^2/\sigma_y^2$ can be also viewed as a relative weight factor.

Carrera & Neuman (1986a) proposed using an automatic algorithm for maximizing L with respect to model parameters and some scaling statistical parameters, while estimating the remaining (hopefully few) statistical parameters by trial and error. This allows simplification of (21), the maximization of which is equivalent to the minimization of $-2\ln L$. After dropping terms that depend only on statistical parameters, this leads to the 'sum of squared residuals' objective function:

$$F=J_h+\lambda J_y \tag{24}$$

where

$$J_h=(\hat{\boldsymbol{h}}-\boldsymbol{h}_m)^t\boldsymbol{V}_h^{-1}(\hat{\boldsymbol{h}}-\boldsymbol{h}_m) \tag{25}$$

and

$$J_y=(\hat{\boldsymbol{y}}-\boldsymbol{y}^*)^t\boldsymbol{V}_y^{-1}(\hat{\boldsymbol{y}}-\boldsymbol{y}^*) \tag{26}$$

where $\hat{\boldsymbol{h}}$ is the vector of heads computed with parameters $\hat{\boldsymbol{y}}$ at observation points.

The trial-and-error estimation of statistical parameters becomes quite simple when one can assume that (22) and (23) are correct. Indeed, if $\boldsymbol{C}_y$ and $\boldsymbol{C}_h$ can be assumed known, except for factors σ_y^2 and σ_h^2, then all one needs to do is to estimate λ. This is can be easily done by minimizing 2 ln L (see eq. (21)). Having found λ, analytical minimization of 2 ln L leads to:

$$\hat{\sigma}_h^2=\frac{F}{N} \tag{27a}$$

$$\hat{\sigma}_y^2=\hat{\sigma}_h^2/\lambda \tag{27b}$$

which are derived assuming that λ is indeed optimal, or to:

$$\hat{\sigma}_h^2=\frac{J_h}{n_h} \tag{28a}$$

$$\hat{\sigma}_y^2=\frac{J_y}{n_y} \tag{28b}$$

which are derived by not assuming that λ is optimal.

The objective function (24) can be minimized using various methods. For the purpose of our later discussion, it is convenient for us to concentrate on the Gauss–Newton family. Gauss–Newton methods are based on iteratively updating the parameter vector by a vector $\boldsymbol{d}^i=\boldsymbol{y}^i-\boldsymbol{y}^{i-1}$ given by

$$\boldsymbol{H}\boldsymbol{d}^i=-\boldsymbol{g}^i \tag{29}$$

where $\boldsymbol{H}$ is a first-order approximation of the Hessian matrix of F and $\boldsymbol{g}^i$ is its gradient, both computed using the parameters at iteration i, $\boldsymbol{y}^i$. Direct derivation of F, neglecting second order derivatives for $\boldsymbol{H}$, leads to:

$$\boldsymbol{g}^i=2[\boldsymbol{J}^t\boldsymbol{C}_h^{-1}(\boldsymbol{h}^i-\boldsymbol{h}_m)+\boldsymbol{C}_y^{-1}(\boldsymbol{y}^i-\boldsymbol{y}^*)] \tag{30}$$

$$\boldsymbol{H}=2[\boldsymbol{J}^t\boldsymbol{C}_h^{-1}\boldsymbol{J}+\boldsymbol{C}_y^{-1}] \tag{31}$$

Solving for $\boldsymbol{d}^i$ from eq. (29) and replacing $\boldsymbol{H}$ by (31) and $\boldsymbol{g}^i$ by (30) allows us to update the parameter vector at iteration $i+1$ from that at iteration i as

$$\boldsymbol{y}^{i+1}=\boldsymbol{y}^i+2\boldsymbol{H}^{-1}[\boldsymbol{J}^t\boldsymbol{C}_h^{-1}(\boldsymbol{h}_m-\boldsymbol{h}^i)+\boldsymbol{C}_y^{-1}(\boldsymbol{y}^*-\boldsymbol{y}^i)] \tag{32}$$

Equation (32) is applied iteratively. That is, $\boldsymbol{y}^1$ is computed from $\boldsymbol{y}^0$ and used for evaluating $\boldsymbol{J},\boldsymbol{H}$, etc. This allows us to apply eq. (32) again, with the matrices updated for computing $\boldsymbol{y}^2$, etc. The process is repeated until convergence.

Error analysis of the estimation is most often performed in terms of the covariance matrix. A lower bound of this matrix is given by the inverse of Fisher's information matrix, the components of which are:

$$f_{kl}=\frac{1}{2}E\left[\frac{\partial^2(-2\ln L)}{\partial y_k \partial y_l}\right] \tag{33}$$

where y_k is the kth component of $\hat{\boldsymbol{y}}$. Assuming that $E(\hat{\boldsymbol{h}}-\boldsymbol{h}_m)=0$, the resulting covariance matrix is precisely

$$\Sigma_y=\boldsymbol{H}^{-1} \tag{34}$$

This would be the exact covariance matrix if heads were a linear function of the parameters. Otherwise, actual uncertainties will be larger than implied by (34). More accurate evaluations of uncertainty are discussed by Vecchia & Cooley (1987).

For the purpose of subsequent comparisons, it is convenient to assume that $\boldsymbol{y}^*$ is estimated by kriging. Therefore, $\boldsymbol{y}^*$ and its covariance matrix are given by (using (2) and (9) for the case of no head data):

$$y^*=\boldsymbol{Q}_{yy_{me}}\boldsymbol{Q}^{-1}_{y_my_{me}}y_{me} \tag{35}$$

$$\boldsymbol{C}_y=\boldsymbol{Q}_{yy}-\boldsymbol{Q}_{yy_{me}}\boldsymbol{Q}^{-1}_{y_my_{me}}\boldsymbol{Q}_{y_my_e} \tag{36}$$

where the subindex e stands for extended, as discussed in Comment 7 (see Carrera & Glorioso (1991) for details).

COMPARISON OF LINEAR AND NON-LINEAR ESTIMATION

The comparison between linear and non-linear methods can be performed either by analyzing the respective formulas or by examining the final results. In general, comparing equations is tedious and may not lead to interesting conclusions. However, we show in this section that there is a nice parallelism between linear and non-linear formulations. Numerical comparisons are discussed in the next section.

In comparing the two formulations, one may examine the equations either at the beginning or at the end of the iterative procedure. Carrera & Glorioso (1991) compared equations (9) and (32) and concluded that for $i=0$ (first iteration) they are identical provided that $\boldsymbol{y}^0$ in (32) is equal to $E(y)$ in (9). That is, if the iterative process is started from the unconditional expectation of $\boldsymbol{y}$, then $\boldsymbol{y}^1$ (computed log T's at the first iteration) are identical to those obtained with cokriging equations. They also showed that, as one should expect, computed covariance matrices are identical. Therefore, the linear formulation can be viewed as a particular case of the non-linear formulation.

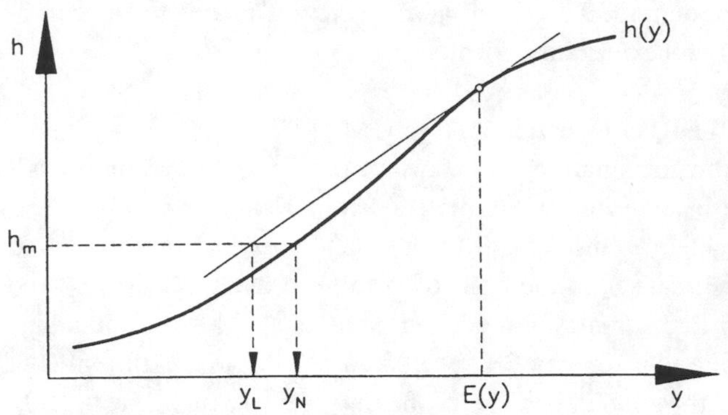

Fig. 2 Illustrating the difference between linear, y_L, and non-linear estimation of y for the case of error-free head measurement. In stationary fields without recharge, $E(y)$ tends to fall very close to the inflection point of h_y, so that, for small variances of y, y_L and y_N may be much closer than implied by the graph.

At the end of the iterative procedure, $\boldsymbol{y}^{i+1}$ equals $\boldsymbol{y}^i$ (in fact, this is one of the stopping criteria). Therefore (32) becomes

$$\boldsymbol{y}^f=\boldsymbol{y}^*+\boldsymbol{C}_{yh}\boldsymbol{C}_h^{-1}[\boldsymbol{h}_m-\boldsymbol{h}(\boldsymbol{y}^f)] \tag{37}$$

where

$$\boldsymbol{C}_{yh}=\boldsymbol{C}_y\boldsymbol{J}^t \tag{38}$$

and $\boldsymbol{y}^f$ stands for the final value of $\boldsymbol{y}$. The important conclusion from (37) is that it is formally identical to (1) and its variations. The same can be said about (36) and (2). Therefore, the conditional expectation formula is still valid. The only difference is that the relationship between head variability and log T variability is performed with estimated $\boldsymbol{y}$'s rather than with the unconditional expectation of $\boldsymbol{y}$. The difference is illustrated in Fig. 2.

2.4 Discussion on linear and non-linear methods

It is clear from the previous section that non-linear methods are generally more accurate than linear methods. On the other hand, the latter do not require iterations, so they are computationally less demanding than the former. In fact, the analytical approach requires not only very little computer time, but also prevents the user from having to discretize, a time-consuming error-prone task. Therefore, the real issue in selecting a linear versus a non-linear approach is the degree of improvement caused by the latter. Carrera & Glorioso (1991) performed extensive comparisons on two problems (one with and one without internal sink/sources) with two levels of head data density over a range of Y and head measurement error variance. For each of these situations, they performed 100 simulations. Comparisons were performed on the basis of both estimation accuracy and consistency between actual errors and estimated variances, as measured by:

$$\epsilon = \frac{1}{N_s}\sum_{j=1}^{N_s}\left(\frac{1}{m}e_j^t\right)^{1/2}$$

$$\epsilon_c = \frac{1}{N_s}\sum_{j=1}^{N_s}\left(\frac{1}{m}e_j^t\Sigma^{-1}e_j\right)^{1/2}$$

where N_s is the number of simulations (100 for each situation, i.e. fixed flow problem, number of head data, Y variogram, etc.); $\boldsymbol{e}$ is the vector of estimation errors (i.e. true minus estimated Y) at each block; and Σ is the estimation covariance matrix. Clearly, good estimation is reflected in small ϵ and consistency between actual errors and Σ is reflected in ϵ_c being close to unity.

Results of these comparisons show that the improvement brought about by non-linearity is very problem-dependent, so that it is difficult to give quantitative guidelines. Qualitatively, the improvement is more significant for large Y variances, large number of head measurements, large head measurement errors and in the presence of internal sink and sources. It is interesting to observe that the relative performance of linear methods degraded when the variance of head measurement errors was decreased. In fact, for large Y variances, reducing head errors led to worsening of linear solutions, probably as a consequence of forcing the linear approximation to be close to the true solution. This suggests that it is important to acknowledge head measurement errors, which had been neglected by most authors. An explanation for this result is given by Figs. 1 and 2: at points of low sensitivity, Y_L may be far from Y_N, an effect which is reduced with large σ_h. That is, when σ_h is small the linear approximation of heads is forced to be close to measured heads.

The improvement caused by the non-linear approach was much more significant in terms of consistency (ϵ_c index) than in terms of actual errors (ϵ index). In fact, for small σ_y and no internal sink/sources, the sizes of errors were virtually identical. On the other hand, for σ_y above unity, Y errors were up to three times smaller in the non-linear approach. Both linear and non-linear approaches were too optimistic ($\epsilon_c>1$). However, while ϵ_c only exceeded 2 in 25% of the non-linear cases, it exceeded 2 in 75% of the linear cases.

Zimmerman & Gallegos (1993) also compared linear and non-linear methods. They simulated flow through a synthetic domain and assumed that heads and log T were measured at a set of points. These measurements were used for characterizing the conditioned log T field using both linear and non-linear methods. Their performance was analyzed by comparing travel times from a prespecified set of points to a given boundary with the actual travel times. Their results showed that linear methods (both numerical and analytical) performed nearly as well as non-linear methods.

This finding is somewhat inconsistent with the results of Carrera & Glorioso (1991). Hence, an explanation should be sought. We see several possible explanations. First, the flow model considered did not have any internal sink/source term which, as discussed above, leads to the closest agreement between linear and non-linear methods. Secondly, the non-linear methods used by Zimmerman & Gallegos (1993) were not identical to the one described earlier. Therefore, the comparisons discussed in the previous section would not be applicable to them. Finally, the entire discussion of Carrera & Glorioso (1991) was based on the assumption that the same stochastic description was used for the linear and the non-linear methods. This was not the case in Zimmerman's comparisons. Any of these three causes, probably a combination of them, may explain the inconsistency.

3 COMPUTATIONAL ASPECTS

3.1 Generalities

Regardless of whether a linear or a non-linear method is used for formulating the inverse problem, one needs to establish the dependence between heads and log T's. In linear methods, this is achieved through the Jacobian matrix (11). Unless an analytical method is available for computing derivatives of heads with respect to log T's, computation of the Jacobian requires solving $\boldsymbol{m}$ systems of equations analogous to the simulation problem (18), $\boldsymbol{m}$ being the number of parameters (dimension of $\boldsymbol{y}$). When using a non-linear method, the problem is formulated as one of optimization, so that the Jacobian is needed only if a Gauss–Newton (see eqs. (29) and (32)) optimization method is used. Other methods do not require the Jacobian matrix. Specifically, quasi-Newton methods (e.g. Broyden, BFGS, DFP, etc.) and conjugate-gradient methods (e.g. Fletcher–Reeves, etc.) need only the gradient of the objective function. Such a gradient can be computed using the adjoint state method, whose CPU requirements are very small compared with the conventional sensitivity equations. Hence, it is not surprising that these methods have been successfully used in groundwater flow problems. Unfortunately, they tend to converge much more slowly than Gauss–Newton methods. In summary, gradient methods are faster per iteration but require more iterations to converge than Gauss–Newton methods.

In comparing different non-linear methods for solving the inverse problem, Cooley (1985) concluded that Marquardt's optimization method is the most efficient for steady-state groundwater flow problems. We have found Cooley's conclusions to be valid for transient problems as well. Difficulties with this method, as with all Gauss–Newton methods, arise when computing the derivatives of state variables with respect to model parameters. Two families of methods are available: direct derivatives and the adjoint state method. The former consists of taking derivatives of the state equations, which leads to a linear system of equations where the unknowns are precisely the columns of the Jacobian matrix (recall eq. (18)). This approach has been used by a large number of authors (e.g. Cooley, 1977; Gavalas, Shah & Seinfeld, 1976; Yoon & Yeh, 1976; Yeh & Yoon, 1981). On the other hand, the adjoint state method can be viewed as a particular case of Lagrange multipliers. However, in its original formulation it was proposed for computing derivatives of scalar functions. While it could be applied to computing the Jacobian matrix (Neuman, 1980a; Sun & Yeh, 1992), it is not competitive whenever the number of observations (in general, the number of scalars whose derivatives are sought) is larger than the number of parameters. What follows is a summary of Carrera & Medina (1994) and is aimed at presenting a revised formulation of the adjoint state equations which is competitive in the use of transient problems whenever the number of observation points (regardless of the number of measurements in each point) is smaller than the number of parameters.

Adjoint state equations were originally proposed by Chavent (1971) in a variational framework. Neuman (1980b) extended these equations, which were applied successfully by Carrera & Neuman (1986b). We will use an alternative approach based on formulating the direct problem in its matrix form (Townley & Wilson, 1985). A thorough presentation of the two methods of computing derivatives (direct and adjoint state) both for linear and non-linear problems is given in Carrera *et al.* (1990).

3.2 General adjoint state equations

General adjoint state equations are thoroughly described by Carrera *et al.* (1990). For the sake of completeness we present here a summary of their equations. Let $f(\boldsymbol{h},\boldsymbol{c},\boldsymbol{p})$ be a scalar function (in general, f is the objective function one wants to minimize); let ψ_i be the state equations in step i which allow state variables at time i to be obtained from those at time $i-1$. It can be shown that the derivatives of f with respect to $\boldsymbol{p}$, while recognizing that both $\boldsymbol{h}$ and $\boldsymbol{c}$ depend on $\boldsymbol{p}$, are given by

$$\frac{df}{d\boldsymbol{p}}=\frac{\partial f}{\partial \boldsymbol{p}}+\sum_{i=0}^{n_t}\lambda_i\frac{\partial\psi_i}{\partial \boldsymbol{p}} \tag{39}$$

where n_t is the total number of time steps (notice that $i=0$ has been included to account for steady-state) and λ_i is the ith adjoint state. This is the solution of

$$\frac{\partial f}{\partial \boldsymbol{x}_j}+\sum_{i=0}^{n_t}\lambda_i\frac{\partial\psi_i}{\partial \boldsymbol{x}_j}=\boldsymbol{0} \qquad j=0,\ldots,n_t \tag{40}$$

where $\boldsymbol{x}_j$ is the vector of state variables (either $\boldsymbol{h}_j$ or $\boldsymbol{c}_j$, depending on the ψ). Therefore, computing the gradient of f involves, first, solving (40) for λ_i and, then, substituting it into (39). The adjoint state λ_i ($i=0,\ldots,n_t$) is linked to f, which appears in both (39) and (40). When obtaining derivatives of a single scalar function df/dp, the computational effort is limited to solving n_t+1 ($i=0,\ldots,n_t$) linear systems of n_n equations (n_n being the dimension of ψ_i and $\boldsymbol{x}_j$). Computational cost is further reduced by the fact that the matrices involved in solving for λ_i (i.e. $\partial\psi_i/\partial\boldsymbol{x}_j$) are identical to those required solving for $\boldsymbol{x}_i$ in the state equations.

However, if $\boldsymbol{f}$ is vectorial (of dimension n_o, number of observations), the above computations would have to be repeated n_o times. It is clear that the corresponding computer cost can become unacceptably large for a large number of observations. In fact, the cost is larger than that of the alternative methods (direct derivation) whenever n_o is larger than m (dimension of $\boldsymbol{p}$). Since, traditionally, one would need a certain degree of observation redundancy (i.e. $n_o \gg n_p$), the above condition would be rarely met, and adjoint state equations were virtually abandoned for the computation of sensitivity matrices. The revitalization of geostatistical inversion has brought back to fashion problems in which the number of parameters can be larger than the number of observations. Hence, it is not surprising that Sun & Yeh (1992) have proposed using adjoint state methods for computing the Jacobian (sensitivity) matrix.

The work of Sun & Yeh (1992) and others is constrained to steady-state problems because the number of observations for transient problems can be very large. The remainder of this section is devoted to examining how adjoint state equations can still be competitive for transient flow possibly with steady-state initial conditions. Carrera & Medina (1994) show that the proposed algorithm can also be applied to transient transport with steady-state flow.

3.3 Improved form of adjoint state equations for transient flow

In this case, state variables, $\boldsymbol{x}_j$, are the vector of nodal heads $\boldsymbol{h}_j$ ($j=0,\ldots,n_t$) and $\boldsymbol{h}_0$ represents initial heads, which may be assumed to represent steady-state flow. State equations are the numerical version of the flow equation:

$$\psi_0=\boldsymbol{A}\boldsymbol{h}_0-\boldsymbol{b}_0=\boldsymbol{0} \tag{41}$$

$$\psi_j=\left(\theta\boldsymbol{A}+\frac{\boldsymbol{D}}{\Delta t}\right)\boldsymbol{h}_j-\left[(\theta-1)\boldsymbol{A}+\frac{\boldsymbol{D}}{\Delta t}\right]\boldsymbol{h}_{j-1}-\boldsymbol{b}_{j-1+\theta}=\boldsymbol{0} \tag{42}$$

where $\boldsymbol{A}$ is the $n_n\times n_n$ symmetric 'conductance' matrix, which depends on transmissivity and leakance factors; $\boldsymbol{D}$ is the storage matrix, which depends on storativity and is often taken as diagonal; and $\boldsymbol{b}_j$ is the vector of nodal sinks and sources at time j. Readers may notice that a θ-weighted time integration scheme is implied in (42). The exact form of $\boldsymbol{A}$, $\boldsymbol{D}$ and $\boldsymbol{b}$ depends on the method for spatial approximation of the flow equation (that is, finite elements, finite differences, etc.). The choice of solution method is unimportant at this stage.

Let us now assume that we want to obtain derivatives of $\boldsymbol{a}^t\boldsymbol{h}_k$ with respect to $\boldsymbol{p}$. We will assume, without loss of generality, that $\boldsymbol{a}^t$ is not a function of $\boldsymbol{p}$ ($\boldsymbol{a}$ may be a function of transmissivities, for example, if $\boldsymbol{a}^t\boldsymbol{h}_k$ represents measurable boundary fluxes). Adjoint state equations are obtained by substituting $f=\boldsymbol{a}^t\boldsymbol{h}_k$ into (40), which leads to

$$\boldsymbol{a}^t\delta_j^k+\lambda_j^k\left(\theta A+\frac{D}{\Delta t}\right)=\lambda_{j+1}^k\left(\frac{D}{\Delta t}+(\theta-1)A\right)=0 \quad j=0,\ldots,n_t \quad (43)$$

where the superindex k is being used to denote that λ_j^k is the adjoint state of $\boldsymbol{a}^t\boldsymbol{h}_k$ (i.e. k is a constant, though it will vary when computing the derivatives of heads at a different time); δ_j^k is Kronecker's delta. Equation (43) is solved backwards in time, from $j=n_t$ through $j=0$. It should be noticed that $\lambda_{n_t+1}^k=\boldsymbol{0}$. Also, $\lambda_j^k=\boldsymbol{0}$ for $j>k$, which is a consequence of $\lambda_{n_t+1}^k=\boldsymbol{0}$.

The most important property of (43) is that k is just a dummy variable. Therefore, adjoint states for different times are identical, though shifted in time:

$$\lambda_j^k=\lambda_{j+m}^{k+m}=\lambda_{j+n_t-k}^{n_t} \quad \text{or} \quad \lambda_j^{n_t}=\lambda_{j+k-n_t}^k \quad (44)$$

The importance of (44) stems from the fact that one does not need to compute a different adjoint state for each k value. A single adjoint state is sufficient. Actually the above needs an additional correction for the case in which initial heads are given by a steady-state. In such cases, the corresponding adjoint state equation would be:

$$\lambda_0^k A=\lambda_1^k\left[\frac{\boldsymbol{D}}{\Delta t}+(\theta-1)\frac{A}{\Delta t}\right] \quad (45)$$

That is, a different λ_0^k is needed for each k value. This is a minor problem, but it can be solved by adding eq. (43) from $j=0$ through n_t, which leads to:

$$\sum_{j=0}^{n_t}\lambda_j^k\boldsymbol{A}=-\boldsymbol{a}^t \quad (46)$$

From eq. (43), for $k=0, j=0$,

$$\lambda_0^0\boldsymbol{A}=-\boldsymbol{a}^t \quad (47)$$

From (47) and (46), one obtains the adjoint state for $j=0$ as

$$\lambda_0^k=\lambda_0^0-\sum_{j=1}^{n_t}\lambda_j^k \quad (48)$$

These equations can now be plugged into (39) to compute the derivatives of f with respect to $\boldsymbol{p}$ for varying k. Let f_k be $\boldsymbol{a}^t\boldsymbol{h}_k$.

The algorithm for computing $df_k/d\boldsymbol{p}$ for $k=0,\ldots,n_t$ can be summarized as follows:

(1) Solve direct problem eqs. (41) and (42).
(2) Solve (47) for λ_0^0.
(3) For $j=n_t$ through $j=0$ perform the following steps:
 (3.1) Solve (43) for $\lambda_j^{n_t}$ (starting with $\lambda_{n_t+1}=0$).
 (3.2) For all observation times between $k=n_t-j+1$ and $k=n_t$ update $df_k/d\boldsymbol{p}$ ($f_k=\boldsymbol{a}^t\boldsymbol{h}_k$) by sequential use of (39), i.e.

$$\frac{df_k}{d\boldsymbol{p}}=\frac{df_k}{d\boldsymbol{p}}+\lambda_{j+k-n_t}^k\frac{\partial\Psi_{j+k-n_t}}{\partial\boldsymbol{p}} \quad (49)$$

 where $\lambda_{j+k-n_t}^k=\lambda_j^{n_t}$, except for $k=0$ (in which case it equals λ_0^0).
(4) Repeat steps (2) and (3) for varying f_k (i.e., by changing $\boldsymbol{a}$ wherever several observation points are available).

Actual implementation of the method is somewhat involved (details are given by Carrera & Medina (1994). The point is that the proposed method requires solving $(n_t+1)n_o$ systems of equations. In geostatistical inversion, this number should be much smaller than either the $(n_t+2)(n_t+1)n_o/2$, required by conventional adjoint state, or the $(n_t+1)\cdot m$, required by the sensitivity method.

4 AN APPLICATION EXAMPLE

4.1 Background

The example we present is part of an exercise formulated by Sandia National Laboratories (SNL) as part of its Performance Assessment (PA) of the Waste Isolation Pilot Project (WIPP).

The WIPP is a project of the US Department of Energy (DOE) aimed at demonstrating the possibility of safe disposal of transuranic radioactive waste at a mined geologic repository near Carlsbad (New Mexico, USA). PA implies identifying the events and processes that might lead to releases of radionuclides to the accessible environment. Some scenarios for such releases involve transport through the Rustler Formation, which overlies the repository and consists of several beds of dolomite, siltstone and evaporitic rocks. Extensive testing has led to the conclusion that the Culebra Dolomite is the most transmissive and laterally continuous unit of the Rustler formation. Therefore, PA efforts have concentrated on this unit.

Early PA exercises made it apparent that transmissivity at the Culebra is extremely variable and that geologic processes controlling this variability are very complex. Complete characterization does not seem plausible both because of the cost of drilling and collecting measurements and because excessive drilling might affect negatively the characteristics

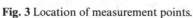

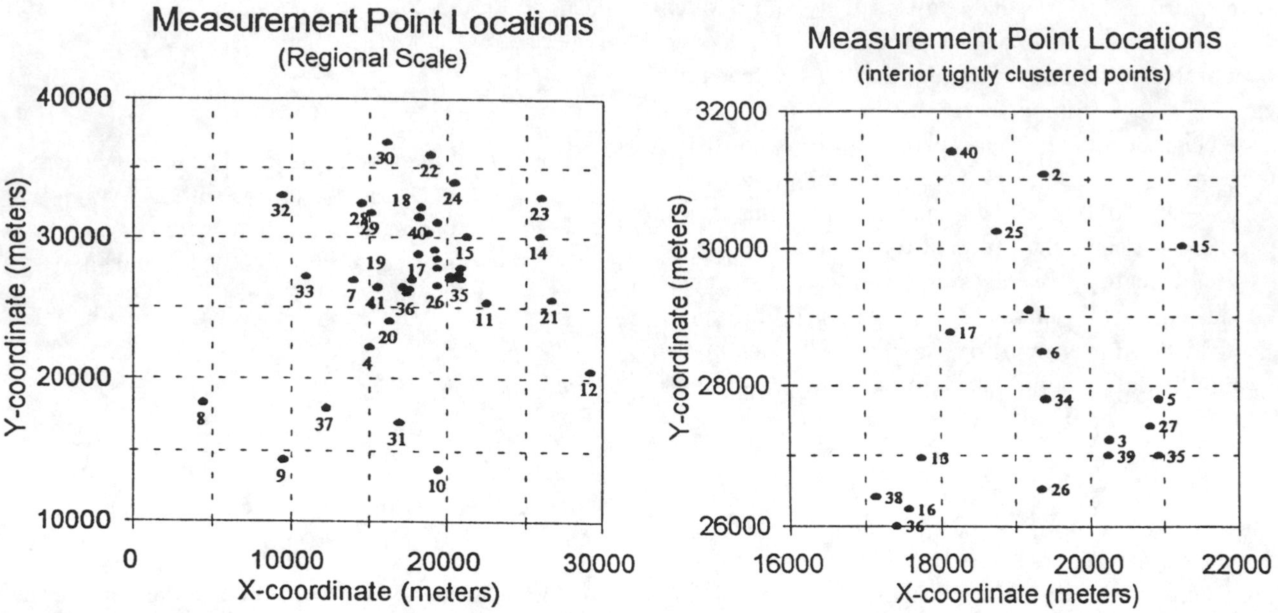

Fig. 3 Location of measurement points.

of the site by increasing its permeability and accessibility to the surface. It seems clear that statistical methods are strongly advisable in this case. The question then becomes: which is the statistical method best suited to yield predictions of arrival groundwater travel times based on available data? In order to address this question, SNL formed a geostatistics expert panel, who set up a sequence of synthetic problems (test problems). Each test problem consisted of a simulated log T field subject to unknown boundary conditions. *T* and *h* were 'measured' at a set of observation points and given to participants, who were expected to provide full log T maps. Actually, rather than a single log T map, participants had to provide a set of 'equally likely' realizations of the log T map.

Performance of each method was then evaluated on the basis of the c.d.f. of travel times computed with this set of log T maps, as compared with the true travel times.

What follows is a brief description of the application of the non-linear maximum likelihood method (Carrera & Neuman, 1986a) to test problem 4. Some of the conclusions from this exercise have been discussed in Section 2.4.

4.2 Description of the example

The log T field was generated so as to ensure the presence of through-going high-T channels, five to six orders of magnitude variations in log T and a small trend. The field was then generated as a multi-Gaussian field with a mean of 5.5, a variance of 0.8 and a Bessel covariance function with an east–west correlation distance of 2050 m and a north–south correlation distance of 1025 m. High-T channels were generated by conditioning this field on several high *T* values. Heads were obtained by applying Dirichlet boundary conditions on all boundaries (boundary heads were generated as an intrinsic random function of order 2). An areal source term was used over an area of high *T* to help direct flow in the desired southwest–northeast direction. Three pumping tests were simulated in order to generate transient data.

Data provided to each participant included:

41 transmissivity measurements at randomly selected points, which are assumed to represent existing boreholes;
steady-state head values at the same locations;
pumping test information, including pumping rate and location and drawdown curves for each test. In all, 40 drawdown curves were provided.

Data locations are shown in Fig. 3.

4.3 Application methodology

STEP 1. CONCEPTUAL MODEL: STRUCTURAL ANALYSIS

This step usually involves an analysis of available hard data (*T*, *h* and hydrochemistry) and soft data (geology, soil use, etc.). The modeler seeks a conceptualization of the system that, hopefully, is qualitatively consistent with all data. In the present case, such analysis had to be limited to *h* and *T* data. Existence of high-T zones could be deduced from the bimodal nature of the log T histogram. Hand contouring of head data suggested the existence of sharp transmissivity contrasts, either high-T channels or low-T barriers. Presence of the former was best derived from pumping tests because

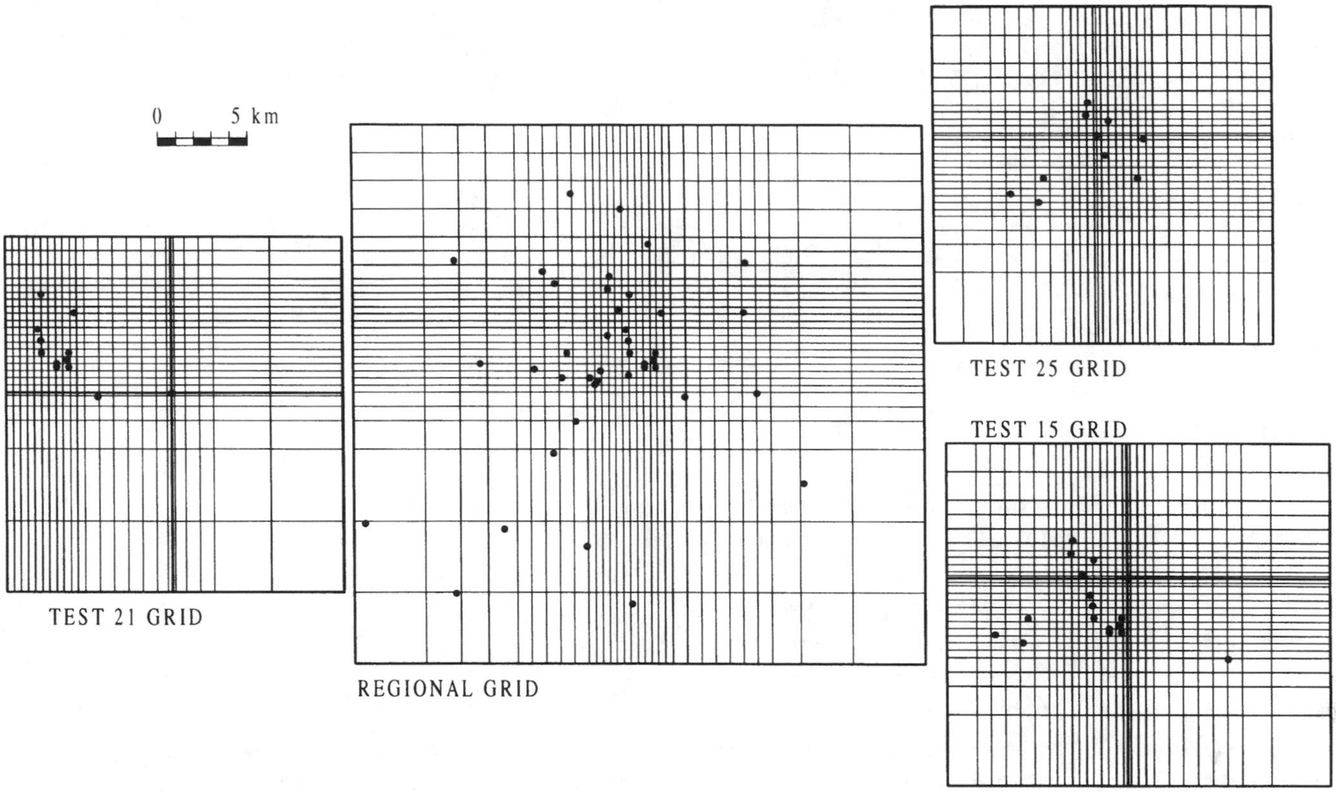

Fig. 4 Finite-element grid. Notice that large portions of the grid are duplicated in simulating the pumping tests. In general, the pumping test grids are identical to the regional grid, except for refinements around the pumping well.

some distant wells showed a quick response while others, much closer, displayed a weak and delayed response.

In summary, the qualitative analysis provided little information of relevance to the conceptual model. Therefore, a stationary random field was assumed for the log T field, in the expectation that any trend or connectivity pattern could be derived through the calibration (conditioning) process. We want to stress that in lacking a geological model, any other assumption would have been gratuituous because it would have implied trying to guess what the model generator had in mind.

A preliminary variogram was obtained by hand-fitting the sample variogram, which did not display any clear directional dependence. We used an exponential variogram with a range parameter of 1500 m and a sill of 2.

STEP 2. DISCRETIZATION: MODEL STRUCTURE

This example has some interesting features: boundary locations and type are unknown, there is no information about overall flow rates (mass balance) and four independent data sets are available (estimation was done using only steady-state data first and the full data set second, but only the latter is described here). The model domain (Fig. 4) was selected so as to include all data points, and it extended towards north and east, so that the area with largest density of data remained more or less centered within the domain. The discretization was made up of rectangles ranging in size from $400 \times 400\,m^2$ in the central area to $4000 \times 4000\,m^2$ at the edges. The increase in size towards the edges was motivated by a desire to reduce the number of nodes and elements. The importance of this reduction becomes apparent when considering the treatment of pumping tests.

There are several approaches in dealing with multiple independent data sets (Carrera, 1994). The most effective one consists of calibrating the model with all the data sets simultaneously. An easy way of doing this was used by Carrera & Neuman (1986c) and consists of repeating the domain for each data set. In this manner, each point in reality is represented by up to four points in the model, one for steady-state and for each test. This explains why four grids are displayed in Fig. 4. The regional grid represents steady-state conditions, while each of the pump test grids will be used for modeling the pumping tests. Obviously, the only difference among the grids is in the boundary conditions and refinements at the pumping well.

Boundary conditions for the regional grid were of the Dirichlet type around the boundary. Actually, boundary heads are unknown, so heads at the corners and mid-points of the sides were estimated. Heads in between were interpolated, which required defining 1-D elements of very large conductivity between each pair of boundary nodes. Prior

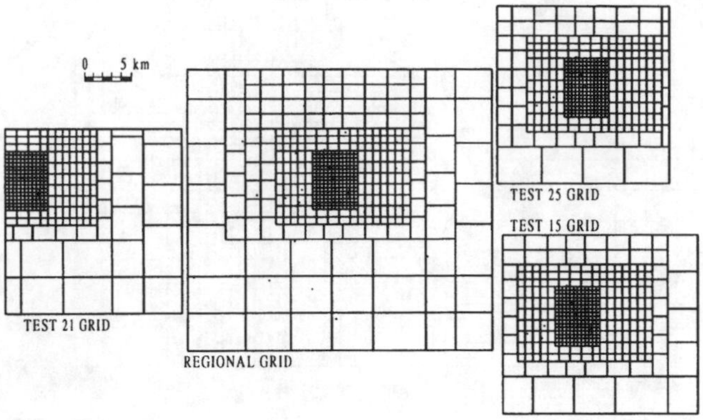

Fig. 5 Geometry of log T blocks.

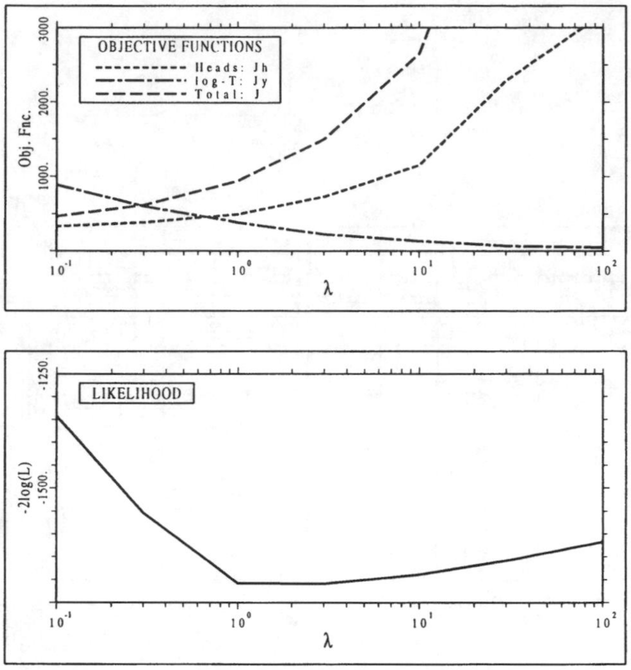

Fig. 6 Estimation of λ. The likelihood function is nearly flat between 1 and 3. A value of λ=1 was chosen because it honors the sample variogram better than λ=3 (other criteria also suggested that the minimum is much closer to 1 than to 3).

estimates of heads at these eight points were first obtained by linear interpolation of head data and later modified after noticing that such an approach would overestimate heads at the northern boundary. No flow boundaries were used for the pumping test grids, but a zero head had to be specified at one node (a corner node) to ensure that the flow matrix during steady-state would not be singular.

In general, transmissivity zones consisted of one element, except where the elements were elongated. In such cases several elements were linked in one zone to ensure that T zones are nearly square in shape. Otherwise, block-T might have to be tensorial, which would have complicated the process. In the end, 433 block-T values had to be estimated. The geometry of the zones is shown in Fig. 5. Prior estimates and the covariance matrix of these values were obtained by block kriging, using the program KRINET (Olivella and Carrera, 1987).

The only parameters remaining to define the estimation problem fully are those controlling the error structure of head data. It was assumed that measurement error was zero, so that only discretization (both of log T and numerical solution) and conceptual errors have to be taken into account. A simple interpolation error analysis suggested a standard deviation of 10 cm for steady-state data and 1 cm for transient data. These values were increased for the pumping wells and for the observations immediately after the start and end of pumping, when numerical errors are expected to be largest.

STEP 3. CALIBRATION

Having fully defined the problem structure, all that remains for estimation is to run the code. Actually, as discussed in Section 2, the error structure of heads and log T may have to be modified at this stage. This involves a trial-and-error procedure to:

(1) Select the best structure among a set of alternatives. This is done by either comparing the likelihood values or, better, the structure identification criteria obtained with the different structures. In our case, a linear variogram was tried during early calibrations, but was soon rejected because it led to poor results (in terms of likelihood) compared with an exponential variogram.

(2) Compute parameters controlling the statistical structure of log T and head errors. In general, this would have involved revising the relative standard deviations assigned to steady-state and transient data, the assumption of stationarity for log T, the range of the variogram, etc. In our case, we simply maximized the likelihood with respect to σ_h^2 and σ_y^2, factors multiplying the covariance matrices V_h and V_y obtained during Step 2. As discussed in Section 2, this is achieved by manually minimizing S with respect to λ. Once the optimum λ has been found, optimal values of σ_h^2 and σ_y^2 can be easily derived.

Optimization with respect to λ is summarized in Fig. 6, where it is shown that the minimum S is found for λ=1. The values of σ_h^2 and σ_y^2 are as follows:

σ_h^2(eq. (27a)=0.99 σ_h^2(eq. (28a)=0.97
σ_y^2(eq. (27b)=0.99 σ_y^2(eq. (28b)=0.90

These values are sufficiently close to unity for us to assume that the prior error structure is correct and should be left untouched. It must be remarked that this is totally coincidental. σ_y^2 is often close to unity because the sample variogram may be sufficient to estimate the true variogram, so that the prior structure of log T may be close to the real one. However, estimation of the head errors structure was rather qualitative, so we find it surprising that we were so close to the optimum.

A last word should be devoted to computational aspects. A typical run took around 3 hours on a DEC ALPHA 7630. Such a run usually consisted of two cycles of Marquardt iterations and one cycle of Broyden (quasi-Newton) iterations. Iteration CPU times are as follows:

Broyden (single adjoint state): 5 s/iter;
Marquardt (conventional sensitivity): 660 s/iter;
Marquardt (improved adjoint state): 260 s/iter.

It is clear that the conventional adjoint state is by far the fastest method. Unfortunately, since it only produces the objective function gradient, it can only be used with conjugate gradients or quasi-Newton methods, which are very slow to converge. On the other hand, it is also clear that in this case (441 parameters, 3347 nodes, 6357 elements, 80 observation points and 81 solution time steps), the modified version of the adjoint state method is much more effective than the conventional sensitivity equations method. It should be added, however, that the Marquardt method failed to produce full convergence in most cases, although it performed better than Broyden's method. That is why both methods were used sequentially on each run.

5 RESULTS

Results are summarized in terms of head maps (Fig. 7), drawdown curves fit (Fig. 8) and transmissivity maps (Fig. 9).

Fig. 7 displays both steady-state heads and drawdowns after ten days of pumping. Considering that head fit is quite good (maximum error is 16 cm), that measurements cover most of the domain with a fairly large density and that, in the absence of sink/source, steady-state head maps tend to be quite smooth, we can assume that the piezometric surface reproduces the true one quite well. A southwest–northeast general trend, which was intended during the definition of the problem, can be observed. The effect of the high-T channel (see Fig. 9) also becomes apparent in Fig. 7. Drawdown maps are much more irregular because they tend to propagate preferentially along high-T zones. It can be observed that the objective of extending the grids far enough, so as to ensure that boundaries did not affect computed drawdowns, has been met only partially. This is because the extent of high-T channels was not known beforehand. In fact, test grids were initially smaller, but had to be extended to reduce boundary effects. As it is, we feel that these do not significantly affect either computed heads or estimated parameters.

Drawdown versus time curves (Fig. 8) show that the fit is excellent, except for a few isolated points. The worst fit cor-

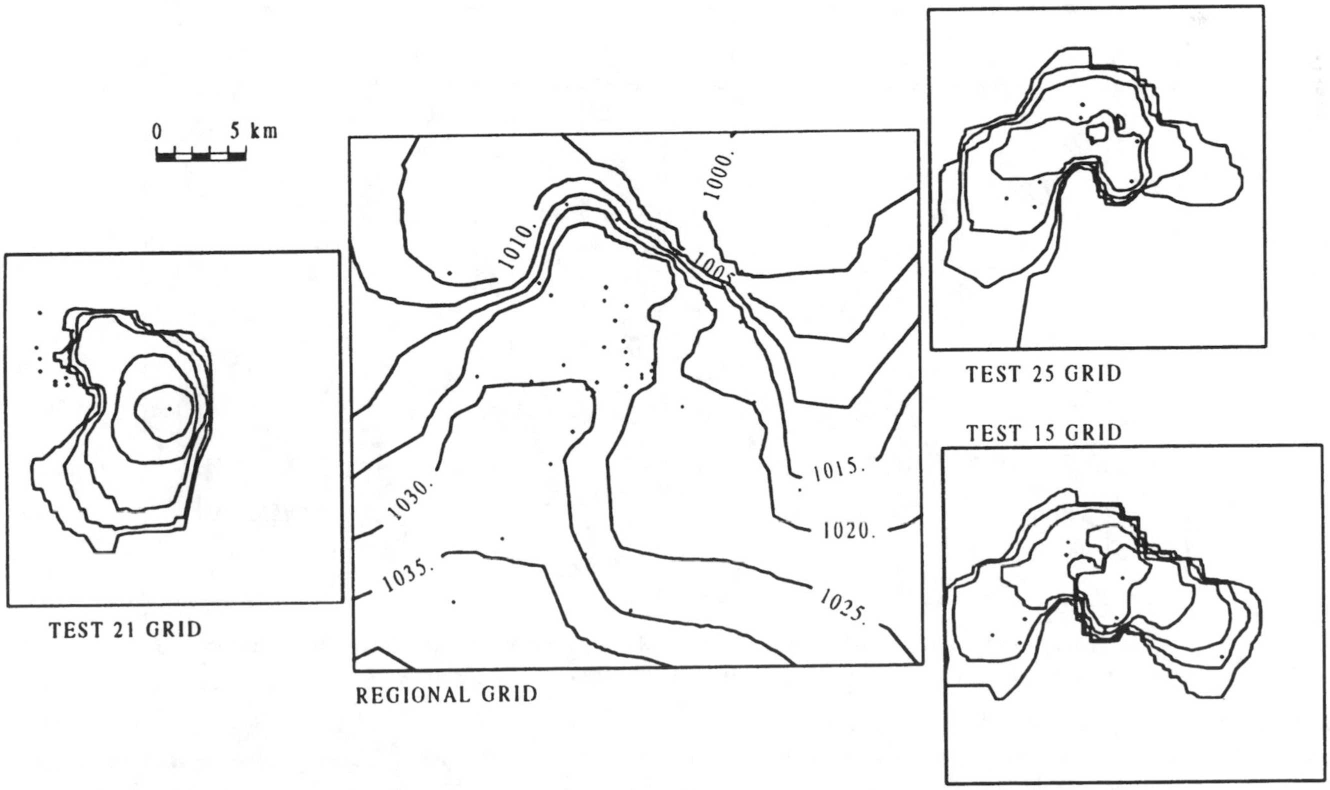

Fig. 7 Computed steady-state heads (regional grids) and drawdowns after ten days of pumping (pump test grids).

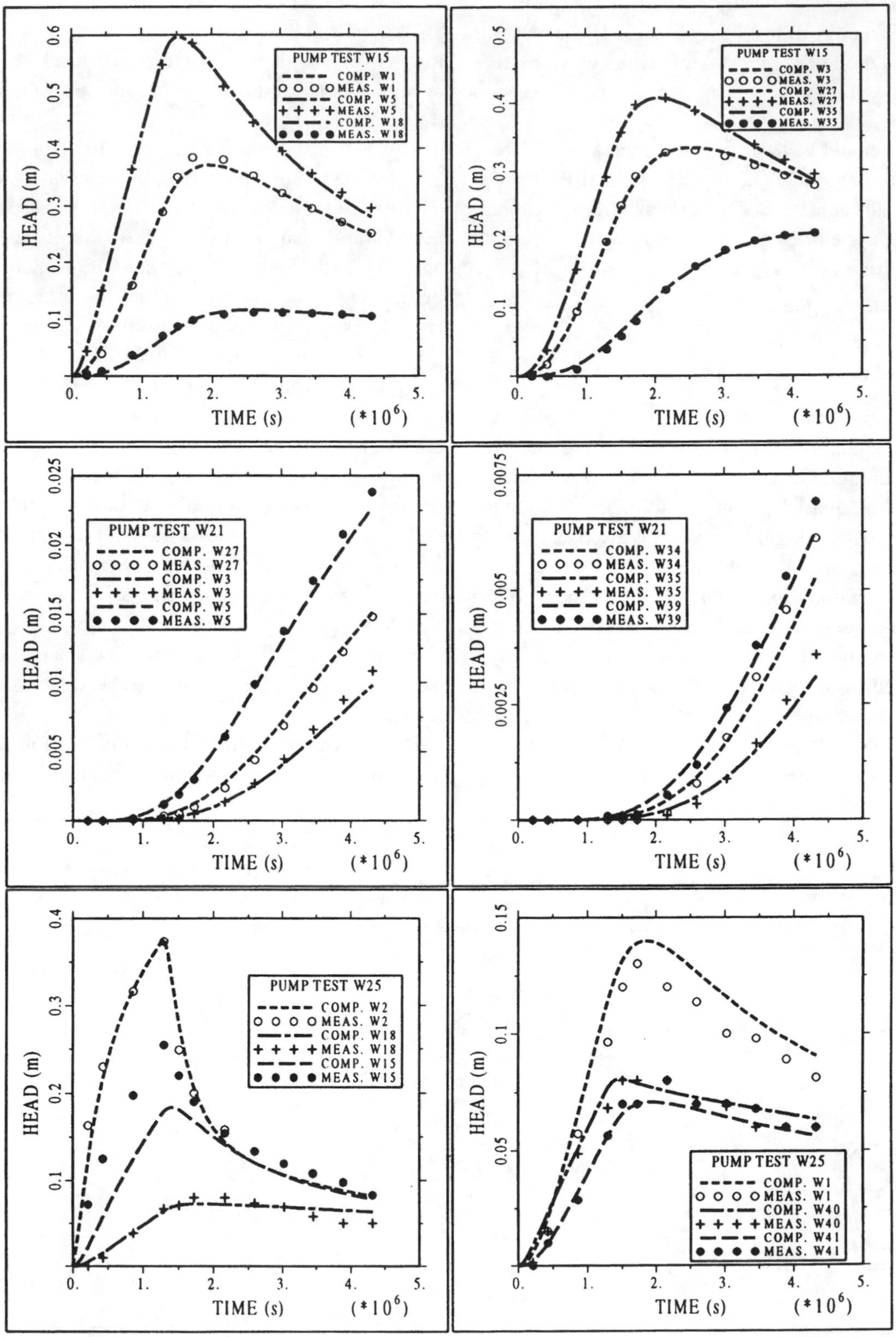

Fig. 8 Drawdown curves at six observation wells for each test (test 15 above, 21 middle and 25 below). Notice that errors are very large at well 15 in response to pumping from 25. This curve accounted for 36% of the objective function.

responds to the response of point 15 to pumping from well 25, which accounts for 36% of the total objective function.

This may be a result of insufficient discretization, in terms of the size of log T blocks for simulating log T variability and in terms of the size of the elements for solving the flow equation. The problem could be solved by either refining the grid or by reducing the weight assigned to this well. Considering that the overall effect would have been an improvement of the

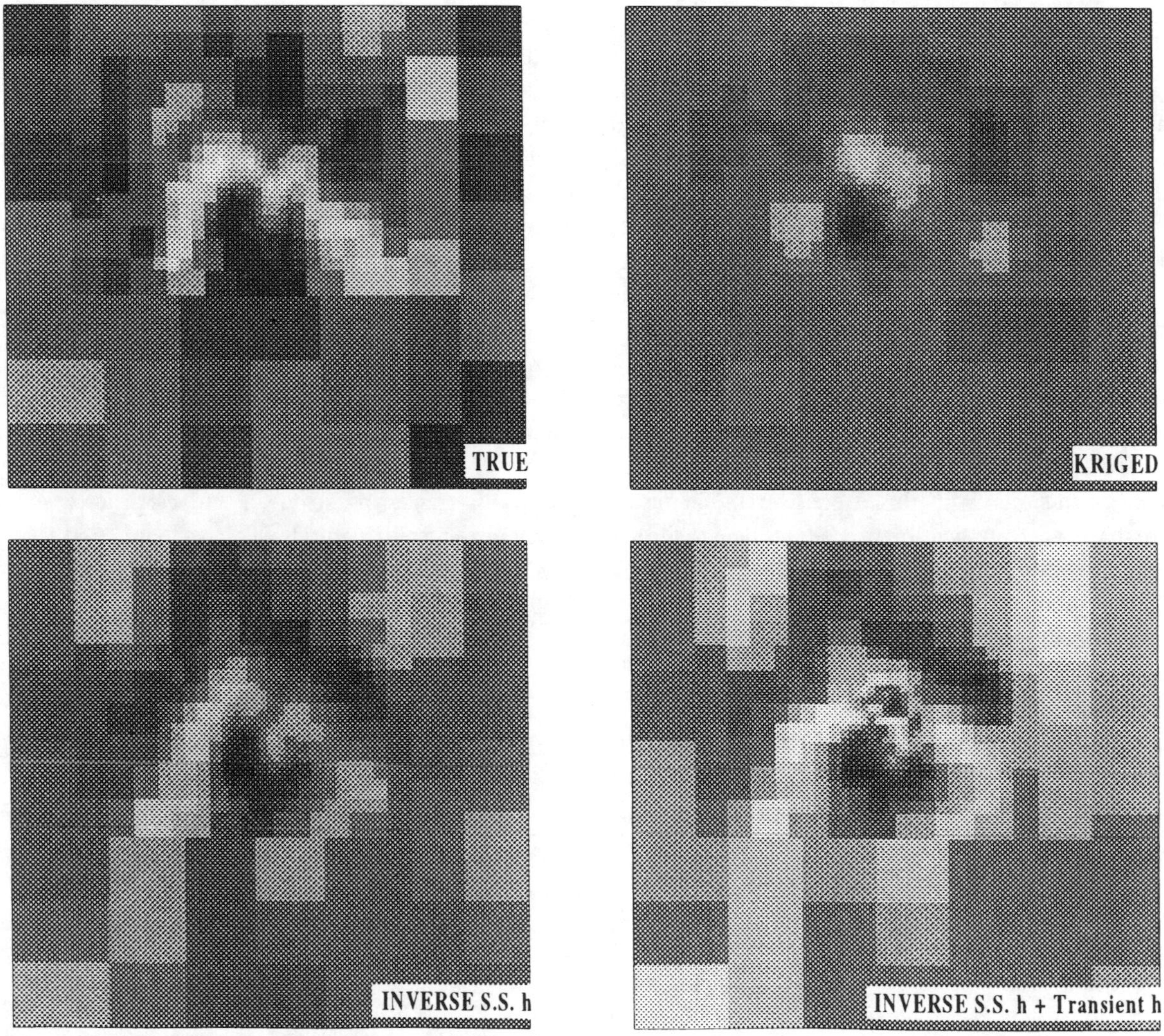

Fig. 9 Computed log T maps at different levels of conditioning. The true log T map (a discretized version of the mean) uses a different gray scale and extends over one area different from the rest. Each gray tone covers 0.5 units of log T (white highest).

head fit and a corresponding reduction of the optimum λ (recall Fig. 6 and eqs. (27) and (28)), we opted for choosing the smallest λ in the interval of indifference in Fig. 6 (1., 3.). In the rest, notably in pump test 15, the fit is excellent.

Transmissivity maps are displayed in Fig. 9. In fact, we show the evolution of the log T map as a function of the level of information. Not surprisingly, the kriging map is extremely smooth except in the area of largest data density. Even there, the map fails to capture the main connectivity features of the true map. The map derived from log T and steady-state head measurements reproduces the sinusoidal channel much better, but is still too smooth. Finally, adding transient measurements allows a much better identification of connectivity pattern and general trends of spatial variability. We find particularly surprising the identification of the high-T zone in the upper right corner. The map is still smooth because, after all, it represents a (conditioned) mean field and it misses some of the features of the true field, but overall it seems to be a fairly good representation of the true field.

6 DISCUSSION AND CONCLUSIONS

Three issues have been examined in this chapter: computational issues, comparison of formulations and an application example. Regarding the first, we have briefly described an improved version of the adjoint state method that should greatly reduce the computational time of geostatistical inverse problems.

Regarding the formulation, we have reviewed linear and non-linear formulations of the inverse problem. The first thing that becomes apparent is that all of them can be viewed as particular cases of the general minimum variance estimator. Therefore, they are not as different as they might look at first sight. Secondly, their main difference stems from the point at which heads are expanded in terms of log T's. In linear methods, heads are expanded around the unconditioned mean log T, while in non-linear methods, they are expanded around the conditioned log T. We argue that this leads to improved estimation and a better and less optimistic evaluation of uncertainty. However, the difference between the two families of methods is small for small variances of log T, large head measurement errors, few head measurements and in the absence of internal sinks and sources.

In fact, in a recent intercomparison exercise led by Sandia National Laboratories, non-linear methods did not perform much better than linear methods (the final conclusions from such an exercise are still pending). Actually, one of the main qualitative conclusions from this exercise was that results were more sensitive to application procedures than to the features of each method.

Application procedures were discussed in an example that was part of the above exercise. One of the findings of this example was that, while capturing most of the features of the true log T field, the final field was still quite different from the true one, despite a very good head fit. This should lead to several reflections. First, information about log T contained in heads is quite limited. Given head and log T data, one may find many different log T fields honoring these data with high accuracy. This is why the above comparisons of methods were not made in terms of mean fields (the only ones we have discussed here) but in terms of many simulated fields. Most simulation methods are based on perturbation of the mean field according to the derived covariance structure (eqs. (2), (34) or similar). Unfortunately, this does not ensure that each of the simulated fields still fits head data. This is what motivated the work of Gómez-Hernández, Sahuquillo & Capilla (1997), who used the opposite approach. That is, they generated fields solely conditioned on log T measurements, but displayed the small-scale log T variability and other desired features, and later perturbed them to ensure that the final field indeed honors head measurements.

A second reflection from the application example is the importance of conceptual issues and the stochastic characterization. One might argue that differences between the true variogram (anisotropic with zero slope at the origin) and the one we used for the model (isotropic with non-zero slope) may be responsible for the differences between the estimated and true fields. Actually, one could go further and contend that connectivity patterns of the true field are not consistent with the assumptions of multi-Gaussianity and stationarity implicit in most stochastic methods. That is, in hindsight, our assumptions about the random field were quite poor. The fact that final results were fairly good can only be attributed to the anomalously large amount of error-free data. Yet, in our experience, well connected high-T zones are frequent, especially in low permeability environments, while data are seldom so abundant and never so accurate. The only way to treat such problems (that is, most problems in low *K* media and many in conventional media), is by relying heavily on geological observations which provide information about density, extent, direction and, often, location of relevant features. Unfortunately, standard geostatistical tools cannot take advantage of this type of information, so in practical problems we rarely use a geostatistical approach (indicator geostatistics is an improvement in this context, but rarely suffices). This leads to the conclusion that the real challenge is to produce geostatistical tools that can take full advantage of qualitative geological information and yet provide the type of results required by most geostatistical inversion methods (basically, trends in the mean and covariance structure). Otherwise, current methods will have to undergo severe conceptual modifications.

ACKNOWLEDGMENTS

Sections 2 and 3 are the result of work funded by ENRESA and the European Union through the RADWAS program. The example in Section 4 was performed with funding from Sandia National Laboratories.

REFERENCES

Carrera, J. (1987). State of the art of inverse problem applied to the flow and solute transport equations. In *Groundwater Flow and Quality Modelling*, eds. E. Custodio *et al.* Dordrecht: Reidel Publishing Co., pp. 273–283.

Carrera, J. (1994). INVERT-4. A Fortran program for solving the groundwater flow inverse problem. User's guide. *CIMNE Technical Report*, 160 pp.+appendices. Barcelona, Spain.

Carrera, J. & Glorioso, L. (1991). On geostatistical formulations of the groundwater flow inverse problem. *Advances in Water Resources*, 14(5), 273–283.

Carrera J. & Medina, A. (1994). An improved form of adjoint-state equations for transient problems. In *X International Conference on Computational Methods in Water Resources*, eds. A. Peters *et al.* Dordrecht: Kluwer Academic Publishers.

Carrera J., Medina, A. & Galarza, G. (1993). Groundwater inverse problem. Discussion on geostatistical formulations and validation. *Hydrogèologie*, no. 4, pp. 313–324.

Carrera, J. & Neuman, S. P. (1986a). Estimation of aquifer parameters under steady-state and transient conditions: I. Background and statistical framework. *Water Resources Research*, 22(2), 199–210.

Carrera, J. & Neuman, S. P. (1986b). Estimation of aquifer parameters under steady-state and transient conditions: II. Uniqueness, stability, and solution algorithms. *Water Resources Research*, 22(2), 211–227.

Carrera, J. & Neuman, S. P. (1986c). Estimation of aquifer parameters under steady-state and transient conditions: III. Applications. *Water Resources Research*, 22(2), 228–242.

Carrera J., Navarrina, F., Vives, L., Heredia, J. & Medina, A. (1990). Computational aspects of the inverse problem. *VIII International Conference on Computational Methods in Water Resources*, eds. G. Gambolati *et al.* Southampton: Computational Mechanics Publications, pp. 513–523.

Chavent, G. (1971). Analyse fonctionelle et identification de coefficients

repartis dans les equations aux derivees partielles. *These de docteur en sciences*, Univ. de Paris, Paris.
Clifton, P. M. &. Neuman, S. P (1982). Effects of kriging and inverse modeling on conditional simulation of the Awa Valley aquifer in southern Arizona. *Water Resources Research*, 18(4), 1215–1234.
Cooley, R. L. (1977). A method of estimating parameters and assessing reliability for models of steady state groundwater flow. 1. Theory and numerical properties. *Water Resources Research*, 133(2), 318–324.
Cooley, R. (1985). Regression modeling of groundwater flow. *U.G.S.S. Open-file Report 85-180*, 450pp.
Dagan, G. (1985). Stochastic modeling of groundwater flow by unconditional and conditional probabilities: The inverse problem. *Water Resources Research*, 21(1), 65–72.
Dagan G. & Rubin, Y. (1988). Stochastic identification of recharge, transmissivity, and storativity in aquifer transient flow: a quasi-steady approach. *Water Resources Research*, 24(10), 1698–1710.
de Marsily, G. (1979). An introduction to geostatistics. Unpublished document, Department of Hydrology, The University of Arizona, Tucson, Arizona.
de Marsily, G., Lavedan, G., Boucher, M. & Fasanino, G. (1984). Interpretation of interference tests in a well field using geostatistical techniques to fit the permeability distribution in a reservoir model. In *Geostatistics for Natural Resources Characterization*, Part 2, eds. G. Verly *et al.* Dordrecht: Reidel, pp. 831–849.
Edwards, A. W. F. (1972). *Likelihood.* Cambridge University Press.
Emsellem, Y. & de Marsily, G. (1971). An automatic solution for the inverse problem. *Water Resources Research*, 7(5), 1264–1283.
Gavalas G. R.,. Shah, P.C. & Seinfeld, T. H. (1976). Reservoir history matching by Bayesian estimation. *Society of Petroleum Engineering Journal,* 261, 337–350.
Gelhar, L. W. & Axness, C. L. (1983). Three-dimensional stochastic analysis of macrodispersion in aquifers. *Water Resources Research*, 19(1), 161–180.
Gómez-Hernández, J., Sahuquillo, A. & Capilla, J. (1997). Stochastic simulation of transmissivity fields conditional to both transmissivity and piezometric data, 1 Theory. *Water Resources Research* (in press).
Graybill, F. A. (1976). *Theory and Application of the Linear Model.* North Scituate, Massachusetts: Duxbury Press.
Hoeksema, R. J. & Clapp, R. B. (1990). Calibration of groundwater flow models using Monte Carlo simulations and geostatistics. In *Calibration and Reliability in Groundwater Modelling,* ed. K. Kovar, IAHS no. 195. The Hague: IAHS.
Hoeksema, R. J. &. Kitanidis, P. K. (1984). An application of the geostatistical approach to the inverse problem in two-dimensional groundwater modeling. *Water Resources Research,* 20(7), 1003–1020.
Hoeksema, R. J. & Kitanidis, P. K. (1985). Comparison of Gaussian conditional mean and kriging estimation in the geostatistical solution of the inverse problem. *Water Resources Research*, 21(6), 825–836.
Journel, A.G. & Huijbregts, Ch.J. (1978). *Mining Geostatistics.* New York: Academic Press.
Kitanidis, P. K. & Vomvoris, E.G. (1983). A geostatistical approach to the inverse problem in groundwater modeling (steady state) and one dimensional simulations. *Water Resources Research,* 19, 677–690.
Kool, J. B. & Parker, J. C. (1988). Analysis of the inverse problem for transient unsaturated flow. *Water Resources Research*, 24(6), 817–830.
La Venue, A. M. & Pickens, J. F. (1992). Application of a coupled adjoint sensitivity and kriging approach to calibrate a groundwater flow model. *Water Resources Research*, 28(6), 1543–1571.
Neuman, S. P. (1973). Calibration of distributed parameter groundwater flow model viewed as a multiple objective decision process under uncertainty. *Water Resources Research*, 9(4), 1006–1021.
Neuman, S. P. (1980a). A statistical approach to the inverse problem of aquifer hydrology, 3, Improved solution methods and added perspective. *Water Resources Research*, 16(2), 331–346.
Neuman, S. P. (1980b). Adjoint state finite elements equations for parameter estimation. In *Finite Elements in Water Resources*, ed. S.Y. Wang. New Orleans: U.S. Department of Agriculture, pp. 2.66–2.75.
Olivella, S. & Carrera, J. (1987). KRINET (Kriging NETwork). Computer code for optimal design of observation networks through geostatistics. User's guide (draft), Barcelona.
Rubin, Y. & Dagan, G. (1987). Stochastic identification of transmissivity and effective recharge in steady groundwater flow: 1. Theory. *Water Resources Research*, 23(7), 1185–1192.
Samper, J. & Carrera, J. (1990). *Geoestadística. Aplicaciones a la Hidrología Subterránea*. Barcelona: Centro Internacional de Métodos Númericos.
Sun, N.-Z. & Yeh, W. W.-G. (1992). A stochastic inverse solution for transient groundwater flow: parameter identification and reliability analysis. *Water Resources Research*, 28(12), 3269–3280.
Townley, L. R. & Wilson, J. L. (1985). Computationally efficient algorithms for parameter estimation and uncertainty propagation in numerical models. *Water Resources Research*, 21(12), 1851–1860.
Vecchia, A. V. & Cooley, R. L. (1987). Simultaneous confidence and predicting intervals for nonlinear regression models, with application to a groundwater flow model. *Water Resources Research*, 23(7), 1237–1250.
Yeh, W. W. G. (1986). Review of parameter identification procedures in groundwater hydrology: The inverse problem. *Water Resources Research*, 22(2), 95–108.
Yeh, W. W. G. & Yoon, Y.S. (1981) Aquifer parameter identification with optimum dimension in parameterization. *Water Resources Research*, 17(3), 664–672.
Yoon, Y. S. & Yeh, W. W. G. (1976). Parameters identification in an homogeneous medium with the finite element method. *Society of Petroleum Engineering Journal,* 16(4), 217–226.
Zimmerman, D. A. & Gallegos, D. P. (1993). A comparison of geostatistically based inverse techniques for use in performance assessment analysis at the WIPP site. Results from test case number 1. *High Level Radioactive, Waste Management Proc. of the IVth Annual Int. Conf.*, Las Vegas, Nevada, Apr. 26–30, 1993. Lagrange Park, Illinois: American Nuclear Society, Inc., pp. 1426–1436.

III

Flow modeling and aquifer management

1 Groundwater flow in heterogeneous formations

PETER K. KITANIDIS
Stanford University

1 INTRODUCTION

1.1 Heterogeneity and uncertainty

Hydrogeologic research in the early seventies emphasized the development and solution of equations that describe the flow of water and the transport of chemicals in geologic formations. It was often maintained that the only obstacles to achieving exact predictions of flow and transport were inadequate process understanding and imprecise and inefficient methods of solution. Stochastic methods in the minds of most hydrogeologists were inextricably linked with empirical statistical models, as in time-series analysis, and consequently inappropriate for subsurface hydrology where the physical and chemical processes are well defined and scientifically studied. Significant progress was then made, and continues to be made, in describing the hydrodynamics of flow in geologic media and the physicochemical transport of soluble and insoluble substances.

Yet, the seventies also witnessed the birth and growth of stochastic groundwater hydrology. As deterministic models were applied in practice, and as the focus switched from problems of resource development to problems of decontamination, it became obvious that medium heterogeneity presented a major roadblock to making accurate predictions of flow and transport (Freeze, 1975; Anderson, 1979). The heterogeneity of the hydrogeologic parameters is so complex and difficult to describe quantitatively that even when the physical process is well understood, as is the Darcian process of saturated flow in a sand, it is impossible to predict deterministically the advective transport of a solute plume in the field. Heterogeneity can be found at all scales, from the Darcy or laboratory (less than a meter) scale to the field (~100 m) and the regional (kilometer) scale.

The high degree of spatial variability in the hydraulic conductivity and the complexity of the causes of this variability are widely recognized as major impediments to making precise predictions. Thus, much of the research in stochastic groundwater hydrology has focused on the evaluation of the effects of variability in hydraulic conductivity on the dependent variables hydraulic head and specific discharge. Furthermore, the resulting variability in the flow velocities is the cause of spreading and dilution of solute plumes. This research is motivated by the need to solve the following general types of practical problems:

- estimation problems, such as inverse problems involving the inference of conductivity from head observations and other information;
- propagation of uncertainty, which means, for example, how uncertainty in the hydraulic conductivity affects the uncertainty in the head;
- problems of scale, such as the questions of 'upscaling', effective conductivity, and macrodispersion, as will be discussed next.

1.2 Scale

The increased awareness of medium heterogeneity also awakened interest in questions of scale that had traditionally (with very few exceptions) been overlooked in hydrogeology. Quantitative hydrogeology followed the fluid-mechanics approach of treating the geologic medium as a continuum with properties that vary gradually. Of course, all quantities studied in fluid mechanics are macroscopic in the sense that they are defined over scales that are very large in comparison to the molecular ones. For example, density is the mass of molecules in a control volume divided by the volume. A cube of water of side equal to 1 μm contains so many molecules that the density is unaffected by the chance movement of molecules in and out of the control volume, and the variation from one small cube to the next is gradual. It is thus justified to replace in fluid mechanics the discrete matter by a continuum medium with the familiar properties (density, viscosity, etc.) By applying conservation principles to infinitesimally small volumes, 'governing' equations in the form of partial differential equations are derived.

This approach has traditionally been followed in flow and transport in geologic formations and forms the basis of analytical formulae and computer-based numerical models used for prediction purposes. The medium is treated as continuous with properties and variables that vary from point to point. Partial differential equations are derived and solved without much attention to issues of scale. For the classical (single-phase, saturated, low-Reynolds number, no physicochemical effects) problem, the flow is studied not at the interstitial ('microscopic') scale but at the Darcy or laboratory scale ('macroscopic'). Darcy's law relates the head gradient to the flow rate, and the coefficient of proportionality is a constituent property of the medium, determined from an experiment, that is called hydraulic conductivity.

Furthermore, in field applications, one is interested in hydraulic conductivities defined over volumes much larger than those relevant in a laboratory setup. In heterogeneous media, the coefficient of hydraulic conductivities and the other properties that affect water storage and flow are defined over a volume and consequently are scale dependent. For example, what conductivity should one assign to the elements of a finite-element model, and how should these parameters change as the volume of the elements changes? What dispersion coefficients are appropriate? Under what conditions are the flow and the advection–dispersion equations with effective parameters applicable? These are important but difficult questions, and answers are emerging slowly.

This chapter highlights progress made and problems remaining in the study of saturated flow in heterogeneous porous formations using stochastic methods. The review does not aspire to be comprehensive. Although this problem is longstanding in stochastic groundwater hydrology, research is continuing as strongly as ever, and new approaches, as well as many refinements, elucidations, and applications of existing methods, are presented every year. The reader should be aware of two recent books on stochastic groundwater hydrology: Dagan (1989) and Gelhar (1993). The chapter starts with the problem of evaluating the moments of the dependent variable (e.g. the head) from the moments of the independent variable (such as log-conductivity) and subsequently discusses problems of upscaling.

2 MONTE CARLO METHODS

The Monte Carlo (MC) method is a conceptually straightforward method that, in a certain sense, is the most general computational approach for evaluating the effects of uncertainty and, as will be seen later, for computing effective parameters. In the MC method, the input variable, which is typically the hydraulic conductivity, is represented as the realization of a stochastic process of spatial coordinates (known as a 'random field') with given probabilistic characterization. The simplest flow equation, obtained from application of the incompressibility condition and Darcy's law for isotropic medium under steady-state conditions without sources or sinks, is:

$$\frac{\partial}{\partial x_i}\left(K\frac{\partial \phi}{\partial x_i}\right)=0 \qquad (1)$$

where K is the spatially variable hydraulic conductivity, ϕ is the hydraulic head, x_i is a spatial coordinate, n is the dimensionality of the flow domain, and $i=1,\ldots,n$. In most applications, the logarithm of the hydraulic conductivity or transmissivity

$$Y=\ln K \qquad (2)$$

called the 'log-conductivity' is modeled as a Gaussian process, for analytical convenience and on the support of data (Freeze, 1975; Hoeksema & Kitanidis, 1985). Then, a set of sample functions or realizations of log-conductivity are synthetically generated using computer-based algorithms (e.g. Gutjahr, 1989; Tompson *et al.*, 1989). The flow equation is solved for each realization separately, with appropriate boundary conditions, and the hydraulic head is computed. Furthermore, through application of Darcy's law, a set of specific discharge realizations may be computed. The principle is that through statistical analysis of these multiple realizations one can characterize the statistical properties of the hydraulic head and the specific discharge as well as the joint probabilistic properties of the log-conductivity, head, and velocity.

Freeze (1975) solved the one-dimensional flow equation with a finite difference scheme, demonstrated that the head variance increases with the log-conductivity variance, and questioned the ability of deterministic models to predict the head due to uncertainty in log-conductivity variability. Freeze also summarized earlier work that used statistical methods in groundwater hydrology. Smith and Freeze (1979a, b) completed this work by accounting for spatial correlation in the log-conductivity and showing how it affects the variability in the head and by presenting simulations of flow in a two-dimensional domain. They also confirmed that the variance in the head is smaller for two-dimensional flow than for one-dimensional flow, for the same log-conductivity variance.

Many applications of the MC method were motivated by interest in evaluating the uncertainty in predicting transport and in quantifying macroscopically the rate of spreading (also known as 'macrodispersion'). Among the earliest works, Schwartz (1977) examined field-scale solute dispersion, and this work was continued in a more comprehensive way by Smith & Schwartz (1980, 1981a). El-Kadi & Brutsaert (1985) examined a simple unsteady flow problem. Black & Freyberg (1987) focused on evaluating the uncertainty in predicting advective solute transport. Hoeksema & Clapp (1990) illustrated the use of the MC method in the geostatistical approach to the inverse problem; i.e., they derived the first two statistical moments of the head using the MC method instead of the linearized approach that had been used in previous works. There have also been many other applications of MC methods in stochastic hydrogeology.

Most of the applications of MC methods used unconditional simulations. That is, typically the probabilistic characterization of the log-conductivity is as a stationary or stationary-increment process without explicitly taking into account the site specific observations of log-conductivity as well as other related information. One among the few exceptions is given in Smith & Schwartz (1981b), who studied the effects of conditioning on conductivity data. Also, the issue of conditional simulations is sometimes addressed in the context of inverse methods. Delhomme (1979) was probably the first to attempt to generate simulations conditional on transmissivity observations and approximately on head observations, using a brute-force approach. Hoeksema & Kitanidis (1989) used a conditional simulations approach within the context of the linear geostatistical approach to the inverse problem. Gutjahr *et al.* (1994) have reviewed the literature on the subject, presenting a method that is a combination of analytical and numerical techniques. A fast Fourier transform (FFT) method allows relatively efficient derivation of log-conductivity and head jointly, and an iterative scheme is also presented.

The MC approach is popular because it appears to be the least restricted of all available methods. Its obvious disadvantage is the computational cost, which should become less of a concern as the unit cost of computing decreases thanks to technological innovations. In reality, however, the MC method is not the cure-all it has been thought to be. For large-contrast cases, i.e. flow problems with log-conductivity that varies a great deal and abruptly, off-the-shelf numerical groundwater flow models may require fine discretization grids, and even then may not be very reliable in computing the hydraulic head and the velocity. Furthermore, a large number of realizations are required to reduce the sampling uncertainty in the estimated moments of the hydraulic head and the velocity. In the classical MC method, accuracy of the estimates is proportional to the inverse of the square root of the number of realizations, $1/\sqrt{N}$.

For example, if the number of realizations is $N=1000$ and it is desired to reduce the error by a factor of ten, one must generate $N=100\,000$ realizations. Also, although the MC method could in theory be used in computing moments higher than the second as well as probability distributions, it must be realized that the variability associated with the sampling of these features may be so large that a massive effort may be required to generate realizations sufficient in number for the results to make sense.

The most promising use of the MC method is in combination with analytical work. That is, the problem should be cast in terms of dimensionless groups, should be pursued analytically as much as possible before employing MC methods, and the results should be summarized cleverly so that they can be used easily without the need to repeat the calculations. One of the most important uses of the technique is to evaluate approximate theories and formulae, such as perturbation methods. In specific applications, the issue of computational efficiency remains central. Work should continue toward developing efficient methods to generate realizations conditional on all direct and related data, a topic related to inverse modeling. Techniques to derive efficiently the moments of dependent variables with large variance are also needed.

3 SMALL-PERTURBATION METHODS

The usefulness of perturbation methods in the solution of stochastic equations has long been recognized in applied mathematics in general and in groundwater hydrology in particular. The essence of the small-perturbation approach is that the stochastic partial differential equation is approximated by expanding the input (log-conductivity) and output (hydraulic head) variables about their mean values:

$$Y=\bar{Y}+Y'$$

$$\phi=\bar{\phi}+\phi' \tag{3}$$

where the bar indicates mean values and the prime indicates fluctuations. Equation (1) using eq. (2) is:

$$\frac{\partial Y}{\partial x_i}\frac{\partial \phi}{\partial x_i}+\frac{\partial^2 \phi}{\partial x_i^2}=0 \tag{4}$$

Then substituting eq. (3) into (4) and keeping up to linear terms of the fluctuations, the following equations are obtained:

$$\frac{\partial \bar{Y}}{\partial x_i}\frac{\partial \bar{\phi}}{\partial x_i}+\frac{\partial^2 \bar{\phi}}{\partial x_i^2}=0 \tag{5}$$

$$\frac{\partial \bar{Y}}{\partial x_i}\frac{\partial \phi'}{\partial x_i}+\frac{\partial^2 \phi'}{\partial x_i^2}=-\frac{\partial Y'}{\partial x_i}\frac{\partial \bar{\phi}}{\partial x_i} \tag{6}$$

The first equation is a deterministic equation that can be solved with the appropriate boundary conditions. The second is a random partial differential equation that relates linearly the output ϕ' with the input Y'. This point becomes clearer if one transforms the partial differential equation (6) into an integral equation (by applying Green's identities); then $\phi'(x)$ is expressed as a volume integral over the whole domain of Y' with appropriate weight function plus boundary integrals. The linearization of the relation between Y' and ϕ' facilitates greatly the derivation of the covariance of the output as well as the input–output cross-covariance. The approach is known by several different names, such as small-perturbation, first-order, linearized, and weak-nonlinearity analysis. We will refer to it as the perturbation method. Early published applications include the numerical approaches of Tang & Pinder (1977), Sagar (1978) and Dettinger & Wilson (1981). The small-perturbation approach has been particularly useful in solving inverse problems (e.g. Hoeksema & Kitanidis, 1984). However, the most important contributions of this methodology have been in deriving analytical solutions.

One of the most successful applications of the perturbation approach is in conjunction with Fourier representations of the stochastic random fields. Under the condition that the log-conductivity is a stationary or stationary-increment random field, it can be represented through an integral transform known as the Fourier–Stieltjes integral. Furthermore, eq. (6) becomes a linear equation with constant coefficients, for which operational-calculus techniques are quite effective. Then, the spectral density of the hydraulic head and the cross-spectral density of the hydraulic head and log-conductivity can be easily derived. An early application of the stochastic spectral analysis to groundwater systems was presented by Gelhar (1974), although the subject was not aquifer heterogeneity in that study. Gelhar (1977) applied the spectral approach to study the effects of spatial variability of the hydraulic conductivity on groundwater flow, pioneering one of the most effective approaches for the derivation of analytical solutions. In that work, Gelhar also demonstrated clearly that the degree of variability of the hydraulic head depends critically on the dimensionality of the flow for a given covariance function of the log-conductivity and made a case for high-dimensional analysis.

The spectral approach was further advanced in the works of Bakr *et al.* (1978), Mizell *et al.* (1982), Gelhar & Axness (1983), and others. The spectral approach is the focus of the recent book by Gelhar (1993). An efficient numerical implementation of the perturbation stochastic spectral approach through fast Fourier transforms was developed by Van Lent & Kitanidis (1989).

A more general method devised to develop analytical perturbation solutions is through the Green's function approach (e.g. Adomian, 1964), which has been advanced in groundwater hydrology for macroscopically uniform flow in infinite-dimensional aquifers by Dagan (1982a,b, 1985a,b) and Rubin & Dagan (1987), and explained in detail in the book by Dagan (1989). The Green's function, easy to obtain for the flow problem discussed in this section, allows one to express the head as a linear integral transform of the log-conductivity. Naff & Vecchia (1986) applied this approach in combination with Fourier series to aquifers of infinite extent along the horizontal but bounded above and below. Rubin & Dagan (1988, 1989) studied the effects of fixed-head or impervious boundary on the head variation in semi-infinite aquifers and showed that the effects of the boundaries are limited to about two to three log-conductivity integral scales. Li & McLaughlin (1991) presented a transfer-function approach, which is similar to the Green's function approach, except that it is carried out in a Fourier (spectral) framework. The transfer function is obtained from the solution of the linearized flow equation. There is considerable interest now in the variable mean case (see Gelhar (1993) and Loaiciga *et al.* (1993)). Rubin & Seong (1994) presented an analytical perturbation method to compute the effects on head variability and flow of log-conductivity which is the superposition of a linear drift and a stationary function. The advantage is that if some of the variability is described through a trend, the small-perturbation assumption is more applicable.

The perturbation analysis is an asymptotic approximation, and its theoretical justification is that it tends to become exact as the log-conductivity variance tends to zero. In practice, of course, one is interested in the accuracy of the method for finite values of the log-conductivity. Dagan (1985b) computed the second-order terms in the head (generalized) covariance and head–log-conductivity cross-covariance for the ideal case of uniform flow in an unbounded aquifer without sources and sinks. He arrived at the conclusion that the results of the perturbation analysis should be sound for variance of log-conductivity up to near unity. The most common method of evaluating the method is through MC simulations, which appears to be a simple matter to those who have not applied it. The difficulty is that the results of MC simulations are affected by (a) the size of the domain and the boundary conditions, (b) the size of the elements used (discretization), and (c) the number of realizations used. For the interesting large-variance cases, many realizations are needed and the computations must be performed on a fine grid. Furthermore, numerical models use Dirichlet and Neuman boundary conditions, which is a disadvantage when trying to evaluate the accuracy of analytical solutions that implicitly assume 'spectral' boundary conditions at infinitely large distances.

An alternative approach is based on the self-consistent approximation, applied in Chirlin & Dagan (1980). Although this does not formally assume small variances, it makes some simplifications about the spatial structure of variability that ought to become more critical as the variance increases.

It is interesting that the perturbation approach has focused so much on the ordinary first-order Taylor-series expansion and that so little has been done in applying other asymptotic methods.

4 OTHER METHODS

The ordinary Monte Carlo approach is computationally expensive and error prone, and it takes great effort to produce and interpret the results in order to develop insights into the controlling factors. Analytical solutions obtained from the perturbation approach are necessarily limited by the requirement for small variances and a number of other idealizations. The limitations of these basic approaches have motivated the search for alternative approaches aimed at: providing a better qualitative understanding of the stochastic flow equation (1), presenting other analytical approximations, and developing more innovative numerical methods of solution. Most of these methods are still under development or have not acquired a large following.

Some of the latest methods start by transforming the differential equation into an integral equation (Serrano, 1988a, b; Unny, 1989; Zeitoun & Braester, 1991). An extensive discussion of the operator and integro-differential representations in stochastic groundwater hydrology is given by Orr & Neuman (1994). One of the advantages of this approach is that the boundary conditions are built in the integral equation right from the beginning instead of being introduced at the end, as is customary in the solution of partial differential equations. This approach, of course, is used by Dagan in the Green's function approach to the linearized equation that has already been discussed. The interest here is in approaches that do not start by linearizing the flow equation. Typically, the differential equation is transformed into an integral equation for the case of the unbounded domain, similar to that in Dagan's perturbation approach, or semi-infinite domains. However, it is not clear why solving this integral equation may be much easier than solving the original partial differential equation, with the exception perhaps of specific cases where specialized integral transforms may provide the solution. It is well known, for example, that the straightforward Neuman-series expansion of the head converges only if the variance is somehow small. In fact, this approach ought to be practically the same with a Taylor-series expansion in the partial differential equation and evaluation of higher-order terms. The difficulty is that the series expansion, in addition to being computationally expensive if many terms are included, diverges when the conductivity contrast exceeds a certain value.

Nevertheless, Zeitoun & Braester (1991) used a Neuman-series expansion of the Laplace transform of the hydraulic head and claimed that, for mean square convergence, the Neuman expansion method may converge for a larger range of variability in conductivity than the classic perturbation method. The semigroup approach (e.g. Serrano & Unny, 1987) is interesting for the insights that it can provide into the mathematical problem, but its usefulness as a computational tool for the derivation of the moments of the head when the conductivity field has large contrast has not been established.

One approach that departs from all others involves the use of diagrammatic or graphic perturbation techniques (Christakos *et al.*, 1993a, b). It is claimed that 'graphic visualizations of the underlying flow processes allow previously undetected features to be seen and can yield more general and accurate results than previous methods' (Christakos *et al.*, 1993a).

Van Lent (1992) developed a numerical spectral approach to solve the stochastic partial differential equation (4) and applied it to the two-dimensional flow case. In this approach, the unbounded domain is replaced by a rectangular domain (or 'cell') with dimension many times the integral scale. Both the log-conductivity and the head are represented as periodic functions within this cell and expanded into Fourier series. Equation (4) is reformulated into a system of linear equations with the Fourier coefficients of the head as the unknowns. Then a realization of the log-conductivity is generated and the corresponding head realization is computed; subsequently, the velocity realization is also computed. The moments of the head and the velocity are computed from ensemble averaging over multiple realizations as well as spatial averaging. For ergodic characteristics, and provided that the cell is large enough, the two types of averaging should yield the same results. As a computational method, the numerical spectral method is expected to be superior in accuracy and efficiency to conventional finite-element or finite-difference methods as the contrast increases and finer grids must be used, provided that the log-conductivity is sufficiently smooth. This method produces highly accurate results, with log-conductivity variances up to about 6 and perhaps higher.

The numerical spectral method is ideally suited for the evaluation of the accuracy of the small-perturbation approximation in the analytical unbounded-domain solutions (e.g. Gelhar, 1977). The boundary conditions in the numerical method are consistent with the condition implicitly assumed in the analytical methods. The only difference is that the numerical spectral method discretizes the spectrum of the log-conductivity, but this discretization error can become as small as desired by increasing the size of the cell. Some of the interesting conclusions from Van Lent's (1992) study of two-dimensional flow are as follows:

(1) Even for cases that the linear theory predicts the head to be a stationary function (Mizell *et al.*, 1982), it is nonstationary, and this nonstationarity becomes more pronounced as the variance increases.
(2) The small-perturbation approximation tends to underpredict the variance of the large-scale fluctuations, and this underprediction increases with the variance.
(3) The perturbation approximation tends to underpredict the specific-discharge variance.
(4) The small-perturbation approximation of the correla-

tion function of the velocity in the direction of flow is reasonably accurate.

(5) The small-perturbation approximation of the correlation function of the velocity component perpendicular to the main-flow direction is not very reliable and the error increases with the variance.

5 EFFECTIVE CONDUCTIVITY

The permeability to saturated flow in a heterogeneous medium may sometimes be described macroscopically through the 'effective conductivity'. That is, a fictitious homogeneous medium is postulated with conductivity that in some average or overall sense is the same as the conductivity of the porous medium. Calculating the effective properties of heterogeneous media is a classical problem in physics which has occupied the most eminent physical scientists and applied mathematicians. The literature on the subject is vast. Even though we will limit our attention to the subsurface-flow literature, it is impossible to do justice to all the methods and their variations that have been tried. In subsurface flow, pioneering works include Warren & Price (1961), Shvidler (1962), and Matheron (1967).

Gelhar (1977) and Gutjahr *et al.* (1978) applied the stochastic perturbation approach to media with stationary log-conductivity. The relationship between head and log-conductivity is linearized and then the ensemble average of the discharge is computed for steady flow in an unbounded domain with unit mean head gradient. This discharge defines the effective conductivity of the medium. An advantage of the perturbation approach is that only the mean and variance of the log-conductivity are required. The result is quite simple for isotropic media: the effective conductivity K^{eff} depends on the conductivity geometric mean K_G, the log-conductivity variance, and the dimensionality of the flow domain n through the formula:

$$K^{eff}=K_G\left(1-\frac{\sigma^2}{n}+\frac{\sigma^2}{2}\right) \tag{7}$$

Dagan (1982c) studied the unsteady flow problem in macroscopically uniform flow in an unbounded domain using the perturbation approximation. The 'relaxation time' was computed. This is the time interval which must elapse after some change in the system so that an effective conductivity can be defined; then, the effective conductivity is the same as the one computed from the steady-flow analysis.

The perturbation approach is a useful analytical tool, but its accuracy for large-variance cases is hard to judge in general terms. Whereas the perturbation approximation to the effective conductivity depends only on moments of the log-conductivity up to the second one, more exact methods applied to large-variance cases are likely to produce results that depend on higher-order moments and probability distributions. Thus, there is no general formula for the effective conductivity of a medium with stationary conductivity, except that some bounds have been established; it is known, for example, that the effective conductivity is between the harmonic and arithmetic mean. The perturbation approximation may be accurate or not depending on the exact geometry of the heterogeneity as described by the variance as well as higher-order moments. If, however, one focuses on Gaussian stationary log-conductivity, then there is a unique effective conductivity that depends on moments only up to second moments. This effective conductivity has been postulated to be:

$$K^{eff}=K_G\exp\left(-\frac{\sigma^2}{n}+\frac{\sigma^2}{2}\right) \tag{8}$$

It is elementary to demonstrate that this formula is correct for $n=1$, and Matheron (1967) has demonstrated that the formula is correct for $n=2$. However, for the three-dimensional case, the result has not been verified analytically, and it is probably inexact. Numerical experiments (Dykaar & Kitanidis, 1992b) have indicated that the difference between the value predicted by this formula and that computed numerically is less than 4% for log-conductivity variance up to 6. Neuman & Orr (1993) found even better agreement between their numerical results and the results of this formula. Thus, eq. (8) is sufficient for most practical purposes, since differences due to non-Gaussianity and other deviations from the assumed conditions may have a more important effect on the value of the effective conductivity than the possible error in this formula.

Dagan (1979, 1981) applied the 'self-consistent approximation' to compute the effective conductivity (as well as moments of head). The analysis is not predicated on a small log-conductivity variance but does depend on some assumptions about the spatial structure. The analysis, which is explained and further developed in the book by Dagan (1989), provides analytical formulae that complement those from perturbation analysis. A numerical evaluation of a result for a two-dimensional two-phase (shale/sandstone) medium was reported in Dykaar & Kitanidis (1992b).

Although most work involves ensemble averaging with stationary random fields, Kitanidis (1990) applied the method of moments which involves spatial averaging for periodic log-conductivity. The method provides insights, such as calculation of the relaxation time, that are often overlooked in the stochastic approach. Perturbation approximations yielded analytical solutions, which, for large periods, as the periodic medium tends to a stationary medium, are the same as those of Gelhar (1977), Gutjahr *et al.* (1978), etc. The main advantage of this approach, however, is that it provides the basis for a relatively efficient and accurate numerical approach to compute effective conductivity in any periodic or stationary medium, Gaussian or not. The method is not limited to small variances. The method (Dykaar & Kitanidis, 1992a, b, 1993) owes its computational efficiency to the use of fast Fourier transforms. In the case of stationarity, the stationary function is approximated as periodic with period L. This is equivalent to approximating the continuous power spectrum of the process with a discrete one, and the approximation improves by increasing L. For the commonly assumed Gaussian log-conductivity, convergence was achieved when the period is 80 integral scales for the two-dimensional case and 30 times for the three-dimensional case.

Many other studies have used straightforward Monte Carlo simulations. Desbarats (1987) computed effective conductivity in sand–shale formations modeled as binary (two possible conductivity values) stationary random functions. The effective conductivity was found to depend critically on the shale volume fraction, the spatial correlation structure, and the flow dimensionality.

There is considerable interest in the hydrologic community in evaluating the effects of deviations from the Gaussian assumption. Desbarats & Srivastava (1991), for a transmissivity function artificially generated over a domain without the Gaussian assumption, solved the steady groundwater flow equations to obtain the distribution of head and specific discharge. The results of these simulations were compared with theoretical predictions using a second-moment characterization of the conductivity field. The simulated field had a log-transmissivity variance of around unity, but embodied features that are perceived as 'non-Gaussian', such as spatial continuity of extreme values, bimodal histogram, strong directional anisotropy, and nested scales of heterogeneity. The results of Desbarats & Srivastava (1991) showed reasonable agreement with the predictions of 'ensemble theoretic models based on perturbation developments' with respect to head covariance (Mizell *et al.*, 1982; Rubin & Dagan, 1988), cross-covariance between head and log-transmissivity (Dagan, 1984, 1985b; Rubin & Dagan, 1988), and effective transmissivity (Gelhar & Axness, 1983). This work shows that moderate deviations from the Gaussian assumption do not affect the results as long as the variance is not large. The issue of non-Gaussian distributions is most important for large-variance cases, a problem that is very difficult to treat rigorously, even for Gaussian random fields.

6 CONDUCTIVITY UPSCALING

In most applications, the applicability of the effective conductivity as defined in the previous section is limited by the fact that there is variability at scales large compared with the scale of interest in the practical problems. The effective conductivity is a macroscopic property defined in essence over a 'representative elementary volume' that is 10 to 100 times the integral scale. This volume of averaging is too large for many applications. Instead, the domain is subdivided into blocks or elements of constant or gradually varying conductivity that form the basis of numerical finite-element or finite-difference codes. Such models can represent large-scale variability but not subgrid variability, i.e. variability at a scale smaller than that of the elements or the grid spacing.

The important questions are as follows:

(1) Under what conditions is it appropriate to treat the blocks as homogeneous and to assign conductivity values to the blocks of the numerical schemes that are properties of the medium only?
(2) What conductivity values should be assigned to the blocks to account properly for the subgrid variability?

Insight into the issue raised in the first question is provided in some works on the effective conductivity problem (Dagan, 1982c; Kitanidis, 1990). In the absence of nearby continuous sources and sinks, sufficient time must elapse from the last time the system was stimulated for the scale of the disturbance to become much larger than the scale of the element. For instance, in the presence of a production well, the elements in its vicinity must be small enough for the hydraulic head to vary gradually, in some sense that can be made mathematically exact. Otherwise, there is probably no unique conductivity value that can be assigned rigorously to the elements, although a value can always be found to reproduce partially the observed behavior for any given boundary conditions.

Regarding what values to assign, Durlofsky (1991) and Dykaar & Kitanidis (1992b) used a periodic formulation. The method, which in these works was applied to rectangular blocks, considers that the particular element is subjected to periodic boundary conditions, and has the important advantage that it produces a whole conductivity tensor. Other methods, which consider that the flow through the element is confined on two opposing sides and subject to a constant head difference on the other two, have the disadvantages that they arbitrarily impose two streamlines and two equipotentials and also that they do not produce tensors. Thus, the periodic boundary conditions, being the least restrictive, are the most appropriate to compute numerically the equivalent conductivity of an element.

Rubin & Gómez-Hernández (1990) aim to derive the statistical moments of the relatively smoothly varying block conductivity (or, in other words, the 'upscaled conductivity') from the point conductivity (such as the highly heterogeneous 'laboratory-scale conductivity'). Their study includes conditioning on local data. Indelman (1993) and Indelman & Dagan (1993a, b) extensively deal with the same question using a more general approach based on variational principles. They study both the question of under what conditions it is appropriate to upscale and the computational issue of how to compute the moments of the upscaled conductivity.

7 CONCLUDING REMARKS

It is obvious from the proliferation of articles in journals and other publications that stochastic hydrology is flourishing and that the classical problem of flow in randomly heterogeneous porous formations is attracting perhaps more attention than ever. In the case of small variance or 'weak nonlinearity', progress has been quite satisfactory and practical methods are available. For large-variance and highly non-Gaussian cases, progress has been slow. Some researchers have chosen to pursue highly sophisticated and complicated approaches, while others have focused on ad hoc methods.

As analysis becomes more advanced, it becomes more important to ask ourselves what we expect to accomplish. When it comes to complex problems, I think that qualitative analysis to identify controlling factors is an important overall objective, and perhaps as much as we can hope to accomplish in terms of practical significance, since more precise quantitative conclusions are conditional on very heavy data requirements and perhaps excessive computational work. For example, I feel that we are coming closer to reaching one of the original goals of stochastic groundwater hydrology, which was to shed light on the conditions under which conventional groundwater models are applicable in the face of heterogeneity. Also, we have at least a qualitative understanding of the spatial correlation structure of hydraulic head and velocity.

REFERENCES

Adomian, G. (1964). Stochastic Green's functions. In *Stochastic Processes in Mathematical Physics and Engineering*, ed. R. Bellman. Providence, RI: American Mathematical Society, pp. 1–39.

Anderson, M. P. (1979). Using models to simulate the movement of contaminants through groundwater flow systems. *CRC Critical Reviews in Environmental Control*, 9(2), 97–156.

Bakr, A. A., Gelhar, L. W., Gutjahr, A. L. & MacMillan, J. R. (1978). Stochastic analysis of spatial variability in subsurface flows, 1. Comparison of one- and three-dimensional flows. *Water Resources Research*, 14(2), 263–271.

Black, T. C. & Freyberg, D. L. (1987). Stochastic modeling of vertically averaged concentration uncertainty in a perfectly stratified aquifer. *Water Resources Research*, 23(6), 997–1004.

Chirlin, G. R. & Dagan, G. (1980). Theoretical head variograms for steady flow in statistically homogeneous aquifers. *Water Resources Research*, 16(6), 1001–1015.

Christakos, G., Miller, C. T. & Oliver, D. (1993a). Cleopatra's nose and the diagrammatic approach to flow modelling in random porous media. In *Geostatistics for the Next Century*. Dordrecht: Kluwer Publishing Co., pp. 341–358.

Christakos, G., Miller, C. T. & Oliver, D. (1993b). Stochastic perturbation analysis of groundwater flow. Spatially variable soils, semi-infinite domains, and large fluctuations. *Stochastic Hydrology and Hydraulics*, 7, 213–239.

Dagan, G. (1979). Models of groundwater flow in statistically homogeneous porus formations. *Water Resources Research*, 15(1), 47–62.

Dagan, G. (1981). Analysis of flow through heterogeneous random aquifers by the method of embedding matrix. 1. Study flow. *Water Resources Research*, 17(1), 107–121.

Dagan, G. (1982a). Stochastic modeling of groundwater flow by unconditional and conditional probabilities 1. Conditional simulations and the direct problems. *Water Resources Research*, 18(4), 813–833.

Dagan, G. (1982b). Stochastic modeling of groundwater flow by unconditional and conditional probabilities 2. The solute transport. *Water Resources Research*, 18(4), 835–848.

Dagan, G. (1982c). Analysis of flow through heterogeneous random aquifers 2. Unsteady flow in confined formations. *Water Resources Research*, 18(5), 1571–1585.

Dagan, G. (1984). Solute transport in heterogeneous porous formations. *Journal of Fluid Mechanics*, 145, 151–177.

Dagan, G. (1985a). Stochastic modeling of groundwater flow by unconditional and conditional probabilities: the inverse problem. *Water Resources Research*, 21(1), 65–73.

Dagan, G. (1985b). A note on higher-order corrections of the head covariances in steady aquifer flow. *Water Resources Research*, 21(4), 573–578.

Dagan, G. (1989). *Flow and Transport in Porous Media*. New York: Springer-Verlag.

Delhomme, J. P. (1979). Spatial variability and uncertainty in groundwater flow parameters: a geostatistical approach. *Water Resources Research*, 15(2), 269–280.

Desbarats, A. J. (1987). Numerical estimation of effective permeability in sand-shale formations. *Water Resources Research*, 23(2), 273–286.

Desbarats, A. J. & Srivastava, R. M. (1991).Geostatistical simulation of groundwater flow parameters in a simulated aquifer. *Water Resources Research*, 27(5), 687–698.

Dettinger, M. D. & Wilson, J. L. (1981). First-order analysis of uncertainty in numerical models of groundwater flow, 1. Mathematical development. *Water Resources Research*, 17(1), 149–161.

Durlofsky, L. J. (1991). Numerical calculation of equivalent grid block permeability tensors for heterogeneous porous media. *Water Resources Research*, 17(5), 699–708.

Dykaar, B. B. & Kitanidis, P. K. (1992a). Determination of effective hydraulic conductivity in heterogeneous porous media using a numerical spectral approach 1. Method. *Water Resources Research*, 28(4), 1155–1166.

Dykaar, B. B. & Kitanidis, P. K. (1992b). Determination of effective hydraulic conductivity in heterogeneous porous media using a numerical spectral approach 2. Results. *Water Resources Research*, 28(4), 1167–1178.

Dykaar, B. B. & Kitanidis, P. K. (1993). Transmissivity of a heterogeneous formation. *Water Resources Research*, 29(4), 985–1001.

El-Kadi, A. I. & Brutsaert, W. (1985). Applicability of effective parameters for unsteady flow in nonuniform aquifers. *Water Resources Research*, 21(2), 183–198.

Freeze, R. A. (1975). A stochastic-conceptual analysis of one-dimensional groundwater flow in nonuniform homogeneous media. *Water Resources Research*, 11(5), 725–741.

Gelhar, L. W. (1974). Stochastic analysis of phreatic aquifers. *Water Resources Research*, 10(3), 539–545.

Gelhar, L. W. (1977). Effects of hydraulic conductivity variations on groundwater flow. In *Hydraulic Problems Solved by Stochastic Methods*, eds. P. Hjorth, L. Joensson & P. Larsen. Fort Collins, CO: Water Resources Publications, chap. 21.

Gelhar, L. W. (1993). *Stochastic Subsurface Hydrology*. Englewood Cliffs, NJ: Prentice Hall.

Gelhar, L. W. & Axness, C. L. (1983). Three-dimensional stochastic analysis of macrodispersion in aquifers. *Water Resources Research*, 19(1), 161–180.

Gutjahr, A. L. (1989). Fast Fourier transform for random field generation. Project Report, Los Alamos National Laboratory.

Gutjahr, A. L., Gelhar, L. W., Bakr, A. A. & MacMillan, J. R. (1978). Stochastic analysis of spatial variability in subsurface flows, 2. Evaluation and applications. *Water Resources Research,* 14(5), 953–959.

Gutjahr, A., Bullard, B., Hatch, S. & Hughson, L. (1994). Joint conditional simulations and the spectral approach for flow modeling. *Stochastic Hydrology and Hydraulics*, 8(1), 79–108.

Hoeksema, R. J. & Clapp, R. B. (1990). Calibration of groundwater flow models using Monte Carlo simulations and geostatistics. In *ModelCARE 90: Calibration and Reliability in Groundwater Modelling*, pp. 33–42. IAHS Publ. 195.

Hoeksema, R. J. & Kitanidis, P. K. (1984). An application of the geostatistical approach to the inverse problem in two-dimensional groundwater modeling. *Water Resources Research,* 20(7), 1003–1020.

Hoeksema, R. J. & Kitanidis, P. K. (1985). Analysis of spatial structure of properties of selected aquifers. *Water Resources Research,* 21(9), 563–572.

Hoeksema, R. J. & Kitanidis, P. K. (1989). Prediction of transmissivities, heads, and seepage velocities using mathematical models and geostatistics. *Advances in Water Resources*, 12(2), 90–102.

Indelman, P. (1993). Upscaling of permeability of anisotropic heterogeneous formations 3. Applications. *Water Resources Research,* 29(4), 935–943.

Indelman, P. & Dagan, G. (1993a). Upscaling of permeability of anisotropic heterogeneous formations 1. The general framework. *Water Resources Research*, 29(4), 917–923.

Indelman, P. & Dagan, G. (1993b). Upscaling of permeability of anisotropic heterogeneous formations 2. General structure and small perturbation analysis. *Water Resources Research,* 29(4), 925–933.

Kitanidis, P. K. (1990). Effective hydraulic conductivity for gradually varying flow. *Water Resources Research*, 26(6), 1197–1208.

Li, S.-G. & McLaughlin, D. (1991). A nonstationary spectral method for solving stochastic groundwater problems: Unconditional analysis. *Water Resources Research,* 7(7), 1589–1605.

Loaiciga, H. A., Leipnik, R. B., Marino, M. A. & Hudak, P. F. (1993). Stochastic groundwater flow analysis in the presence of trends in heterogeneous hydraulic conductivity fields. *Mathematical Geology*, 25(2), 161–176.

Matheron, G. (1967). *Elements Pour une Theorie des Milieux Poreux.* Paris: Masson et cie.

Mizell, S. A., Gutjahr, A. L. & Gelhar, L. W. (1982). Stochastic analysis of spatial variability in two-dimensional steady groundwater flow assuming stationary and nonstationary fields. *Water Resources Research,* 18(4), 1053–1067.

Naff, R. L. & Vecchia, A. V. (1986). Stochastic analysis of three-dimensional flow in a bounded domain. *Water Resources Research,* 22(5), 695–704.

Neuman, S. P. & Orr, S. (1993). Prediction of steady state flow in nonuniform geologic media by conditional moments: Exact nonlocal formalism, effective conductivities, and weak approximation. *Water Resources Research,* 29(2), 341–364.

Orr, S. & Neuman, S. P. (1994). Operator and integro-differential representations of conditional and unconditional stochastic subsurface flow. *Stochastic Hydrology and Hydraulics*, 8, 157–172.

Rubin, Y. & Dagan, G. (1987). Stochastic identification of transmissivity and effective recharge in steady groundwater flow. *Water Resources Research,* 23(7), 1185–1192.

Rubin, Y. & Dagan, G. (1988). Stochastic analysis of boundaries effects on head spatial variability in heterogeneous aquifers, 1. Constant head boundary. *Water Resources Research,* 24(10), 1689–1697.

Rubin, Y. & Dagan, G. (1989). Stochastic analysis of boundaries effects on head spatial variability in heterogeneous aquifers, 1. Impervious boundary. *Water Resources Research,* 25(4), 707–712.

Rubin, Y. & Gómez-Hernández, J. J. (1990). A stochastic approach to the problem of upscaling of conductivity in disordered media: Theory and unconditional numerical simulations. *Water Resources Research,* 26(4), 691–701.

Rubin, Y. & Seong, K. (1994). Investigation of flow and transport in certain cases of nonstationary conductivity fields. *Water Resources Research,* 30(11), 2901–2911.

Sagar, B. (1978). Galerkin finite element procedure for analyzing flow through random media. *Water Resources Research,* 14(6), 1035–1044.

Schwartz, F. W. (1977). Macroscopic dispersion in porous media: The controlling factors. *Water Resources Research,* 13(4), 743–752.

Serrano, S. E. (1988a). General solution to random advective-dispersive transport equation in porous media, 1. Stochasticity in the sources and in the boundaries. *Stochastic Hydrology and Hydraulics,* 2, 79–98.

Serrano, S. E. (1988b). General solution to random advective-dispersive transport equation in porous media, 2. Stochasticity in parameters. *Stochastic Hydrology and Hydraulics,* 2, 99–112.

Serrano, S. E. & Unny, T. E. (1987). Semigroup solutions of the unsteady groundwater flow equation with stochastic parameters. *Stochastic Hydrology and Hydraulics,* 1, 281–296.

Shvidler, M. I. (1962). Flow in heterogeneous media. *Izv Akad. Nauk SSSR*, 3, 185–190. (In Russian.)

Smith, L. & Freeze, R. A. (1979a). Stochastic analysis of steady state groundwater flow in a bounded domain, 1. One-dimensional simulations. *Water Resources Research*, 15(3), 521–528.

Smith, L. & Freeze, R. A. (1979b). Stochastic analysis of steady state groundwater flow in a bounded domain, 2. Two-dimensional simulations. *Water Resources Research*, 15(6), 1543–1559.

Smith, L. & Schwartz, F. W. (1980). Mass transport, 1. A stochastic analysis of macroscopic dispersion. *Water Resources Research*, 16(2), 303–313.

Smith, L. & Schwartz, F. W. (1981a). Mass transport: Analysis of uncertainty in prediction. *Water Resources Research*, 17(2), 351–369.

Smith, L. & Schwartz, F. W. (1981b). Mass transport, 3. Role of hydraulic conductivity measurements. *Water Resources Research,* 17(5), 1463–1479.

Tang, P. H. & Pinder, G. F. (1977). Simulation of groundwater flow and mass transport. *Advances in Water Resources*, 13(1), 25–30.

Tompson, A. F. B., Ababou, R. & Gelhar, L. W. (1989). Implementation of the three-dimensional turning bands random field generator. *Water Resources Research*, 25(10), 2227–2243.

Warren, J. E. & Price, H. S. (1961). Flow in heterogeneous porous media. *Society Petroleum Engineering Journal*, 1, 153–167.

Unny, T. E. (1989). Stochastic partial differential equations in groundwater hydrology. *Stochastic Hydrology and Hydraulics,* 3, 135–153.

Van Lent, T. (1992). Numerical spectral methods applied to flow in highly heterogeneous aquifers. Ph.D. thesis, Stanford University.

Van Lent, T. & Kitanidis, P. K. (1989). A numerical spectral approach for the derivation of piezometric head covariance functions. *Water Resources Research*, 25(11), 2287–2298.

Zeitoun, D. G. & Braester, C. (1991). A Neuman expansion approach to flow through heterogeneous formations. *Stochastic Hydrology and Hydraulics*, 5, 207–226.

2 Aspects of numerical methods in multiphase flows

RICHARD E. EWING
Texas A & M University

ABSTRACT The ability to simulate numerically single-phase and multiphase flow of fluids in porous media is extremely important in developing an understanding of the complex phenomena governing the flow. The flow is complicated by the presence of heterogeneities in the reservoir at many different length scales and by phenomena such as diffusion and dispersion. These effects must be effectively modeled by terms in coupled systems of nonlinear partial differential equations which form the basis of the simulator. The simulator must be able to model both single and multiphase flows and the transition regimes between the two in unsaturated flow applications. A discussion of some of the aspects of modeling unsaturated and multiphase flows in the presence of heterogeneities and channeling is presented along with directions for future work. Simulators are severely hampered by the lack of knowledge of reservoir properties, heterogeneities, and relevant length scales and important mechanisms like diffusion and dispersion. Simulations can be performed either deterministically, to predict the outcome of a single realization of reservoir and flow properties, or via stochastic techniques to incorporate uncertainties of flow directly. Due to the extreme difficulties in using stochastic differential equation models for nonlinear multiphase flows, we will concentrate on the potential of deterministic models. Recent developments have been made in homogenization, scaled averaging, and the use of the simulator as an experimental tool to develop methods to model the interrelations between localized and larger-scale media effects. Monte Carlo techniques using simulators with effective parameters can generate statistics for multiphase flow. Understanding the functional forms of the model parameters that result from these scaling operations is also important in devising parameter estimation techniques for parameters needed for models at different length scales. We will briefly discuss aspects of parameter estimation for multiphase flow.

1 INTRODUCTION

The understanding and prediction of the behavior of the flow of multiphase or multicomponent fluids through porous media are often strongly influenced by heterogeneities, either large-scale lithological discontinuities or quite localized phenomena. Considerable information can be gained about the physics of multiphase flow of fluids through porous media via laboratory experiments and pore-scale models; however, the length scales of these data are quite different from those required from field-scale understanding. The coupled fluid–fluid interactions are highly nonlinear and quite complex. The presence of heterogeneities in the medium greatly complicates this flow. We must understand the effects of heterogeneities coupled with nonlinear parameters and functions on different length scales. We can then use the simulators as 'experimental tools' in the laboratory of high performance computing to simulate the process on increasingly larger length scales to develop intuition on how to model the effects of heterogeneities at various levels.

The partial differential equation models used in the simulators are convection dominated. Mixed-finite-element methods are described to treat the strong variation in coefficients arising from heterogeneities. An operator-splitting technique is then used to address the disparate temporal scales of convection, diffusion, and reaction terms. Convection is treated by time-stepping along the characteristics of the associated pure convection problem, and diffusion is modeled via a Galerkin method for single-phase flow and a Petrov–Galerkin technique for multiphase regimes. Eulerian–Lagrangian techniques, MMOC (modified method of characteristics) described by Douglas & Russell (1982) or Ewing, Russell & Wheeler (1983), or ELLAM (Eulerian–Lagrangian localized adjoint methods) introduced by Celia *et al.* (1990), effectively treat the advection-dominated processes. Extensions of ELLAM to the multiphase regime appear in Ewing (1991). Accurate approximations of the fluid velocities needed in the Eulerian–Lagrangian time-stepping procedure are obtained by mixed-finite-element methods. When reaction terms are present, their rapid effects in relation to convection or diffusion must be treated carefully since they strongly affect the conditioning of the resulting systems.

In order to scale the highly localized behavior of fine-scale fingering generated by heterogeneous media up to computational and field scales, we must develop techniques to obtain effective parameters for coarse-grid models which match fine-grid simulations. Ewing, Russell & Young (1989a) presented a coarse-grid dispersion model of heterogeneity and viscous fingering to match fine-grid simulation of miscible displacement processes. They adjusted longitudinal and transverse dispersivities in a dispersion tensor to match recovery curves for simulations of viscous fingering on fine grids. Although they were able to match production from various simulations, they pointed out that permeability averages, variances, and standard deviations alone are not able to determine dispersivities, since the specific permeability distribution in each realization can have significant impact upon the flow. Espedal *et al.* (1991) considered similar dispersion models to describe fingering processes in immiscible, two-phase flow. Neuman, Winter & Newman (1987) and Neuman & Zhang (1990) also presented various models for both Fickian and non-Fickian dispersion. In this paper we combine these ideas for multiphase and multicomponent flow, using dispersion models coupled with accurate treatment of first-order transport effects for both models. This coupling is very important for saturated–unsaturated models, which possess aspects of each process. The dispersion models are presented for both single and multiphase cases. Then accurate high-resolution numerical simulators are introduced and used as our experimental tools. Numerical results have illustrated the success of dispersion models for all of these problems.

Even at the finest grid level available using effectively scaled models, many phenomena cannot be resolved. Understanding and modeling these localized phenomena require the use of adaptive or local grid refinement. Usual implementation of local grid refinement techniques destroys the efficiency of large-scale simulation codes. Techniques which involve a relatively coarse macro-mesh are the basis for domain decomposition methods (Bramble *et al.*, 1988; Ewing, 1989a) and associated parallel solution algorithms. Accuracy, efficiency of implementation, and adaptivity of these techniques are discussed (see also Ewing, 1989b; Ewing, Lazarov & Vassilevski, 1991, 1994). Experiences obtained while using these ideas in three-dimensional multiphase industrial petroleum simulators are briefly described in Boyett, El-Mandouh & Ewing (1992) and Ewing *et al.* (1989a).

In Section 2, we discuss some of the difficulties involved in describing a reservoir with features of many different length scales. In Section 3, we present model equations for field-scale simulations. In Section 4, we discuss mixed-finite-element discretization techniques to approximate fluid velocities in the presence of heterogeneities as accurately as possible. Section 5 contains a description of operator-splitting techniques. Sections 6 and 7 present concepts of local grid refinement and upscaling. Some numerical experiments are discussed in Section 8. Finally, in Section 9 we present some conclusions and directions for future study.

2 RESERVOIR CHARACTERIZATION AND DESCRIPTION

The processes of both single and multiphase flow involve convection, or physical transport, of the fluids through a heterogeneous porous medium. The equations used to simulate this flow at a macroscopic level are variations of Darcy's law. Darcy's law has been derived for both single and multiphase regimes by Slattery (1969, 1970) via a volume averaging of the Navier–Stokes equations, which govern flow through the porous medium at a microscopic or pore-volume level. The length scale for Navier–Stokes flow (10^{-4}–10^{-3} meters) is very different from the scale required by normal reservoir simulation (10–10^{-3} meters). Reservoirs themselves have scales of heterogeneity ranging from pore level to field scale. In the standard averaging process for Darcy's law, many important physical phenomena which may eventually govern the macroscopic flow could be lost. The continued averaging of reservoir and fluid properties necessary to use grid blocks of the size of 10–10^2 meters in field-scale simulators further complicates the modeling process. We discuss certain techniques to try to address these scaling problems.

Diffusion and dispersion are often critical to the flow processes and must be understood and modeled. Molecular diffusion is typically fairly small. However, dispersion, or the mechanical mixing caused by velocity variations and flow through heterogeneous rock, can be extremely important and should be incorporated in some way in our models.

Since the mixing and velocity variations are influenced at all relevant length scales by the heterogeneous properties of the reservoir, much work must be done in the volume averaging of terms like porosity and permeability. Neuman & Yakowitz (1979), Neuman (1980), Neuman, Fogg & Jacobson (1980), Gelhar & Axness (1983), Furtado *et al.* (1991), have shown that statistical methods have promise in this area. Statistical techniques are currently being considered to obtain effective permeability tensors for large-scale models of flow through anisotropic or fractured media.

Due to the scarcity of direct data measurements, many researchers (Kitanidis & Vomvoris, 1983; Carrera & Neuman, 1986; Yeh, 1986; Ginn & Cushman, 1990) have included pore pressure measurements as part of the inverse problem for parameter estimation. In order to enhance these techniques, newer parameter estimation techniques (Rubin, Mavko & Harris, 1992) also incorporate seismic velocity data in addition to the hydrologic flow data. Stochastic models have been used effectively to determine various properties for single-phase flow (Dagan, 1985; Rubin & Dagan, 1987a,b; Dagan & Rubin, 1988).

The effects of dispersion in various flow processes have been discussed extensively in the literature (e.g., see Perkins & Johnston (1963) and Warren & Skiba (1964)). Russell & Wheeler (1983) and Young (1984) have given excellent surveys of the influence of dispersion and the attempts to incorporate it in present reservoir simulators. Various terms which affect the length of the dispersive mixing zone include reservoir heterogeneity and viscosity and velocity variations. Much work is needed to quantify these effects and to obtain useful effective dispersion coefficients for field-scale simulators. The dispersion tensor has strong velocity dependence. The longitudinal dispersion is often an order of magnitude larger than the transverse dispersion. This variation enhances unstable flow regimes induced by viscosity differences and reservoir heterogeneity. Initial work on correlation of dispersion coefficients presented with statistical simulations by Ewing *et al.* (1989b) will be discussed below.

In both single and multiphase regimes, the process of flows of fluids with very different viscosities through a heterogeneous porous medium can be unstable. If the flow rate is sufficiently high, then the interface between the contaminant and the water may become unstable and may form fingers which grow in length in a nonlinear fashion. This phenomenon, termed *viscous fingering,* is well known; different techniques for understanding and modeling it were surveyed by Ewing & George (1984).

Ewing *et al.* (1989b) indicated the ability of coarse-grid dispersion models to match results of both laboratory experiments and fine-grid simulations on highly heterogeneous meshes in a single-phase flow context. The use of a dispersion tensor in petroleum recovery applications avoided the optimistic recovery predictions often attributed to standard convection–diffusion models. Heterogeneity by itself, when not highly correlated, was shown to be less of an influence on recovery than viscous fingering from adverse mobility ratios; however, effects of heterogeneity are important and must be incorporated. The mixing parameter approach of Todd & Longstaff (1972) predicted premature breakthroughs and optimistic ultimate recoveries compared with experiments and dispersion models. More research is needed in this area, which is currently underway. Studies are being carried out to extend the global effective dispersion concept to multiphase flow in an analogous manner.

3 DISPERSION MODELS

In order to ensure that the information passed from scale to scale is dependent upon the physical properties of the flow and not upon the numerics of the specific simulator, we have extensively studied the codes used in Ewing *et al.* (1983) and Dahle *et al.* (1990) and have shown them to be essentially free of numerical dispersion and grid orientation effects. There are no Courant number restrictions, and the only grid restrictions are sufficiently fine to resolve waves of length $2\Delta x$. Otherwise, interpolation errors could cause difficulties. In Dahle *et al.* (1990), local grid refinement around the interface allowed significantly greater grid spacings elsewhere. The codes utilize mixed-finite-element methods for accurate fluid velocities in the presence of heterogeneities and Eulerian–Lagrangian techniques for accurate fluid transport without numerical dispersion.

These techniques apply well to the transport of contaminants through the saturated or unsaturated soil zones. We first consider for simplicity a two-dimensional horizontal reservoir where gravity effects are negligible. The single-phase flow of an incompressible fluid with a dissolved solute in a horizontal porous reservoir $\Omega \subset \mathcal{R}^2$, over a time period $J=[T_0, T_1]$, is given by

$$-\nabla\cdot\left(\frac{\mathbf{K}}{\mu}\nabla p\right)\equiv\nabla\cdot\mathbf{u}=q,\ x\in\Omega,\ t\in J \qquad (1)$$

$$\phi\frac{\partial c}{\partial t}-\nabla\cdot(\mathbf{D}\nabla c-\mathbf{u}c)=q\tilde{c},\ x\in\Omega,\ t\in J \qquad (2)$$

where p and $\mathbf{u}$ are the pressure and Darcy velocity of the fluid mixture, ϕ and $\mathbf{K}$ are the porosity and the permeability of the medium, μ is the concentration-dependent viscosity of the mixture, c is the concentration of the contaminant solute, q is the external rate of flow, and $\tilde{c}$ is the measurable inlet or outlet concentration. The form of the diffusion-dispersion tensor $\mathbf{D}$ is given by

$$\mathbf{D}=\phi(x)\{d_m I=|\mathbf{u}|(d_\ell E(\mathbf{u})+d_t E^{\perp}(\mathbf{u}))\} \qquad (3)$$

where

$$E_{ij}(\mathbf{u})=\frac{\mathbf{u}_i\mathbf{u}_j}{|\mathbf{u}|^2} \qquad (4)$$

$E^{\perp}=I-E$, d_m is the molecular diffusion coefficient, d_ℓ and d_t are the longitudinal and transverse dispersion coefficients, respectively. In general, $d_\ell \approx 10 d_t$, but this may vary greatly with different soils, fractured media, etc. The viscosity μ in eq. (1) is assumed to be determined by some empirical relationship or mixing rule based on contaminant concentration. In addition to eqs. (1) and (2), initial and boundary conditions are specified. The flow at wells is modeled in eqs. (1) and (2) via point or line sources and sinks.

When either an air phase or a nonaqueous phase liquid contaminant (NAPL) is present, the equations describing two-phase, immiscible flow in a horizontal porous medium are given by

$$\frac{\partial(\phi\rho_w S_w)}{\partial t}-\nabla\cdot\left(\mathbf{K}\frac{\rho_w k_{wi}}{\mu_w}\nabla p_w\right)=q_w\rho_w,\ x\in\Omega,\ t\in J \tag{5}$$

$$\frac{\partial(\phi\rho_a S_a)}{\partial t}-\nabla\cdot\left(\mathbf{K}\frac{\rho_a k_{ai}}{\mu_a}\nabla p_a\right)=q_a\rho_a,\ x\in\Omega,\ t\in J \tag{6}$$

where the subscripts w and a refer to water and air contaminant, respectively, S_i is the saturation, p_i is the pressure, ρ_i is the density, k_{ri} is the relative permeability, μ_i is the viscosity, and q_i is the external flow rate, each with respect to the ith phase. The saturations sum to unity.

One of the saturations can be eliminated; let $S=S_w=1-S_a$. The pressure between the two phases is described by the capillary pressure

$$p_c(S)=p_a-p_w \tag{7}$$

Although formally the equations presented in (1) and (2) seem quite different from those in (5) and (6), the latter system may be rearranged in a form which very closely resembles the former system. In order to use the same basic simulation techniques in our sample computations to treat both miscible and immiscible displacement, we will follow the ideas of Chavent (1976), and utilize a miscible/immiscible flow analogy.

Let Ω in $\mathbf{R}^3$ represent a porous medium. The global pressure p and total velocity $\mathbf{v}$ formulation of a two-phase water (w) and air (a) flow model in Ω are given by the following equations:

$$S_a c_a\frac{dp}{dt}+\nabla\cdot\mathbf{v}=-\frac{\partial\phi(p)}{\partial t}+q(x,S_w),\ x\in\Omega,\ t>0 \tag{8}$$

$$\mathbf{v}=-\mathbf{K}\lambda(\nabla p-\mathbf{G}_\lambda),\ x\in\Omega,\ t>0 \tag{9}$$

$$\phi\frac{\partial S_w}{\partial t}+\nabla\cdot(f_w\mathbf{v}-\mathbf{K}\lambda_a q_w\delta\rho\mathbf{g}+\mathbf{D}(S_w)\cdot\nabla S_w)=$$

$$-S_w\frac{\partial\phi(p)}{\partial t}+q_w,\ x\in\Omega,\ t>0 \tag{10}$$

The global pressure and total velocity are defined by.

$$p=\frac{1}{2}(p_w+p_a)+\frac{1}{2}\int_{S_c}^{S}\frac{\lambda_a-\lambda_w}{\lambda}\frac{dp_c}{d\xi}d\xi \text{ and } \mathbf{v}=\mathbf{v}_w+\mathbf{v}_a \tag{11}$$

Further, $d/dt\equiv[\phi(\delta/\delta t)]+[(\mathbf{v}_a/S_a)\cdot\nabla]$, $\lambda=\lambda_w+\lambda_\alpha$ is the total mobility, $\lambda_i=k_{ri}/\mu_i$, $i=w,a$, is the mobility for water and air, and $\mathbf{K}$ is the absolute permeability tensor.

The gravitational forces G_λ and capillary diffusion term $D(S)$ are expressed as

$$\boldsymbol{G}_\lambda=\frac{\lambda_w\rho_w+\lambda_a\rho_a}{\lambda}\mathbf{g} \quad\text{and}\quad \mathbf{D}(S)=-\mathbf{K}\lambda_a f_w\frac{dp_c}{dS} \tag{12}$$

and the compressibility c_a and fractional flow of water f_w are defined by

$$c_a=\frac{1}{\rho_a}\frac{d\rho_a}{dp_a} \quad\text{and}\quad f_w=\frac{\lambda_w}{\lambda} \tag{13}$$

We note that, in this formulation, the only diffusion/dispersion term is capillary mixing described by (12).

The phase velocities for water and air, which are needed in transport calculations, are given by

$$\mathbf{v}_w=f_w\mathbf{v}+\mathbf{K}\lambda_a f_w\nabla p_c-\mathbf{K}\lambda_a f_w\delta\rho\mathbf{g}$$

$$\mathbf{v}_a=f_a\mathbf{v}+\mathbf{K}\lambda_w f_a\nabla p_c-\mathbf{K}\lambda_w f_a\delta\rho\mathbf{g} \tag{14}$$

where $f_\alpha=\lambda_\alpha/\lambda$, $\alpha=w$, a, and $\delta\rho=\rho_a\rho_w$. Within the groundwater literature, the pressure is normally scaled by the gravity potential function. Equation (8) would then be given in terms of the pressure head. We should also note that if the Richards approximation, infinite mobility of air, or $p_a=0$ are valid, eq. (10) can be replaced by $p_c(S_w)=-p_w$. We may note that the phase velocity for air is given by eqs. (14) even if the Richards approximation is used.

If Γ is the boundary of Ω, general boundary conditions for eqs. (8)–(10) can be given by a combination of the following expressions:

$$p=p_\Gamma(x,t),\ \ x\in\Gamma_1,\ \ t>0 \tag{15}$$

$$\mathbf{v}\cdot\nu+b(x,t,S_w)p=G(x,t,S_w),\ \ x\in\Gamma_2,\ \ t>0 \tag{16}$$

$$\int_{\Gamma_3}\mathbf{v}\cdot\nu=g(t),\ \text{and}\ \ p=p_\Gamma(x,t)+d(t),\ \ x\in\Gamma_3,\ \ t>0 \tag{17}$$

$$S_w=S_\Gamma(x,t),\ \ x\in\Gamma_4 \tag{18}$$

$$(f_w\mathbf{v}+\mathbf{K}\lambda_a f_w(\nabla p_c-\delta\rho\mathbf{g}))\cdot\nu+b_w(x,t,S_w)p=G_w(x,t,S_w),$$

$$x\in\Gamma_5,\ \ t>0 \tag{19}$$

where Γ_i, $i=1,...,5$ are given partitions of Γ.

Normally the boundary conditions will be nonlinear functions of the physical boundary conditions for the original two-pressure formulation (Chen, Ewing & Espedal, 1994). This means that we have to iterate on the boundary conditions as a part of the solution process. Our experience is that this does not cause problems.

Since the transport term and the diffusion/dispersion term in (10) are governed by the fluid velocity, accurate simulation requires an accurate approximation of the velocity **v**. Because the lithology in the reservoir can change abruptly, causing rapid changes in the flow capacities of the rock, the tensor **K** in eqs. (9) and (10) can have discontinuous entries. In this case, in order for the flow to remain relatively smooth, the pressure changes extremely rapidly. Thus, standard procedures of solving eqs. (8) and (9) as a parabolic partial differential equation for pressure, differentiating or differencing the result to approximate the pressure gradient, and then multiplying by the discontinuous $\mathbf{K}\lambda$ can produce very poor approximations to the velocity **v**. A mixed-finite-element method for approximating **v** and p simultaneously, via a coupled system of first-order differential equations, has been used. This formulation allows the removal of singular terms as in Ewing *et al.* (1985), and accurately treats the problems of rapidly changing flow properties in the reservoir.

Both single-phase and multiphase codes used in our simulation utilize a physical dispersion tensor with different longitudinal and transverse terms. Although this is clearly natural for single-phase-contaminant modeling, the local physics of multiphase flows does not normally involve a dispersion phenomenon. However, via perturbation analysis, M.S. Espedal and P. Langlo have developed a natural dispersion tensor arising from heterogeneous flow at larger length scales following the single-phase work of Dagan (1989). Descriptions of these concepts appear in Espedal *et al.* (1990, 1991). Furtado *et al.* (1991) have arrived stochastically at a dispersion phenomenon with effects somewhere between transport and diffusion in origin. We feel that this corresponds to the need to match the gross permeability effects with first-order transport concepts and the finer-scale fingering with dispersion models.

4 MIXED METHODS FOR ACCURATE VELOCITY APPROXIMATIONS

The system in eqs. (8)–(10) is solved sequentially. An approximation for **v** is first obtained at time level $t=t^n$ from a solution of eq. (9) with the mobility λ evaluated from the value of S_w at time level t^{n-1}. Equations (8) and (9) can be solved via a mixed-finite-element method for the fluid velocity.

There are two major sources of error in the methods currently being utilized for finite-difference discretizations of eqs. (8)–(10). The first occurs in the approximation of the fluid pressure and velocity. The second comes from the techniques for upstream weighting to stabilize eq. (10). We first describe mixed-finite-element methods for the accurate approximation of the total velocity **v**. We then discuss some alternative upstream-weighting techniques in the next section for use in eq. (10).

Since the transport term in eq. (10) is governed by the fluid velocity, accurate simulation requires an accurate approximation of the velocity **v**. Because the lithology in the reservoir can change abruptly, causing rapid changes in the flow capabilities of the rock, the coefficient **K** in eqs. (9) and (10) can be discontinuous. In this case, in order for the flow to remain relatively smooth, the pressure changes extremely rapidly. Standard procedures of solving eqs. (8) and (9) are to eliminate the velocity and solve the remaining second-order equation as a parabolic partial differential equation for pressure; the differentiation of $\mathbf{K}\lambda$ can produce very poor approximations to the velocity **v**. In this section, a mixed-finite-element method for approximating **v** and p simultaneously, via the coupled system of first-order partial differential equations (8) and (9), will be discussed.

We define certain function spaces and notation. Let $W- L^2(\Omega)$ be the set of all functions on Ω whose square is finitely integrable. Let $H(div;\Omega)$ be the set of vector functions $\mathbf{v} \in [L^2(\Omega)]^2$ such that $\nabla\cdot\mathbf{v} \in L^2(\Omega)$ and let

$$V=H(div;\Omega)\cap\{\mathbf{v}\cdot\mathbf{n}=0 \text{ on } \Gamma\}$$

Let

$$(v,w)=\int_\Omega vw\,dx, \quad <v,w>=\int_\Gamma wv\,ds \quad \text{and} \quad \|v\|^2=(v,v)$$

be the standard L^2 inner products and norm on Ω and Γ. We obtain the weak solution form of eqs. (8) and (9) by dividing each side of eq. (9) by $\mathbf{K}\lambda$ (inverting the tensor $\mathbf{K}\lambda$), multiplying by a test function $\mathbf{u} \in V$, and integrating the result to obtain

$$((\mathbf{K}\lambda)^{-1}\mathbf{v},u)=(p,\nabla u)+(G\lambda,u), \quad u\in V \tag{20}$$

The right-hand side of eq. (20) was obtained by further integration by parts and use of the boundary conditions.

Next, multiplying eq. (8) by $w \in W$ and integrating the result, we complete our weak formulation, obtaining

$$(\nabla\cdot\mathbf{v},w)=(q,w)-\left(S_aC_a\frac{dp}{dt},w\right)-\left(\frac{\partial\phi(p)}{\partial t},w\right) \tag{21}$$

For a decreasing sequence of mesh parameters $h>0$, we choose finite-dimensional subspaces V_h and W_h with $V_h\subset V$ and $W_h\subset W$ and seek a solution pair $(\mathbf{U}_h;P_h)\in V_h\times W_h$ satisfying

$$((\mathbf{K}\lambda)^{-1}\mathbf{V}_h,\mathbf{u}_h)=(P_h,div\mathbf{u}_h)=(G\lambda,\mathbf{u}_h), \quad \mathbf{u}_h\in V_h \tag{22}$$

$$(div\,\mathbf{V}_h,w_h)=(q,w_h)-\left(S_aC_a\frac{dp}{dt}+\frac{\partial\phi(p)}{\partial t},w_h\right), \quad w_h\in W_h \tag{23}$$

Equations (22) and (23) lead to a saddle-point problem when the compressibility is small, requiring care in solution. Preconditioning or efficient iterative methods are essential. Effective block preconditioners are presented in Ewing *et al.* (1985), and efficient multigrid techniques are being developed.

5 OPERATOR-SPLITTING TECHNIQUES

In finite-difference simulators, the convection is stabilized via upstream-weighting techniques. In a finite-element setting, we use a possible combination of a modified method of characteristics and Petrov–Galerkin techniques to treat the transport separately in an operator-splitting mode.

In miscible or multicomponent flow models, the convective part is a linear function of the velocity. An operator-splitting technique has been developed to solve the purely hyperbolic part by time-stepping along the associated characteristics (Douglas & Russell, 1982; Ewing, Russell & Wheeler, 1984; Ewing *et al.*, 1985; Russell, 1985).

In immiscible or multiphase flow, the convective part is nonlinear. A similar operator-splitting technique to solve this equation needs reduced time steps because the pure hyperbolic part may develop shocks. Recently, an operator-splitting technique has been developed for immiscible flows (Espedal & Ewing, 1987; Dahle *et al.*, 1990), which retains the long time-steps in the characteristic solution without introducing serious discretization errors.

The splitting of the convective part of eq. (10) into two parts, $\mathbf{f}^m(S)+\mathbf{b}(S)S$, is constructed (Espedal & Ewing, 1987) such that $\mathbf{f}^m(S)$ is linear in the shock region, $0\leq S\leq S_1\leq 1$, and $\mathbf{b}(S)\equiv 0$ for $S_1\leq S\leq 1$.

The operator splitting is defined by the following set of equations:

$$\phi\frac{\partial \bar{S}_w}{\partial t}+\frac{d}{dS}\mathbf{f}^m(\bar{S}_w)\cdot\nabla\bar{S}_w\equiv\phi\frac{d}{d\tau}\bar{S}_w=0 \tag{24}$$

$$\phi\frac{\partial S_w}{\partial t}+\nabla\cdot(\mathbf{b}^m(S_w)S_w)-\nabla\cdot(D(S_w)\nabla S_w)=-S_w\frac{\partial\phi(p)}{\partial t}+q_w \tag{25}$$

$t_m\leq t\leq t_{m+1}$, together with proper initial and boundary conditions. As noted earlier, the saturation S_w is coupled to the pressure/velocity equations, which will be solved by mixed-finite-element methods (Douglas *et al.*, 1983; Ewing & Heinemann, 1984; Ewing *et al.*, 1984, 1985).

For a fully developed shock, the characteristic solution of eq. (24) will always produce a unique solution and, as in the miscible or single-phase case, we may use long time steps Δt without loss of accuracy. Equation (25) is solved by Petrov–Galerkin variational methods, where the time derivative and the nonlinear constants are approximated by the solution from eq. (24) (Espedal & Ewing, 1987). An iterative solution procedure based on domain decomposition methods (Bramble *et al.*, 1988) is used in the solution of the variational form of eq. (24).

6 LOCAL GRID REFINEMENT

It seems natural to relate the size of the coarse domains to the solution of the pressure–velocity equation (Espedal & Ewing, 1987), since the velocity varies slowly and defines a natural long length scale compared with the variation of the saturation S at a front. A simple local error estimate, which determines if a coarse-grid block must be refined, is given in Espedal & Ewing (1987). Normally, local refinement must be performed if a fluid interface is located within the coarse-grid block in order to resolve the solution there. A slightly different strategy is to make the region of local refinement big enough such that we can use the same refinements for several of the large time steps allowed by the method. The local grid refinement strategy combined with operator splitting is defined in the literature (Espedal & Ewing, 1987; Dahle *et al.*, 1990). The solution at groups of the coarse-grid vertices and the local refinement calculations may be sent to separate processors to achieve a high level of parallelism in the solution process.

The difficulty with these techniques is the communication of the solution between the fine and coarse grids. The use of local grid refinement in large-scale simulators often destroys the vectorization and efficient solution capabilities of the codes. Patch approximation techniques coupled with domain decomposition iterative solution methods (Bramble *et al.*, 1988) have proven to be very effective in developing accurate and efficient local grid refinement in the context of existing simulators. Mass balance considerations are very important for accuracy. Approximation techniques for cell-centered finite-difference methods appearing in Ewing, Lazarov & Vassilevski (1991, 1994) are discussed and compared with other methods.

These techniques can be extended to local time-stepping schemes (Ewing *et al.*, 1989b) and to algorithms for mixed finite-element methods (Ewing *et al.*, 1990). Mixed methods are being incorporated into existing finite-difference simulators to address the need for accurate approximation of fluid velocities in the context of heterogeneous media.

7 UPSCALING

The solution procedure described above represents an excellent tool for handling multiscale phenomena. Often data such as permeability, porosity, and capillary forces will have a multiscale dependence. Our multilevel solution procedure fits into this very well. However, the local refinement capabilities also mean that we must be able to give the appropriate model equations for different computational scales. Given a local computational grid, subgrid information has to be incorporated properly into the data representation.

Large-scale groundwater or oil reservoirs may have a very complex structure, and the geological description is normally a subject of great uncertainty. A two-stage geostatistical model has been proposed (Haldorsen & Damseth, 1990):

> Large-scale heterogeneities associated with facies are modeled from the information achieved from seismic data, well data, and analogous outcrops.
>
> Rock properties of the facies are modeled by a continuous multivariate Gaussian field or other statistical models. Seismic and well data can be used by a conditioning technique, and core data and other available data should be used to determine the statistical properties of the random field (mean, variance, correlation, etc.).

The coarsest computational domains should coincide with the facies of the model. The level of refinement of these domains and the grid within a given domain have to be decided from the knowledge of geometry, permeability variation, pressure gradients, etc.

In both groundwater and petroleum modeling, a substantial amount of research has been done on the upscaling of the permeability field to give a grid-block permeability (Dagan, 1989; Rubin, 1990), which could be used within our computational framework. The homogenization type of upscaling (Amaziane & Bourgeat, 1988; Amaziane, Bourgeat & Koebbe, 1991), which leads to a symmetric block-tensor for the permeability, seems to be especially well suited. The additive Schwarz type of domain decomposition methods leads to zero-Dirichlet boundary conditions for the local computations, consistent with the periodic boundary condition needed for the homogenization technique. One should note that within our computational framework, we need only the assumption of a periodic media locally on a given domain. We want to extend this single-scale homogenization technique to a multiscale model. Based on a wavelet representation of permeability, we have started research within this area, and so far the results look promising.

The upscaling of the saturation equation gives a new macrodispersion term in eq. (10), originating from the subgrid permeability variation. For a single-phase model this has been successfully studied (Lasseter, Waggoner & Lake, 1986; Dagan, 1989; Ewing *et al.*, 1989b; Rubin, 1990; Binning, 1994).

Within two-phase models little work has been done. Using a multifractal hypothesis, scaling laws for macrodispersion terms have recently been presented in the literature (Glimm & Lindquist, 1992; Glimm *et al.*, 1993). Also, macrodispersion models have recently been derived for a model where the permeability has a lognormal distribution (Langlo & Espedal, 1994). The derivation is based on the solution technique given above. It gives a saturation-dependent block-tensor dispersion coefficient. From the numerical experiments that are performed, we can conclude that the weakly correlated saturation fluctuations, on average, can be adequately described by this dispersion term.

Upscaling, leading to block-tensor dispersion terms, falls naturally into our computational setup, and we will continue our work based on this kind of modeling.

8 NUMERICAL EXPERIMENTS

A wide variety of numerical experiments have been performed using the techniques and concepts described in this chapter. A variety of numerical results for single-phase problems were presented by Ewing *et al.* (1989b). There the entire heterogeneity distribution and fingering phenomenon were described, with some lack of success, by the dispersion tensor alone. We feel that the highly correlated heterogeneities must be described by first-order effective permeability via homogenization techniques and the subgrid effects by dispersion methods. Continued research is underway for these problems.

In the multiphase flow case, experiments have been described in Espedal *et al.* (1990, 1991). These computations indicate that fingering instabilities initiated by fine-grid heterogeneities can effectively be modeled via a dispersion tensor. However, highly correlated heterogeneities must be treated by effective coarse-grid permeabilities. The nonlinearities limit the growth of the mixing zone in this study so that an isotropic tensor can be used for the fine-grid effects. Future work will consider a continuum of scales of heterogeneity.

In all the numerical experiments, we systematically varied mobility ratio, longitudinal and transverse dispersivity, and heterogeneity on fine grids. We used lognormal permeability distributions, considering the effect of variance. We also simulated several different randomly generated permeabilities with the same statistical properties to see whether the gross fingering behavior and recovery are similar. Then we sought relationships between the fine-grid parameters and those in the coarse-grid models to use effective parameters which match 'averaged' properties of many fine-grid simulations. The computations for both the multicomponent and multiphase models on fine grids have been matched to some extent via dispersion models for limited examples. Additional work for several scales of heterogeneity is essential.

9 CONCLUSIONS AND DIRECTIONS FOR FURTHER STUDY

Given the importance of dispersion and fingering in the modeling of many flow processes, research must progress in several directions. First, the averaging processes used to change length scales must be improved, perhaps via more statistical techniques, to obtain better effective reservoir coefficients for the macroscopic models. Simultaneously, the effective length scales of dispersion and its effect upon dispersive phenomena must be better understood. Also, better macroscopic techniques for including the effects (perhaps statistical) of channeling and dispersion must be developed.

Even if the information known about the reservoir properties in a highly heterogeneous medium is complete, the problem of how to represent this medium on coarse-grid blocks of different length scales still remains. The power of supercomputers must be brought to bear for simulation studies using homogenization and statistical averaging to represent fine-scale phenomena on coarser grids. The estimated viscosity, permeability, or dispersion coefficients must be modified to incorporate the important effects of fingering. Describing fingering on a grid size that resolves the fingers is impossible, even on the largest supercomputers, but since its presence can dominate flow, its effects must be included in large-scale simulators on coarse grids. We are developing effective equations and effective parameters to model this important effect. Much work is needed in this research area. Additional directions for research have been given in Section 8. Progress will depend upon the combination of stochastic techniques and deterministic simulation with accurate numerical techniques that do not possess artificial grid-size related diffusion or dispersion phenomena that may mask the true phenomena.

REFERENCES

Amaziane, B. & Bourgeat, A. (1988). Effective behavior of two-phase flow in heterogeneous reservoir. In *Numerical Simulation in Oil Recovery*, ed. M.F. Wheeler. IMA Volumes in Mathematics and Its Application 11. Berlin: Springer Verlag, pp. 1–22.

Amaziane, B., Bourgeat, A. & Koebbe, J. (1991). Numerical simulation and homogenization of two-phase flow in heterogeneous porous media. In *Transport in Porous Media II*, eds. U. Hornung, Z.M. Dogan & P. Knaber. Dordrecht: Kluwer Academic Publishers, pp. 519–548.

Binning, P.J. (1994). *Modeling Unsaturated Zone Flow and Contaminant Transport in the Air and Water Phases.* Ph.D. Dissertation, Department of Civil Engineering and Operations Research, Princeton University.

Boyett, B.A., El-Mandouh, M.S. & Ewing, R.E. (1992). Local grid refinement for reservoir simulation. In *Computational Methods in Geosciences, Frontiers in Applied Mathematics*, eds. W.E. Fitzgibbon & M.F. Wheeler. Philadelphia, PA: SIAM, pp. 15–28.

Bramble, J.H., Ewing, R.E., Pasciak, J.E. & Schatz, A.H. (1988). A preconditioning technique for the efficient solution of problems with local grid refinement. *Computer Methods in Applied Mechanics and Engineering*, 67, 149–159.

Carrera, J. & Neuman, S.P. (1986). Estimation of aquifer parameters under transient and steady state conditions, 1. Maximum likelihood method incorporating prior information. *Water Resources Research*, 22(2), 199–210.

Celia, M.A., Russell, T.F., Herrera, I. & Ewing, R.E. (1990). An Eulerian-Lagrangian localized adjoint method for the advection-diffusion equation. *Advances in Water Resources*, 13, 187–206.

Chavent, G. (1976). A new formulation of diphasic incompressible flows in porous media. In *Applications of Methods of Functional Analysis to Problems in Mechanics*, Lecture Notes in Mathematics 503. Berlin: Springer-Verlag.

Chen, Z., Ewing, R.E. & Espedal, M. (1994). Multiphase flow simulation with various boundary conditions. In *Computational Methods in Water Resources*, eds. A. Peters, G. Wittum, B. Herrling, U. Meissner, C.A. Brebbia, W. G. Gray & G.F. Pinder. Dordrecht: Kluwer Academic Publishers, pp. 925–932.

Dagan, G. (1985). Stochastic modeling of groundwater flow by unconditional and conditional probabilities: The inverse problem. *Water Resources Research*, 21(1), 165–72.

Dagan, G. (1989). *Flow and Transport in Porous Formations.* Berlin: Springer-Verlag.

Dagan, G. & Rubin, Y. (1988). Stochastic identification of recharge, transmissivity and storativity in aquifer unsteady flow: A quasi-steady approach. *Water Resources Research*, 24(10), 1698–1710.

Dahle, H.K., Espedal, M.S., Ewing, R.E. & Sævareid, O. (1990). Characteristic adaptive subdomain methods for reservoir flow problems. *Numerical Methods for Partial Differential Equations*, 6, 279–309.

Douglas, J. Jr., Ewing, R.E. &. Wheeler, M.F. (1983). A time-discretization procedure for a mixed finite element approximation of miscible displacement in porous media. *R.A.I.R.O. Analyse Numerique*, 17, 249–265.

Douglas, J. Jr., & Russell, T.F. (1982). Numerical methods for convection dominated diffusion problems based on combining the method of characteristics with finite element or finite difference procedures. *SIAM Journal on Numerical Analysis*, 19, 871–885.

Espedal, M.S. & Ewing, R.E. (1987). Characteristic Petrov-Galerkin subdomain methods for two-phase immiscible flow. *Computer Methods in Applied Mechanics and Engineering*, 64, 113–135.

Espedal, M.S., Hansen, R., Langlo, P., Sævareid, O. & Ewing, R.E. (1990). Heterogeneous porous media and domain decomposition methods. *Proceedings 2nd European Conference on the Mathematics of Oil Recovery*. Paris: Technip, pp. 157–163.

Espedal, M.S., Langlo, P., Sævareid, O., Gislefoss, E. & Hansen, R. (1991). Heterogeneous reservoir models, local refinements, and effective parameters. *SPE 21231, Proceedings of Eleventh SPE Symposium on Reservoir Simulation*, Anaheim, CA, pp. 307–316.

Ewing, R.E. (1989a). Domain decomposition techniques for efficient adaptive local grid refinement. In *Domain Decomposition Methods*, eds. T.F. Chan, R. Glowinski, J. Periaux & O.B. Widlund. Philadelphia, PA: SIAM, pp. 192–206.

Ewing, R.E. (1989b). Adaptive grid refinements for transient flow problems. In *Adaptive Methods for Partial Differential Equations*, eds. J.E. Flaherty, P.J. Paslow, M.S. Shephard & J.D. Vasilakis. Philadelphia, PA: SIAM, pp. 194–205.

Ewing, R.E. (1991). Operator splitting and Eulerian-Lagrangian localized adjoint methods for multiphase flow. In *The Mathematics of Finite Elements and Applications VII MAFELAP 1990*, ed. J. Whiteman. San Diego, CA: Academic Press Inc., pp. 215–237.

Ewing, R.E. & George, J.H. (1984). Viscous fingering in hydrocarbon and recovery processes. In *Mathematical Methods in Energy Research*, ed. K.I. Gross. Philadelphia, PA: SIAM, pp. 194–213.

Ewing, R.E. & Heinemann, R.F. (1984). Mixed finite element approximation of phase velocities in compositional reservoir simulation. *Computer Methods in Applied Mechanical Engineering*, 47, 161–176.

Ewing, R.E., Lazarov, R.D. & Vassilevski, P.S. (1991). Local refinement techniques for elliptic problems on cell-centered grids. I: Error analysis. *Mathematical Computation*, 56(194), 437–462.

Ewing, R.E., Lazarov, R.D. & Vassilevski, P.S. (1994). Local refinement techniques for elliptic problems on cell-centered grids. II: Optimal order two-grid iterative methods. *Numerical Linear Algebra Application*, 1(4), pp. 337–368.

Ewing, R.E., Russell, T.F. &. Wheeler, M.F. (1983). Simulation of miscible displacement using mixed methods and a modified method of characteristics. *SPE 12241, Proceedings of Seventh SPE Symposium on Reservoir Simulation*, San Francisco, CA, pp. 71–82.

Ewing, R.E., Russell, T.F. & Wheeler, M.F. (1984). Convergence analysis of an approximation of miscible displacement in porous media by mixed finite elements and a modified method of characteristics. *Computer Methods in Applied Mechanical Engineering*, 47, 73–92.

Ewing, R.E., Russell, T.F. & Young, L.C. (1989a). An anisotropic coarse-grid dispersion model of heterogeneity and viscous fingering in five-spot miscible displacement that matches experiments and fine-grid simulations. *SPE 18441, Proceedings of Tenth SPE Symposium Reservoir Simulation,* Houston, Texas, February 6–8, pp. 447–466.

Ewing, R.E., Boyett, B.A., Babu, D.K. &. Heinemann, R.F. (1989b). Efficient use of locally refined grids for multiphase reservoir simulations. *SPE 18413, Proceedings of Tenth SPE Symposium on Reservoir Simulation*, Houston, TX, February 6–8, 1989, pp. 55–70.

Ewing, R.E., Koebbe, J.V., Gonzalez, R. & Wheeler, M.F. (1985). Mixed finite element methods for accurate fluid velocities. In *Finite Elements in Fluids*, vol. 6, eds. R.H. Gallagher, G.F. Carey, J.T. Oden & O.C. Zienkiewicz. New York: Wiley, pp. 233–249.

Ewing, R.E., Lazarov, R.D., Russell, T.F. & Vassilevski, P.S. (1990). Local refinement via domain decomposition techniques for mixed finite element methods with rectangular Raviart-Thomas elements. In *Domain Decomposition Methods for Partial Differential Equations,* eds. T.F. Chan, R. Glowinski, J. Periaux & O.B. Widlund. Philadelphia, PA: SIAM, pp. 98–114,

Furtado, J., Glimm, J., Lindquist, W.B. & Pereira, L.F. (1991). Characterization of mixing length growth for flow in heterogeneous porous media. *Proceedings of Eleventh SPE Symposium on Reservoir Simulation*, Anaheim, CA, pp. 317–322.

Gelhar, L.W. & Axness, C.L. (1983). Three-dimensional stochastic analysis of macro-dispersion in aquifers. *Water Resources Research,* 19(1), 161–180.

Ginn, T.R. & Cushman, J.H. (1990). Inverse methods for subsurface flow: A critical review of stochastic techniques. *Stochastic Hydrology Hydraulics*, 4, 1–26.

Glimm, J. & Lindquist, W.B. (1992). Scaling laws for macrodispersion. *Proceedings of Ninth International Conference on Computer Methods in Water Resources*, vol. 2, pp. 35–49.

Glimm, J., Lindquist, W.B., Pereira, F. & Zhang, Q. (1993). A theory of macrodispersion for the scale up problem. *Transport in Porous Media*, 13, 97–122.

Haldorsen, H. & Damseth, E. (1990). Stochastic modeling. *Journal of Petroleum Technology,* 42, 404–413.

Kitanidis, P.K. & Vomvoris, E.G. (1983). A geostatistical approach to the inverse problem in groundwater modeling (steady state) and one-dimensional simulations. *Water Resources Research,* 19(3), 677–690.

Langlo, P. & Espedal, M.S. (1994). Macrodispersion for two-phase, immiscible flow in porous media. *Advances in Water Resources,* 17, 297–316.

Lasseter, T.J., Waggoner, J.R. & Lake, L.W. (1986). Reservoir heterogeneities and their influence on ultimate recovery. In *Reservoir Characterization,* eds. LvW. Lake & H.B. Carroll. New York: Academic Press, pp. 545–560.

Neuman, S.P. (1980). A statistical approach to the inverse problem of aquifer hydrology, III. Improved solution method and added perspective. *Water Resources Research,* 16(2), 331–346.

Neuman, S.P., Fogg, A. & Jacobson, E. (1980). A statistical approach to the inverse problem of aquifer hydrology, II. Case study. *Water Resources Research,* 16, 33–58.

Neuman, S.P., Winter, C.L. & Newman, C.M. (1987). Stochastic theory of field-scale dispersion in anisotropic porous media. *Water Resources Research,* 23(3), 453–466.

Neuman, S.P. & Yakowitz, S. (1979). A statistical approach to the inverse problem of aquifer hydrology, I. Theory. *Water Resources Research,* 15(4), 845–860.

Neuman, S.P. & Zhang, K. (1990). A quasi-linear theory of non-Fickian and Fickian subsurface dispersion, 1. Theoretical analysis with application to isotropic media. *Water Resources Research,* 26(5), 887–902.

Perkins, T.K. & Johnston, O.C. (1963). A review of diffusion and dispersion in porous media. *Society of Petroleum Engineers Journal,* 3, 70–84.

Rubin, Y. (1990). Stochastic modeling of macrodispersion in heterogeneous porous media. *Water Resources Research,* 26, 133–141.

Rubin, Y. & Dagan, G. (1987a). Stochastic identification of transmissivity and effective recharge in steady state groundwater flow, 1. Theory. *Water Resources Research,* 23(7), 1185–1192.

Rubin, Y. & Dagan, G. (1987b). Stochastic identification of transmissivity and effective recharge in steady state groundwater flow, 2. Case study. *Water Resources Research,* 23(7), 1192–1200.

Rubin, Y., Mavko, G. & Harris, J. (1992). Mapping permeability in heterogeneous aquifers using hydrological and seismic data. *Water Resources Research,* 28(7), 1809–1816.

Russell, T.F. (1985). The time-stepping along characteristics with incomplete iteration for Galerkin approximation of miscible displacement in porous media. *Society of Industrial and Applied Mathematics Journal of Numerical Analysis,* 22, 970–1013.

Russell, T.F. & Wheeler, M.F. (1983). Finite element and finite difference methods for continuous flows in porous media. In *The Mathematics of Reservoir Simulation,* ed. R.E. Ewing, Frontiers in Applied Mathematics. Philadelphia, PA: SIAM, pp. 35–106.

Slattery, J.C. (1969). Single-phase flow through porous media. *American Institute of Chemical Engineers Journal,* 15, 866–872.

Slattery, J.C. (1970). Two-phase flow through porous media. *American Institute of Chemical Engineers Journal,* 15, 345–352.

Todd, M.R. & Longstaff, W.J. (1972). The development, testing, and application of a numerical simulator for predicting miscible flood performance. *Journal of Petroleum Technology,* 253, 874–882.

Warren, J.E. & Skiba, F.F. (1964). Macroscopic dispersion. *Society of Petroleum Engineering Journal,* 4, 215–230.

Yeh, W.W.-G. (1986). Review of parameter identification procedures in groundwater hydrology: The inverse problem. *Water Resources Research,* 22(2), 95–108.

Young, L.C. (1984). A study of spatial approximations for simulating fluid displacements in petroleum reservoirs. *Computer Methods in Applied Mechanical Engineering,* 47, 3–46.

3 Incorporating uncertainty into aquifer management models

STEVEN M. GORELICK
Stanford University

1 INTRODUCTION

The aim of aquifer simulation is to predict how hydraulic heads and solute concentrations respond to system stresses such as pumping and recharge. In many cases, simulation alone does not provide the results necessary to manage groundwater resources. What is needed is a design tool to find the best arrangement of pumping and recharge in order to control the heads and concentrations over space and time. One might want the managed head declines to be limited to some target value, or to ensure that solute concentrations never exceed water quality standards at supply wells. These controls must be maintained while achieving some objective, such as minimizing cost or risk. The recognition that simulation is not an end in itself has led to the development of aquifer management models.

During the past 30 years, the field of aquifer management modeling has developed as a distinct discipline. It has provided a framework which replaces trial and error simulations. Modern aquifer management models combine simulation tools with optimization techniques. The optimization techniques were developed in fields such as operations research and applied physics, and were adapted to unite with aquifer simulation models. This combination of formal optimization and aquifer simulation provides a framework that forces the engineer or hydrologist to formulate carefully the groundwater management problem. The problem must contain a series of constraints on heads, drawdowns, pumping rates, hydraulic gradients, groundwater velocities, and solute concentrations. In addition, an objective, usually involving costs, cost surrogates, or risk, must be stated. The power underlying this union of highly computational technologies was unleashed when computers became fast and cheap. On-the-shelf programs for aquifer management modeling are now available.

Initial aquifer management models were deterministic and restricted to simulation involving linear systems, such as those described by the equation for confined aquifers (Gorelick, 1983). During the past decade, the discipline has grown to handle broad classes of nonlinear problems as well as uncertainty from a variety of sources. Simulation-management models that were once linear and deterministic are now nonlinear and stochastic (Gorelick, 1990; Wagner, 1995). In real-world problems uncertainty exists in our ability to predict behavior. The consequences of this uncertainty for managing groundwater resources are obvious; severe uncertainty forces us to overdesign our aquifer management schemes and/or to accept greater risk that our design policy will not be successful. Therefore, it is not surprising that recent developments in aquifer simulation-optimization have focused heavily on formulating problems in probabilistic terms, and on developing innovative techniques to overdesign cost effectively groundwater management schemes in order to accommodate our predictive uncertainty.

This chapter will discuss the categories of uncertainty relevant to aquifer management models and some of the key contributions in each area. Because of the tremendous concern over groundwater quality, much of the work in simulation-optimization modeling has focused on the design of aquifer remediation systems (Ahlfeld *et al.*, 1988; Gorelick *et al.*, 1993). This chapter is not intended to be a complete literature review, and will focus on problems of groundwater quality and aquifer remediation. There have been several recent methodological advances that have greatly improved the efficiency of solving simulation-optimization problems. They are discussed in an excellent review by Wagner (1995).

2 THE NATURE OF UNCERTAINTY

A model is a simplified representation of a system. Uncertainty is a property of a model whether it is explicitly recognized or not. Uncertainty stems from our failure or inability to measure, understand, or represent all of the features of a true system. With any model we get uncertainty for free.

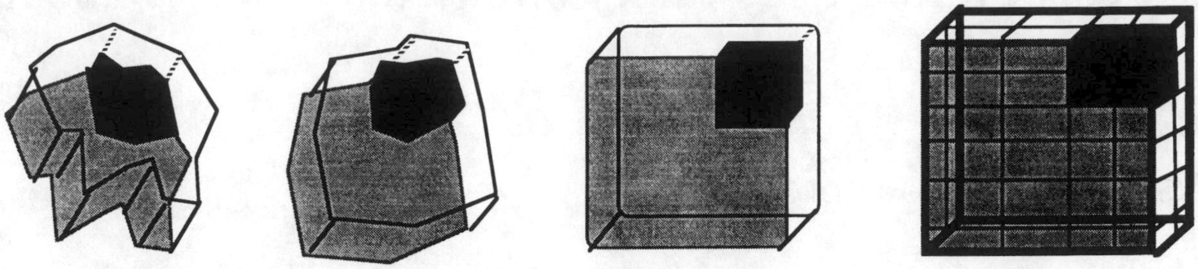

Fig. 1 Procedure for model development showing heterogeneity and its representation.

The specific origins of uncertainty, and how it has been dealt with, have guided the development of stochastic simulation-optimization modeling. Before getting into specific approaches, it is useful to consider the schematic representation of the modeling process shown in Fig. 1. Given a true system, we make observations and conduct tests in order to develop a conceptual model. The conceptual model is where our understanding of the key processes, structures, and excitations are delineated. Going from the limited observations of the true system to the conceptual model is the most crucial step in simulation model development. This is the stage where we either capture the essence of system behavior or introduce the greatest uncertainty. If we fail to recognize primary physical and chemical processes in the development of the *conceptual model*, then model uncertainty will overshadow our predictive capacity. On the other hand, if reasonable simplifications of the true system were made in developing the conceptual model, then the uncertainty may be properly represented in the *structural and mathematical model.*

The representation of spatial structure is a critical feature of aquifer models. Complicated geologic, hydrologic, and geochemical forces are responsible for complex geometries and spatial patterns of hydraulic parameters. Although we may have a reasonable conceptual model, we must make some decisions about how to represent spatial variability. The easiest thing to do is to ignore spatial variability and call the modeled system homogeneous. This is commonly done when constructing models used to analyze aquifer pump and slug tests. Immediately, one can see that the assumption of homogeneity is tied to the scale represented by the model. At some small scale no real medium is homogeneous, but this approximation may suffice over the volume of influence of a well test. At scales beyond that of the pump test, spatial variability in hydraulic conductivity and other model parameters must be dealt with. The structural model may be zonal or continuously variable. Our decision here is one of appropriate model parameterization. If a zonal model is adopted, then physical representation of spatial variability within each zone is lost, and that neglected complexity becomes a component of model uncertainty. Even if a model of continuous spatial variation is adopted, we may be forced into an overly simplistic mathematical representation of spatial connectivity and natural variability. Proper representation of our conceptual model may be jeopardized. Aquifer tests measure spatially averaged aquifer properties, and this connection to common field test values is one attraction of zonal models. Our *mathematical model* contains all of our modeling decisions that originated from the conceptual model and the structural model. It is generally a partial differential equation, which may be deterministic or stochastic, accompanied by boundary and initial conditions, which may also be variable and uncertain. During the past two decades our ability to solve complex mathematical models using numerical methods has grown significantly. At some scale *numerical models* will be discrete, and parameter values can only be represented as an area or volume average.

3 DETERMINISTIC FORMULATION OF SIMULATION-OPTIMIZATION PROBLEM

Before discussing simulation-optimization approaches that incorporate uncertainty, let us consider a generic deterministic formulation. A simulation-optimization problem includes an objective function that is minimized or maximized. In the generic formulation, the cost of pumping is to be minimized, where p_1 is a cost function that depends upon the vector of pumping rates $\mathbf{q}$. In addition, the total cost objective can include the minimization of fixed costs, p_2, which may involve fixed costs of well and treatment plant installation. The second component of a simulation-optimization formulation is the constraint set. The constraints can include limitations on a variety of head-dependent variables, $\mathbf{H}$, such as heads, drawdowns, or hydraulic gradients. Constraints can also include limitations on solute concentrations, $\mathbf{C}$, at any particular location or time. The variables $\mathbf{H}$

and **C** depend on the vector of pumping rates, **q**, which are called decision variables. The values of **H** in a particular constraint equation are obtained directly or indirectly through simulation. Heads and hydraulic gradients are calculated for any particular set of pumping rates, **q**, by solving a flow model, $f(\mathbf{q})$. Concentrations are calculated by solving both flow and transport models, $g(\mathbf{q})$. Finally, each decision variable or dependent variable is bounded above by $\mathbf{q}^*,\mathbf{H}^*,\mathbf{C}^*$ and below $\mathbf{q}_l,\mathbf{H}_l,\mathbf{C}_l$.

The generic formulation is:

Minimize Cost of pumping [+ fixed installations]	$p_1(\mathbf{q})[+p_2(\mathbf{q})]$
Subject to:	
Heads, drawdowns, hydraulic gradients, are typically linear functions of pumping rates.	$f(\mathbf{q})=\mathrm{H}$
Concentrations are nonlinear functions of pumping rates	$g(\mathbf{q})=\mathrm{C}$
Pumping rates, **q**, are bounded	$\mathbf{q}_l \le \mathbf{q} \le \mathbf{q}^*$
Head-dependent variables, **H**, are bounded	$\mathbf{H}_l \le \mathbf{H} \le \mathbf{H}^*$
Concentrations, **C**, are restricted to meet water quality standards	$\mathbf{C} \le \mathbf{C}^*$

Solving this simulation-optimization problem results in the selection of optimal pumping rates at specified locations. Where values of pumping are set to zero, no well is needed. In this sense, the optimal-pumping well locations are determined. The solution will identify the least-cost set of pumping rates such that the constraints are obeyed.

For our purposes, the simulation components can be thought of as any (numerical) model that has been carefully developed to predict heads and concentrations for a particular aquifer. In the simulation-optimization literature the governing equations for heads and concentrations are usually given in 2D as:

$$f(\mathbf{q}) \equiv S\frac{\partial h}{\partial t} = \frac{\partial}{\partial x_i}\left[T_{ij}\frac{\partial h}{\partial x_j}\right] - q$$

$$g(\mathbf{q}) \equiv R\frac{\partial c}{\partial t} = \frac{\partial}{\partial x_i}\left[D_{ij}\frac{\partial c}{\partial x_j}\right] - v_i\frac{\partial c}{\partial x_i} - \frac{q}{\theta b}(c_s - c) \quad i,j=1,2$$

where

- $f(\mathbf{q})$ is the function that relates pumping to heads;
- $g(\mathbf{q})$ is the function that relates pumping to concentrations and depends on h which is given by $f(\mathbf{q})$;
- h is hydraulic head, L;
- T_{ij} is the transmissivity tensor, L^2/T;
- q represents fluid sinks (+) or sources (−) L/T;
- c is concentration, M/L^3;
- D_{ij} is the hydrodynamic dispersion tensor, L^2/T;
- v_i represents the groundwater velocity vector components, L/T;
- R is a retardation factor;
- c_s is the concentration in a fluid source, and is equal to c for a fluid sink, M/L^3;
- θ is the effective porosity, L^3/L^3;
- b is aquifer thickness, L;
- x_i are spatial coordinates, L and,
- t is time, T

The most heavily computational part of simulation-optimization modeling is the simulation of heads and concentrations, $f(\mathbf{q})$ and $g(\mathbf{q})$, for different vectors of pumping rates, **q**, while searching for an optimal set of pumping rates that meets the constraints.

4 INCORPORATING UNCERTAINTY

Most aquifer management models assume that the correct conceptual model was selected and that all of the primary physical and chemical processes are adequately represented in the mathematical model. Therefore, all uncertainty resides in the model parameters. Two approaches have been developed to represent uncertainty due to variability in model parameters. The first is the zonal approach, and the second is a geostatistical approach. In both of these approaches model uncertainty is fully contained in the model parameters, such as hydraulic conductivity, or unmanaged fluxes, such as recharge. In both approaches, there are two stages: parameter estimation and simulation-optimization under uncertainty. The formulations differ from the deterministic one above by recognizing that uncertainty is at the heart of the problem.

The parameters in the simulation model are uncertain, and therefore in order to guarantee that the constraints are met, the system must be overdesigned. Fig. 2 is a schematic representation of the roles of simulation, optimization, and uncertainty. The overall problem is couched in an optimization framework with an objective function and a series of constraints. The optimization model contains a full simulation model to represent the physical and chemical processes as well as possible. At the heart of the simulation-optimization model is uncertainty which is embedded in the simulation model parameters.

5 ZONAL APPROACH

The zonal approach assumes that uncertainty is fully contained within the model parameters and that heads and concentrations are random variables. It is assumed that their

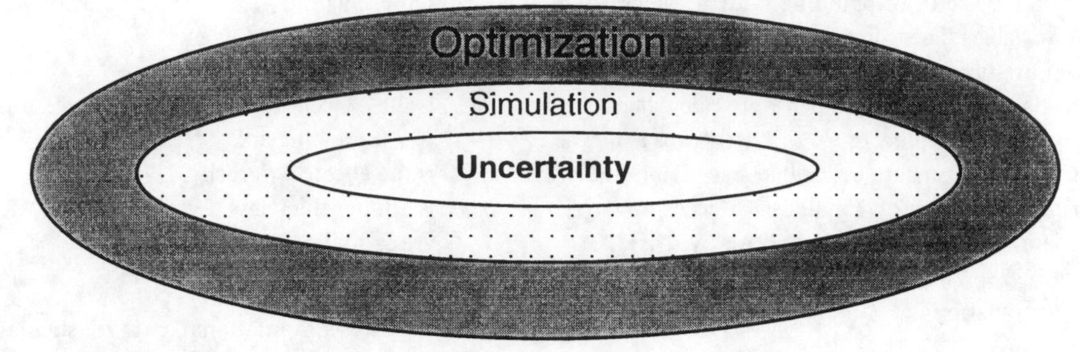

Fig. 2 Role of uncertainty in simulation-optimization models.

random behavior can be explained if the uncertainty of the model parameters is quantified. Each uncertain parameter in the simulation model, for example transmissivity or recharge, is constant over a zone, but its value is unknown. The formulation of the stochastic simulation-optimization problem begins with the deterministic formulation shown above, but considers the dependent variables **H** and **C** which are described by a probability distribution at any location and time, *i*. One writes a stochastic constraint on a particular dependent variable that depends on head. It states that the *probability must be greater than some specified reliability*, r_h, that a head, drawdown, or hydraulic gradient, h_i, is below a specified limit h_i*:

$$\text{prob } \{h_i \leq h_i^*\} \geq r_h \quad \text{for all } i$$

A stochastic constraint on solute concentration states that the probability must be greater than some specified reliability, r_c, that a concentration will not exceed the water quality standard c_i*:

$$\text{prob } \{c_i \leq c_i^*\} \geq r_c \quad \text{for all } i$$

These probabilistic constraints have been included in simulation-optimization models by converting them to chance constraints (Wagner & Gorelick, 1987; Tiedeman & Gorelick, 1993; Chan, 1994). A chance constraint is a deterministic equivalent where the state variable is a function of the uncertain model parameters (Charnes & Cooper, 1962). For example, **H** might represent the hydraulic gradient. Let us assume that the only uncertain parameter is transmissivity, T; $h=h(T)$. The chance constraint for a hydraulic gradient at any specified location and time, *i*, is

$$\underset{\text{expected value}}{E[h_i]} + \underset{\text{stochastic component}}{N^{-1}(r_h)\sigma[h_i]} \leq \underset{\text{gradient limit}}{h_i^*}$$

The first term is the expected value of the hydraulic gradient and is the predicted gradient when the mean value of the uncertain parameter, transmissivity, is used in the simulation model. The second term is the stochastic component, where N^{-1} is the value of the standard normal cumulative distribution corresponding to a specified reliability level r_h, and $\sigma[h_i]$ is the standard deviation of the simulated hydraulic gradient as a function of uncertainty in transmissivity. This term reflects the uncertainty in the hydraulic gradient due to the fact that transmissivity is unknown. If the transmissivity is known and equal to the estimated value, then the stochastic component is zero. Note that the value of the standard normal cumulative distribution corresponding to $r_h=50\%$ is zero. In that case there is only the mean value of the transmissivity and the stochastic constraint is identical to the deterministic constraint shown earlier.

In order to obtain the statistical information required in the chance constraints, a relationship must be obtained between the uncertain model parameter values and the dependent variables. The first-order Taylor series approximation to the mean and covariance of head are based completely on the mean and covariance of the model parameters. This relationship was given by Dettinger & Wilson (1981) as

$$E[h(T)] \cong h(\tilde{T})$$

$$\text{Cov}(h(\tilde{T})) = \mathbf{J}\ \text{Cov}(\tilde{T})\mathbf{J}^{t}$$

where E is the expected value, Cov is the covariance, **J** is the Jacobian of head with respect to transmissivity, $(\partial h/\partial \tilde{T})$, and the superscript t denotes transpose. $\tilde{T}$ is the estimated value of transmissivity. This relationship was extended to concentrations by Wagner & Gorelick (1987) and to hydraulic gradients for unconfined flow conditions with uncertain recharge by Tiedeman & Gorelick (1993).

An application of the zonal approach is found in Tiedeman & Gorelick (1993). They considered a contaminant capture design problem in southwest Michigan involving a vinyl chloride plume that is migrating toward Lake Michigan. The area that was modeled is approximately 1.6 km by 2.1 km. A three-dimensional model of the groundwater flow system was developed in which the aquifer was divided into four layers, each approximately 5 m thick. The upper layer is unconfined. The region was divided into two zones, each with a different hydraulic conductivity. There were limited aquifer test values and the zonal conductivity values were calibrated parameters. In addition, the flow system is dominated by recharge from precipitation which was modeled by developing a time series of monthly recharge values. With the relative monthly values fixed, a recharge multiplier was the sole calibration parameter for recharge. Model calibration was based upon transient three-dimensional flow simulation in which initial conditions were established for September 1987. Simulated and observed heads were matched for 30 heads during March 1988 and 31 heads during September 1988. The calibration procedure used a least-squares approach (Cooley, 1979), which resulted in estimated values of the two zonal conductivity values and the recharge multiplier. In addition, the simulation-regression procedure gave an estimate of the covariance of the three parameter estimates. A capture design based on chance constraints written for hydraulic gradients was constructed and solved.

Fig. 3 shows the groundwater flow system near Lake Michigan for which a plume capture system was developed. The shaded region delineates the 10 μg/ℓ plume margin for vinyl chloride. The open circles show seven potential remediation wells that were optimally selected from ten possible locations. Using the parameter estimates alone, the 50% reliability capture curve is shown. This corresponds to a total pumping rate of 5.1 ℓ/s. In order to overcome the uncertainty in the estimates of hydraulic conductivity and recharge, a 90% reliability capture curve was designed using the zonal stochastic formulation described earlier. In order to obtain 90% reliability, the pumping rate must increase to 7.1 ℓ/s. It is interesting to note that Tiedeman and Gorelick found that, for a two-well system, capture could be achieved for a reliability of 50% and required 8.7 ℓ/s. However, it was infeasible to design a 90% reliability capture system with just two wells because additional pumping, beyond about 10 ℓ/s, results in excessive drawdowns in the wells; that is, the wells become dry. If uncertainty in parameters is ignored, one obtains a very misleading solution.

It is clear that uncertainty in model parameter values forces one to increase pumping in order to guarantee plume containment. Tiedeman and Gorelick (1993) found that it was necessary to increase pumping by about 40% to achieve 90% reliability containment. The zonal approach was also used by Gailey & Gorelick (1993) who applied it to design reliable capture schemes for the Gloucester Landfill problem in Ottawa, Canada. They found that, in order to increase the design reliability from 50% to 90%, pumping must be increased by a safety factor of about 25%. Because both the stochastic and expected value components of the chance constraint depend on the optimal value of the decision variables (the pumping rates), the safety factor could not be determined ahead of time. Rather, in each case safety factors are given by the solution of the optimization problem.

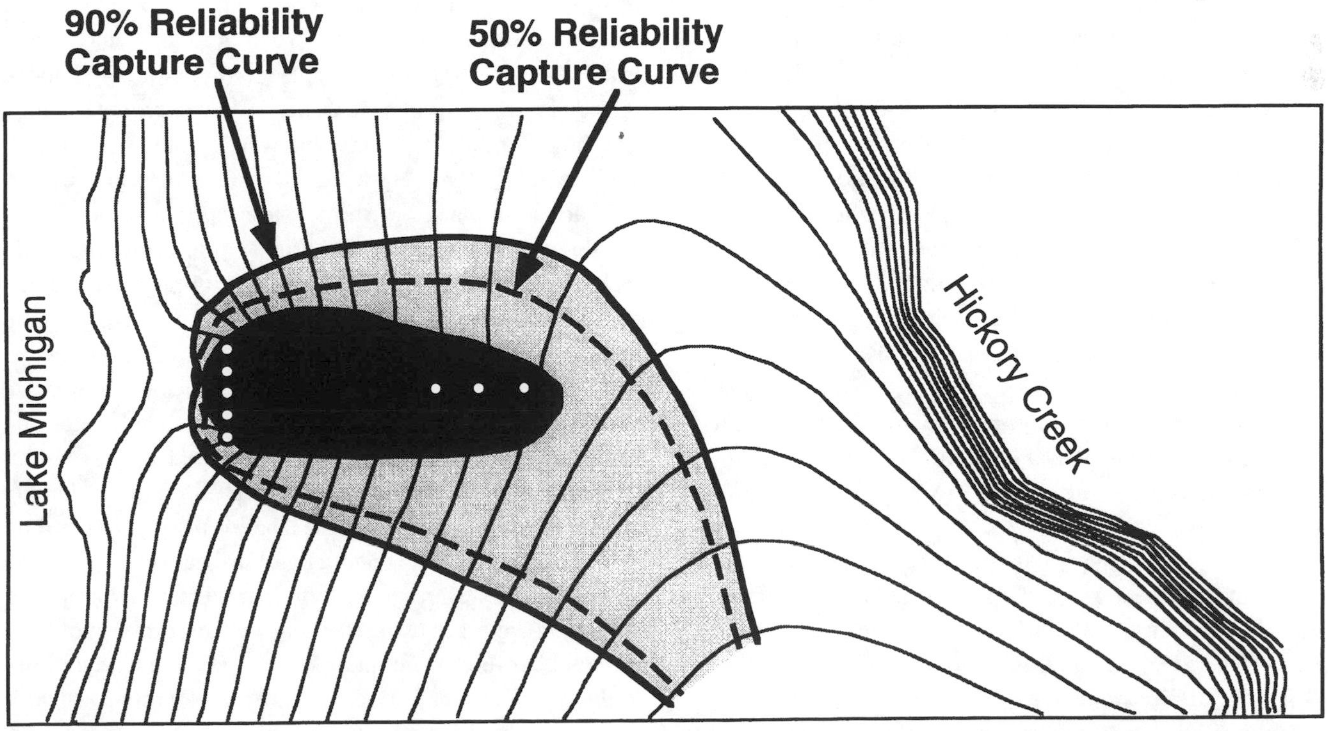

Fig. 3 Capture curves for 50% and 90% reliability based on uncertainty in zonal hydraulic conductivities and recharge from precipitation (Tiedeman & Gorelick, 1993).

6 GEOSTATISTICAL APPROACH

In the zonal approach, the view of the real system is that the aquifer consists of a number of different zones each with its own uniform hydraulic conductivity value. The value for each zone is uncertain. Including that uncertainty in a simulation model results in a statistical distribution of hydraulic heads and solute concentrations. This consequent uncertainty of dependent variables forces one to overdesign any particular management scheme in order to guarantee that the constraints are actually met in the real system.

In a somewhat different view of uncertainty, the geostatistical approach sees the real system as an aquifer with continuously spatially variable properties. The values are not zonal, they are unknown, and therefore the complete map of parameter values of the true system is virtually unknowable. At best one can produce a smoothed rendition of the true system that is based upon some form of interpolation and a statistical model of spatial correlation. One can also characterize and then reproduce the variable and highly uncertain texture that may be added to the smoothed field. This view is appealing because geologists recognize that sedimentary, erosional, tectonic, and geochemical processes are primary agents that shape hydraulic conductivity variations. Such processes are complex and do, in fact, result in media properties that vary dramatically through space, yet maintain spatial correlation. This view is also appealing to engineers because tools of modern geostatistics can be applied to estimate, describe, and reproduce the spatial variability of any field. In the geostatistical approach the parameters are not ones that give the hydraulic conductivity at each location. Rather they describe key statistical properties of the media, for example the mean, variance, and spatial correlation of values. Geostatistics can be quite good at producing images that maintain the texture of mapped features and can be conditioned so that they reproduce the sparse and spatially variable data that we commonly have. In between data points, the interpolated values are uncertain. The view of heterogeneity as continuous and not zonal is undoubtedly a superior conceptual model for problems involving contaminant transport. While variations in hydraulic heads are relatively insensitive to variations in transmissivity, solute concentrations are quite sensitive to heterogeneity in transmissivity.

Given that any map of hydraulic conductivity is highly uncertain because it is based upon a small amount of data, a second approach to simulation-optimization was developed that is based upon multiple equally likely realizations (Gorelick, 1987; Wagner & Gorelick, 1989; Chan, 1993). Assuming that one can estimate the spatial statistics associated with transmissivity variations, it is straightforward to generate maps or realizations of transmissivity values. The formulation of the optimization problem for heads, **h**, and concentrations, **c**, only considering pumping costs, *p*, now becomes:

Minimize Cost of pumping	$p(\mathbf{q})$
Subject to	
Heads as functions of pumping rates for	
Realization 1	$f_1(\mathbf{q})=\mathbf{h}$
Realization 2	$f_2(\mathbf{q})=\mathbf{h}$
Realization 3	$f_3(\mathbf{q})=\mathbf{h}$
⋮	⋮
Realization *k*	$f_k(\mathbf{q})=\mathbf{h}$
⋮	⋮
Realization *n*	$f_n(\mathbf{q})=\mathbf{h}$
Concentrations as functions of pumping rates for	
Realization 1	$g_1(\mathbf{q})=\mathbf{c}$
Realization 2	$g_2(\mathbf{q})=\mathbf{c}$
Realization 3	$g_3(\mathbf{q})=\mathbf{c}$
⋮	⋮
Realization *k*	$g_k(\mathbf{q})=\mathbf{c}$
⋮	⋮
Realization *n*	$g_n(\mathbf{q})=\mathbf{c}$
Pumping rates, **q**, are bounded	$\mathbf{q}_l\leq\mathbf{q}\leq\mathbf{q}^*$
Heads are bounded	$\mathbf{h}_l\leq\mathbf{h}\leq\mathbf{h}^*$
Concentrations, **c**, are restricted to meet water quality standards	$\mathbf{c}\leq\mathbf{c}^*$

This formulation says the following: Find the least-cost pumping rates so that head and concentration constraints are never violated for each and every map (realization) of the hydraulic conductivity field. In an aquifer remediation problem the goal is to find a single set of pumping well locations and rates that will provide the desired degree of cleanup for each potential reality, and that will be optimal for the set of all realities as represented in the stack of *n* maps. This approach is similar to a Monte Carlo approach, but it has one significant difference. In a true Monte Carlo approach one would obtain a feasible pumping solution for each *individual* map, but the feasible solution for one map might violate the constraints associated with any other map. In the above formulation, the solution must be robust and must never violate any constraint for any map. This is accomplished by stacking the constraints where each element in the stack (subset of the constraint set) corresponds to the simulation for one realization. Together there will be *n* simulations based upon *n* hydraulic conductivity realizations.

Fig. 4 shows a graphical representation of the multiple realization approach using stacking. For this problem of contaminant capture, the binding constraints will be hydraulic gradients surrounding the capture zone. In the simplest case, the solution will merely identify the worst-case realization and design a set of pumping rates to meet the constraints for that map. That is, the solution for that worst-case map is so

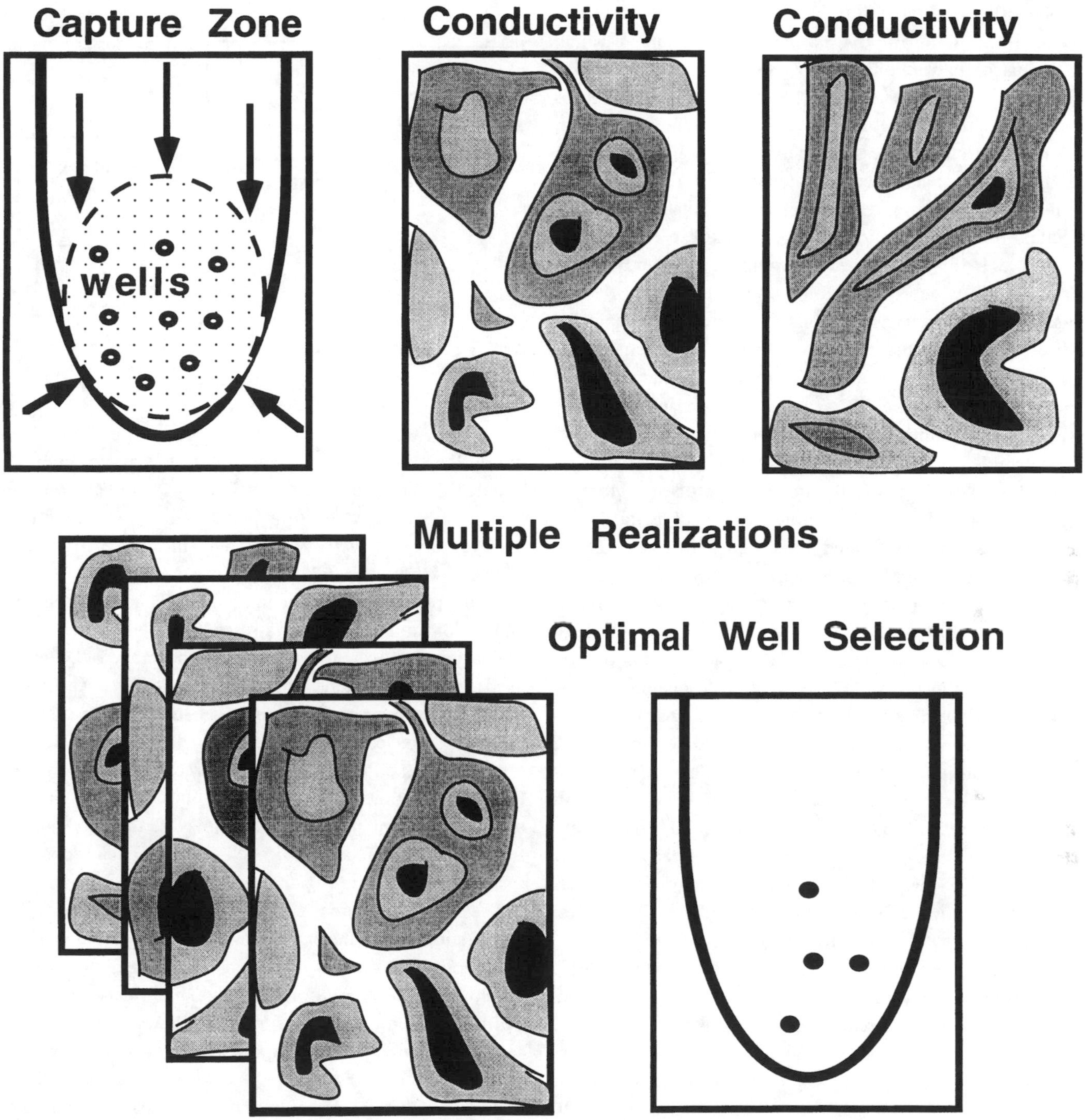

Fig. 4 Multiple realizations of hydraulic conductivity and the stacking process which results in an optimal well selection.

overdesigned that the pumping rates associated with it are sufficient to capture the contaminant in every realization. However, in general, there may be no single realization that represents a worst case. Rather there may be parts of one hydraulic conductivity field that dictate pumping in one region and there may be parts of another hydraulic conductivity field that dictate pumping in another region. The solution may be based upon several realizations, and the optimal solution will meet each constraint in every part of every realization in the stack.

Some work has been carried out to determine the relationship between the number of realizations in the stack and the reliability of the solution. Imagine that there were only five realizations in the stack and a solution was found such that all of the constraints in those five realizations were met. The obvious question is: Given many realizations, 1000 perhaps, to which that (five realization) solution was applied, how many of the 1000 would be expected to be cleaned up? It is a clear problem that can be addressed with direct post-optimization Monte Carlo simulation. This is done by simply applying the optimal solution to 1000 simulated conductivity fields and counting the number of realizations in which the constraints were or were not violated. Such a study was conducted by Wagner & Gorelick (1989), and it was shown that with as few as 10 to 30 stacked constraint sets, design reliability or effectiveness was between 92% and 99%. Chan (1993) showed that, based on order statistics, the system reliability for a stack of n realizations is approximately $n/(n+1)$. That is, for nine realizations in the stack the expected reliability is 90%. He also developed an argument based on Bayesian principles in which the reliability is $(n+1)/(n+2)$. In either case, his nonparametric relations show good agreement with Monte Carlo results, especially when $n>10$. Chan (1993) conducted extensive sensitivity analyses to spatial statistical model parameters, such as variance and correlation length of transmissivity, and to the type of covariance structure. Results indicated only modest sensitivity.

In a similar view of the stochastic problem, Morgan, Eheart & Valocchi (1993) generated a series of spatially variable realizations and aimed to design a reliable remedial containment system. In their approach they include an indicator that flags constraint sets in which constraints are violated. They define full reliability as a solution for which all constraint sets are satisfied, and zero reliability for a solution where all the constraint sets are violated. A solution that violates 50% of the constraint sets would be 50% reliable. In this manner they were able to construct a curve showing the minimal pumping rate versus reliability. Chan (1994) followed a similar approach in which the number of constraint violations is related to the system reliability. Identifying constraining realizations was the subject of Ranjithan, Eheart & Garrett (1993). They used artificial neural networks to determine the relationship between pumping for contaminant capture and patterns of transmissivity. Once the neural network screened out the critical realizations, the multiple realization formulation as shown previously could be used, but now with far fewer realizations. They were able to discard 75% of the realizations as inconsequential (clearly nonbinding), and therefore show a savings in solution time for the multiple realization problem. Wagner, Shamir & Nemati (1992) used a multiple-realization approach to target remedial strategies given an uncertain conductivity field. Their formulation differs from others in that it permits constraint violations on plume escape, but induces a penalty in the objective function for such violations known as a 'recourse cost'.

7 FEEDBACK CONTROL SYSTEMS

Another line of research has adopted the notion that uncertainty can be reduced during remediation if data gathered during the early stages of aquifer management are used to update the management scheme. This can be done in one of two ways: by simply adjusting pumping or by re-estimating the hydraulic conductivity field, reducing the level of predictive uncertainty, and then redetermining the optimal pumping plan. Typically, the initial field is based on a small amount of data and minimal stresses to the system. The preliminary hydraulic conductivity field and range of unmanaged flow stresses may be estimated, but the uncertainty bounds will be large. Updating procedures can be used to obtain a better estimate of the parameter field, reduce the uncertainty, and fine tune the management scheme. Some of the methods that employ feedback and updating of the simulation model also attempt to optimize the data collection network. A procedural flow chart for a feedback control system is displayed in Fig. 5.

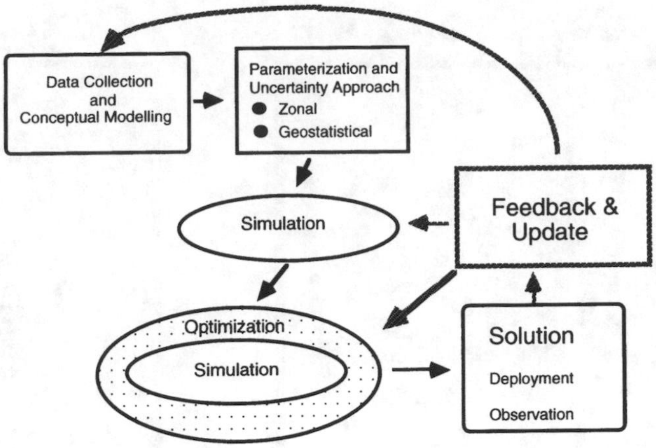

Fig. 5 Flow chart of feedback control in simulation-optimization problems.

The notion of feedback implies that there is opportunity to re-estimate parameters based on data collected during operation of an aquifer management system. Jones (1992) developed an optimization procedure in which the zonal hydraulic conductivities in the simulation model were sequentially updated based upon head data collected at the end of each pumping stage. Lee and Kitanidis (1991) employed a dynamic programming approach to couple the parameter estimation and optimization-management problems. The idea is to find the minimum-cost pumping scheme and to estimate the unknown transmissivities. They approached the problem as one of dual control. They recognized that the data collection effort can be tailored to provide the greatest amount of information to reduce costs of the pumping scheme, and, at the same time, that the managed pumping rates and well locations can be adjusted to provide the most information from future data collection. For example, high pumping rates at the initial stage can be established to stress and probe the system in order to achieve a better initial estimate of the transmissivities.

The problem of combining optimal data collection and minimum-cost aquifer remediation was the topic of some recent studies. Tucciarelli & Pinder (1991) adopted the philosophy that one has a budget that must accommodate both data collection and remediation costs. If additional data cannot reduce the remediation costs, then those data are not worthwhile. In their work, they make the assumption that additional data will reduce uncertainty in the transmissivity field and thereby reduce remediation costs. This is not always the case because new data may in fact find a previously unidentified flow path (high conductivity channel), and therefore the remedial system may become more expensive. Andricevic & Kitanidis (1990) and Andricevic (1993) considered minimizing costs while minimizing uncertainty in model predictions by developing a sequential sampling network of head measurements. They showed that the optimal sampling and management objective is not necessarily one that reduces predictive uncertainty of hydraulic heads. The optimal sequencing of new observation wells that minimizes the expected cost of remediation plus sampling was discussed by James & Gorelick (1994). They employed a Bayesian data worth framework in which a plume must be located and hydraulically contained in a cost-effective manner. Sources of uncertainty are aquifer heterogeneity, source uncertainty, and source loading history. One important notion is that reducing uncertainty in hydraulic conductivity, while pleasing, may be almost irrelevant to reducing the cost of remediation. The issue is whether or not additional monitoring or test data, when combined with existing data, result in reduced remedial costs. The work by James & Gorelick builds upon the pioneering work of Freeze *et al.* (1992), who applied decision analysis to select the best alternative strategy for site investigation from among several predetermined alternatives.

The philosophy of employing a best design and then updating the managed system to overcome constraint violations was the topic of Whiffen & Shoemaker (1993). Given estimates of model parameters, the optimal strategy is determined but constraints are treated as targets without fixed bounds. That is, the constraints are written into the objective function so that deviation of desired concentrations and heads from target values are minimized. They suggest that the initial solution be deployed in the field and then actual constraint violations are noted. Using feedback laws, which reweight the pumping and penalty portions of the objective function, they determine an updated pumping scheme. No attempt is made to re-estimate the uncertain model parameters. Rather, the idea is to adjust pumping rates to compensate for constraint violations. Even simple measures, such as adjusting pumping during remediation, can result in significantly lower remediation costs (Culver & Shoemaker, 1992).

8 ALTERNATIVE CONCEPTUAL MODELS

All of the studies cited above assume that the simulation model is indeed the correct conceptual model. If this assumption is incorrect, then it is unlikely that lumping this major source of uncertainty into existing model parameters, like transmissivity and recharge, will be sufficient to result in an optimal or even feasible management strategy. The problem of alternative conceptual models remains an open area in the literature. One example of optimization under alternative conceptual models appears in Haggerty & Gorelick (1994). This study considered the remediation of multiple contaminant plumes subject to rate-limited transfer. In one model of the system, local chemical equilibrium was assumed to exist for each of the sorbing solutes. In an alternative model, rate-limited mass transfer was assumed to exist such that slow desorption occurs and contaminants remain relatively immobile. For a classic down-gradient contaminant capture scheme, the optimization model identified a solution with the minimum pumping rate when the local equilibrium model was used. However, when the kinetic model was used, with a slow first-order mass transfer rate, cleanup within a 15-year horizon was infeasible.

There are many processes that control contaminant migration that may be significant, and yet they are not reflected in our simulation models. At the core of simulation-optimization models is the simulation model. If it is incorrect there is little point in attempting to optimize a strategy for aquifer management. If one can identify alternative models, then stochastic simulation-optimization models can incorporate them. For example, the multiple realization approach can include any number of models for flow, $f(\mathbf{q})$, and solute transport, $g(\mathbf{q})$, as well as any underlying sets of model parameters. The procedure will identify an optimal solution that satisfies the constraints for all alternative models. Another useful option is to employ decision analysis to identify the most cost-effective alternative, if one can clearly identify different alternative conceptual models.

9 FUTURE DIRECTIONS

The combined use of simulation and optimization methods for groundwater management under uncertainty provides an important tool for hydrologists, engineers, and water managers. In order to compensate for predictive model uncertainty it is necessary to employ management schemes that are overdesigned and therefore rather immune to our inadequate knowledge of model parameter values.

Conceptual model uncertainty is perhaps the greatest form of doubt about our predictive capability and therefore the one that leads to the most significant concern in aquifer management. In many cases simple linear simulation models for groundwater flow have distinct advantages over complex simulation models involving both flow and transport. For aquifer remediation, the consequence of making a mistake is so severe that overdesign is essential. Much work has been done that relies upon processes, such as dispersion and retardation, that are difficult to quantify with the sparse data typical of hydrogeologic investigations. In such cases, management models that guarantee plume capture rather than achieving a particular target concentration should be used. Given that continuous spatial variability has such a strong influence on solute migration, the zonal approach to parameter estimation and simulation-optimization should only be used in very homogeneous environments where a nonzero concentration target exists.

There are several techniques that have the capability of determining proper overdesign in aquifer management problems. The simplest is the multiple realization approach. By generating a stack of realizations whose properties and geometries span the true system, and by guaranteeing that the optimal pumping scheme is sufficient to manage every realization, the engineer can produce pumping and recharge policies that are robust and that have a high probability of success. To date, the multiple realization approach has been developed for use with geostatistically generated random fields. This is not a requirement of the method. Any set of maps may be included as members of the stack. If there is controversy about the use of a particular geostatistical model or range of parameter values, then realizations using additional models of spatial variability and parameter values can be included through the use of supplementary maps that are added to the stack. The small price paid is the generation of the realizations and solving a simulation-optimization problem with a larger stack.

The philosophy of feedback and updating is clearly valuable. However, some of the methods used to accomplish this are very complicated, and their use is often accompanied by a temptation to ignore complexities in aquifer heterogeneity, transport, mixing, and chemical reactions. For aquifer contamination problems, when beginning an investigation of a new site, the monitoring network should be thought through in light of the potential remedial system. In some cases the investigator loses sight of the fact that the idea of site characterization is not to delineate all features of the site. For example, it may be that identifying a plume peak concentration is irrelevant to the design of a contaminant capture system because all of the contaminant will end up at the pumping/treatment well anyway.

Perhaps the most important feature of the simulation-optimization approach is that it forces one to come to terms with several key goals of aquifer management, for any particular application, and the role of uncertainty in management strategy development. One must formalize the constraints and cope with those that are contradictory. One must ask how much risk one is willing to assume that some of the constraints may not be met because of model uncertainty. Alternatively, one must identify the additional value of guaranteeing that a management scheme will work. Overdesign is a tool used to overcome uncertainty, and is a form of potentially costly insurance. In aquifer management, uncertainty breeds risk, and risk mitigation breeds overdesign.

ACKNOWLEDGMENTS

The author would like to thank Dr Brian Wagner of the USGS for providing a preprint of his fine review of research published in simulation-optimization modeling during the past four years. In addition, the author gratefully acknowledges the support of NSF grant BCS-8957186 and the Hewlett Packard company.

REFERENCES

Ahlfeld, D. P., Mulvey, J. M., Pinder, G. F. & Wood, E. F. (1988). Contaminated groundwater remediation design using simulation, optimization, and sensitivity theory, 1. Model development. *Water Resources Research*, 24(3), 431–441.

Andricevic, R. (1993). Coupled withdrawal and sampling designs for groundwater supply models. *Water Resources Research*, 29(1), 5–16.

Andricevic, R. & Kitanidis, P. K. (1990). Optimization of the pumping schedule in aquifer remediation under uncertainty. *Water Resources Research*, 26(5), 875–885.

Chan, N. (1993). Robustness of the multiple realization method for stochastic hydraulic aquifer management. *Water Resources Research*, 29(9), 3159–3167.

Chan, N. (1994). Partial infeasibility method for chance-constrained aquifer management. *ASCE Journal of Water Resources Planning and Management*, 120(1), 70–89.

Charnes, A. & Cooper, W. W. (1962). Chance-constraints and normal deviates. *Journal of American Statistics Association*, 57, 134–148.

Cooley, R. L. (1979). A method of estimating parameters and assessing reliability for models of steady state groundwater flow, 2. Application of statistical analysis. *Water Resources Research*, 15(3), 603–617.

Culver, T. B. & Shoemaker, C. A. (1992). Dynamic optimal control for groundwater remediation with flexible management periods. *Water Resources Research*, 28(3), 629–641.

Dettinger, M. D. & Wilson, J. L. (1981). First order analysis of uncertainty in numerical models of groundwater flow, Part 1. Mathematical development. *Water Resources Research*, 17(1), 149–161.

Freeze, R. A., James, B., Massmann, J., Sperling, T. & Smith, L. (1992). Hydrogeological decision analysis, 4: The concept of data worth and its use in the development of site investigation strategies. *Groundwater*, 30(4), 574–588.

Gailey, R. M. & Gorelick, S. M. (1993). Design of optimal, reliable plume capture schemes: Application to the Gloucester Landfill groundwater contamination problem. *Groundwater*, 31(1), 107–114.

Gorelick, S. M. (1983). A review of distributed parameter groundwater management modeling methods. *Water Resources Research*, 19(2), 305–319.

Gorelick, S. M. (1987). Sensitivity analysis of optimal groundwater contaminant capture curves: spatial variability and robust solutions. In *Proceedings of the Conference Solving Groundwater Problems with Models.* Denver, Colorado: NWWA, pp. 133–146.

Gorelick, S. M. (1990). Large scale nonlinear deterministic and stochastic optimization: Formulations involving simulation of subsurface contamination. *Mathematical Programming*, 48, 19–39.
Gorelick, S. M., Freeze, R. A., Donohue, D. & Keely, J. F. (1993). *Groundwater Contamination: Optimal Capture and Containment.* Ann Arbor, Michigan: Lewis Publishers.
Haggerty, R. & Gorelick, S. M. (1994). Design of multiple contaminant remediation: Sensitivity to rate-limited mass transfer. *Water Resources Research*, 30(2), 435–446.
James, B. R. & Gorelick, S. M. (1994). When enough is enough: The worth of monitoring data in aquifer remediation design. *Water Resources Research*, 30(12), 3499–3513.
Jones, L. (1989). Some results comparing Monte Carlo simulation and first order Taylor series approximation for steady groundwater flow. *Stochastic Hydrology Hydraulics*, 3, 179–190.
Jones, L. (1990). Explicit Monte Carlo simulation head moment estimates for stochastic confined groundwater flow. *Water Resources Research*, 26(6), 1145–1153.
Jones, L. (1992). Adaptive control of groundwater hydraulics. *ASCE Journal of Water Resources Planning and Management*, 118(1), 1–17.
Lee, S-L. & Kitanidis, P. K. (1991). Optimal estimation and scheduling in aquifer remediation with incomplete information. *Water Resources Research,* 27(9), 2203–2217.
Morgan, D. R., Eheart, J. W. & Valocchi, A. J. (1993). Aquifer remediation design under uncertainty using a new chance constrained programming technique. *Water Resources Research,* 29(3), 551–561.
Ranjithan, S., Eheart, J. W. & Garrett, J. H. Jr (1993). Neural network-based screening for groundwater reclamation under uncertainty. *Water Resources Research*, 29(3), 563–574.
Tiedeman, C. &. Gorelick, S. M. (1993). Analysis of uncertainty in optimal contaminant capture design. *Water Resources Research*, 29(7), 2139–2153.
Tucciarelli, T. & Pinder, G. (1991). Optimal data acquisition strategy for the development of a transport model for groundwater remediation. *Water Resources Research*, 27(4), 577–588.
Wagner, B. J. (1995). Recent advances in simulation-optimization groundwater management modeling. In *U.S. National Report to the IUGG, 1991–1994, Reviews of Geophysics,* vol. 33, ed. R. Vogel, pp. 1021–1028.
Wagner, B. J. & Gorelick, S. M. (1987). Optimal groundwater quality management under parameter uncertainty. *Water Resources Research*, 23(7), 1162–1174.
Wagner, B. J. & Gorelick, S. M. (1989). Reliable aquifer remediation in the presence of spatially variable hydraulic conductivity: From data to design. *Water Resources Research*, 25(10), 2211–2225.
Wagner, J. M., Shamir, U. & Nemati, H. R. (1992). Groundwater quality management under uncertainty: Stochastic programming approaches and the value of information. *Water Resources Research*, 28(5), 1233–1246.
Whiffen, G. J. & Shoemaker, C. A. (1993). Nonlinear weighted feedback control of groundwater remediation under uncertainty. *Water Resources Research*, 29(9), 3277–3289.

IV

Transport in heterogeneous aquifers

1 Transport of inert solutes by groundwater: recent developments and current issues

YORAM RUBIN
University of California at Berkeley

ABSTRACT Some recent developments in stochastic modeling of transport of inert solutes are discussed. Rather than tabulating all that has been done, we focus on some central issues. We explore the attempts to depart from the limited model for dispersion of the mean concentrations toward a complete stochastic description, and continue by evaluating the means for narrowing the ensemble of all physically plausible realizations toward the one which will eventually become a reality. We discuss some scaling issues: starting from the relationship between the scale of the plume and the scale of the heterogeneity, and ending by analyzing the behavior of finite plumes in domains with evolving scales of heterogeneity. We conclude by exploring the issue of the concentration variability. The limitations of the two-moment characterization are evaluated, and we present a method for computing the entire concentration probability distribution function (pdf) as well as some recent ideas on how to use tracer data for Bayesian updating of the concentration moments.

1 INTRODUCTION

Application of the stochastic paradigm to the problems of contaminant transport in porous media has been a major center of research activity over the last few years. This surge of activity reflects the growing recognition of this concept as a viable problem solving tool, as well as the increase in the efforts to solve environmental problems.

It is not the goal of this chapter to take stock of all that has been achieved in recent years. The alternative pursued here is to identify and analyze some central areas of activity within the discipline with a view toward recording problems that were resolved or became better understood, as well as others that still await resolution. Some issues are omitted to prevent overlap with the other contributions in this book, and will be mentioned only briefly.

In order to gain a perspective on recent developments, we refer to the study of Sposito, Jury & Gupta (1986) for a list of fundamental problems at that time, which will be used as a benchmark. That study focused on the stochastic convection-dispersion equation (CDE) of solute transport, which was considered then to be the main vehicle for stochastic analysis, and listed the following as problems or deficiencies:

Transport is modeled through an equation for the mean concentration which lacks the means to describe the properties of the ensemble of realizations. Furthermore, the mean concentration is not accessible to experimental measurements.

The main physical interest is not in the ensemble behavior unless it is relevant to prediction for a single aquifer, yet very few rigorous results are available for predicting the single realization out of the ensemble, which would be useful for site-specific analysis.

No rigorous definition is available of the classes of spatially variable coefficients which allow the modeling of transport as a Fickian process. Related issues: the conditions which lead to Fickian processes, the validity of the Fickian approximation in a single aquifer, and an explanation for the apparent increase of the dispersion coefficients with the overall dimension of the region through which solute transport occurs, coined by Sposito *et al.* (1986) as the 'scale effect'.

Sposito *et al.* (1986) continued to identify several promising areas of research: they suggested the concept of 'evolving spatial heterogeneities' as a potential explanation of the scale effect, and they pointed to the statistical procedure of conditioning as a possible bridge between the ensemble mean statistics and the need to predict transport in a single realization. They also pointed out the need for experimental verification of the assumptions of homogeneity and ergodicity of the random hydraulic conductivity functions, which are cornerstones of the stochastic approach to solute transport.

In order to assess the progress that has been made regarding the questions raised by Sposito *et al.* (1986), this chapter is built as follows: Section 2 presents and discusses the Eulerian and Lagrangian approaches to stochastic solute transport; Section 3 discusses the problems associated with expressing the Lagrangian analysis in Eulerian coordinates, and in doing that explores the transition from the analysis of the ensemble to the prediction of the single realization; Section 4 discusses the evolution of non-ergodic plumes and its relation to the scale of the plume and the scale of heterogeneity, and ventures into the area of transport in formations of evolving scales of heterogeneity, with the goal of exploring the 'scale effect'; Section 5 is concerned with the variability of the concentration and the concentration's variance. In our summary, the developments of the various sections will be contrasted with the issues raised by Sposito *et al.* (1986).

2 ON THE EULERIAN AND LAGRANGIAN PICTURES

Work in the area of transport has been carried out in recent years using both Eulerian and Lagrangian frameworks. It is proper to introduce these frameworks as the starting point for a discussion on recent developments.

2.1 The Eulerian picture

The Eulerian approach concerns the solution of the stochastic partial differential equation for the random concentration C:

$$\frac{\partial C}{\partial t}+\mathbf{U}\cdot\nabla C=-\mathbf{u}\cdot\nabla C+\nabla\cdot(\underline{\underline{D}}_m\nabla C) \qquad (1)$$

where $\mathbf{V}(\mathbf{x})=\mathbf{U}+\mathbf{u}(\mathbf{x})$ is the velocity, $\mathbf{U}=\langle\mathbf{V}\rangle$ and $\mathbf{u}$ is the local fluctuation. Here and subsequently, boldfaced variables are vectors. Angle brackets denote the expected value operator; $\underline{\underline{D}}_m$ is the pore scale dispersion tensor. Solutions to (1) have been sought either in the form of Monte-Carlo simulations, where the entire pdf of C can be derived, or otherwise through analytic solutions leading to the low-order moments of C. In both cases one seeks to relate C to the variability of the hydraulic conductivity and to other physical parameters controlling flow.

To derive the moments of C, we write $C(\mathbf{x}, t)=\langle C(\mathbf{x}, t)\rangle+c(\mathbf{x}, t)$. Upon substitution in (1), an equation for the ensemble mean concentration is obtained:

$$\frac{\partial\langle C\rangle}{\partial t}+U_i\frac{\partial\langle C\rangle}{\partial x_i}=-\frac{\langle\partial u_i c\rangle}{\partial x_i}+\underline{\underline{D}}_m\nabla^2\langle C\rangle \qquad (2)$$

and for c:

$$\frac{\partial c}{\partial t}+U_i\frac{\partial c}{\partial x_i}=-u_i\frac{\partial\langle C\rangle}{\partial x_i}-\frac{\partial}{\partial x_i}(u_i c)+\frac{\partial}{\partial x_i}\langle u_i c\rangle+\underline{\underline{D}}_m\nabla^2 c \qquad (3)$$

In (2) and subsequently, the rule of repeated indices applies, $i=1,\ldots,m$, where m denotes the number of space dimensions.

Equations (2) and (3) are the basis for the Eulerian approach to transport. Equations (3) can be used to obtain expressions for $\langle u_i c\rangle$ (2), as well as for the higher-order moments of C. For example, multiplying (3) by c and taking the expected value, we get an equation for the concentration variance σ_c^2:

$$\frac{\partial\sigma_c^2}{\partial t}+U_i\frac{\partial\sigma_c^2}{\partial x_i}=-\frac{\partial}{\partial x_\beta}\left\{\langle c^2u_\beta\rangle-2D_m\frac{\partial^2\sigma_c^2}{\partial x_\beta}\right\}-2\langle c\,u_\beta\rangle\frac{\partial\langle C\rangle}{\partial x_\beta}$$
$$-2D_m\left\langle\left(\frac{\partial c}{\partial x_\beta}\right)^2\right\rangle \qquad (4)$$

The first two terms on the r.h.s. of (4) are written as a divergence, and consequently can be taken as representing the diffusion of the concentration fluctuations in space through non-linear effects and as a consequence of pore-scale dispersion. The third term, which expresses the interaction between the flux $\langle cu_\beta\rangle$ and the mean concentration gradient, is the generation term for the concentration fluctuations. The last term represents the irreversible destruction of the C fluctuations by pore-scale dispersion.

Similar to the way in which (4) was derived, one can derive an equation for the flux term $\langle u_i c\rangle$ of (2) by multiplying (3) by u_i and averaging. This task is often carried out by assuming that transport is dominated by the gradient of the mean concentration:

$$-\langle c\mathbf{u}\rangle=\underline{\underline{D}}\nabla\langle c\rangle \qquad (5)$$

where $\underline{\underline{D}}$ is the macrodispersion tensor, and one then seeks an expression for $\underline{\underline{D}}$. The major thrust of the analytical Eulerian methods has been along the lines of (5). Such an approach was adopted in the context of groundwater by Gelhar & Axness (1983). Assuming $\nabla\langle C\rangle$ to be constant, and taking $\partial C/\partial t=0$, these authors derived $\underline{\underline{D}}$ for small variance of the logconductivity. $\underline{\underline{D}}$ should then be viewed as an asymptotic large-time limit, applicable for domains of small variability and for plumes of large spatial extent.

Given the non-local nature of the transport problem, the adoption of (5), although justified for pragmatic reasons, may appear to be counter-intuitive. However, Lagrangian analysis shows that transport becomes Fickian at large travel times. For small variance in the logconductivity, particle displacements are also Gaussian at early travel times; hence (5) becomes valid also at pre-asymptotic times provided that $\underline{\underline{D}}$ depends initially on time. Attempts to derive a time-dependent $\underline{\underline{D}}$ using an Eulerian framework have been reported, but this is a formidable task that requires some simplifying assumptions.

Koch & Brady (1988), Graham & McLaughlin (1989a, b) and Naff (1990) have attempted to obtain a time dependent solution to (3) and (5) by assuming that the second and third terms on the r.h.s. of (3) can be dropped out based on order relationships. This assumption leads to the cancellation of the term $\langle c^2 u_\beta \rangle$ on the r.h.s. on (4), and hence to simplification of the solution. Dagan & Neuman (1991) showed that the dropped term is actually of the same order as the retained terms at early to medium-range travel times. Given the nature of the divergence, this approach is expected to lead to an underestimation of high-order spatial moments. Dagan & Neuman (1991) showed that in the case of a finite velocity integral scale, this approach leads to fourth-order spatial moments that are correct only at large time. They, and, later, Neuman (1993), further showed that application of the solution which is based on the truncated equation to the case of slug injection leads to a prediction that the mean concentration bifurcates at early travel times, which is not in line with exact solutions developed in the field of turbulent diffusion. A solution for (5) that avoids the truncation of (4) was reported recently by Deng, Cushman & Delluer (1993).

Equation (5) extends the concept of mixing length to scalar diffusion. Similar attempts in the field of turbulent diffusion have generally ended with indeterminate results (McComb, 1990). Corrsin (1973) pointed out that attempts to model turbulent diffusion by analogies with gas kinetic theory fail to meet many of the underlying restrictions imposed upon the latter. The important requirement that mean concentrations should vary by only a negligible amount over the equivalent of one mean free path, i.e., the Lagrangian integral length scale, are difficult to satisfy in most turbulent flows, and we may add that the same is true in most cases of flow in heterogeneous media.

The introduction of a macrodispersion tensor leads to an equation identical to the equation of molecular diffusion and hence to a fundamental question regarding scale. When dealing with molecular diffusion, the scale of heterogeneity is the molecular mean free path. This length scale, the number of molecules moving in space, the nature of their motion in space and the scale of the sampling devices allow the modeling of single realizations using an ensemble mean equation such as (2). $\underline{\underline{D}}$ of (5) models heterogeneities at a different scale, and in situations that are not analogous to the case of molecular diffusion. This has an immediate bearing on the interpretation of field studies: a point measurement of C cannot be used for inference of $\underline{\underline{D}}$. Similarly, the actual concentration in the field cannot be modeled through its ensemble mean (Sposito *et al.*, 1986). Putting aside the goal of modeling concentration using $\underline{\underline{D}}$, we are left with the less ambitious, but perhaps more realistic, goal of modeling spatial moments of the mean concentration. Indeed, a major success of the stochastic methods is in dealing with spatial moments.

A fundamental solution of the molecular diffusion equation is that the spatial moments grow linearly in time. It is important then to understand when the Fickian regime is obtained which permits one to use asymptotic $\underline{\underline{D}}$ as a parameter for growth of spatial moments. If we are dealing with a cloud of particles originating from a point, Taylor's (1921) analysis shows that the travel time elapsed must be larger than the Lagrangian time scale. In many cases, however, the time of origin of the cloud from the point source may not be known. It is useful then to define the applicability of (2) in terms of the size of the cloud of particles. Simple analysis from the field of turbulent diffusion (Fischer *et al.*, 1975) shows that the cloud size needs to be substantially larger than the length scale over which fluid velocities are correlated along the mean flow direction. Using a covariance model such as that developed by Rubin (1990), this length scale should be of the order of a distance of at least ten integral scales of the logconductivity. How much larger? For practical reasons, this subject needs a firm resolution, and will be addressed in Section 4, using the Lagrangian framework.

2.2 The Lagrangian picture

A resolution of the issues of time and scale involved with the application of (1) was obtained using Taylor's (1921) Lagrangian framework, which was outlined first by Dagan (1982) in the context of transport in porous media. The fundamental concept in the Lagrangian approach is that of a fluid particle. This particle is modeled as an individual entity. In order to be modeled using Darcy's law, this particle needs to be slightly larger than the pore scale, and in order to capture the effects of heterogeneity, it should be much smaller than the integral scale of the conductivity. In the Lagrangian terminology, $\mathbf{X}(t;\mathbf{x}_0, t_0)$ denotes the position at time t of a particle that is at $\mathbf{x}_0$ at time t_0 (i.e., $\mathbf{X}(t_0;\mathbf{x}_0, t_0)=\mathbf{x}_0$). For each Eulerian variable, for example the velocity $\mathbf{V}(\mathbf{x}, t)$, the corresponding Lagrangian variable is defined in conjunction with its trajectory, e.g. $\mathbf{V}(t;\mathbf{x}_0)=\mathbf{V}[\mathbf{X}(t;\mathbf{x}_0, t_0)]$, and, additionally $\mathbf{V}(t_0)=\mathbf{V}_0$.

The Lagrangian approach investigates the concentration field associated with the particle of mass M

$$C(\mathbf{x}, t)=\frac{M}{n}\delta(\mathbf{x}-\mathbf{X}) \tag{6}$$

where n denotes the porosity and δ is the Dirac delta. $\mathbf{X}(t;\mathbf{x}_0, t_0)$ is expressed in terms of the velocity field as:

$$\mathbf{X}(t;\mathbf{x}_0,t_0)=\int_{t_0}^{t}\mathbf{V}[\mathbf{X}(t';\mathbf{x}_0,t_0)]dt' \qquad (7)$$

Hence a random function model for the velocity field and subsequently for **X** are prerequisites for the solution of (6). In the stochastic approach, the primary interest is in the Lagrangian pdf $f_L(\mathbf{V}, \mathbf{X}; t|\mathbf{V}_0=\mathbf{v}_0, \mathbf{x}_0)$ since it allows the computation of the various moments of **X**, using (7). Note that here **x** denotes the sample space corresponding to $\mathbf{X}(t; \mathbf{x}_0)$, while in the Eulerian approach it is just a parameter. A comprehensive review of results based on the coupling of (6) and (7) is given in Dagan (1987). Most significant there is the derivation of time-dependent, pre-asymptotic macrodispersion coefficients.

A direct numerical implementation of the Lagrangian concept is to release a large number of particles at the source ($\mathbf{X}(t_0)=\mathbf{x}_0$) with initial velocities $\mathbf{V}(t_0)$ distributed according to the Eulerian pdf $f(\mathbf{V}; \mathbf{x}_0, t_0)$. The stochastic model equations are integrated to obtain $\mathbf{V}(t)$ and $\mathbf{X}(t)$, and the expected particle number density is determined and used in conjunction with (6) to determine the concentration field.

An alternative is to derive both Lagrangian and Eulerian pdf's, which will then be used to derive the various moments of **X** using (7). In general, this can be achieved through the relationship (Pope, 1994):

$$f(\mathbf{V};\mathbf{x},t)=\int\int f(\mathbf{V}_0;\mathbf{x}_0,t_0)f_L(\mathbf{V};\mathbf{x},t|\mathbf{V}_0;\mathbf{x}_0)d\mathbf{V}_0 d\mathbf{x}_0 \qquad (8)$$

where integration is over all velocities and over the entire flow domain at t_0. The Lagrangian pdf f_L is thus the transition density for the Eulerian velocity, since it determines the transition to the Eulerian pdf $f(\mathbf{V})$ from t_0 to t. Since f_L determines f, it also determines simple Eulerian means, such as the mean and variance of the velocity, which can be used as coefficients in stochastic Eulerian models.

General solutions for f and f_L are not available. However, some significant results were obtained for situations where two-moment characterization is statistically exhaustive, and f_L is determined in terms of hydrogeological parameters. These results, which are in line with the ideas outlined in Dagan (1984), are limited to small variance in the logconductivity, $\sigma_y^2<1$. In this case, $f=f_L$ when considering terms of order up to σ_y^2. It is important to note that the requirement of $\sigma_y^2<1$ does not imply that the logconductivity Y needs to be Gaussian in order to derive the moment of **X** (7).

Simple, closed form analytical expressions for the velocity covariances have been obtained for two-dimensional (Rubin, 1990) and for isotropic three-dimensional flows (Zhang & Neuman, 1992). Numerical quadrature is needed for three-dimensional, anisotropic domains (Rubin & Dagan, 1992; Zhang & Neuman, 1992). A faster algorithm was recently reported by Ezzedine (1997). The above results were obtained using perturbation techniques, and the terms retained are of the order of σ_y^2, hence limiting their applicability to formations of small variances of the logconductivity. It is further assumed that the conductivity field is stationary and that the mean head gradient is constant. More recent works departed from this framework. Rubin & Bellin (1994) derived velocity covariances for a uniformly recharged field. In Rubin & Seong (1994), the velocity covariances were derived for non-stationary logconductivity fields, where the non-stationarity manifests itself as a linear trend in the mean logconductivity.

While Gaussianity of the displacements is a precondition for an application of (5) into (2), the assumption of Gaussianity is not inherent to the Lagrangian approach: the challenge here is to characterize exhaustively and parsimoniously the Lagrangian velocity field such that tracking particles will capture reliably the effects of heterogeneity. Numerical experiments by Bellin, Salandin & Rinaldo (1992) showed that the velocity covariances derived for $\sigma_y^2<1$ are fairly accurate even at large variabilities. This work shows that the deficiency of the low-order derivation is in the small separation distances of the covariances, and that at the medium to large separation distances the low-order derivations are quite sufficient. However, two-moment characterization of the velocity may not be sufficient for non-Gaussian Y unless some normal score transforms for the velocity can be developed, and more work is needed to determine the pdf's of the velocity at this range. Non-Gaussianity of Y is not a prerequisite of the linear theory of Dagan (1984) for computing spatial moments.

The Lagrangian approach proved successful for the reason that the derivation of the macrodispersion coefficients at the pre-asymptotic regime addresses a fundamental need in field applications, and additionally it provided the means to address theoretical problems, such as the 'scale effect'. From a mathematical point of view, the success stemmed from the approximation $f=f_L$, which is justified at $\sigma_y^2<1$: at this range, it is easy to see that approximating the Lagrangian trajectory by its expected value, which in turn can be expressed using the Eulerian mean velocity, leads to consistent results, and allows the replacement of the Lagrangian covariances by the Eulerian ones. This aspect of the Lagrangian approach is not without its pitfalls, and will be addressed further in the next section.

Time-dependent macrodispersion coefficients are not the modeler's best friends when it comes to applications. In some situations the initial travel time of the plume may not be known. Another problem is as follows: consider two plumes of the same species, released at different times, yet both occupying at a given time the same volume in space. Are these two plumes going to evolve differently?

The first question was dealt with by Dagan & Sposito (1994). They showed that any evolved state of the plume that occurs after the initial appearance of solute can be used as a reference state on which to base predictions, provided that spatial interdependence between the initial state and the velocity field is accounted for. The second question appears now to be resolved: if the two different plumes occupy the same volume in space, their subsequent evolutions will be identical.

2.3 An interim summary

The significant achievement of both approaches is in relating field-scale geological heterogeneity to macrodispersion by capitalizing on the correlation between the fluid velocity on the one hand, and hydraulic conductivity and hydraulic head on the other. The main products of both methods have been the macrodispersion coefficients, or otherwise an expression for the spatial moments. It is emphasized that the macrodispersion coefficients are not equivalent to those used for molecular diffusion. For once, molecular diffusion is a monotonous growth process, and the concentration in the single realization, at the scale of the measurement, is practically identical to its expected value, while concentrations and spatial moments of actual plumes in the ground can differ significantly from their expected values (see Smith & Schwartz, 1980; Quinodoz & Valocchi, 1990). Additionally, the spatial moments, in the case of molecular diffusion, measure space which is 'filled' with the tracer, and the concentration is smoothly varying at the measurement scale. Macrodispersion coefficients, on the other hand, are not 'space-filling' measures. Consequently, as pointed out recently by Kitanidis (1994), they cannot be used to estimate dilution.

The pursuit of macrodispersion coefficients by both the Lagrangian and Eulerian techniques is more in line with tradition and much less a reflection of necessity. Recently, however, we have been able to see a departure from this tradition in order to be more in line with the true nature of the two concepts. The Eulerian concept, being 'static', is ideal for the control volume type of analysis, and hence for computing concentrations. The Lagrangian concept is focused on the displacement of solute particles, and is better suited for derivation of travel times and solute fluxes. There is a growing recognition of this distinction, and indeed in recent years the Lagrangian method is less focused on the derivation of macrodispersion coefficients (see Dagan & Nguyen, 1989; Desbarats, 1990; Dagan, Cvetkovic & Shapiro, 1992; Cushey and Rubin, (1997)).

3 ON THE PROBLEM OF EXPRESSING THE LAGRANGIAN ANALYSIS IN EULERIAN COORDINATES

The use of the Lagrangian approach allows the establishment of some simple and powerful results. For applications, however, we need information from an Eulerian coordinate system. The goal of relating the two coordinate systems for flow in random fields has not yet been achieved, at least not in a general sense. Several attempts to achieve this goal have been reported, and are reviewed here.

The Lagrangian approach requires the evaluation of $\langle \mathbf{u}(t_0)\mathbf{u}(t)\rangle$ (see Fischer *et al.*, 1975; McComb, 1990), where $\mathbf{u}(t)$ is the Lagrangian velocity fluctuation at time t, which is related to the Eulerian one by

$$\mathbf{u}(t)=\mathbf{u}[\mathbf{X}(t)] \tag{9}$$

From (9), the Lagrangian covariance becomes

$$\langle \mathbf{u}(t_0)\mathbf{u}(t)\rangle=\langle \mathbf{u}[\mathbf{x}(t_0)]\mathbf{u}[\mathbf{X}(t)]\rangle \tag{10}$$

where the average is taken over many particle paths $\mathbf{X}(t)$. The average on the l.h.s. of (10) is quite straightforward, but it is the one on the r.h.s. which is more difficult, because it involves two unaveraged random variables. This averaging cannot be carried out as though $\mathbf{u}[\mathbf{X}(t)]$ and $\mathbf{X}(t)$ are independent variables.

A more formal statement of the l.h.s. of (10) is (McComb, 1990, chap. 12):

$$\langle \mathbf{u}(t_0)\mathbf{u}(t)\rangle=\int\int \mathbf{u}(t_0)\mathbf{u}[\mathbf{X}(t)]f_L\{\mathbf{u}[\mathbf{X}(t)|\mathbf{X}(t)]\}f[\mathbf{X}(t)]d\mathbf{u}[\mathbf{X}(t)]d\mathbf{X}(t) \tag{11}$$

and the main difficulty here is that $f_L\{\mathbf{u}[\mathbf{X}(t)]\}$ and $f[\mathbf{X}(t)]$ are not independent of each other, as is clear if we recall that

$$\mathbf{X}(t)=\int_0^t \mathbf{V}[\mathbf{X}(\tau)]d\tau \tag{12}$$

The problem can be simplified if one supposes that the path followed by any particle is not a random variable, but is rather a prescribed path $\mathbf{x}^*$. In that case, $f[\mathbf{X}(t)]=\delta[\mathbf{X}(t)-\mathbf{x}^*(t)]$, and the integral (11) simplifies to

$$\langle \mathbf{u}(t_0)\mathbf{u}(t)\rangle=\int \mathbf{u}(t_0)\mathbf{u}[\mathbf{x}^*(t)]f_L[\mathbf{u}(\mathbf{x}^*(t))]d\mathbf{u}[\mathbf{x}^*(t)] \tag{13}$$

which contains only one unaveraged quantity.

Along these lines, Dagan (1984) suggested the use of $\mathbf{x}^*=\langle\mathbf{X}\rangle$, where $\langle\mathbf{X}\rangle$ is the unconditional mean displacement of the ensemble of trajectories. This choice is consistent with the first-order approach adopted in that study. An alternative suggested by Dagan (1984) and Rubin (1991a) is to use $\mathbf{x}^*=\langle\mathbf{X}^c\rangle$, where $\langle\mathbf{X}^c\rangle$ is the conditional mean trajectory of the ensemble of particle trajectories:

$$\langle\mathbf{X}^c\rangle=\langle\mathbf{X}|\{N\}\rangle \tag{14}$$

and $\{N\}$ is the set of data available for conditioning, such as measurements of conductivity, hydraulic head, and perhaps other types of data (see Hyndman, Harris & Gorelick, 1994). Using this approach, the Lagrangian covariance (11) becomes:

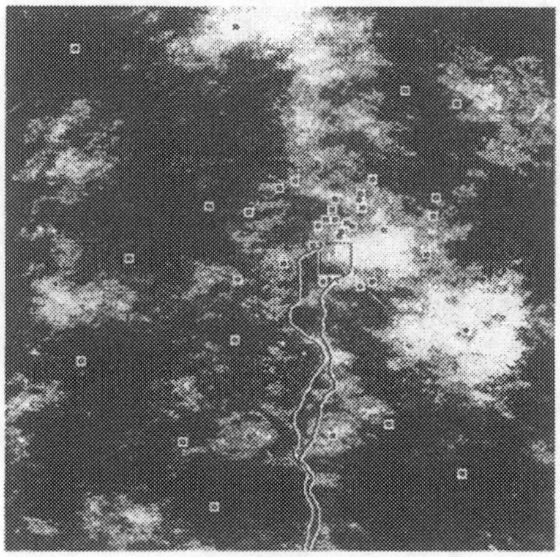

Test Problem 1 true log(T) field
(20km × 20km area)

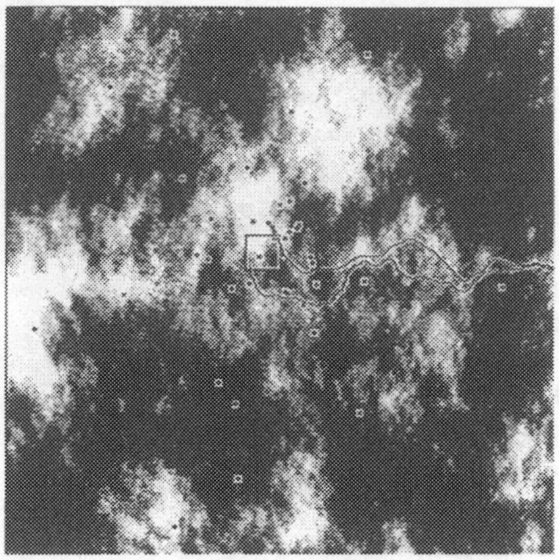

Test Problem 2 true log(T) field
(20km × 20km area)

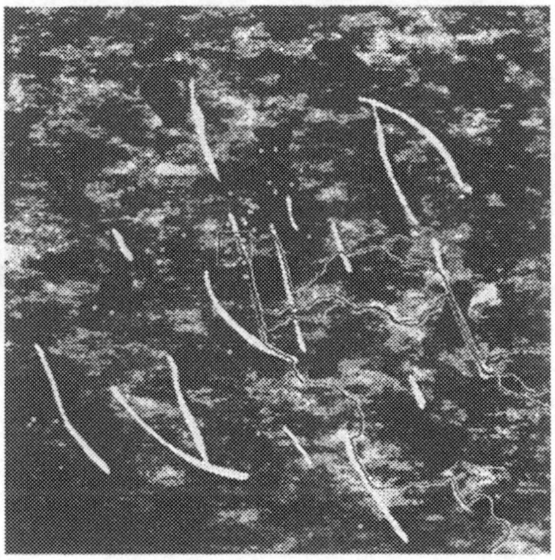

Test Problem 3 true log(T) field
(20km × 20km area)

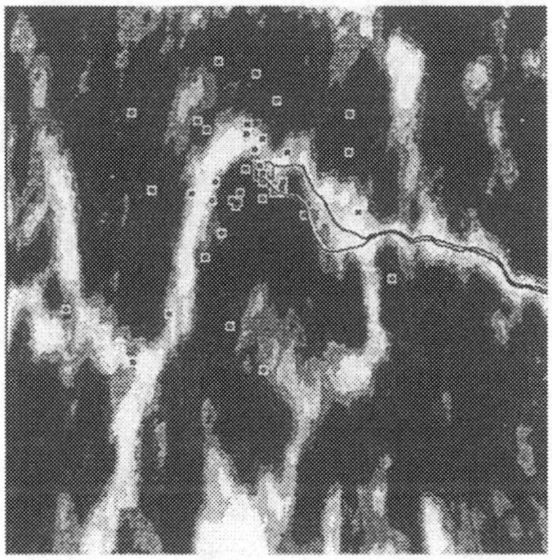

Test Problem 4 True log(T) field
(30km × 30km area)

Fig. 1 A gray-scale map showing a logconductivity map and particle trajectories used by Sandia National Laboratories to evaluate inverse and transport codes. The different methods are evaluated by their capability of reconstructing the trajectories and travel times when given only a limited amount of information.

$$\langle \mathbf{u}(t_0)\mathbf{u}(t)\rangle = \int \mathbf{u}(t_0)\mathbf{u}[\langle \mathbf{X}^c(t)\rangle] f_L^c\{\mathbf{u}[\langle \mathbf{X}^c(t)\rangle]\}\, d\mathbf{u}[\langle \mathbf{X}^c(t)\rangle] \quad (15)$$

where f_L^c is the conditional pdf. The merit of this option is that it limits the ensemble of $\mathbf{X}(t)$ to the more likely realizations given $\{N\}$.

The choice of either $\mathbf{x}^* = \langle \mathbf{X}\rangle$ or $\mathbf{x}^* = \langle \mathbf{X}^c\rangle$ should be dictated by the parameters of the transport problem. It is of consequence when dealing with the displacement of finite-size plumes over short distances. In all other situations, either the plume or the trajectory become ergodic, and the conditional means approach the unconditional ones. Rubin & Dagan (1992) showed in a series of figures how the conditional mean displacement approaches the unconditional one at large travel time: at such distances the ensemble mean of all local subensembles of velocities along the trajectory approaches the unconditional ensemble mean. A similar case can be made for the displacement of ergodic plumes, and this will be discussed in Section 4.

A recent investigation into the efficiency of using $\mathbf{x}^* = \langle \mathbf{X}^c\rangle$ as the alternative was conducted as part of the performance assessment of the proposed low level nuclear waste repository site WIPP in New Mexico. The investigation consisted of the following steps:

A heterogeneous conductivity field was generated and the hydraulic head was solved for given boundary conditions (see Fig. 1).

A series of imaginary particles was released from different locations, and the travel time to some distance from the source was computed (see Fig. 1).

A small number of head and conductivity measurements were then used to identify the hydraulic parameters of the aquifer and to predict travel time and trajectories in the form of cumulative distribution functions (CDFs).

The predicted statistics were then evaluated through a comparison with the information obtained by simulation based on the entire complete sets.

A sample result obtained using (15) and following the method outlined in Rubin (1991b) is given in Fig. 2. A preliminary report of this study is given in Zimmerman *et al.* (1997). This study concludes that the choice of $\mathbf{x}^*=\langle\mathbf{X}^c\rangle$ is unequivocally better than the unconditional alternative, and furthermore that (15) offers a very attractive method for dealing with heterogeneity and uncertainty.

A different approach to the integration of (11) is based on Corrsin's (1973) independence hypothesis, which was developed for turbulent flow. It consists in rewriting (11) as follows:

$$\langle\mathbf{u}(t_0)\mathbf{u}(t)\rangle=\int\langle\langle\mathbf{u}(t_0)\mathbf{u}(\mathbf{x},t)\delta[\mathbf{x}-\mathbf{X}(t)]\rangle\rangle d\mathbf{x} \quad (16)$$

where the double angled brackets denote the joint average over the velocity field and the particle paths. Corrsin's hypothesis proceeds by assuming that for large travel times the joint average (16) is separable into two independent ones:

$$\langle\mathbf{u}(t_0)\mathbf{u}(t)\rangle=\int\langle\mathbf{u}(t_0)\mathbf{u}(\mathbf{x},t)\rangle\langle\delta[\mathbf{x}-\mathbf{X}(t)]\rangle d\mathbf{x}$$
$$=\int\langle\mathbf{u}(t_0)\mathbf{u}(\mathbf{x},t)\rangle f[\mathbf{x}(t)]d\mathbf{x} \quad (17)$$

or, in other words, that $\mathbf{X}$ and $\mathbf{U}$ are independent. In an application to turbulent flow, Saffman (1963) assumed $\mathbf{X}$ to be Gaussian, and this assumption was later adopted in groundwater applications.

The main thrust in applying Corrsin's hypothesis to the problem of transport in aquifers can be found in the works of Dagan (1988), Neuman & Zhang (1990), Zhang & Neuman (1990), and Glimm *et al.* (1993). Application of Corrsin's hypothesis is meant to account for the non-linear terms arising from the deviation of $\mathbf{X}(t)$ from $\langle\mathbf{X}(t)\rangle$, which are neglected in Dagan's (1984) approach based on order relationship.

The main claim of the Corrsin-based works on the non-linear, i.e. large σ_Y^2, range is based on the non-linear relationship between the Lagrangian and Eulerian time scales, which follows from (17). In fact, application of (17) to the case of turbulent flow showed that the Lagrangian time scale, T_L, and the Eulerian one, T_E, are related through (Saffman, 1963; see also McComb, 1990)

$$T_L=\beta\, T_E \quad (18)$$

and consequently that the Lagrangian and Eulerian correlation functions are related by

$$R_L(\beta t)=\beta\, R_E(t) \quad (19)$$

with $\beta \sim V/\sigma_{u_1}$, where σ_{u_1} is the r.m.s. velocity fluctuation in the mean flow direction.

Equations (18) and (19) are known also as the Hay and Pasquill conjecture. The Hay and Pasquill conjecture is based on the observation that the Eulerian and Lagrangian correlation functions have similar shapes but different scales. If we apply the Hay and Pasquill assumption to the groundwater flows, the dependence on β suggests that, at a large variability in σ_Y^2, the longitudinal spread is inversely proportional to σ_Y^2, an element which of course cannot be captured by the linear theories. This type of behavior was observed in numerical simulations (see Kitanidis, 1988). Works in the area of turbulence suggest a value of $1.1<\beta<8.5$. Numerical simulations (A. Bellin, 1994, personal communication) suggest that a value of $\beta>1$ can be expected also in groundwater applications.

While both Dagan's (1984) and Neuman & Zhang's (1990) macrodispersion models are similar in every sense but in the way $\mathbf{X}$ is treated, this element alone introduces significant differences. Neuman and Zhang's model depends non-linearly on an effective Peclet number P_e which is inversely proportional to the longitudinal macrodispersion coefficient. Here, pore-scale dispersion and macrodispersion are not additive. Dagan's model was developed for $P_e \to \infty$ and the effects of pore-scale dispersion and macrodispersion on mixing are taken as additive.

Both models display qualitatively similar results when it comes to longitudinal mixing, but there is a profound difference in the lateral macrodispersion coefficient D_T. Dagan's model indicates that D_T approaches zero at very large travel times, while the Neuman and Zhang model indicates that D_T has a finite asymptotic value. Dagan's result is a straightforward outcome from the hole-type covariance (zero integral scale) that the lateral velocity covariance displays along the mean flow direction, and the approximation of $\mathbf{X}(t)-\mathbf{X}(t')$ by the projection of its expected value along the mean flow direction. Neuman and Zhang's model allows some deviation from the mean flow direction, and hence the hole effect does not automatically kick in.

Neuman and Zhang found the asymptotic limit of D_T to be proportional to σ_Y^4. Dagan's initial analysis (1984) is carried out up to order σ_Y^2 and indicates the zero limit. More recently, Dagan (1994a) pursued a higher-order analysis while employing the same inconsistency as the Neuman and Zhang analysis: the Eulerian velocity covariances are derived at order σ_Y^2, yet the deviations of the particles from their trajectories are accounted for at higher-order accuracy. Because of that similarity, these models should become identical at

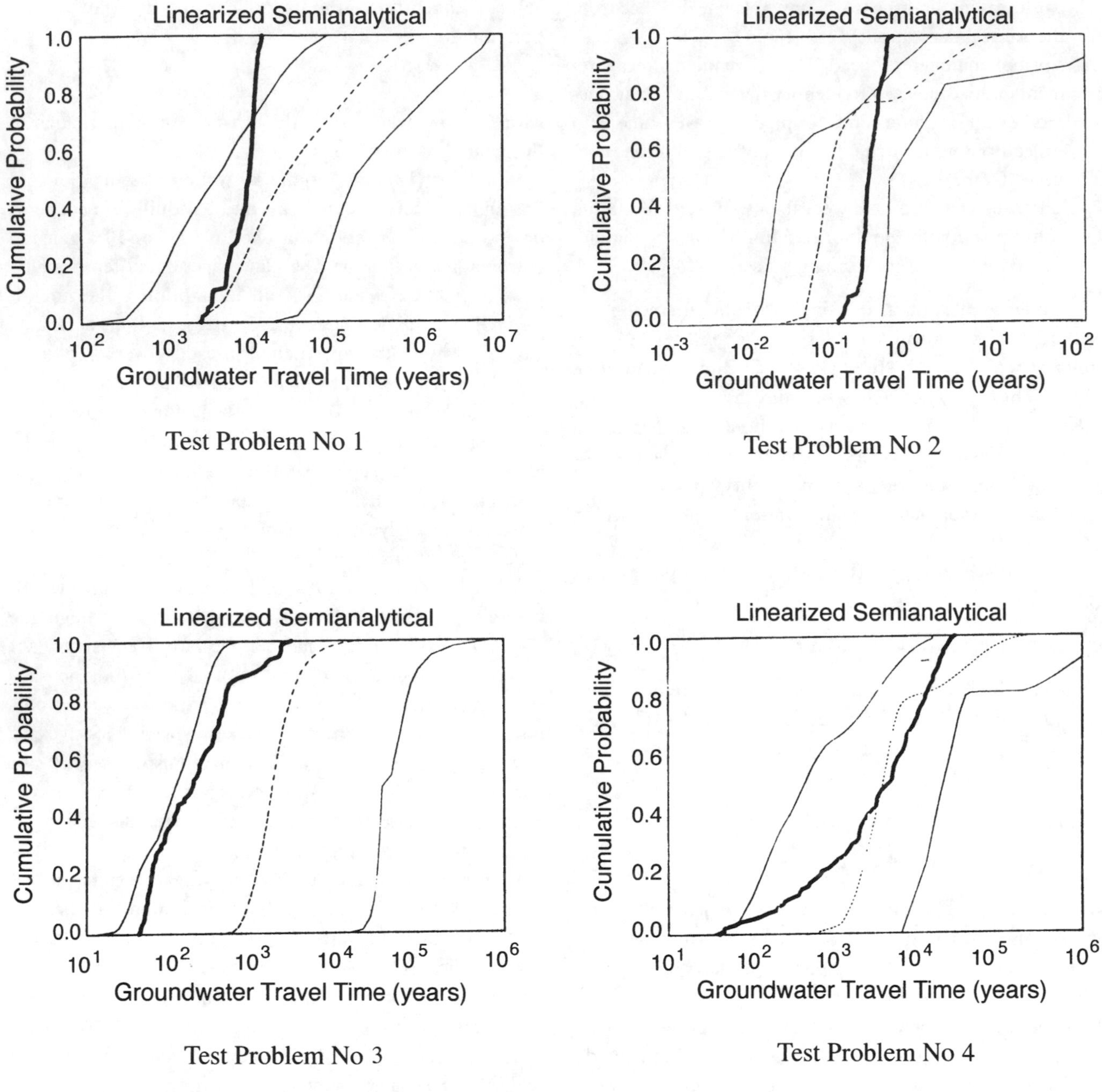

Fig. 2 Estimated groundwater travel times cumulative distribution functions (CDFs) for the four test problems shown in Fig. 1. The dashed lines show the mean CDF for the waste placement panel shown as a square in Fig. 1. The thin solid lines show the 95% intervals of confidence. The bold solid lines show the actual average CDF as computed by Sandia given the complete data set. The estimated CDFs were computed using eqs. (14) and (15) based on a partial data set.

high P_e numbers, which unfortunately does not happen. Clearly Corrsin's hypothesis leads to profound differences which are not recovered by Taylor's analysis, and it awaits further verification.

It is interesting to analyze these models vis-à-vis numerical simulations. Since both models were developed under the assumption of small variability of the logconductivity $\sigma_Y^2<1$, numerical simulations at small σ_Y^2 can only reveal internal inconsistencies, but those carried out at higher σ_Y^2 can reveal the range of applicability of the linear theories.

We start by looking at longitudinal spread. The Bellin *et al.* (1992) and Chin & Wang (1992) studies found a surprisingly good agreement between the linear theories and the numerical results, even at large σ_Y^2. Glimm *et al.* (1993) compared a Corrsin-based perturbation method with numerical results, and they considered, in addition to σ_Y^2, the rate of decay of the logconductivity correlation at large separation distances. They found that linear methods gave accurate estimates of mixing lengths for σ_Y^2 as high as 5.0, provided that the rate of decay is such that the conductivity has a finite integral scale. For a slow rate of decay, the linear approximations seem to fail at smaller variances. Follin (1992) simulations, conducted for the case of a finite integral scale, also confirm the findings of the previous studies, yet suggest extending the applicability range of the linear methods to variances as large as 16.

One should be careful in projecting this good agreement to high values of σ_Y^2. Kitanidis (1988) found, using numerical simulations, a complete breakdown of the linear theories at σ_Y^2 of the order of 20, and he showed a dramatic reduction in the longitudinal spread compared to the one predicted by the linear theories.

While Dagan's model is strictly linear, and hence does not have any formal claims on the non-linear range, Neuman and Zhang refer to their work as quasi-linear, alluding to the non-linear fashion in which the particle trajectories are treated. Still, a task yet unfulfilled is that of trying to explain the surprising success of the linear theories at such large variabilities.

This success is all the more surprising given the reliance on the Gaussianity of **X**, which is common to the above-mentioned methods. Note that, however, at small σ_Y^2 and for a Gaussian *Y* this is rigorous at any travel distance, while for a non-Gaussian *Y*, **X** will become Gaussian at some finite travel distance, based on considerations coming from the central limit theorem. It is also not uncommon to assume Gaussian **X** in turbulent diffusion, even at early times, based on experimental evidence. An analysis of early travel times in Gaussian and non-Gaussian logconductivity fields based on the work of Scheibe & Cole (1994) reveals a very good proximity to analytical models which assume Gaussian displacements (see Fig. 3). Desbarats & Srivastava (1991) is a remarkable example to the success of the linear theories in non-Gaussian fields. They digitized a mountainous terrain and scaled it for conductivities. The outcome was a bimodal logconductivity field with σ_Y^2 around 1. They found that the results of their numerical transport simulations were matched successfully by Dagan's (1984) model.

One explanation is the relatively small variability of the velocity field observed even at relatively large σ_Y^2. In the linear range, Dagan (1984) derived the coefficient of variation of the longitudinal velocities in the form $\rho_{u1} = {}^3\!/_8\,\sigma_Y^2$. Glimm *et al.* (1993) found the coefficient to vary between 0.2 at small σ_Y^2 to 0.1 at $\sigma_Y^2 = 5.1$, which suggests that a perturbation expansion in the velocities is applicable even at large variances in the logconductivity, due to its lower variability. This smaller variability reflects the fact that the velocity is constrained by boundary conditions as well as mass conservation.

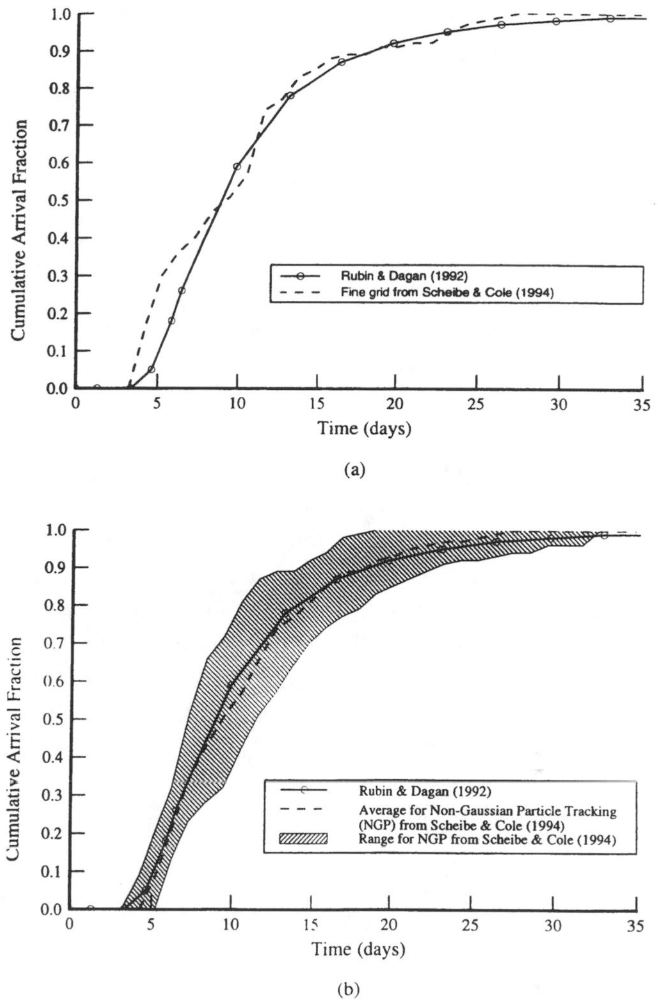

Fig. 3 A comparison of travel time CDFs obtained by the Lagrangian technique (Dagan & Nguyen, 1989; Rubin & Dagan, 1992) with numerical results obtained by Scheibe & Cole (1994) (a) for a Gaussian logconductivity field and (b) for a non-Gaussian field. 'Cumulative Arrival Fraction' is the term used by Scheibe & Cole (1994) and is equivalent to CDF.

Glimm *et al.* (1993) concluded that as the correlations decay, and become small, they enter the range of applicability of the perturbation theory. This suggests that the agreement would be asymptotically correct, improving with increase in the travel distance, which is indeed what their numerical results indicate.

Chin & Wang (1992) suggest that the surprisingly good performance of the linear model of Dagan (1984) at large σ_Y^2 is a fortuitous outcome from mutual cancellation of errors: errors in estimating the velocity covariance are cancelled by errors in estimating the mean velocity.

Some experience on this issue has been gained in turbulent flow. Adrian (1979) investigated flow patterns in turbulent flow, and showed that the mean flow pattern, which is associated with slight deviations from the mean, is very similar to the one associated with very large deviations. Furthermore, the use of a low-order expansion in terms of the velocity fluctuation led to results almost identical to those obtained using a high-order expansion.

The story of success is very different when it comes to lateral mixing: there is not much supporting evidence to encourage the applicability of the low-order solutions beyond their nominal range. Follin (1992, fig. 118) found a good match for σ_Y^2 up to 4, but Chin & Wang (1992, fig. 13) found that for σ_Y^2~2 the agreement is not favorable. However, this situation is consistent with the order of the derivation.

When it comes to comparing the theoretical models with actual field studies, the linear and quasi-linear models predict the longitudinal mixing quite well, in line with the theoretical models. With regard to lateral mixing, these models do quite well at early times, but fail at large travel times, as was indicated by the few recent large-scale field tests. Although problems with the experimental field procedures cannot be entirely discounted, this finding is not surprising given the conclusions drawn from the numerical simulations. Recent work (Rehfeldt & Gelhar, 1992; Farell *et al.*, 1994), suggests that the gap in lateral mixing may be attributed to flow unsteadiness, but Dagan, Bellin & Rubin (1996) showed that this is not necessarily the case. Rubin & Bellin (1994) showed that the presence of distributed recharge leads to enhancement of the lateral speed. An application to Cape Cod (Ezzedine & Rubin, 1997) showed that by accounting for recharge good agreement between theory and observations was obtained (see Fig. 4).

4 SCALING ISSUES

Replacing $\mathbf{x}^*=\langle\mathbf{X}\rangle$ by $\mathbf{x}^*=\langle\mathbf{X}^c\rangle$ as suggested in Section 3 is an attempt to capture more accurately the Lagrangian trajectory through conditioning. A salient question is whether for certain plume sizes conditioning becomes redundant due to ergodicity. This brings to the fore the question of how to deal with non-ergodic, finite-size solute bodies, or better, with the entire spectrum of possible plume sizes.

For a better definition of the problem, consider the fundamental relation (Fischer *et al.*, 1975, Kitanidis, 1988; Dagan, 1990) written for space of dimension m:

$$\langle S_{ij}(t)\rangle = S_{ij}(0) + X_{ij}(t) - R_{ij}(t) \quad i,j=1,\ldots,m \tag{20}$$

between the expected value of the spatial moments of a finite plume about its centroid $\langle S_{ij}\rangle$ and its initial value, $S_{ij}(0)$, the single particle's displacement variance–covariance tensor X_{ij},

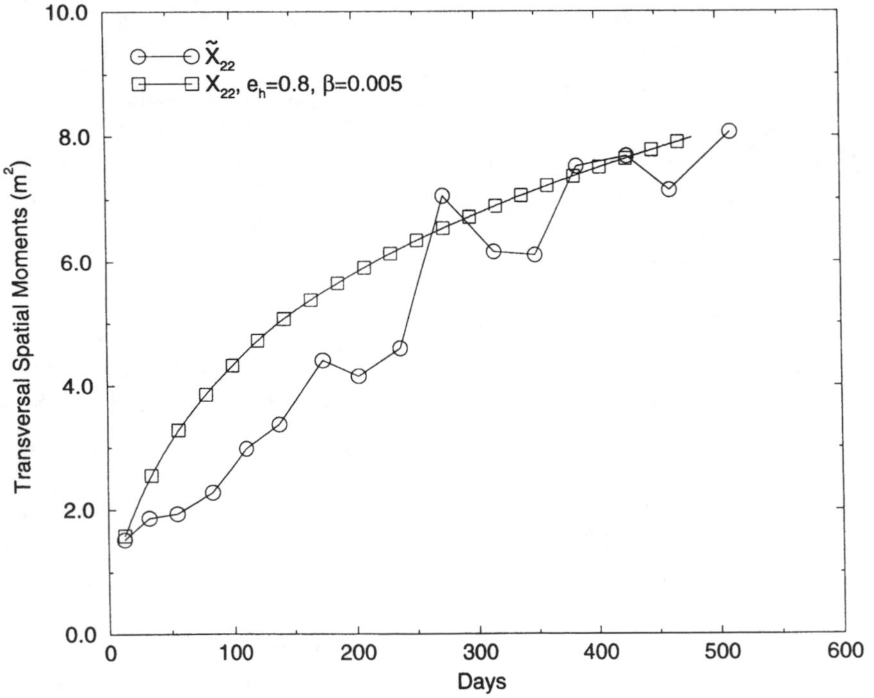

Fig. 4 A comparison between the lateral spatial moment $\tilde{X}_{22}$ of the experimental plume at Cape Cod and a theoretical model. The theoretical $\tilde{X}_{22}$ was computed following Ezzedine & Rubin (1997) based on first-order velocity covariances which were developed for a horizontal ratio of anisotropy e_h=0.8. The dependence on $\beta=RI_yT_GJ_0$ was derived in Rubin & Bellin (1994). R is the natural, disturbed recharge, I_y is the horizontal integral scale of the logconductivity, T_G is the geometric mean transmissivity and J_0 is the gradient at the release point. In the above analysis, a small β is assumed.

and the variance–covariance tensor of the plume's centroid displacement, R_{ij}. R_{ij} is defined as the variance of the space average

$$\frac{1}{V_0}\int_{V_0} \mathbf{X}(t;\mathbf{a})da \tag{21}$$

with V_0 being the injection volume. Equation (21) assumes that the initial concentration C_0 is constant over V_0. When dealing with a single particle's displacement, (R_{ij}) simplifies to

$$R_{ij}(t)=X_{ij}(t) \tag{22}$$

with the results for X_{ij} already presented in Section 3. When dealing with an ergodic plume, V_0 is very large and $R_{ij}(t)=0$, with the simple outcome that $D_{ij}=\frac{1}{2}\,\partial X_{ij}/\partial t$ describes the rate of growth of ergodic plumes only, or alternatively the rate of growth of the ensemble average of all non-ergodic plumes (see also Fischer *et al.*, 1975), which can be interpreted as an imaginary envelope of performance determined by all possible realizations. For field applications, however, it is important to analyze the behavior of the single, possibly non-ergodic realization, with appropriate tools, since this behavior is different from the pattern followed by ergodic plumes. (See Quinodoz and Valocchi, 1990.)

As an alternative to D_{ij}, Dagan (1990) defined an actual dispersion coefficient

$$\bar{D}_{ij}=\frac{1}{2}\frac{dS_{ij}}{dt} \qquad i,j=1,\ldots,m \tag{23}$$

which describes the growth rate of the plume's spatial moments around its centroid. Since the fluctuations of the centroid are filtered out, $\bar{D}_{ij}$ is better suited than D_{ij} to analyze the spread of non-ergodic plumes. Being non-ergodic, however, makes $\bar{D}_{ij}$ a random function. Its expected value is obtained from (20):

$$\langle\bar{D}_{ij}\rangle=\frac{1}{2}\frac{dX_{ij}}{dt}-\frac{1}{2}\frac{dR_{ij}}{dt} \tag{24}$$

Since $0\leq R_{ij}\leq X_{ij}$, we find that $\langle\bar{D}\rangle\leq D_{ij}$, hence the use of D_{ij} for modeling of non- ergodic plumes is non-conservative. As the dimension of the initial solute body V_0 increases, R_{ij} approaches zero and $\langle\bar{D}_{ij}\rangle$ approaches D_{ij}. For non-ergodic conditions, $\langle\bar{D}_{ij}\rangle$ needs to be characterized by its variance and perhaps even higher-order moments.

Since $\langle\bar{D}_{ij}\rangle$ depends on V_0 and not just on the travel time, it is less meaningful than D_{ij}: it depends on the parameters of the transport problem, not just on physical parameters. Besides traditional value, there is no compelling reason for working with $\langle\bar{D}_{ij}\rangle$ rather than $\langle S_{ij}\rangle$. It is important to notice that $\langle\bar{D}_{ij}\rangle$ depends strongly on l_T, the dimension of the plume that is orthogonal to the mean flow direction. Dagan (1990, fig. 4) showed that in the case of stratified flow the ratio $\langle\bar{D}_{ij}\rangle/D_{11}$, referring to longitudinal spread coefficients, approaches unity when l_T/I_Y is of the order of 30. For two-dimensional flows, a ratio of 10 to 20 is more likely (Dagan, 1991). The convergence of $\langle\bar{D}_{ij}\rangle$ to D_{ij} with the size of V_0 signifies the approach of the single plume behavior to that of the ensemble mean, as well as the reduction of R_{ij}. This means that the sensitivity of the plume's displacement to local heterogeneities reduces, and so does the role of conditioning.

As becomes evident, the applicability of D_{ij} is closely related to the scale of the plume: the plume needs to be much larger than the scale of heterogeneity. Recently, attention has been focused on formations with multiple scales of heterogeneity: some geological phenomena were found to display more and more scales of heterogeneity as the scale of observation increases. It has been suggested that, in some cases, an infinite number of scales co-exist. Evidence in support of such phenomena in hydrogeology is quite limited. Measurements of core permeability taken along deep wells indicate the presence of increasing scales of heterogeneity along the vertical (Ababou & Gelhar, 1990). Similar arguments were also raised in a more recent study by Desbarats & Bachu (1994), although no attempt was made here to establish the presence of a trend which might lead to similar observations. More intriguing evidence is contained in the works of Lallemand-Barres & Peaudecerf (1978) and Gelhar, Welty & Rehfeldt (1992), who analyzed dozens of transport experiments and pointed to a trend of increasing macrodispersivity with the scale of the experiment. These data were not taken all from a single site, and hence should not be interpreted as a definite indication to the presence of evolving scales; however, there was enough of an impetus here for researchers to check into this possibility. Of direct relevance to our previous discussion is to define macrodispersion in such formations, the problem being that the applicability of macrodispersion coefficients and the notion of ergodicity are inseparable, yet ergodicity in formations of evolving scales cannot be attained, due to the lack of a finite integral scale.

A regression analysis by Neuman (1990, 1994) of the compendium of experiments mentioned above suggested that

$$\alpha_L=CL^{\beta},\quad \beta=\begin{matrix}1.5 \text{ for } L\leq 100\text{ m}\\ 0.75 \text{ otherwise}\end{matrix} \tag{25}$$

where C and β are constants, α_L is the longitudinal macrodispersivity and L is the travel distance. This implies that the mixing length varies as L^{ρ} with $\rho=(1+\beta)/2$. Neuman's work assumes that the logconductivity field is characterized by a variogram $\gamma_Y(r)\sim L^{1/2}$.

Numerical and analytical work by Glimm *et al.* (1993) along the same lines focused on formations where the logconductivity is characterized by the covariance

$$C_Y(r)=br^{-\eta} \tag{26}$$

for large r, where b and η are slowly varying functions of r. For rapidly decaying correlations, $\eta>1$ the dispersion is found to be Fickian, but for $0<\eta<1$, C_Y has an infinite integral scale, and the mixing length exponent becomes $\rho=1-(\eta/2)$. The exponent η in (26) defines the rate of decay of the covariance. But since

$$\frac{\partial \ln C_Y}{\partial \ln r}=-\frac{\partial \ln \gamma_Y}{\partial \ln r} \tag{27}$$

by definition, we find that the formations investigated by Glimm *et al.* (1993) and those investigated by Neuman (1990, 1994) become statistically identical for $\eta=\frac{1}{2}$, which leads to $\rho=0.75$ according to Glimm *et al.* (1993). This value agrees well with Neuman's (1990) value of ρ for transport in domains with $L>100$ m.

The agreement between Glimm *et al.* (1993) and Neuman's model is quite surprising since they are conceptually different: Neuman on the one hand analyzed a collection of single realizations, which do not constitute a statistical ensemble, and Glimm *et al.* (1993) analyzed a statistical ensemble. We recall, however, that ensemble averages and single plumes' moments are interchangeable only in the case of a finite scale, and this suggests that, in both cases, a finite length scale, which is determined by the dimension of the solute body, does exist.

A plume of infinite extent samples the heterogeneity at all scales, but in realistic applications plumes are of finite extent, and variability at scales much larger than the scale of the plume does not enhance spreading. At most we can expect that large-scale variability would affect, at least initially, the mean displacement of the plume much more than it would affect spreading.

This aspect was demonstrated recently in studies of transport in two different formations. The first study concerns heterogeneous formations where the mean logconductivity has a linear trend (Rubin & Seong, 1994). The linear trend in this study is taken as a manifestation of large-scale variability, much larger than the scale of the observed heterogeneity. It was found that the effect of the trend on spreading becomes noticeable only after a very large travel time, while its effect on the mean velocity is evident at all times. More significantly, this study showed that the presence of the trend translates into shear flow that cannot be modeled as a function of the total variance. The second study (Rubin, 1995) concerns bimodal heterogeneity, such as that observed in sand–shale formations, or fractured rocks, or unconsolidated formations embedded with shoe-string sands. Here also it is found that the effect of large-scale heterogeneity becomes noticeable only after the travel distance becomes of the order of the larger scale of heterogeneity.

A very vivid demonstration of the role of large-scale heterogeneities on mixing, albeit from the field of turbulent diffusion, is given by List, Gartrell & Winant (1990, fig. 7). These results are taken from an actual experiment conducted in the ocean, and they are in line with the previous, theoretically derived findings, namely that the effects of large-scale heterogeneity kick in only after a large travel time.

These studies all suggest that the concept of macrodispersion coefficients may be less applicable in formations of evolving scales of heterogeneity than in formations of finite scale since ergodicity of finite-size plumes with regard to spatial moments can never be attained here.

Ababou & Gelhar (1990), as well as Cushman & Ginn (1993), recognized the significance of the plume's scale in determining the spread of finite-size plumes in formations of evolving scales. A recent study (Dagan, 1994b) investigated macrodispersibility in such formations using (26), with $0<\eta<1$, as a model for the heterogeneity. Employing the concept of actual dispersion coefficients (23), this study shows that, for plumes of finite length scale, actual spreading in space attains a Fickian limit despite the lack of a finite heterogeneity scale. This Fickian limit is proportional to $\ell^{\eta+1}$, where ℓ is the initial transversal dimension of the plume. Since all the experiments leading to (25) start with a finite ℓ, there is obviously a disagreement between (25) and Dagan (1994b), with one possible explanation being that (25) is based on pre-asymptotic regimes.

A different case is that of domains with stationary increments; the domain is characterized by an unbounded variogram:

$$\gamma_Y=\alpha r^{\beta} \tag{28}$$

Dagan (1994b) showed that the actual dispersion is a function of γ_Y, irrespective of the existence of σ_Y^2. This implies that the heterogeneity which affects the actual dispersion is only the one that is characterized by scales smaller than the scale of the plume. It is also shown in that study that for $\beta>1$ the actual dispersion (24) is anomalous and the mixing length grows as $L^{(\beta-1)/2}$. For $\beta<1$ the actual dispersion is Fickian, unlike the behavior indicated by (25), which was obtained for $\beta=\frac{1}{2}$. The maximum possible growth rate of the mixing length according to Dagan is linear with the travel distance, and it corresponds to $\beta=2$ in (28).

Dagan's distinction between the macrodispersion coefficients, representing ergodic situations, and the effective macrodispersion coefficients, which pertain to the non-ergodic situations, gives rise to a strange breed of coefficients which depend on the definition of the transport problem in addition to the physical parameters of the aquifer. The problem is fundamental, not in the sense that it requires departure from traditional methods, but because of related issues. Non-ergodic plumes are sensitive to local parameter configurations, and hence their motion can be better defined by conditioning. It follows then that the macrodispersion coefficients can be expanded to account for the available measurements, and hence will modify with each additional piece of information. Lack of information is offset by allowing for larger dispersion. Hence we end up with coefficients that reflect physical reality as well as the extent of our familiarity with this reality. An attempt to deal with this problem is reported in Neuman (1993). The complexity of the issue suggests that the concept of dispersion coefficients may be out of place in most field applications, and we may be better off by discarding it in favor of more robust concepts.

5 ON THE CONCENTRATION VARIABILITY AND THE CONCENTRATION VARIANCE

Both Eulerian and Lagrangian methods are concerned with the variance of the concentration, for the simple reason that quantification of the uncertainty associated with prediction of the concentration is of paramount importance in environmental studies: decisions under conditions of uncertainty require means for assessing risks. Most of the work to date has been done on derivation of the concentration variance, and very little work has been done on higher-order moments, let alone on the entire concentration pdf. Since a general case for the concentration to be Gaussian cannot be made, a two-moment characterization is not exhaustive. Despite the limitations of the two-moment characterization, significant work on the concentration variance has been reported, which is worth exploring.

Using (6) as a starting point, it was shown (see Dagan, 1989; Rubin, 1991b) that in the case of slug injection through a source of size V_0 at time t_0, the variance of the volume-averaged concentration detected by a sampler of volume V at time t is given by:

$$\sigma_c^2(t)=\frac{C_0^2}{V^2}\int_V\int_V\int_{V_0}\int_{V_0}[f_L(\mathbf{x}', \mathbf{x}''; t, \mathbf{a}, \mathbf{b}, t_0)-f_L(\mathbf{x}', t, \mathbf{a}, t_0)\, f_L(\mathbf{x}'', t, \mathbf{b}, t_0)]d\mathbf{a}d\mathbf{b}\, d\mathbf{x}'d\mathbf{x}'' \tag{29}$$

where $f_L(\mathbf{x}', \mathbf{x}''; t, \mathbf{a}, \mathbf{b}, t_0)$ is the two-particle displacement pdf: $f_L(\mathbf{x}', \mathbf{x}''; t, \mathbf{a}, \mathbf{b}, t_0)d\mathbf{x}'d\mathbf{x}''$ denotes the probability of two particles released at **a** and **b** at time t_0 to be at points $\mathbf{x}'$ and $\mathbf{x}''$, respectively, at time t. C_0 denotes the initial concentration at V_0, assumed here to be constant for simplicity.

From (29) it appears that σ_c^2 can be reduced by spatial averaging. This was confirmed and demonstrated in Bellin, Rubin & Rinaldo *et al.* (1994, fig. 8). However, for a significant reduction, averaging must be done over very large volumes (Rubin, 1991b), and the downside is the loss of resolution. Nevertheless, tracers are sampled by different devices, and (29) offers a useful method for interpretation. A different method of variance reduction is through conditioning of the various pdf's on field data.

Simplification of (29) can be obtained for small V. In the case of a very small sampler, $V \to 0$, and, neglecting dilution, the concentration at any point in the field is either C_0 or zero, and hence the concentration pdf becomes

$$f(C)=\left(1-\frac{\langle C\rangle}{C_0}\right)\delta(C)+\frac{\langle C\rangle}{C_0}\delta(C-C_0) \tag{30}$$

The concentration variance and coefficient of variation are obtained from (30) in the form (Dagan, 1989)

$$\sigma_c^2=\langle C\rangle(C_0-\langle C\rangle); \qquad \frac{\sigma_c^2}{\langle C\rangle^2}=\frac{C_0}{\langle C\rangle}-1 \tag{31}$$

Several significant observations can be made at this point: the coefficient of variation can become exceedingly large when $\langle C\rangle$ is small, and hence the prediction of the concentration at the fringes of the plume or at large travel times is subjected to large uncertainty; since pore-scale dispersion leads to destruction of the concentration fluctuations and to the smoothing of the concentration field, the relations in (31) constitute upper bounds; the concentration variance is always finite but the coefficient of variation is not; for a fixed point the concentration variance may increase or decrease with time, depending on its position relative to the plume's centroid.

Following (31), the variance σ_c^2 has a stationary point when

$$\frac{\partial\sigma_c^2}{\partial x}=\frac{\partial\langle C\rangle}{\partial x}[C_0-2\langle C\rangle]=0 \tag{32}$$

For a bell-shape distribution of $\langle C\rangle$, stationary points will occur at peak concentration and where $\langle C\rangle=C_0/2$. From the second derivative

$$\frac{\partial^2\sigma_c^2}{\partial x^2}=\frac{\partial^2\langle C\rangle}{\partial x^2}[C_0-2\langle C\rangle]-2\left[\frac{\partial\langle C\rangle}{\partial x}\right]^2 \tag{33}$$

we find that at the plume center (where $\partial\langle C\rangle/\partial x=0$ and $\partial^2\langle C\rangle/\partial x^2$ is negative), the variance will attain a maximum if $\langle C\rangle$ is less than $C_0/2$ and a minimum otherwise. Hence the location of the maximum variance varies with time, and it will be at the plume center only after the expected value of the concentration there becomes smaller than $C_0/2$. Prior to this time, the maximum variance is located along a 'halo' around the centroid. Due to the non-stationary nature of the field, the variance, unlike the coefficient of variation, is not an indicator of uncertainty.

The source of the concentration variance σ_c^2 in (31) is the variability in the velocity field. This uncertainty is related to the probability that a particle's trajectory may or may not be at a certain point at a given time, and the only mechanism considered here is the variability in the conductivity field.

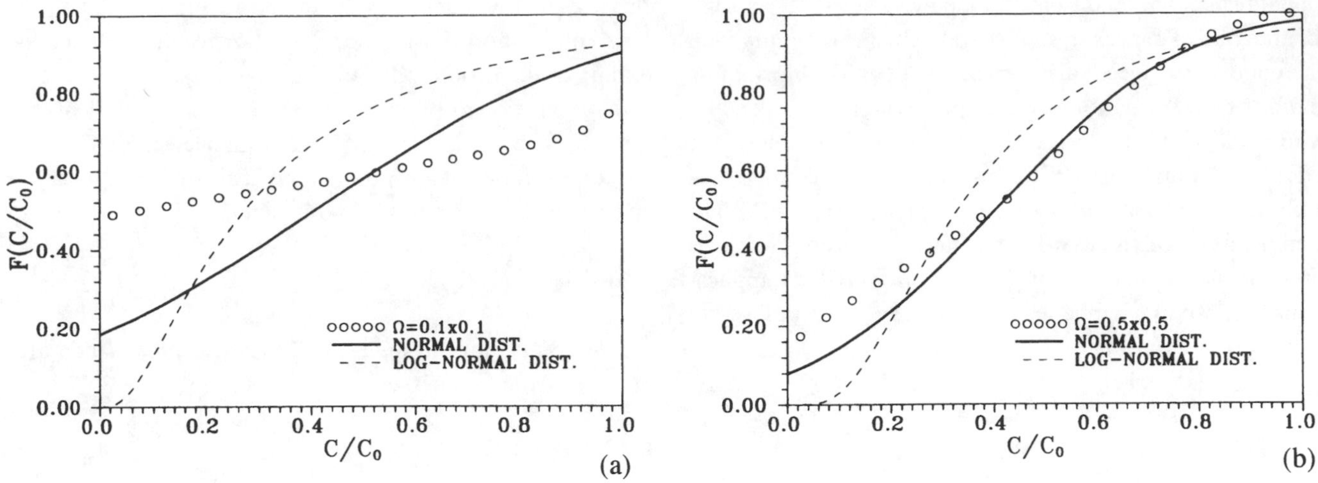

Fig. 5 Cumulative probability distribution function of the concentration for sampling volumes. (a) 0.1×0.1 integral scales, (b) 0.5×0.5 integral scales. The figure is from Bellin *et al.* (1994). The samples are located 2 logconductivity integral scale downstream from the source.

Under the conditions leading to (30), i.e., for a basically divergence-free flow and in the absence of dilution, the space integral of σ_c^2 is constant in time, regardless of local variations in σ_c^2. This is clearly related to the exclusion of pore-scale dispersion, which is the source of variance reduction mechanism $2D_m\langle(\partial c'/\partial x_\beta)^2\rangle$ in (4).

Kapoor & Gelhar (1994) investigated the concentration variance reduction using an Eulerian formalism. They considered (4) while assuming that the mean concentration gradient is constant and finite and that the mean concentration field is smooth. One of the problems with (4) is that the presence of sharp fronts when coupled with the assumption of gradient-based dispersion leads to infinite concentration variances, which can never occur for a finite initial concentration. This limits the applicability of (4) to transport regimes characterized by small Peclet numbers ($P_e = I/\alpha$). In the absence of sharp fronts, Kapoor & Gelhar (1994) hypothesize that the slope of the logconductivity covariance near the origin, characterized by the microscale, determines the rate of variance reduction. Smaller microscale is indicative of larger small-scale variability, and hence of a larger interface area between zones of different conductivity, which in turn enhances the smoothing of the fluctuations in the concentration, and to variance reduction. The role of the microscale in explaining dilution is questionable because it sometimes does not even exist, such as in the case of an exponential correlation function. Work in progress suggests that dilution can be related to pore-scale dispersion rather than to the logconductivity microscale.

A formula or an equation for the variance is of limited value for predictive purposes because the concentration, in general, is neither Gaussian nor log-Gaussian. Under conditions which are often met in field-scale situations it is close to being bimodal (see (30)). Even when departing from bimodality, the concentration is bounded by zero and the initial concentration C_0, and therefore it cannot be described by a Gaussian distribution. Hence the variance is a measure of variability, but not a measure of uncertainty.

A more complete description can be obtained by deriving the entire concentration pdf. Generally, it is a tedious task that can be accomplished only through an extensive series of Monte Carlo runs. In a recent study, Bellin *et al.* (1994) and Rubin, Cushey & Bellin (1994) suggested a fast way to compute the concentration pdf. The approach is based on particle tracking, and the savings come from a fast and economic way of generating the fluid velocity fields which will not be explored here. The concentration is found to be non-Gaussian and the availability of the first two moments is clearly insufficient for predictive purposes. Pore-scale dispersion is shown to lead to departure from strict bimodality but still does not lead to Gaussian distributions. The sampling volume has a dramatic effect on the concentration pdf. Not surprisingly, for large volumes the Gaussian approximation works fairly well, especially around and above the median, where the effect of the zero lower bound is less accentuated. This behavior is demonstrated in Fig. 5 (from Bellin *et al.*, 1994). The concentration pdf is computed for different sampling volumes, and an attempt is made to model it using a normal and log-normal model. In the case of the larger sampling volume, the normal model works reasonably well over a large range of values, and its performance improves for high concentration values.

The formulation of the concentration estimation problem in a stochastic framework also allows us to derive conditional estimates of the concentration moments. In the Lagrangian approach conditioning can be done by replacing the displacement pdf's in (29) by their conditional counterparts, for example:

$$f_L(\mathbf{x}';t,\mathbf{a},t_0) \rightarrow f_L^c(\mathbf{x}',t,\mathbf{a},t_0)|\{N\} \tag{34}$$

where f_L^c is the conditional pdf and $\{N\}$ denotes the set of data available for conditioning. In the Eulerian formalism conditioning is done by replacing unconditional expected values with conditional ones (Graham & McLaughlin, 1989a,b; Neuman, 1993); for example,

$$\frac{\partial\langle C\rangle}{\partial t} \rightarrow \frac{\partial\langle C|\{N\}\rangle}{\partial t} \tag{35}$$

This approach requires that conditional nonstationary macrodispersion coefficients are derived at considerable computational cost. Graham and McLaughlin avoided this pitfall by conditioning directly the velocity fields using the Kalman filter method, and thus freeing themselves from the need to lump the effects of heterogeneities on one hand, and measurements on the other, into a single coefficient.

Of particular interest is the diversity of measurements that can be included in $\{N\}$. Several works reported using conductivity and/or hydraulic head measurements to condition transport (Smith & Schwartz, 1980; Dagan, 1984; Graham & McLaughlin, 1989b; Rubin, 1991a,c). Graham & McLaughlin (1989b), Rubin (1991c) and more recently Ezzedine & Rubin (1996) also explored the possibility of using concentration data for conditioning. Rubin (1991c) showed that the concentration has a well defined spatial correlation structure that can be used for conditioning using a kriging formalism. Fig. 6 (Rubin, 1991c) demonstrates the correlation coefficient of variation $\rho_c(\mathbf{x},\mathbf{x}')$ for $\mathbf{x}$ at the plume centroid and $\mathbf{x}'$ varying all over the flow domain.

ρ_c is non-stationary and also time-dependent. It is anisotropic with a larger correlation distance along the mean flow direction and it typically displays positive correlations over short separation distances, followed by negative correlations at larger distances, a fact simply reflective of mass conservation. Using the correlation coefficients, a dramatic reduction in the estimation uncertainty can be obtained, see Fig. 7 (Rubin, 1991c).

More recent work (Ezzedine & Rubin, 1996) discusses the potential of conditioning the concentration estimates on measurements of hydraulic conductivity and hydraulic head. Using linear estimation methods, such a task would require derivation of the cross-covariances between the concentration, the conductivity and the hydraulic head, but it is shown to lead to very favorable results. This study also explores the possibility of using tracer data for conditioning. Conditioning on tracer data can be done in several ways: one

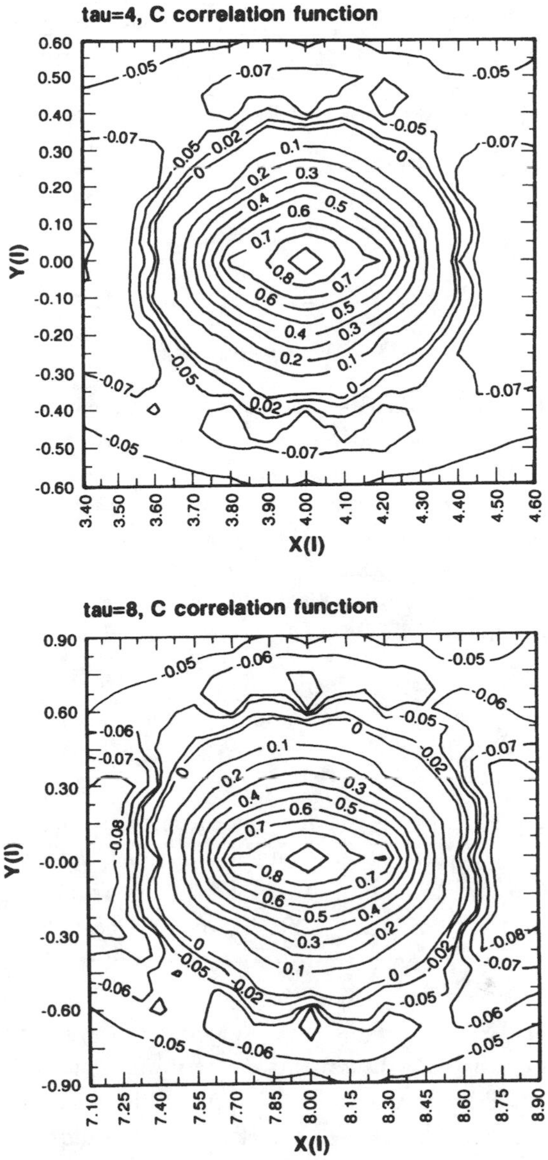

Fig. 6 The spatial correlation function of the concentration $\rho_c(\mathbf{x}, \tau, \mathbf{x}', \tau)$: it represents the correlation between the concentration at point $\mathbf{x}$, which is taken here at the plume's centroid, at time τ, and the concentration at $\mathbf{x}'$, which we vary over the entire domain. ρ_c is shown for τ=4 and τ=8 (non-dimensional times). Note that ρ_c becomes negative away from $\mathbf{x}$, which is an outcome from mass conservation: ρ_c is a hole-type covariance. ρ_c is also time-dependent. The figure is taken from Rubin (1991c).

can condition on actual concentration measurements, but also on travel times or actual displacements. (Rubin and Ezzedine, 1997).

6 SUMMARY

Some of the issues raised by Sposito *et al.* (1986) have been explored in depth in the last few years, and although not all are completely resolved, research in recent years has defi-

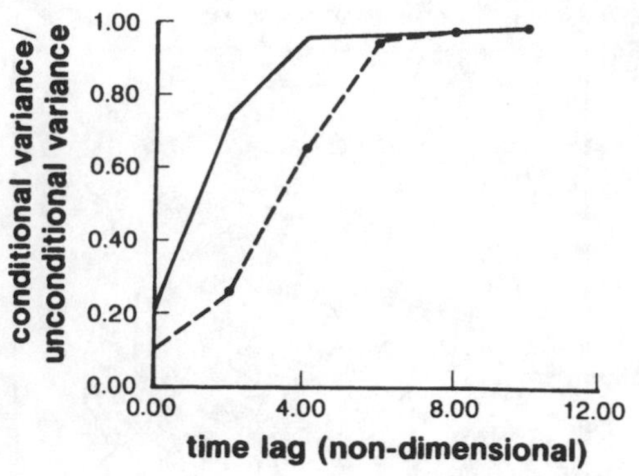

Fig. 7 The ratio between the conditional concentration variance and the unconditional one for $C(\mathbf{x}, \tau)$, the concentration at the plume's centroid at $\tau = 12$ (from Rubin, 1991c). Conditioning is on concentration measurements taken at a time log $|\tau = \tau'|$ before τ. The solid line describes the case where the data are located symmetrically upstream and downstream from **x**, and the dashed line describes the case where the two data are located symmetrically around **x** and across the mean flow direction.

nitely led to a better understanding of transport in heterogeneous media, and of what can be reasonably expected from available tools.

The CDE for the mean concentration is of limited applicability. Nothing has changed with regard to that since 1986. Only the limitations are better understood: the mean concentration is not accessible for measurements, and the ensemble mean concentration is not relevant for environmental studies. Theoretical and field studies have shown, however, that the CDE is useful, under certain conditions, for predicting the spatial moments of the mean concentration. These conditions are expressed in terms of the dimensions of the plume and the elapsed travel time, and are quite prohibitive. However, once these conditions are met, the spatial moments become also the moments of the single realizations, and hence measurable quantities. Substantial effort has been devoted, however, to developing methods or techniques to evaluate the higher-order moments of the concentration.

Derivation of macrodispersion coefficients using the CDE and following an Eulerian formalism has been accomplished under restrictive conditions. The Lagrangian approach offered an easy alternative and allowed the derivation of pre-asymptotic coefficients, and consequently, the definition of the conditions which lead to a Fickian transport regime. Unlike the case of molecular diffusion, the spatial moments in the single realization do not constitute a monotonous growth process, and since the mean concentration gradient is rarely smooth, they cannot be used to measure the actual volume which is filled with solute.

It is recognized that plumes do not sample heterogeneities which are characterized by a length scale larger than its dimensions: the large-scale variability affects the displacement of the centroid, and the small-scale variability contributes the most to mixing. That led to the concept of actual, or effective, macrodispersivities, which depend on the plume dimensions. These coefficients measure mixing around the finite-size plume's centroid, and avoid confusing that process with the departure of that centroid from the ensemble mean displacement. By doing so, these coefficients are accessible to measurements, unlike the coefficients that measure spread around the theoretical mean displacement, which are just a theoretical concept. Using the concept of actual macrodispersivities allows investigation of the scale effect. In the case of stationary formations with no finite integral scale, transport is Fickian for plumes of finite lateral dimension. Similar to the case of a finite integral scale, transport becomes Fickian after a certain period, which can be interpreted as scale effect, but should not be confused with anomalous diffusion. In both cases, the plume's finite dimension limits the range of scales of variability which affect mixing. Anomalous diffusion can be observed also in the case of stationary increments of the logconductivity, but only when the variogram of the increments grows at a rate faster than linear with distance.

The role of conditioning as a means to distinguish a single realization out of an ensemble for site-specific applications has been established. Conditioning defines a subensemble which is in agreement with the data and at the same time is physically plausible. Conditioning has been employed for prediction of spatial moments, and more recently in the estimation of travel times and the moments of the concentration.

ACKNOWLEDGMENT

This study was supported through NSF Grant 9304481.

REFERENCES

Ababou, R. & Gelhar, L.W. (1990). Self-similar randomness and spectral conditioning: analysis of scale effects in subsurface hydrology. In *Dynamics of Fluids in Hierarchical Porous Media*, ed. J. Cushman. New York: Academic Press.

Adrian, R.J. (1979). Conditional eddies in isotropic turbulence. *Physics of Fluids*, 22(11), 2065–2070.

Bellin, A., Rubin, Y. & Rinaldo, A. (1994). Eulerian-Lagrangian approach for modeling of flow and transport in heterogeneous geological formations. *Water Resources Research*, 30(11), 2913–2924.

Bellin, A., Salandin, P. & Rinaldo, A. (1992). Simulation of dispersion in heterogeneous porous formations: statistics, first-order theories, convergence of computations. *Water Resources Research*, 28(9), 2211–2228.

Chin, D.A. & Wang, T. (1992). An investigation of the validity of first-order stochastic dispersion theories in isotropic porous media. *Water Resources Research*, 28(6), 1531–1542.

Corrsin, S. (1973). Comment on transport equations in turbulence. *Physics of Fluids*, 16(1), 157–158.

Cushey, M. A. & Rubin, Y. (1997). Field-scale transport of nonpolar organic solutes in 3-D heterogeneous aquifers. *Environmental Science Technology*, in press.

Cushman, J.H. & Ginn, T.R. (1993). Nonlocal dispersion in media with continuously evolving scales of heterogeneity. *Transport in Porous Media*, 13, 123–138.

Dagan, G. (1982). Stochastic modeling of groundwater flow by unconditional and conditional probabilities 2. The solute transport. *Water Resources Research*, 18, 835–848.

Dagan, G. (1984). Solute transport in heterogeneous porous formations. *Journal of Fluid Mechanics*, 145, 151–177.

Dagan, G. (1987). Theory of solute transport by groundwater. *Annual Review of Fluid Mechanics*, 19, 183–215.

Dagan, G. (1988). Time-dependent macrodispersion for solute transport in anisotropic heterogeneous aquifers. *Water Resources Research*, 24, 1491–1500.

Dagan, G. (1989). *Flow and Transport in Porous Formations*. New York: Springer Verlag.

Dagan, G. (1990). Transport in heterogeneous formations: spatial moments, ergodicity and effective dispersion. *Water Resources Research*, 26(6), 1281–1290.

Dagan, G. (1991). Dispersion of a passive solute in non-ergodic transport by steady velocity fields in heterogeneous formations. *Journal of Fluid Mechanics*, 233, 197–210.

Dagan, G. (1994a). An exact nonlinear correction to transverse macrodispersivity for transport in heterogeneous formations. *Water Resources Research*, 30(10), 2699–2706.

Dagan, G. (1994b). The significance of heterogeneity of evolving scales to transport in porous formations. *Water Resources Research*, 30(12), 3322–3336.

Dagan, G. & Neuman, S.P. (1991). Nonasymptotic behavior of a common Eulerian approximation for transport in random velocity fields. *Water Resources Research*, 27(12), 3249–3257.

Dagan, G. & Nguyen, V. (1989). A comparison of travel time and concentration approaches to modeling transport by groundwater. *Journal of Contaminant Hydrology*, 4, 79–81.

Dagan, G. & Sposito, G. (1994). Predicting solute plume evolution in heterogeneous porous formations (Technical Report). *Water Resources Research*, 30 (2), 585–589.

Dagan, G., Bellin, A. & Rubin, Y. (1996). Lagrangian analysis of transport in heterogeneous formations under transient flow conditions. *Water Resources Research*, 32(4), 891–901.

Dagan, G., Cvetkovic V. & Shapiro, A. (1992). A solute-flux approach to transport in heterogeneous formations 1. The general framework. *Water Resources Research*, 28(5), 1369–1376.

Deng, F.W., Cushman, J.H. & Delleur, J.W. (1993). A fast Fourier transform analysis of the contaminant transport problem. *Water Resources Research*, 29(9), 3241–3247.

Desbarats, A. (1990). Macrodispersion in sand-shale sequences. *Water Resources Research*, 26(1), 142–150.

Desbarats, A. & Bachu, S. (1994). Geostatistical analysis of aquifer heterogeneity from the core scale to the basin scale: A case study. *Water Resources Research*, 30(3), 673–684.

Desbarats, A.J. & Srivastava, R.M. (1991). Geostatistical characterization of groundwater flow parameters in a simulated aquifer. *Water Resources Research*, 27(5), 687–698.

Ezzedine, S. (1997). Fast computation of head and velocity covariances in three-dimensional statistically axisymmetric heterogeneous porous media. *Water Resources Research*, 33(1), 267–270.

Ezzedine, S. & Rubin, Y. (1996). A geostatistical approach to the conditional estimation of specially distributed solute concentration. *Water Resources Research*, 32(4), 853–862.

Ezzedine, S. & Rubin, Y. (1997). Analysis of the Cape Cod tracer data. *Water Resources Research*, 33(1), 1–12.

Farell, D.A., Woodbury, A.D., Sudicky, E.A. & Rivett, M. O. (1994). Stochastic and deterministic analysis of dispersion in unsteady flow at the Borden tracer-test site. *Journal of Contaminant Hydrology*, 15, 159–185,

Fischer, H.B., List, E. J., Koh, R.C.Y., Imberger, J. & Brooks, N.H. (1975). *Mixing in Inland and Coastal Waters*. New York: Academic Press.

Follin, S. (1992). *Numerical Calculations on Heterogeneity of Groundwater Flow*. Ph.D. Dissertation. Stockholm: Royal Institute of Technology.

Gelhar, L.W. & Axness, C.L. (1983). Three-dimensional stochastic analysis of macrodispersion in aquifers. *Water Resources Research*, 19(7), 161–180.

Gelhar, L.W., Welty, C. & Rehfeldt, K.R. (1992). A critical review of data on field-scale dispersion in aquifers. *Water Resources Research*, 28, 1955–1974.

Glimm, J., Lindquist, W.B., Pereira, F. & Zhang, Q. (1993). A theory of macrodispersion for the scale-up problem. *Transport in Porous Media*, 13, 97–122.

Graham, W. & McLaughlin, D. (1989a). Stochastic analysis of nonstationary subsurface solute transport 1. Unconditional moments. *Water Resources Research*, 25(2), 215–232.

Graham, W. & McLaughlin, D. (1989b). Stochastic analysis of nonstationary subsurface solute transport 2. Conditional moments. *Water Resources Research*, 25(11), 2331–2335.

Hyndman, D., Harris, J.M. & Gorelick, S.M. (1994). Coupled seismic and tracer test inversion for aquifer property characterization. *Water Resources Research*, 30(7), 1965–1978.

Kapoor, V. & Gelhar, L.W. (1994). Transport in three-dimensional heterogeneous aquifers: 1. Dynamics of concentration fluctuations. *Water Resources Research*, 30(6), 1775–1788.

Kitanidis, P.K. (1988). Prediction by the method of moments of transport in a heterogeneous formation. *Journal of Hydrology*, 102, 453–473.

Kitanidis, P.K. (1994). The concept of the dilution index. *Water Resources Research*, 30(7), 2011–2026.

Koch, D.L. & Brady, J.F. (1988). Anomalous diffusion in heterogeneous porous media. *Physics of Fluids*, 31, 965–973.

Lallemand-Barres, P. & Peaudecerf, P. (1978). Recherche des relations entre la valeur de la dispersivite macroscopique d'un milieu aquifere, ses autres caracteristiques et les conditions de mesure. Etude bibliographique. *Bulletin BRGM*, III, no. 4, 277–284.

List, J., Gartrell, G. & Winant, C.D. (1990). Diffusion and dispersion in coastal waters. *Journal of Hydraulic Engineering*, 116(10), 1158–1178.

McComb (1990). *The Physics of Fluid Turbulence*. Oxford University Press.

Naff, R.L. (1990). On the nature of dispersive flux in saturated heterogeneous porous media. *Water Resources Research*, 26(5), 1013–1026.

Neuman, S.P. (1990). Universal scaling of hydraulic conductivities and dispersivities in geologic media. *Water Resources Research*, 26(8), 1749–1758.

Neuman, S.P. (1993). Eulerian-Lagrangian theory of transport in space-time nonstationary velocity fields: exact nonlocal formalism by conditional moments and weak approximation. *Water Resources Research*, 29(3), 633–645.

Neuman, S.P. (1994). Generalized scaling of permeabilities: validation and effect of support scale. *Geophysical Research Letters*, 21(5), 349–352.

Neuman, S.P. & Zhang, Y.K. (1990). A quasi-linear theory of non-Fickian and Fickian subsurface dispersion, 1, Theoretical analysis with application to isotropic media. *Water Resources Research*, 26(5), 887–902.

Pope, S.B. (1994). Lagrangian pdf methods for turbulent flows. *Annual Review of Fluid Mechanics*, 26, 23–63.

Quinodoz, H.A.M. & Valocchi, A.J. (1990). Macrodispersion in heterogeneous aquifers: numerical experiments. In *Proceedings of the Conference on Transport and Mass Exchange Processes in Sand and Gravel Aquifers*, vol. 1, ed. G. Moltyaner. Ottawa: Atomic Energy of Canada, pp. 465–468.

Rehfeldt, K.R., & Gelhar, L.W. (1992). Stochastic analysis of dispersion in unsteady flow in heterogeneous aquifers. *Water Resources Research*, 28, 2085–2099.

Rubin, Y. (1990). Stochastic modeling of macrodispersion in heterogeneous porous media. *Water Resources Research*, 26(1), 133–142.

Rubin, Y. (1991a). Prediction of tracer plume migration in disordered porous media by the method of conditional probabilities. *Water Resources Research*, 27(6), 1291–1308.

Rubin, Y. (1991b). Transport in heterogeneous porous media: prediction and uncertainty. *Water Resources Research*, 27(7), 1723–1738.

Rubin, Y. (1991c). The spatial and temporal moments of tracer concentration in disordered porous media. *Water Resources Research*, 27(11), 2845–2854.

Rubin, Y. (1995). Transport in bimodal, multiple scale heterogeneous formations. *Water Resources Research*, 31(10), 2461–2468.

Rubin, Y. & Bellin, A. (1994). The effects of recharge on flow nonuniformity and macrodispersion. *Water Resources Research*, 30(4), 939–948.

Rubin, Y. & Dagan, G. (1992). Conditional estimation of solute travel time in heterogeneous formations 1. Impact of transmissivity measurements. *Water Resources Research*, 28(4), 1033–1040.

Rubin, Y. & Ezzedine, S. (1997). The travel times of solutes at the Cape Cod tracer experiment: Data analysis, modeling and structural parameters inference. *Water Resources Research*, in press.

Rubin, Y. & Seong, K. (1994). Investigation of flow and transport in certain cases of nonstationary heterogeneous formations. *Water Resources Research*, 30(11), 2901–2911.

Rubin, Y., Cushey, M. A. & Bellin, A. (1994). Modeling of transport in groundwater for environmental risk assessment. *Journal of Stochastic Hydraulics and Hydrology*, 8(1), 57–77.

Saffman, P. G. (1963). *Applied Scientific Research*, A11, 245.

Scheibe, T. D. & Cole, C. R. (1994). Non Gaussian particle tracking: Application to scaling of transport processes in heterogeneous porous media. *Water Resources Research*, 30(7), 2027–2040.

Smith, L. & Schwartz, F. W. (1980). Mass transport, 1. Stochastic analysis of macrodispersion. *Water Resources Research*, 16(2), 303–313.

Sposito, G., Jury, W. A. & Gupta, V. (1986). Fundamental problems in the stochastic convection-dispersion model of solute transport in aquifers and field soils. *Water Resources Research*, 22(1) 77–88.

Taylor, G. I. (1921). Diffusion by continuous movements. *Proceedings of the London Mathematical Society*, 2(20), 196–214.

Zhang, D. & Neuman, S. P. (1992). Comment on 'A note on head and velocity covariances in three-dimensional flow through heterogeneous anisotropic porous media' by Y. Rubin and G. Dagan. *Water Resources Research*, 28(12), 3343–3344.

Zhang, Y. K. & Neuman, S. P. (1990). A quasi-linear theory of non-Fickian and Fickian subsurface dispersion: 2. Application to anisotropic media and the Borden site. *Water Resources Research*, 26(5), 903–913.

Zimmerman, D.A., de Masily, G., Gotway, C.A., Marietta, M.G., Axness, C.L., Beauheim, R., Bras, R., Carrera, J., Dagan, G., Davies, P.B., Gallegos, D.P., Galli, A., Gómez-Hernandez, J., Gorelick, S.M., Grindrod, P., Gutjahr, A.L., Kitanidis, P.K., Lavenue, A.M., McLaughlin, D., Neuman, S.P., Ramarao, B.S. & Rubin, Y. (1997). A comparison of seven geostationally-based inverse approaches to estimate transmissivities for modeling advective transport by groundwater flow. *Water Resources Research*, submitted.

2 Transport of reactive solutes

VLADIMIR CVETKOVIC
Royal Institute of Technology, Stockholm

1 INTRODUCTION

Modelling coupled reactions and flow in subsurface formations is of importance in *geochemistry* for interpreting phenomena such as weathering, diagenesis, ore deposition, etc., in reservoir *engineering* for predicting displacement of oil by chemical flooding, in *contaminant hydrology* for predicting the fate of pollutants in soil, groundwater and deep rock formations, and in *biogeochemistry* for quantifying fluxes along flow paths that control element cycling.

Many theoretical and experimental studies over the past decade have focused on understanding flow and nonreactive transport in heterogeneous aquifers (e.g. Dagan, 1982, 1984, 1989; Gelhar & Axness, 1983; Shapiro & Cvetkovic, 1988; Rubin, 1990; Dagan, Cvetkovic & Shapiro, 1992; Neuman, 1993). The use of geostatistical methods for hydraulic data evaluation, and analytical and/or numerical models for calculating the statistics of Eulerian and Lagrangian fluid velocity, have notably increased our confidence in predictions of nonreactive transport in heterogeneous aquifers. A number of relatively simple analytical models that may account for non-Fickian effects, nonergodic transport conditions, etc., are currently available for macrodispersivity as well as for spatial and temporal moments.

In a parallel development, significant progress in simulations of complex reaction systems has improved our understanding of how, under idealized conditions, reactions influence transport, and how transport influences reactions. Numerical models for many simultaneous reactions are currently available, where reaction rates may vary over several orders of magnitude. Models of varying degree of complexity have been used for generic studies (e.g. Liu & Narasimhan, 1989b; Yeh & Tripathi, 1991; McNab & Narasimhan, 1994; Walter *et al.*, 1994b), as well as for case studies (e.g. Valocchi, Street & Roberts, 1981; Kinzelbach, Schäfer & Herzer, 1991). In spite of this progress, however, implementation of models for simulating complex reaction systems to real field conditions is only in its infancy (e.g. Kent *et al.*, 1994).

Comprehensive field experiments on reactive transport in aquifers are relatively few. The experiment at the Borden site, Ontario, was conducted in a sandy aquifer on a scale <60 m where the sorption of five organic compounds was quantified (Roberts, Goltz & Mackay, 1986; Thorbjarnarson & Mackay, 1994). A comprehensive experiment at Cape Cod, Massachusetts, was carried out in a sand and gravel aquifer and involved Li^+ and molybdate (MoO_4^{2-}) on scales <200 m (Le Blanc *et al.*, 1991), and also chromate and selenate on scales <30 m (Kent *et al.*, 1994). An experiment in a sand and gravel aquifer at Vejen, Denmark, involved Na^+ and K^+ and was focused on ion exchange on scales <100 m (Bjerg & Christensen, 1993). A comprehensive field experiment in a single fracture of granite rock was conducted at Grimsel, Switzerland; it involved isotopes of Na, Sr and Cs on scales <5 m and was focused on sorption and diffusive mass transfer relevant for radionuclides (Frick *et al.*, 1992).

This chapter focuses on the conceptual framework for modelling reactive subsurface transport in view of the random heterogeneity observed for physical and chemical properties. Various currently available modelling approaches are mentioned, although this chapter is not a review. Most of the discussed concepts are general; however, our main interest is within contaminant hydrology, the particular focus being on groundwater. We present the Lagrangian formulation of reactive transport that has been developed specifically for describing reactions along advection flow paths. Recognizing the necessity of simplifications in modelling of reactive transport, we discuss possibilities for implementing relatively simple analytical and/or (semi)analytical models for process identification and characterization, as well as for predictive purposes.

2 PROBLEM DESCRIPTION

2.1 General configuration

We consider a two-phase heterogeneous porous formation. The fluid is a mobile phase in the form of an aqueous solution entirely filling the pore space. The immobile phase is

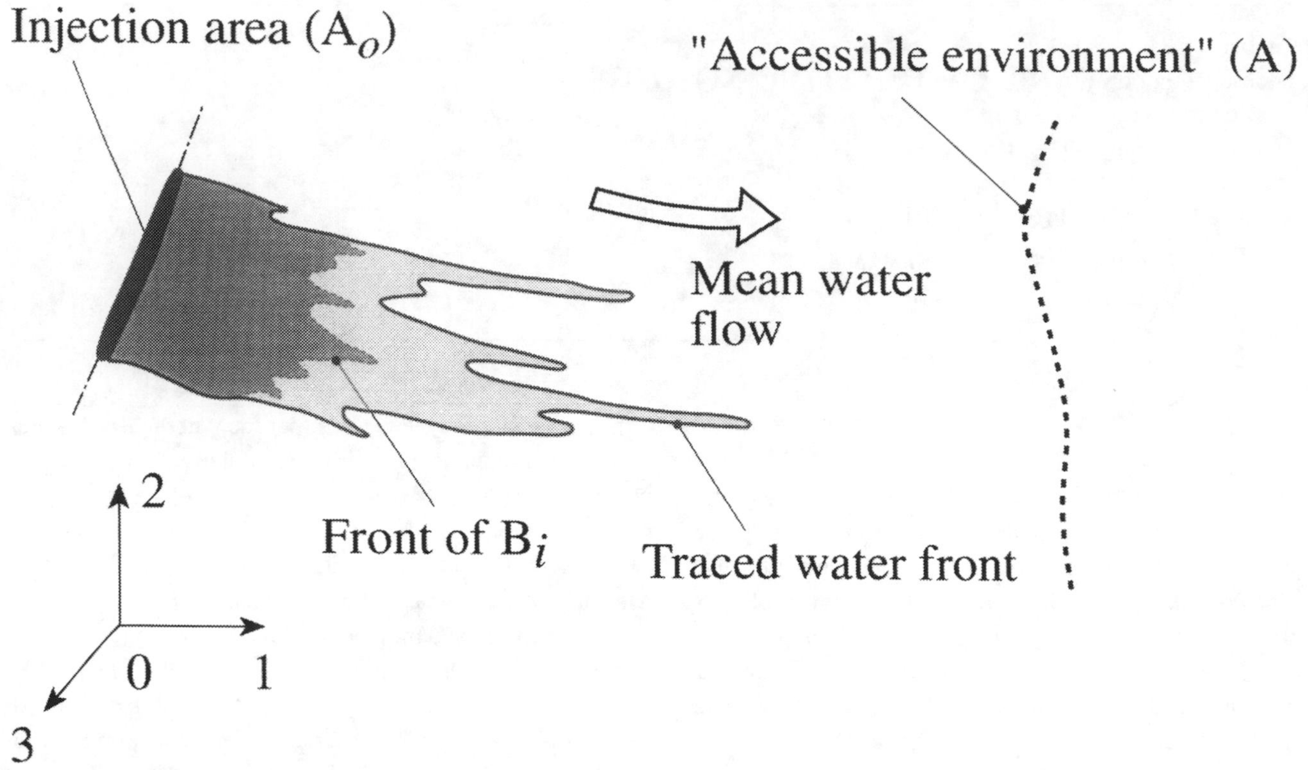

Fig. 1 Configuration sketch; B_i denotes a given species.

essentially solid grains composed of minerals, organic material, etc. that may contain immobile fluid within, say, intra-aggregate pores; also, regions of essentially immobile water may exist in the flow field.

The water is moving due to a prescribed hydraulic head, or a given hydraulic gradient. The hydraulic conductivity, K, is a locally isotropic, random space function (RSF) with given statistical properties (mean and covariance); consequently the water velocity, $\mathbf{v}$, is a RSF. For simplicity, we assume the effective porosity, θ, to be constant. The water velocity field, $\mathbf{v}=\mathbf{v}(\mathbf{x},t)$, is obtained from the mass balance equation combined with Darcy's law. Time dependence of the water velocity is possible due to temporal variations in the boundary conditions, which we assume such that the prevailing flow pattern does not change, only the magnitude of the velocity may change in time.

A mixture of dissolved species with composition different from that of the groundwater enters the porous formation over a given surface A_0 at $t=0$ (Fig. 1). The temporal variation of the component concentrations at A_0 is assumed to be given. Due to the variable properties of the medium, the injected mixture develops an erratic front relative to the ambient water and follows an irregular pattern of flow paths (Fig. 1); in general, different components/species of the injected mixture will develop different propagation fronts. We shall assume that the mixture itself and the ongoing reactions do not influence groundwater flow.

2.2 Reactions

At $t>0$, a total $M+M^*$ of species undergo M_R reactions in a porous formation. The mobile species B_i $(i=1,M)$ are transported by the water movement and undergo both homogeneous reactions and heterogeneous reactions with the immobile species B_j^* $(j=1,M^*)$. The term 'heterogeneous reactions' is used for general exchanges between the mobile water and immobile water/solid, such that even mass transfer reactions due to, e.g., diffusion into intra-aggregate pores and stagnant water (Brusseau & Rao, 1989; Wood, Kraemer & Hearn, 1990) may be accounted for.

Stoichiometric equations for the reaction system are (Friedly & Rubin, 1992)

$$0 \Leftrightarrow \sum_{i=1}^{M} \nu_{ri} B_i + \sum_{j=1}^{M^*} \nu_{rj}^* B_j^* \qquad r=1,M_R \tag{1}$$

where ν,ν^* are stoichiometric coefficient matrices, and the formulation is such that all the M_R reactions are linearly independent. By convention, the stoichiometric coefficients of the reactants are negative. Each reaction $r=1,M_R$ will have a net rate (forward minus reverse) denoted by I_r; for the equilibrium limit of the rth reaction, $I_r=0$, which yields the thermodynamic equilibrium relationship.

The symbolic expression (1) is general and may represent complex reactions (dissolution, precipitation, oxidation, reduction, hydrolysis, complexation), as well as simple (nonspecific) exchange and/or transformation reactions that are most often used in engineering applications (sorption–desorption, diffusive mass transfer, decay, degradation). For instance, a two-component reversible exchange with essentially infinite exchange capacity is written as $B \leftrightarrow B^*$; if the exchange capacity is limited, we write $B_1 + B_2^* \leftrightarrow B_1^*$ where B_2^* is the sorbent and B_1^* is the immobilized product. The simplest irreversible reaction is $B_1 \rightarrow B_2$, for instance, due to transmutation, or $B \rightarrow B^*$ applicable to, say, degradation.

In the following, we discuss the dynamic modelling of the propagation front evolution for any given species B_i. Because of the irregularity of the transition zone (Fig. 1), we generally have to adopt some type of averaging. For instance, we may define the first and second spatial moment of the transition zone for species B_i, and describe its evolution in time. We may also be interested in quantifying the first two moments (mean and variance) of the concentration of species B_i at any point in the domain. Alternatively, we may be interested in the first two moments of the mass flux (mean and variance) of species B_i as it is discharged into the accessible environment over a given surface A (Fig. 1).

3 EULERIAN FORMULATION

3.1 Governing equations

The coupled transport-reaction equations for the concentration of species B_i are written in the form

$$\frac{\partial}{\partial t}\begin{pmatrix}\theta C_i\\ C_j^*\end{pmatrix}+\nabla\cdot\begin{pmatrix}\mathbf{J}_i\\ \mathbf{0}\end{pmatrix}=\begin{pmatrix}\Im_i\\ \Im_j^*\end{pmatrix}\qquad i=1,M;\quad j=1,M^* \tag{2}$$

where θ is the effective porosity. The rate functions are defined by

$$\Im_i=\sum_{r=1}^{M_R}\nu_{ir}I_r[C_1,\dots,C_M,C_1^*,\dots,C_{M^*}^*;P(\mathbf{x},t)]$$

$$\Im_j^*=\sum_{r=1}^{M_R}\nu_{jr}^*I_r[C_1,\dots,C_M,C_1^*,\dots,C_{M^*}^*;P(\mathbf{x},t)]$$

where P denotes a set of reaction parameters (e.g. rate coefficients) which may depend on space and time. In (2), $\mathbf{J}_i$ is the mass flux of B_i defined as $\mathbf{J}_i=\theta\mathbf{v}C_i-(\theta\mathbf{D}\cdot\nabla C_i)$, where $\mathbf{v}$ is the groundwater advection velocity and $\mathbf{D}$ is the tensor of hydrodynamic dispersion (Bear, 1972). The immobile concentrations in (2) are defined per unit bulk volume of the representative elementary volume (REV). Derivation of the transport-reaction equation of the type (2) is discussed, for example, by Lichtner (1985). A classification of the different types of reactions represented by (2) that are relevant for groundwater is given, for instance, by Rubin (1983) and Cherry, Gillham & Barker (1984).

3.2 Special cases

If steady-state is established with respect to the mobile concentrations C_i, eq. (2) may be written as

$$\frac{\partial}{\partial t}\begin{pmatrix}0\\ C_j^*\end{pmatrix}+\nabla\cdot\begin{pmatrix}\mathbf{J}_i\\ \mathbf{0}\end{pmatrix}=\begin{pmatrix}\Im_i\\ \Im_j^*\end{pmatrix}\qquad i=1,M;\quad j=1,M^* \tag{3}$$

The system (3) is the basis for what is referred to as a 'quasi-stationary' approximation (Lichtner, 1988) and is applicable for modelling fluid–rock interactions over geologic time. Although the system is steady-state with respect to C_i, it is not in equilibrium; however, (3) is the equivalent of equilibrium for an open system.

Integrating (2) over the entire domain (volume) of interest, we obtain for a zero-dimensional open system

$$\frac{d}{dt}\begin{pmatrix}\theta C_i\\ C_j^*\end{pmatrix}+\begin{pmatrix}F_i\\ 0\end{pmatrix}=\begin{pmatrix}\Im_i\\ \Im_j^*\end{pmatrix}\qquad i=1,M;\quad j=1,M^* \tag{4}$$

where F_i is the net flux of species B_i and is related to the water flux into and out of the control volume; the quantities in (4) are defined per unit control volume. Although (4) cannot be used for modelling front propagation (fully mixed conditions are assumed), it is relatively simple to solve and is useful for a first assessment of possible reactions in systems with many interacting species. A steady-state form of (4) has been used for modelling geochemical systems (see, e.g., Furrer, Westall & Sollins, 1989).

If we close the system (4) by setting $F_i=0$, we recover the standard model for batch conditions, i.e. a flux-free system that has not yet reached equilibrium. Once steady-state is reached, (4) yields the equation system of thermodynamic equilibrium as $\Im_i=0$ and $\Im_j^*=0$.

Open and closed zero-dimensional models are significant since they are most often used for interpretation of laboratory data.

3.3 Solutions

For given boundary conditions $C_i(\mathbf{a},t)$, where $\mathbf{a}\in A_0$, initial conditions $C_i(\mathbf{x},0)$ and $C_j^*(\mathbf{x},0)$, and for given rate functions and reaction parameters, $P=P(\mathbf{x},t)$, eq. (2) can in principle be solved for the mobile concentration $C_i=C_i(\mathbf{x},t)$; the solution quantifies the propagation of a contaminant front, a redox front, a dissolution front, a biomass production front, etc., through the porous formation.

The interpretation of the quantities in (2) will essentially depend on the scale of the reactive transport problem. For instance, if the mean velocity is unidirectional and (2) is written in one dimension, where $\mathbf{v}$ represents a mean velocity, then the concentration is essentially an expected residence concentration, spatially averaged across the propagation front, i.e. orthogonal to the mean flow. A multidimensional simulation of (2) can in principle be applied to heterogeneous formations, where both the advection velocity and reaction parameters are spatially and/or temporally variable, provided that sufficient data for a deterministic simulation are available.

Different solution techniques and models for the system (2) with P=*const*. have been proposed and implemented for a variety of transport problems in the subsurface (e.g. Jennings, Kirkner & Theis, 1982; Miller & Benson, 1983; Schulz & Reardon, 1983; Liu & Narasimhan, 1989a,b; Yeh & Tripathi, 1991; Engesgaard & Kipp, 1992; Friedly & Rubin, 1992; Walter *et al.*, 1994a,b). The system (2) has also been applied to reactions mediated by microbial activity (e.g. Molz, Widdowson & Benefield, 1986; Frind *et al.*, 1990; Kinzelbach *et al.*, 1991; McNab & Narasimhan, 1994; Zysset, Stauffer & Deacos, 1994).

Changes in species concentration can be described statistically by Monte Carlo simulations using (2), provided that the statistics of flow velocity and reaction parameters are known. Such simulation would yield both expected concentrations and the associated variances that quantify the concentration fluctuations. Analytical and/or simulation results for the system (2) that explicitly account for the heterogeneity of subsurface formation properties have been provided only for relatively simple, two-component, sorption–desorption reactions.

In particular, linear equilibrium sorption where the fluid velocity and the distribution coefficient are spatially variable has been analysed analytically (e.g. Kabala & Sposito, 1991; Chrysikopoulos, Kitanidis & Roberts, 1992) and using simulations (e.g. Tompson, 1993; Burr, Sudicky & Naff, 1994). Nonlinear equilibrium sorption with physical and/or chemical heterogeneity has been considered analytically (Kabala & Sposito, 1994) and by numerical simulations (Bosma & Van der Zee, 1993; Tompson, 1993; Tompson, Schafer & Smith, 1996). Rate-limiting sorption with first-order kinetics and random fluid velocity and rate coefficients ratio has been investigated analytically (Hu, Deng & Cushman, 1995) and by numerical simulations (Burr *et al.*, 1994). The results of these studies have shown that the effect of spatial variability in flow and reaction parameters may be significant for quantifying transport on the field scale, in particular if a strong correlation exists between a sorption parameter and the hydraulic conductivity.

4 LAGRANGIAN FORMULATION

Field-scale transport of nonreactive solutes through heterogeneous formations is advection dominated. A significant fraction of the water flow may be along more or less distinct ('preferential') flow paths, the distinction being more apparent for stronger heterogeneity. As a consequence, solute transport often takes place over relatively limited portions of a formation, a fact that may significantly influence reactions (see e.g., Kent *et al.*, 1994).

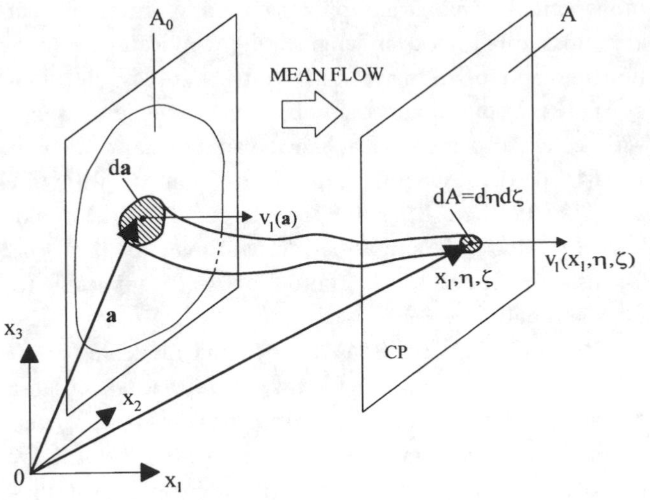

Fig. 2 Advection flow path.

In the following, we discuss the Lagrangian formulation of reactive transport that is specifically focused on advection flow paths in groundwater (Cvetkovic & Dagan, 1994; Dagan & Cvetkovic, 1996). The Lagrangian approach to reactive transport in the subsurface has been applied for multi-component ion-exchange (Charbeneau, 1988), fluid–rock interactions (Lichtner, 1988), linear equilibrium sorption (Dagan, 1989; Bellin *et al.*, 1993), rate-limiting sorption with first-order kinetics (e.g. Cvetkovic & Shapiro, 1990; Selroos & Cvetkovic, 1994), parallel linear equilibrium and non-equilibrium sorption (Destouni & Cvetkovic, 1991) and biodegradation (Simmons, Ginn & Wood, 1995; Ginn, Simmons & Wood, 1995)

4.1 Advection flow paths

For simplicity, we shall assume that the flow field is steady-state, and that the average velocity field, $\mathbf{U}=E(\mathbf{v})$, is uniform and parallel to the x_1-direction of a Cartesian coordinate system $\mathbf{x}(x_1,x_2,x_3)$ (Fig. 2), i.e. $\mathbf{U}(U,0,0)$, where $E(\ )$ is the ensemble average operator. The statistical properties of $\mathbf{v}$, given the statistical properties of K, have been investigated by analytical and numerical methods (e.g. Dagan 1982, 1984; Gelhar & Axness, 1983; Rubin, 1990) and are here assumed known.

The Lagrangian representation of the movement of a marked fluid particle originating from $\mathbf{x}=\mathbf{a}$ is based on its trajectory $\mathbf{x}=\mathbf{X}(t;\mathbf{a})$, which satisfies the kinematic equation:

$$\frac{d\mathbf{X}}{dt}=\mathbf{v}(\mathbf{X}), \qquad \mathbf{X}(0;\mathbf{a})=\mathbf{a} \tag{5}$$

The evaluation of the first two statistical moments of $\mathbf{X}$ has been the basis for a number of investigations of solute transport by groundwater (e.g. Dagan, 1984, 1989). The field-scale macrodispersion tensor is obtained from the rate of change of the second moment of **X**.

An alternative Lagrangian representation of the marked fluid particle is derived from **X** as (Dagan *et al.*, 1992)

$$t=\tau(x_1,\mathbf{a}) \text{ solution of } x_1-X_1(t;\mathbf{a})=0$$
$$\eta(x_1;\mathbf{a})=X_2(t;\mathbf{a});\ \zeta(x_1;\mathbf{a})=X_3(t;\mathbf{a})$$

where $X_1(t;\mathbf{a})$ is assumed to be a monotonically increasing (unique) function of t. Although isolated stagnation points with $\mathbf{v}=\mathbf{0}$ are possible (Sposito, 1994), these are not considered significant for the advection of species. A modified formulation is required to strictly account for possible non-uniqueness of $X_1(t;\mathbf{a})$; for simplicity, we assume in the present discussion no stagnation and/or 'backward' flow.

The function $\tau(x_1;\mathbf{a})$ is the travel (residence) time of the advecting fluid particle from **a** to a 'control plane' normal to the mean flow at x_1 (Fig. 2). Similarly, $x_2=\eta(x_1;\mathbf{a})$ and $x_3=\zeta(x_1;\mathbf{a})$ are the equations of a streamline passing through $\mathbf{x}=\mathbf{a}$. These random functions can be related to the velocity field through

$$\frac{d\tau}{dx_1}=\frac{1}{v_1(x_1,\eta,\zeta)};\quad \frac{d\eta}{dx_1}=\frac{v_2(x_1,\eta,\zeta)}{v_1(x_1,\eta,\zeta)};\quad \frac{d\zeta}{dx_1}=\frac{v_3(x_1,\eta,\zeta)}{v_1(x_1,\eta,\zeta)} \qquad (6)$$

A joint probability density function (pdf) for τ,η,ζ that quantifies the three-dimensional particle spreading can be defined (Dagan *et al.*, 1992). The marginal pdf $g(\tau;x_1-a_1)$ quantifies the probability of a particle originating at $\mathbf{x}=\mathbf{a}$ crossing the control plane at x_1 at time $t=\tau$; various forms of the pdf g for transport in soils and groundwater have been discussed (e.g. Jury, 1982; Simmons, 1982; Shapiro & Cvetkovic, 1988; Dagan & Nguyen, 1989; Bellin, Salandin & Rinaldo, 1992; Destouni, 1993; Russo, 1993; Destouni, Sassner & Jensen, 1994).

4.2 Reactions along flow paths

In the following, we shall neglect the mass transfer *between* adjacent flow paths due to diffusion and pore-scale dispersion and focus on advection only. Setting $\mathbf{J}_i=\theta\mathbf{v}C_i$ in (2), we obtain the governing transport equations as

$$\frac{\partial}{\partial t}\begin{pmatrix}C_i\\C_j^*\end{pmatrix}+\left[\mathbf{v}\cdot\nabla\begin{pmatrix}C_i\\0\end{pmatrix}\right]=\begin{pmatrix}\Im_i\\\Im_j^*\end{pmatrix} \qquad i=1,M;\ \ j=1,M^* \qquad (7)$$

where the water is assumed incompressible, and effective porosity has been incorporated into the source term. Following the procedure of Cvetkovic & Dagan (1994) and Dagan & Cvetkovic (1996), we use a transformation replacing the independent variables $\mathbf{x},t$ with $\boldsymbol{\xi},t$ in (7), where the components of $\boldsymbol{\xi}$ are defined as

$$\xi_1=\tau(x_1;\mathbf{a});\ \xi_2=x_2-\eta(x_1;\mathbf{a});\ \xi_3=x_3-\zeta(x_1;\mathbf{a}) \qquad \mathbf{a}\in A_0 \qquad (8)$$

Setting $\xi_2=\xi_3=0$ yields transport equations along the advection flow path associated with the streamline through **a** as

$$\frac{\partial}{\partial t}\begin{pmatrix}C_i\\C_j^*\end{pmatrix}+\frac{\partial}{\partial\tau}\begin{pmatrix}C_i\\0\end{pmatrix}=\begin{pmatrix}\Im_i\\\Im_j^*\end{pmatrix}\begin{matrix}[C_1,\ldots,C_M,C_1^*,\ldots,C_{M^*}^*;\ P(\tau)]\\ [C_1,\ldots,C_M,C_1^*,\ldots,C_{M^*}^*;\ P(\tau)]\end{matrix} \qquad (9)$$

for $\xi_1=\tau$, $\xi_2=\xi_3=0$ $(i=1,M;\ j=1,M^*)$ where $P(\tau)\equiv P[\mathbf{X}(t;\mathbf{a}),t]\|_{t=\tau}$. If the reaction parameters do not change in time, only in space, then $P=P(\mathbf{x})$, and $P(\tau)\equiv P[\mathbf{X}(\tau)]$. Note that C_i, C_j^*, $\Im_i$, $\Im_j^*$ and P are different functions in (7) and (9), although we retain the same notation for simplicity.

Equation (9) is written for advection only; however, the potentially significant effect of dispersion and diffusion of solutes into essentially immobile regions *within* advection flow paths (intra-granular pores) are accounted for in the exchange terms, i.e. as mass transfer reactions. Hence, the (non)reactive species (tracer) cannot propagate faster than advected by the groundwater on the Darcy scale; the retardation of species due to any type of mass transfer into immobile zones is accounted for in the mass exchange term.

4.3 Solutions

For initial conditions $C_i(0,\tau)=\widehat{C}_i(\tau)$ and $C_j^*(0,\tau)=\widehat{C}_j^*(\tau)$, boundary conditions $C_i(t,\mathbf{a})=C_i^0(t)$, for given rate functions and for given $P=P(\tau)$, the advection–reaction system (9) can in principle be solved to yield the evolution of the injected mixture propagation front for a given flow path as $C_i=C_i(t,\tau)$, valid for $\mathbf{a}\in A_0, x_2=\eta(x_1;\mathbf{a})$, $x_3=\zeta(x_1;\mathbf{a})$. The three-dimensional nature of the solution of (9) stems from the dependence of τ,η,ζ upon x_1 and **a** within A_0 (Fig. 1). Since the streamline and the travel time differ from realization to realization of the velocity field, τ,η,ζ are RSFs, whereby C_i is also a RSF. In principle, any statistical moment of C_i can be computed if the joint pdf of these RSFs are known for different **a** and x_1. A considerable simplification is achieved if $P(\tau)$ in (9) is substituted for by constant, as yet unspecified, reaction parameters, P_e, applicable for a given advection flow path. If $P=P(\mathbf{x})$ is correlated to the hydraulic conductivity, the average ('equivalent') effective P_e will be correlated to the water travel/residence time τ. In a subsequent section we shall discuss the correlation between P_e and τ.

In general, (9) needs to be solved numerically. Lichtner (1988) developed a simulation model for fluid–rock interactions by considering pure advection and invoking a quasi-stationary assumption. For several types of two-component reactions, and for simpler multicomponent reactions, analytical solutions have been obtained for analyzing chromatographic effects. In particular, for multicomponent exchange with $P=P_e=const.$, (9) can be reduced to the Reimann problem and solved by analytical means (e.g. Rhee, Aris & Amundson, 1970; Charbeneau, 1988). Similarly, analytical solutions of (9) for two-component reactions with $P=P_e=const.$ are available for ion-exchange with second-order kinetics (e.g. Thomas, 1944), for kinetically controlled linear and bilinear sorption (e.g. Amundson, 1950), and for mineral dissolution with nonlinear kinetics (e.g. Lund & Fogler, 1976).

Equation (9) may be generally simpler to solve than (2). However, the solution is applicable for an advection flow path that is unspecified. In addition, if $P=P_e=const.$ is assumed along a flow path, we need to define the average reaction parameters, i.e. relate P_e to the statistics of measurable $P=P(\mathbf{x})$.

5 FIELD-SCALE DESCRIPTION

In the following, we shall assume that the reaction parameters are constant effective values, possibly correlated to the water residence/travel time, τ. We wish to derive expressions for global quantities that describe transport of species B_i (e.g. the first two moments of the mass flux, expected spatial moments etc.) as functions of the solution $C_i=C_i(t,\tau)$ and the statistics of the water velocity and reaction parameters. The simplifying assumptions in this section are identical to those in Section 4. In the following, we apply the methodology of Cvetkovic & Dagan (1994) and of Dagan & Cvetkovic (1996).

5.1 Mass flux

From $C_i=C_i(t,\tau)$, obtained by solving (9), the mass flux of species B_i across an elementary area $dA=d\eta d\zeta$ of the control plane A (Fig. 2) can be evaluated as

$$q_i=\theta C_i v_1(x_1,\eta,\zeta) \tag{10}$$

The field-scale integrated mass flux (or discharge) across the control plane A at x_1 of the species B_i, which is injected at time $t=0$ into the porous formation over an area A_0 at $x_1=0$, is obtained from (10) as

$$Q_i(t,x_1)=\int_A q_i dA=\theta\int_{A_0} C_i(t,\tau)v_0 d\mathbf{a} \tag{11}$$

where C_i depends on $\mathbf{a}\in A_0$ through τ. In (11), we have used the water mass balance equation in the form $v_1(x_1,\eta,\zeta)dA=v_0 d\mathbf{a}$, where $v_0\equiv v_1(\mathbf{a})$ and $d\mathbf{a}$ is an elementary surface area of A_0. The concentration C_i, and consequently Q_i, is a RSF. We characterize the field-scale mass flow of species B_i by the first two moments:

$$E(Q_i)\equiv\overline{Q}_i(t,x_1)=\theta A_0\int_0^\infty\int_0^\infty C_i(t,\tau,P_e)\tilde{g}(\tau,P_e;x_1)d\tau dP_e \tag{12}$$

and

$$E(Q_i^2-\overline{Q}_i^2)\equiv\sigma_{Qi}^2(t,x_1)=\theta^2\int_{A_0}\int_{A_0}\int_0^\infty\int_0^\infty\int_0^\infty\int_0^\infty C_i(t,\tau';P_e')\,C_i(t,\tau'';P_e'')$$

$$\times\tilde{g}_2(\tau',\tau'',P_e',P_e'';x_1,\mathbf{a}',\mathbf{a}'')dP_e'dP_e''d\tau'd\tau''d\mathbf{a}'d\mathbf{a}''-\overline{Q}_i^2 \tag{13}$$

where we retain the same symbol, A_0, for the measure of the area over which the species is injected into the formation. The functions $\tilde{g}$ and $\tilde{g}_2$ are defined by

$$\tilde{g}(\tau,P_e;x_1)\equiv\int v_0 g(\tau,v_0,P_e;x_1)dv_0 \tag{14}$$

$$\tilde{g}_2(\tau',\tau'',P_e',P_e'';x_1,\mathbf{a}',\mathbf{a}'')$$

$$\equiv\int\int v_0',v_0''g_2(\tau',\tau'',v_0',v_0'',P_e',P_e'';x_1,\mathbf{a}',\mathbf{a}'')dv_0'dv_0'' \tag{15}$$

where $\mathbf{a}'$, $\mathbf{a}''\in A_0$ and g and g_2 are the one- and two-particle joint pdfs. The functions g and g_2 depend on the flow field and the reaction parameter field. If P_e is a fixed parameter, (12) and (13) simplify to

$$\overline{Q}_i(t,x_1)=\theta A_0\int_0^\infty C_i(t,\tau,P_e)\tilde{g}(\tau;x_1)d\tau \tag{16}$$

and

$$\sigma_{Qi}^2(t,x_1)=\theta^2\int_{A_0}\int_{A_0}\int_0^\infty\int_0^\infty C_i(t,\tau';P_e)\,C_i(t,\tau'';P_e)$$

$$\times\tilde{g}_2(\tau',\tau'',x_1,\mathbf{a}',\mathbf{a}'')d\tau'd\tau''d\mathbf{a}'d\mathbf{a}''-\overline{Q}_i^2 \tag{17}$$

where $\tilde{g}(\tau;x_1)$ and $\tilde{g}_2(\tau',\tau'';x_1,\mathbf{a}',\mathbf{a}'')$ are defined analogous to (14) and (15).

If the correlation of τ and P_e with v_0 is negligible, then we have $\tilde{g}=Ug$ and $\tilde{g}_2=U^2g_2$. For field-scale transport governed by the advection-dispersion equation, g takes specific forms (e.g. Kreft & Zuber, 1978; Shapiro & Cvetkovic, 1988). The function $\tilde{g}$ for correlated v_0 and τ, and for fixed (given) P_e, which is consistent with a Gaussian distribution for $X_1(t)$, has been derived recently (Cvetkovic & Dagan, 1996). In principle, g and g_2 can be computed using standard *non-reactive* particle tracking Monte Carlo simulations, provided that the statistics of the reaction parameter field is known and that the groundwater flow field $\mathbf{v}(\mathbf{x})$ has been determined.

Equation (16) is similar in form to expressions that have been used in chromatographic theory (e.g. Villermaux, 1974) and in soils (e.g. Sardin *et al.*, 1991) for linear non-equilibrium sorption, where g is referred to as a 'transfer function'. For linear irreversible sorption, (17) reduces to the form derived by Destouni (1992), and for reversible non-equilibrium sorption by Selroos & Cvetkovic (1994); for nonreactive solute, (17) reduces to the form given by Dagan *et al.* (1992).

From $\overline{Q}_i$, the expected flux averaged concentration can be evaluated as $\overline{Q}_i/\overline{Q}_0$ (Kreft & Zuber, 1978); $\overline{Q}_0$[L^3/T] is the total water flux in which the species mass is mixed, thereby accounting for dilution. The two moments $\overline{Q}_i$ and σ^2_{Qi} may be considered as conditional moments if the parameters are uncertain; unconditional values are obtained through integration, assuming the pdfs (or the corresponding moments) of the uncertain parameters (e.g. Dagan, 1989; Andricevic & Cvetkovic, 1996). The integrated mass flux $\overline{Q}_i$ quantifies the evolution of the propagating mixture front in the direction of the mean flow only. Following the procedure of Dagan & Cvetkovic (1996), transverse extent of the mixture front can also be evaluated. Temporal moments for nonergodic transport of reacting solute can be evaluated as an alternative to the mass flux representation (Selroos, 1995).

5.2 Spatial moments

For some applications, it is of interest to quantify the spatial extent of the propagating mixture front in the longitudinal direction by means of spatial moments (e.g. Roberts *et al.*, 1986). Recent analytical and simulation efforts have been directed toward evaluating spatial moments of a solute advected by a random velocity field and subject to sorption reactions. Specifically, linear equilibrium sorption has been considered (Chrysikopoulos *et al.*, 1992; Bellin *et al.*, 1993; Bosma *et al.*, 1993), linear nonequilibrium sorption (Valocchi, 1989; Quinodoz & Valocchi, 1993; Cvetkovic & Dagan, 1994; Hu *et al.*, 1995), and nonlinear equilibrium sorption (Bosma & Van der Zee, 1993; Tompson, 1993).

In the following, we give expressions for longitudinal spatial moments under additional simplifying assumptions: (i) species are injected at constant concentration $C_i(t,0)=C_i^0$, with uniform initial conditions $C_i(0,\tau)=\overline{C}_i$; (ii) C_i is a monotonically increasing or decreasing function; (iii) transport conditions are ergodic; and (iv) $P=const.$

The first two longitudinal spatial moments are defined and evaluated as (Dagan & Cvetkovic, 1996)

$$E(R_i)=R_i(t)=\frac{1}{\theta|C_i^0-\overline{C}_i|A_0}\int_0^\infty \frac{\partial M_i(x_1,t)}{\partial x_1}dx_1=U\int_0^\infty \Gamma_i(t,\tau)d\tau \quad (18)$$

and

$$E(S_i)=S_i(t)=\frac{1}{\theta|C_i^0-\overline{C}_i|A_0}\int_0^\infty (x_1-R_i)^2\frac{\partial^2 M_i}{\partial x_1^2}dx_1$$

$$=2U^2\int_0^\infty \tau\Gamma_i(t,\tau)d\tau-R_i^2+\int_0^\infty \frac{dX_{11}}{d\tau}\Gamma_i(t,\tau)d\tau \quad (19)$$

where X_{11} is the longitudinal variance of nonreactive particle displacement (e.g. Dagan, 1984) and $M_i(x_1,t)$ is the total mass of species B_i from the injection plane ($x_1=0$) to the control plane at x_1. The function Γ_i is defined from (Dagan & Cvetkovic, 1996)

$$C_i(t,\tau)=\overline{C}_i+(C_i^0-\overline{C}_i)\Gamma_i(t,\tau;C_j^0,\overline{C}_j) \quad (20)$$

where we emphasize that, in general, Γ_i is a nonlinear function of $\overline{C}_j$ and C_j^0. Higher-order spatial moments can also be evaluated following the procedure of Dagan & Cvetkovic (1996). For pulse injection, (18) and (19) take a modified form (Cvetkovic & Dagan, 1994).

6 EQUIVALENT REACTION PARAMETERS

Combined laboratory and field investigations indicate a discrepancy between reaction parameters measured in batch and column tests, and those estimated from field-scale experiments; it has been observed for sorption reactions (e.g. Ptacek & Gillham, 1992) as well as for other types of reactions (e.g. Kent *et al.*, 1994). The discrepancy between laboratory and field reaction parameters may be due to their spatial variability, such that data from laboratory experiments do not provide representative values. In addition, due to heterogeneity, mixing will be different in the laboratory and the field, which may have a significant impact on reactions (e.g. Kent *et al.*, 1994).

A number of investigations have focused on the effect of spatial variability in hydraulic and/or reaction parameters on transport. The considered reactions are linear equilibrium sorption using analytical (e.g. Garabedian, 1987; Dagan, 1989; Cvetkovic & Shapiro, 1990; Kabala & Sposito, 1991; Chrysikopoulos *et al.*, 1992; Bellin *et al.*, 1993) and simulation methods (e.g. Bosma *et al.*, 1993), nonlinear equilibrium sorption using analytical models (e.g. Van der Zee & Van Riemsdijk, 1987; Wise, 1993) and simulations (e.g. Bosma & Van der Zee, 1993; Tompson, 1993), and linear nonequilibrium sorption using analytical methods (e.g. Destouni & Cvetkovic, 1991; Hu *et al.*, 1995) and numerical simulations (e.g. Burr *et al.*, 1994). These investigations have shown that the effects on transport may be significant, depending on the degree and type of correlation between the hydraulic conductivity, *K*, and the reaction parameters, *P*.

6.1 Analytical result

If $P(\mathbf{x})$, and thereby $P(\tau)=P[\mathbf{X}(\tau)]$, are RSFs, then eqs. (9) are stochastic partial differential equations (e.g. Beran, 1968) and a general solution is not available. As noted in Section 4.3, a considerable simplification for solving (9) is achieved if $P=P(\tau)$ in (9) is replaced by an average (equivalent) value, $P_e=const.$, valid for a given flow path. In the following, we shall discuss the definition and nature of P_e.

Cvetkovic & Dagan (1997) have investigated the effect of spatial variability in P for two-component exchange (sorption) reactions: nonlinear equilibrium sorption (where P corresponds to the maximum sorption capacity), and linear nonequilibrium sorption (where P corresponds to the ratio of rate coefficients). For these cases, where $P(\mathbf{x})$ is a given stationary RSF, the solution of (9) is obtained as $C_i=C_i(t,\tau,\mu)$, where μ is a Lagrangian functional that depends on $P(\mathbf{x})$ and on the trajectory $\mathbf{X}$ and can be used to define P_e as:

$$\mu(\tau;\mathbf{a})=\int_0^\tau P[\mathbf{X}(t;\mathbf{a})]dt \equiv P_e\tau \qquad (21)$$

Thus P is weighted by the infinitesimal residence times along the flow path $\mathbf{X}$.

For the abovementioned class of sorption reactions, (21) provides a definition of an average ('equivalent') reaction parameter, P_e, for a given flow path. In general, P_e is a function of τ. If $P(\mathbf{x})$ is not correlated with the hydraulic conductivity, then P_e will become constant for $t\to\infty$. If $P(\mathbf{x})$ is correlated to the hydraulic conductivity, and hence to the advection velocity $\mathbf{v}=\mathbf{v}(\mathbf{x})$, correlation between τ and P_e may persist even for $\tau\to\infty$. Although (21) provides a rigorous definition of P_e only for a limited class of sorption–desorption reactions, we may assume as a working hypothesis that (21) is also applicable for other reactions. In the following, we shall evaluate P_e and assess the correlation between τ and P_e using a first-order perturbation expansion.

6.2 Approximative result

Let P and K be RSFs defined as $K(\mathbf{x})=K_G\exp[Y(\mathbf{x})]$ and $P(\mathbf{x})=P_G\exp[\beta Y(\mathbf{x})+W(\mathbf{x})]$, where $Y{:}N(0,\sigma_Y^2)$ and $W{:}N(0,\sigma_W^2)$, the subscript G denotes geometric means of K and P, and β is a parameter characterizing the degree and type of correlation between K and P, (e.g. Robin *et al.*, 1991). The covariance functions C_Y and C_W for Y and W are assumed to be negative exponentials with integral scales ℓ_Y and ℓ_W. We now wish to use the definition (21) and evaluate the mean and variance of the equivalent reaction parameter, P_e, and the degree of correlation between τ and P_e, as functions of β, σ_Y^2, σ_W^2, ℓ_Y and ℓ_W.

We approximate in (21) $P[\mathbf{X}(t)]\approx P(X_1,0,0)=P(Ut,0,0)$, and $\tau=\bar{\tau}=x_1/U$ in the limit of the integral where $U=K_GJ/\theta$ and J is the mean hydraulic gradient assumed parallel to x_1; these approximations are consistent with the first-order approximation (e.g. Dagan, 1984). The mean of P_e is the arithmetic mean, i.e. $\bar{P}_e=\bar{P}$. To evaluate the variance of P_e, we expand the exponentials and retain first-order terms; the fluctuations of P_e and τ are then approximated by

$$P_e'\equiv P_e-\bar{P}_e\approx\frac{P_G}{x_1}\int_0^{x_1}(\beta Y+W)dx \quad \text{and} \quad \tau'\equiv\tau-\bar{\tau}\approx-\frac{1}{U}\int_0^{x_1}Ydx$$

where, for simplicity, the effect of fluctuations in the hydraulic gradient on τ has also been neglected. We have the following limiting cases. For $x_1/\ell_Y\to 0$ and $x_1/\ell_W\to 0$

$$\sigma_{P_e}^2=P_G^2\sigma_Y^2\left(\beta^2+\frac{\sigma_W^2}{\sigma_Y^2}\right);\quad \rho_{\tau P_e}=\frac{-\beta}{\left(\beta^2+\dfrac{\sigma_W^2}{\sigma_Y^2}\right)^{1/2}} \qquad (22)$$

and for sufficiently large x_1/ℓ_Y (assuming $\ell_W/\ell_Y\le 1$)

$$\sigma_{P_e}^2=\frac{2P_G^2\sigma_Y^2}{(x_1/\ell_Y)}\left(\beta^2+\frac{\sigma_W^2}{\sigma_Y^2}\frac{\ell_W}{\ell_Y}\right);\quad \rho_{\tau P_e}=\frac{-\beta}{\left(\beta^2+\dfrac{\sigma_W^2}{\sigma_Y^2}\dfrac{\ell_W}{\ell_Y}\right)^{1/2}} \qquad (23)$$

where the correlation coefficient is $\rho_{\tau P_e}\equiv E(\tau' P_e')/\sigma_\tau\sigma_{P_e}$. For cases where the correlation between v_0 and τ is negligible, we may approximate $g(\tau,P_e)=\tilde{g}(\tau,P_e)/U$ in (12) by a joint pdf (say lognormal) where (22) and (23) provide the additional parameters. For $\sigma_{P_e}^2\to 0$, $g(\tau,P_e)\to g(\tau)\delta(P_e-\bar{P})$, and the equivalent reaction parameter is exactly the arithmetic mean of $P(\mathbf{x})$.

Expressions (22) and (23) indicate that the variance of the equivalent reaction parameter $\sigma_{P_e}^2$ is a function of scale. For a transport scale that is comparable to the heterogeneity scale (case (22)), $\sigma_{P_e}^2$ may be large for $|\beta|$ close to unity, and/or for σ_W^2 large. For a transport scale large in comparison to the heterogeneity scale (case (23)), the variance $\sigma_{P_e}^2$ diminishes as $x_1/\ell_Y\to\infty$. Also, for fluctuations in the reaction parameter that are over short distances ('white noise' type), the uncorrelated part of P fluctuations (quantified by σ_W^2) has a small effect on $\sigma_{P_e}^2$. The correlation coefficient $\rho_{\tau P_e}$ is independent of the transport scale.

Expressions (22) and (23) have been obtained using simplifications and are strictly valid for small fluctuations only. However, if the statistics of $K(\mathbf{x})$ and $P(\mathbf{x})$ are known, then the statistics of P_e, as well as the degree of its correlation with τ, can be evaluated using standard flow and nonreactive particle-tracking simulations.

7 DISCUSSION

Mathematical models for reactive subsurface transport are used in solving essentially two types of problems: the *inverse* and the *direct* problem. Interpreting field and/or laboratory data for the purpose of identifying reactive processes, and characterizing them (i.e. determining the controlling parameters, such at reaction rates, solubility constants, etc.) is referred to as the inverse problem. A direct problem refers to predictions, i.e. extrapolation in space and/or time, of the reactive transport process in general described as the evolution of species propagation fronts.

In subsurface contaminant hydrology, the focus is on the movement and transformation (due to reactions) of contaminants along flow paths, from the source area to the accessible environment (Fig. 1). Hence, analysis of reactive transport is frequently limited to identifying and characterizing the reversible and irreversible mass transfer/sorption reactions that retard and/or 'remove' the contaminant from the groundwater. In the general case, however, a complex reaction system may have to be considered to account for all the changes (in pH, redox potential, microbial activity, etc.) that control contaminant movement and transformation.

7.1 Process identification and characterization

In order to identify reactions and estimate reaction parameters, experiments may be performed on a variety of scales. The simplest experiment is of the batch type, on scales typically of the order $10^{-2}-10^{-1}$ m, where the resident concentration is measured as a function of time. Column experiments account for the fluid movement on the laboratory scale, say up to 10^0 m. Field experiments are on scales of the order 10^0-10^2 m, or larger. For routine column and field-scale experiments involving pumping, reactive and nonreactive solutes are injected, and generally the breakthrough curves are monitored as functions of time at one or several locations (e.g. Wise & Charbeneau, 1994); these essentially yield the integrated flux-averaged concentration. In unconsolidated formations and for more comprehensive field studies under natural gradients, reactive solute plumes can be monitored at many locations where the resident concentration is measured as a function of time (e.g. Kent *et al.*, 1994; Thorbjarnarson & Mackay, 1994).

From data collected on various scales (batch, column, field), it is a nontrivial task to identify uniquely the prevailing reactive processes and relate estimated reaction parameters. Different reactions (including physical mass transfer) may yield similar effects. For instance, extended tailing (asymmetry), which is frequently observed in breakthrough curves, may be due to diffusive mass transfer, kinetically controlled sorption, or preferential flow and non-Fickian transport. In order to discriminate between these different processes and effects, a combination of reactive and nonreactive tracers may have to be used, as well as nonreactive tracers with different diffusivities (Frick *et al.*, 1992; Heer, Hadermann & Jacob, 1994). For contaminants that are found in relatively high concentrations, sorption (equilibrium or non-equilibrium) may be nonlinear, requiring tests with different initial and/or boundary conditions. A significant complication is due to possible microbial activity in the field, which may be critical for reaction rates and is very difficult to duplicate in the laboratory.

Even if equivalent reaction parameters can be defined analytically, the application of these results is not straightforward. Batch experiments are generally done on small, refined samples that may have little relevance for reactive processes on the field scale. Undisturbed column experiments in the laboratory or the field (e.g. Ptacek & Gillham, 1992), may be more representative; however, this will depend on the sample size relative to the heterogeneity scale. For many applications, an important problem appears to be how to identify a *minimum* scale (batch, column or field) that is qualitatively representative of a particular class of reactions, and type of formation, and can be used for upscaling and predictions.

7.2 (De)coupling flow and reactions

The most general framework for modelling reactive transport in groundwater is given by the Eulerian formulation (2). The Lagrangian formulation is derived from (2) based on the assumption that the flow pattern is steady-state, and that the transport is on a scale where advection is dominant. Using the transformation (8), solutions of the advection–reaction system (7), and of the flow and nonreactive transport through a three-dimensional formation, can be solved separately; expressions (12), (13) or (19) analytically couple these solutions. The probabilistic features of the flow and the reaction parameter field are included in g and g_2 in (14) and (15), whereas $C_i(t,\tau)$ accounts for reactions. For spatial moment analysis using, say, (19), the flow and nonreactive transport are characterized by X_{11}, whereas reactions are accounted for through Γ_i. In the following, we focus on the species mass flux, and we discuss different ways of implementing (12)–(17).

If the boundary conditions of the considered domain are complex, one can use standard simulations of nonreactive transport (particle-tracking) for determining g and g_2; the equivalent reaction parameters are evaluated as (21). The reactions, on the other hand, are solved in the (τ,t)-domain from (9), for fixed P_e. The results are then integrated using (12)–(17), where (9) needs to be solved for a suitable number of P_e values over a finite interval, determined by the form of g and g_2.

A considerable simplification can be achieved if analytical models for advective transport and/or reactions are applicable. Analytical solutions of (9) are, in fact, available for a relatively large class of sorption–desorption reactions, linear and nonlinear, equilibrium and kinetically controlled. If the flow and nonreactive transport are solved numerically, then the reactive transport can be evaluated using, e.g., (12) and (16); this has been illustrated for linear nonequilibrium sorption (Selroos & Cvetkovic, 1992, 1994).

The functions g and g_2 are essentially dependent on the flow and the reaction parameter fields. These functions can be approximated as analytical functions. Standard simulations of flow and particle-tracking can be used for more accurate evaluation of g and g_2, or of their corresponding moments. Particular attention to the choice of g and g_2 should be given if extreme events, such as the 'first-arrival' of a reactive contaminant, are important.

For large-scale predictions, the formation may exhibit statistical nonstationarity, for instance if the mean hydraulic conductivity increases with the transport scale. Possible self-similarity of the formation would imply that a finite integral scale cannot be defined. Also, the boundary conditions for flow may be time dependent, such that the mean fluid velocity changes in time. Analytical stochastic models for fluid flow and nonreactive transport that take into account the above complexities are increasingly available (e.g. Neuman, 1990; Dagan, 1994; Indelman & Abramovich, 1994); the results of these studies can, in principle, be incorporated into estimates of the functions g and g_2.

7.3 Evaluation of fluxes: a few examples

For many applications, a common requirement is to predict the fluxes of contaminants from a given source area (surface deposition and infiltration ponds, surface waste deposits, accidental spills, deep subsurface waste repositories) to a given discharge area (water supply wells, a river, lake, sea or the biosphere in general) referred to as the accessible environment, A (Fig. 1). The predictive models should, on the one hand, capture the relevant transport and transformation processes, and on the other be sufficiently simple such that a large number of different scenarios can be evaluated with reasonable effort. In the following, we mention a few recent examples where analytical models for reactive solute flux were developed and implemented.

Mass flux of solute subject to kinetically controlled sorption–desorption through the integrated soil–groundwater system has been considered by Destouni & Graham (1995); a solute is injected over a given soil surface area, and is advected through the soil and groundwater to a given discharge area. The recharge of both solute and water from the unsaturated zone into the groundwater was shown to have a potentially significant effect on the solute flux through the discharge area, by inducing significantly earlier and less dispersed breakthrough for increasing groundwater recharge.

Based on the evaluation of chemical field data and modelling of geochemical processes reported by Strömberg & Banwart (1994), the Lagrangian framework for reactive solute transport has been applied for estimating large-scale, long-term leaching of Cu from waste rock of the Aitik copper mine in northern Sweden (Eriksson & Destouni, 1997). This approach to coupling geochemical reactions with field-scale transport modelling is an alternative to more comprehensive coupled simulations (e.g. Walter *et al.*, 1994b).

The health risk from radionuclides at the Nevada Test Site was evaluated by Andricevic, Daniels & Jacobson (1994) using the mass flux approach. An analytical model for assessment of risk from radioactive tracers H^{3+} and Sr, injected by the US DOE in 1968 into the groundwater at the Gnome site, New Mexico, was applied by Andricevic & Cvetkovic (1996). Nonergodic conditions were assumed in view of the relatively small source size compared with the scale of the transport problem. The analysis showed how the health risk for individuals consuming the groundwater at the closest inhabited area downstream depends on hydraulic and reaction parameters and their associated uncertainty. This framework is currently being extended and implemented for risk and safety assessment studies within the Swedish nuclear waste isolation program (Selroos, 1997).

Pump-and-treat remediation of aquifers is a special case of the configuration in Fig. 1, where A_0 would correspond to the injection and A would correspond to the extraction wells; in between is a contaminant plume that is displaced from the aquifer by clean water. Analytical models for evaluating the efficiency of contaminant extraction by pumping with different rates have been derived by Berglund (1995), and these show the influence of aquifer heterogeneity and the type of sorption model assumed; this analysis has currently been extended to a broader class of sorption reactions and aquifer heterogeneity (Berglund & Cvetkovic, 1996).

8 SUMMARY

The Eulerian and Lagrangian approaches to modelling reactive subsurface transport are complementary. The three-dimensional Eulerian system (2) is most general and suitable for detailed and comprehensive simulations toward a better basic understanding of the coupling between reactions and transport under various conditions; it has been used in most applications to date, and its usefulness, as well as its limitations, are fairly well established. The Lagrangian approach decouples reactions from pure advection along flow paths by neglecting mass transfer between flow paths, and may simplify the modelling considerably; it has been used in comparatively few applications, mainly because its conceptual basis has not been apparent. Recent work has provided a clearer theoretical ground for the Lagrangian approach, hence increasing the possibility of its wider application.

For modelling reactive transport in groundwater, extensive information may be required. First, one needs to identify the dominant reactive processes for a given system: physical mass transfer, sorption, possible microbial mediation, redox reactions, complexation, etc. Next, the parameters for controlling reactions (rate coefficients, solubility constants, etc.) need to be estimated. Furthermore, assessment of heterogeneity is required, i.e. estimation of mean, variance, integral scales, etc. The source of contamination also needs to be characterized chemically (to establish boundary conditions for reactive transport modelling) and physically (source scale relative to heterogeneity scale for assessing uncertainty of predictions). If the transport problem is on a relatively large (regional) scale, possible statistical nonstationarity, self-similarity and temporal changes of the regional boundary conditions for flow may need to be identified and accounted for.

In view of the system complexity and inaccessibility, modelling reactive transport in groundwater always has limited accuracy. Hence, the problem for any application is to somehow identify a minimum, or *acceptable*, accuracy for both system data and model predictions. Rational criteria for acceptable accuracy can be provided only if reactive transport modelling is set in an appropriate environmental context. Thus, the increasing demand is for relatively simple models that account for dominant processes of subsurface reactive transport and can readily be used in sensitivity studies within a broader environmental impact evaluation.

For contamination problems where the focus is on a few species subject to sorption and/or transformation, relatively simple (semi)analytical models can be obtained using the Lagrangian approach. These are suitable for scoping calculations when planning field and laboratory experiments, and analysis of sensitivity to different reactions, parameters, boundary conditions ('scenarios'), etc. Models for solute mass flux are often of direct practical interest, and can provide the link between different subsystems of the subsurface (soil–groundwater–surface water). In addition, the statistics of the flux-averaged concentration (mean and variance) are often the quantities required as input to 'modules' that account for the effect of contaminants on human health, ecosystems, etc.

Stochastic modelling of subsurface reactive transport is currently under development, with increasing possibilities for engineering and earth science applications. One of the most important problem areas is how to combine field and laboratory experiments efficiently with the stochastic and deterministic modelling such that fairly general and routine methodologies for field-scale process identification and characterization may eventually be developed. In this context, the few comprehensive field studies carried out to date that were specifically designed to address the effect of heterogeneity on reactive transport have provided the first guidelines.

Another increasingly important problem area is that of describing fluxes through the surface–subsurface–surface system (i.e. from the soil–atmosphere interface, through the soil and groundwater, to surface water) for quantifying key components of biogeochemical cycles. The challenge in this context is to combine effectively stochastic modelling with short-term and possibly long-term, field experiments, complemented by data from natural and/or anthropogenic isotope studies; the purpose of this integration is to characterize dominant flow paths from local to regional scales, to identify the most likely (bio)chemical reactions along them, and to provide estimates of dominant biogeochemical fluxes.

ACKNOWLEDGMENTS

The author is grateful to Georgia Destouni at the Royal Institute of Technology, Stockholm, Sweden and Steven Banwart at the University of Bradford, Bradford, UK, for reading the manuscript and providing useful comments. This work was supported by the Royal Institute of Technology, Stockholm, Sweden.

REFERENCES

Amundson, N. R. (1950). Mathematics of adsorption in beds. II. *Journal of Physical Colloid Chemistry*, 54, 812–820.

Andricevic, R. & Cvetkovic V. (1996). Evaluation of risk from contaminants migrating by groundwater. *Water Resources Research* (in press).

Andricevic, R., Daniels, J. & Jacobson, R. (1994). Radionuclide migration using travel time transport approach and its application to risk analysis. *Journal of Hydrology*, 163, 125–145.

Bear, J. (1972). *Dynamics of Fluids in Porous Media*. New York: Elsevier.

Bellin, A., Salandin, P. & Rinaldo, A. (1992). Simulation of dispersion in heterogeneous porous formations: Statistics, first-order theories, convergence of computations. *Water Resources Research*, 28, 2211–2228.

Bellin, A., Rinaldo, A., Bosma, W. J. P., Van der Zee, S. E. A. T. M. & Rubin, Y. (1993). Linear equilibrium adsorbing solute transport in chemically and physically heterogeneous porous formations, 1. Analytical solutions. *Water Resources Research*, 29, 4019–4030.

Beran, M. J. (1968). *Statistical Continuum Theory*. New York: Interscience.

Berglund, S. (1995). The effect of Langmuir sorption on pump-and-treat remediation of a stratified aquifer. *Journal of Contaminant Hydrology*, 18, 199–220.

Berglund, S. & Cvetkovic, V. (1996). Contaminant displacement in aquifers: Coupled effects of flow heterogeneity and nonlinear sorption. *Water Resources Research*, 32, 23–32.

Bjerg, P. & Christensen, T. H. (1993). A field experiment on cation-exchange affected multicomponent transport in a sandy aquifer. *Journal of Contaminant Hydrology*, 12, 269–290.

Bosma, W. J. & Van der Zee, S. E. A. T. M. (1993). Transport of reacting solute in a one-dimensional, chemically heterogeneous porous medium. *Water Resources Research*, 29, 117–131.

Bosma, W. J., Bellin, A., Van der Zee, S. E. A. T. M. & Rinaldo, A. (1993). Linear equilibrium adsorbing solute transport in chemically and physically heterogeneous porous formations, 2. Numerical results. *Water Resources Research*, 29, 4031–4043.

Brusseau, M. L. & Rao, P. S. C. (1989). Sorption nonideality during organic contaminant transport in porous media. *CRC Critical Reviews in Environmental Control*, 19, 33–99.

Burr, D. T., Sudicky, E. A. & Naff, R. L. (1994). Nonreactive and reactive solute transport in three-dimensional heterogeneous porous media: Mean displacement plume spreading, and uncertainty. *Water Resources Research*, 30, 791–815.

Charbeneau, R. J. (1988). Multicomponent exchange and subsurface solute transport: Characteristics, coherence, and the Riemann problem. *Water Resources Research*, 24, 57–64.

Cherry, J. A., Gillham, R. W. & Barker, J. F. (1984). Contaminants in groundwater: Chemical processes. In *Groundwater Contamination*. Washington, D.C.: National Academy Press.

Chrysikopoulos, V. D., Kitanidis, P. K. & Roberts, P. V. (1992). Macrodispersion of sorbing solutes in heterogeneous porous formations with spatially periodic retardation factor and velocity field. *Water Resources Research*, 28, 1517–1530.

Cvetkovic, V. & Dagan, G. (1994). Transport of kinetically sorbing solute by steady random velocity in heterogeneous porous formations. *Journal of Fluid Mechanics*, 265, 189–215.

Cvetkovic, V. & Dagan, G. (1996). Reactive transport and immiscible flow in geological media. 2 Applications. *Proceedings of the Royal Society of London A*, 452, 303–328.

Cvetkovic, V. & Dagan, G. (1997). Contaminant transport in aquifers with spatially variable hydraulic and sorption parameters, 1. Theory (submitted).

Cvetkovic, V. & Shapiro, A. M. (1990). Mass arrival of sorptive solute in heterogeneous porous media. *Water Resources Research*, 26, 2057–2067.

Dagan, G. (1982). Stochastic modeling of groundwater flow by conditional and unconditional probabilities, 2. The solute transport. *Water Resources Research*, 18, 835–848.

Dagan, G. (1984). Solute transport in heterogeneous porous formations. *Journal of Fluid Mechanics*, 145, 151–177.

Dagan, G. (1989). *Flow and Transport in Porous Formations*. New York: Springer-Verlag.

Dagan, G. (1991). Dispersion of a passive solute in non-ergodic transport by steady velocity fields in heterogeneous formations. *Journal of Fluid Mechanics*, 233, 197–210.

Dagan, G. (1994). The significance of heterogeneity of evolving scales to transport in porous formations. *Water Resources Research*, 30, 3327–3336.

Dagan, G. & Cvetkovic, V. (1996). Reactive transport and immiscible flow in geological media. 1 General theory. *Proceedings of the Royal Society of London A*, 452, 285–301.

Dagan, G., Cvetkovic, V. & Shapiro, A. (1992). A solute-flux approach to transport in heterogeneous formations 1. The general framework. *Water Resources Research*, 28, 1369–1376.

Dagan, G. & Nguyen, V. (1989). A comparison of travel time and concentration approaches to modelling transport by groundwater. *Journal of Contaminant Hydrology*, 4, 79–91.

Destouni, G. (1992). Prediction uncertainty in solute flux through heterogeneous soil. *Water Resources Research*, 28, 793–801.

Destouni, G. (1993). Field-scale solute flux through macroporous soils. In *Water Flow and Solute Transport in Soil*, eds. D. Russo & G. Dagan. Heidelberg: Springer-Verlag, pp. 33–44.

Destouni, G. & Cvetkovic, V. (1991). Field scale mass arrival of sorptive solute into the groundwater. *Water Resources Research*, 27, 1315–1325.

Destouni, G. & Graham, W. (1995). Solute transport through an integrated heterogeneous soil-groundwater system. *Water Resources Research*, 31, 1935–1944.

Destouni, G., Sassner, M. & Jensen, K. H. (1994). Chloride migration in heterogeneous soil, 2. Stochastic modeling. *Water Resources Research*, 30, 747–758.

Engesgaard, P. & Kipp, K. L. (1992). A geochemical transport model for redox-controlled movement of mineral fronts in groundwater-flow systems: A case of nitrate removal by oxidation of pyrite. *Water Resources Research*, 28, 2829–2843.

Eriksson, N. & Destouni, G. (1997). Combined effects of dissolution kinetics, secondary mineral precipitation, and preferential flow on copper leaching from mining waste rock. *Water Resources Research* (in press).

Frick, U. *et al.* (1992). Grimsel Test Site, The radionuclide migration experiment – Overview of investigations 1985–1990. *NTB 91-04, NAGRA*, Wettingen.

Friedly, J. C. & Rubin, J. (1992). Solute transport with multiple equilibrium-controlled or kinetically controlled chemical reactions. *Water Resources Research*, 28, 1935–1953.

Frind, E. O., Duynisveld, W. H. M., Strebel, O. & Boettcher, J. (1990). Modeling of multicomponent transport with microbial transformation in groundwater: The Fuhrberg case. *Water Resources Research*, 26, 1707–1719.

Furrer, G., Westall, J. & Sollins, P. (1989). The study of soil chemistry through quasi-steady-state models: I. Mathematical definition of model. *Geochimica Cosmochimica Acta*, 53, 595–601.

Garabedian, S. P. (1987). *Large-scale dispersive transport in aquifers: Field experiments and reactive transport theory*. Ph.D. dissertation.

Gelhar, L. J. & Axness, C. L. (1983). Three-dimensional stochastic analysis of macrodispersion in aquifers. *Water Resources Research*, 19, 161–180.

Ginn, T. R., Simmons, C. S. & Wood, B. D. (1995). Stochastic-convective transport with nonlinear reaction: Biodegradation with microbial growth. *Water Resources Research*, 31, 2689–2700.

Goltz, M. N. & Roberts, P. V. (1987). Using the method of moments to analyze three-dimensional diffusion-limited solute transport from temporal and spatial perspectives. *Water Resources Research*, 23, 1575–1585.

Heer, W., Hadermann, J. & Jacob, A. (1994). Modelling the radionuclide migration experiments at the Grimsel Test Site. In *Transport and Reactive Processes in Aquifers*, eds. T. Dracos & F. Stauffer. Rotterdam: Balkema.

Hu, X., Deng, F. W. & Cushman, J. H. (1995). Nonlocal reactive transport with physical and chemical heterogeneity, 2. Linear nonequilibrium sorption. *Water Resources Research*, 31, 2239–2252.

Indelman, P. & Abramovich, B. (1994). Nonlocal properties of nonuniform averaged flows in heterogeneous media. *Water Resources Research*, 30, 3385–3394.

Jennings, A. A., Kirkner, D. J. & Theis, T. L. (1982). Multicomponent equilibrium chemistry in groundwater quality models. *Water Resources Research*, 18, 1089–1096.

Jury, W. A. (1982). Simulation of solute transport using a transfer function model. *Water Resources Research*, 18, 363–368.

Kabala, Z. J. & Sposito, G. (1991). A stochastic model of reactive solute transport with time-varying velocity in a heterogeneous aquifer. *Water Resources Research*, 27, 341–350.

Kabala, Z. J. & Sposito, G. (1994). Statistical moments of reactive solute concentration in a heterogeneous aquifer. *Water Resources Research*, 30, 759–768.

Kent, D. B., Davis, J. A., Anderson, L. C. D., Rea, B. A. & Waite, T. D. (1994). Transport of chromium and selenium in the suboxic zone of a shallow aquifer: Influence of redox and adsorption reactions. *Water Resources Research*, 30, 1099–1114.

Kinzelbach, W., Schäfer, W. & Herzer, J. (1991). Numerical modeling of natural and enhanced denitrification processes in aquifers. *Water Resources Research*, 27, 1123–1135.

Kreft, A. & Zuber, A. (1978). On the physical meaning of the dispersion equation and its solution for different initial and boundary conditions. *Chemical Engineering Science*, 33, 1471–1480.

LeBlanc, D .R., Garabedian, S. P., Hess, K. M., Gelhar, L. W., Quadri, R. D., Stollenwerk, K. G. & Wood, W. W. (1991). Large-scale natural gradient test in sand and gravel, Cape Cod, Massachusetts, 1. Experimental design and observed tracer movement. *Water Resources Research*, 27, 895–910.

Lichtner, P. C. (1985). Continuum model for simultaneous chemical reactions and mass transport in hydrothermal systems. *Geochimica Cosmochimica Acta*, 49, 779–800.

Lichtner, P. C. (1988). The quasi-stationary state approximation to coupled mass transport and fluid-rock interaction in a porous medium. *Geochimica Cosmochimica Acta*, 52, 143–165.

Liu, C. W. & Narasimhan, T. N. (1989a). Redox-controlled multiple-species reactive chemical transport, 1. Model development. *Water Resources Research*, 25, 869–882.

Liu, C. W. & Narasimhan, T. N. (1989b). Redox-controlled multiple-species reactive chemical transport, 2. Verification and application. *Water Resources Research*, 25, 883–910.

Lund, K. & Fogler, H. S. (1976). Acidization – V, The Prediction of the movement of acid and permeability fronts in sandstone. *Chemical Engineering Science*, 31, 381–392.

McNab, W. W. & Narasimhan, T. N. (1994). Modeling reactive transport of organic compounds in groundwater using a partial redox disequilibrium approach. *Water Resources Research*, 30, 2619–2635.

Miller, C. W. & Benson L. V. (1983). Simulation of solute transport in a chemically reactive heterogeneous system: Model development and application. *Water Resources Research*, 19, 381–391.

Molz, F. J., Widdowson, M. A. & Benefield, L. D. (1986). Simulation of microbial growth dynamics coupled to nutrient and oxygen transport in porous media. *Water Resources Research*, 22, 1207–1216.

Neuman, S. P. (1990). Universal scaling of hydraulic conductivities and dispersivities in geological media. *Water Resources Research*, 26, 1749–1758.

Neuman, S. P. (1993). Eulerian-Lagrangian theory of transport in space-time nonstationary velocity fields: Exact nonlocal formalism by conditional moments and weak approximation. *Water Resources Research*, 29, 633–645.

Ptacek, C. J. & Gillham, R. W. (1992). Laboratory and field measurements of non-equilibrium transport in the Borden aquifer, Ontario, Canada. *Journal of Contaminant Hydrology*, 10, 119–158.

Quinodoz, H. A. M. & Valocchi, A. J. (1993). Stochastic analysis of the transport of kinetically sorbing solutes in aquifers with randomly heterogeneous hydraulic conductivity. *Water Resources Research*, 29, 3227–3240.

Rhee, H. K., Aris, R. & Amundson, N. R. (1970). On the theory of multicomponent chromatography. *Philosophical Transactions of the Royal Society of London*, A267, 419–455.

Roberts, P. V., Goltz, N. M. & Mackay, D. M. (1986). Natural-gradient experiment on solute transport in a sand aquifer, 3. Retardation estimates and mass balances of organic solutes. *Water Resources Research*, 22, 2047–2058.

Robin, M. J. L., Sudicky, E., Gillham R. & Kachanoski, R. (1991). Spatial variability of strontium distribution coefficients and their correlation with hydraulic conductivity in the Canadian Forces Base Borden aquifer. *Water Resources Research*, 27, 2619–2632.

Rubin, J. (1983). Transport of reacting solutes in porous media: relation between mathematical nature of problem formulation and chemical nature of reactions. *Water Resources Research*, 19, 1231–1252.

Rubin, Y. (1990). Stochastic modeling of macrodispersion in heterogeneous porous media. *Water Resources Research*, 26, 133–141.

Russo, D. (1993). Stochastic modeling of solute flux in a heterogeneous partially saturated porous formation. *Water Resources Research*, 29, 1731–1744.

Sardin, M., Schweich, D., Leij, F. J. & van Genuchten, M. Th. (1991). Modeling the nonequilibrium transport of linearly interacting solutes in porous media: A review. *Water Resources Research*, 27, 2287–2307.

Schulz, H. D. & Reardon, E. J. (1983). A combined mixing cell/analytical model to describe two-dimensional reactive solute transport for unidirectional groundwater flow. *Water Resources Research*, 19, 493–502.

Selroos, J. O. (1995). Temporal moments for nonergodic solute transport in heterogeneous aquifers. *Water Resources Research*, 31, 1705–1712.

Selroos, J. O. (1997). A stochastic analytical framework for safety assessment of waste repositories. 1. Theory. *Groundwater* (in press).

Selroos, J. O. & Cvetkovic, V. (1992). Modeling solute advection coupled with sorption kinetics in heterogeneous formations. *Water Resources Research*, 28, 1271–1278.

Selroos, J. O. & Cvetkovic, V. (1994). Mass flux statistics of kinetically sorbing solute in heterogeneous aquifers: Analytical solution and comparison with simulations. *Water Resources Research*, 30, 63–69.

Shapiro, A. M. & Cvetkovic, V. D. (1988). Stochastic analysis of solute arrival time in heterogeneous porous media. *Water Resources Research*, 24, 1711–1718.

Simmons, C. S. (1982). A stochastic-convective transport representation of dispersion in one-dimensional porous media systems. *Water Resources Research*, 18, 1193–1214.

Simmons, C. S., Ginn, T. R. & Wood, B. D. (1995). Stochastic-convective transport with nonlinear reaction: Mathematical framework. *Water Resources Research*, 31, 2675–2688.

Sposito, G. (1994). Steady groundwater flow as a dynamical system. *Water Resources Research*, 30, 2395–2401.

Strömberg, B. & Banwart, S. (1994). Kinetic modelling of geochemical processes at the Aitik mining waste rock site in northern Sweden. *Applied Geochemistry*, 9, 583–595.

Thomas, H. C. (1944). Heterogeneous ion exchange in a flowing system. *Journal of American Chemical Society*, 66, 1664–1666.

Thorbjarnarson, K. W. & Mackay, D. M. (1994). A forced-gradient experiment on the solute transport in the Borden aquifer, 3. Nonequilibrium transport of the sorbing organic compounds. *Water Resources Research*, 30, 401–419.

Tompson, A. F. B. (1993). Numerical simulation of chemical migration in physically and chemically heterogeneous porous media. *Water Resources Research*, 29, 3709–3726.

Tompson, A. F. B., Schafer, A. L. & Smith, R. W. (1996). Impacts of physical and chemical heterogeneity on cocontaminant transport in a sandy porous medium. *Water Resources Research*, 32, 801–818.

Valocchi, A. (1989). Spatial moment analysis of the transport of kinetically adsorbing solute through stratified aquifers. *Water Resources Research*, 25, 273–279.

Valocchi, A. J., Street, R. L. & Roberts, P. V. (1981). Transport of ion-exchanging solutes in groundwater: Chromatographic theory and field simulation. *Water Resources Research*, 17, 1517–1527.

Van der Zee, S. E. A. T. M. & Van Riemsdijk, W. H. (1987). Transport of reactive solute in spatially variable soil systems. *Water Resources Research*, 23, 2059–2069.

Villermaux, J. (1974). Deformation of chromatographic peaks under the influence of mass transfer phenomena. *Journal of Chromatographic Science*, 12, 822–831.

Walter, A. L., Frind, E. O., Blowes, D. W., Ptacek, C. J. & Molson, J. W. (1994a). Modeling of multicomponent reactive transport in groundwater, 1. Model development and evaluation. *Water Resources Research*, 30, 3137–3148.

Walter, A. L., Frind, E. O., Blowes, D. W., Ptacek, C. J. & Molson, J. W. (1994b). Modeling of multicomponent reactive transport in groundwater, 2. Metal mobility in aquifers impacted by acidic mine tailings discharge. *Water Resources Research*, 30, 3149–3158.

Wise, W. R. (1993). Effects of laboratory-scale variability upon batch and column determinations of nonlinearly sorptive behavior in porous media. *Water Resources Research*, 29, 2983–2992.

Wise, W. R. & Charbeneau, R. J. (1994). In situ estimation of transport parameters: A field demonstration. *Groundwater*, 23, 420–430.

Wood, W. W., Kraemer, T. F. & Hearn, P. P. (1990). Intragranular diffusion: An important mechanism influencing solute transport in clastic aquifers? *Science*, 247, 1569–1572.

Yeh, G. T. & Tripathi V. S. (1991). A model for simulating transport of reactive multispecies components: Model development and demonstration. *Water Resources Research*, 27, 3075–3094.

Zysset, A., Stauffer, F. & Deacos, Th. (1994). Modeling of reactive groundwater transport governed by degradation. *Water Resources Research*, 30, 2423–2434.

3 Nonlocal reactive transport with physical and chemical heterogeneity: linear nonequilibrium sorption with random rate coefficients

BILL X. HU, FEI-WEN DENG AND JOHN H. CUSHMAN

Purdue University

ABSTRACT A nonlocal, first-order, Eulerian, stochastic theory was developed for the mean concentration of a conservative tracer. This was extended to account for nonequilibrium linear sorption with random partition coefficient, K_d, but deterministic rate constant, K_r. Here we extend these results to account for both random K_d and K_r. The basic governing balance law for mean solution-phase concentration is nonlocal in space and time. Nonlocality is manifest in a dispersive flux, an effective convective flux and in sources and/or sinks. The mean concentration balance law is solved exactly in Fourier–Laplace space and numerically inverted to real space. Mean concentration contours and various spatial moments are presented graphically. All results simplify to earlier results under appropriate conditions. The results indicate that there are gaps in existing data sets at major reactive-chemical study sites. The model also suggests the need to design new experiments, and it further suggests that a number of novel correlation functions should be obtained.

1 INTRODUCTION

Our goals in this work are two-fold. First we highlight the need for nonlocality in transport theory, and secondly we extend our previous reactive transport (Hu, Deng & Cushman, 1995) model to account for both random forward and backward rate coefficients. We limit our attention here to porous media that have two natural scales: a scale on which Darcy's and Fick's laws hold locally (see Dagan, 1989; Gelhar, 1993) and a larger scale defined in terms of the integral scale of the log-fluctuating conductivity. These latter assumptions, while restrictive, are consistent with the main body of literature on transport in porous media. Elsewhere, the authors' group (Cushman & Ginn, 1993; Cushman, Hu & Ginn 1994) has made significant progress in eliminating the need for these assumptions. These more general theories, however, have kernels that at present are hard to interpret and measure experimentally. The model presented herein suggests that new experiments must be designed to examine reactive chemical transport.

Before proceeding further, it is advantageous to discuss the concept of nonlocality within the framework of theories for transport in incompletely characterized porous media. A constitutive theory is said to be nonlocal if it involves integrals (over space and or time) or derivatives of order higher than the first (the degree of the highest-order derivative is a measure of the extent of the nonlocality). A good example of a nonlocal constitutive theory is provided by Deng, Cushman & Delleur (1993). These authors develop a convolution Fickian model for dispersion of a conservative solute:

$$\mathbf{q}=\int_0^t \int_{R^3} \mathbf{D}(\mathbf{x}-\mathbf{y},t-t')\cdot\nabla_{\mathbf{y}}\overline{C}(\mathbf{y},t')d\mathbf{y}dt$$

where $\mathbf{q}$ is the dispersive flux, $\mathbf{D}$ is a dispersion kernel and $\overline{C}$ is the mean concentration. This constitutive relation is nonlocal in both space and time. The nonlocal flux was derived from a small-scale local dispersive process. The main question that we must pose here is: why does a process that appears inherently local on a small scale become nonlocal on a larger scale? A concise answer to this question is that small-scale boundary information is suppressed on the macroscale. We illustrate the emergence of nonlocality on the large scale via Fig. 1. To determine the flow field at any point $\mathbf{x}$ in each

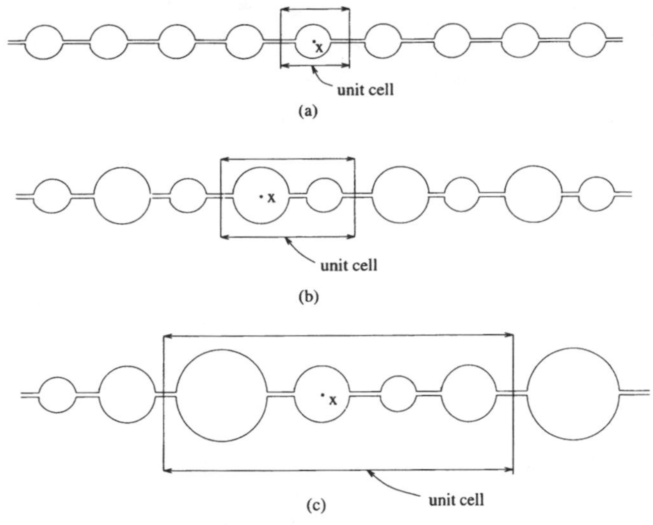

Fig. 1 Unit cells for model capillary tubes.

capillary in Figs. 1(a)–(c), information throughout a unit cell is needed. The governing small-scale equation with slow flow is Stokes, which is a local equation that is subject to boundary conditions throughout the unit cell. To determine the flow field at x, information 'far' from x must be used (via boundary conditions (BCs) over the unit cell). 'Far' here is quantified by the size of the unit cell. In many instances, one seeks only the average flux. For the average flux to be local, the unit cell must be viewed from a scale on which it appears as a point, so that the small-scale boundary data are irrelevant. If, however, the scale of observation is such that the unit cell remains discernible, then small-scale boundary data are important. If small-scale BCs are neglected in the averaging process, then nonlocality is manifest in the appearance of a macroscale nonlocal flux.

In transport theory we are interested in the trajectory of particles, or plume evolution. For a conservative tracer, the trajectory of a particle is dictated by both the mean flow velocity and by the fluctuating flow velocity. The fluctuating velocity is dependent on the degree of heterogeneity associated with the porous matrix, i.e. the complexity of the pore-scale boundary value problem. In many problems there is local homogeneity in which the hydraulic conductivity is well defined and locally uniform. In this case, the pore-scale problem, which dictates the velocity field, is often replaced by a larger-scale flow problem defined via mass continuity, Darcy's law, and regions of constant K. The boundaries between regions of constant K play the same role on this larger scale as did the pore boundaries on the pore scale. Unit cells are now constructed from the conductivity field. When upscaling to the 'reservoir' scale, if the conductivity unit cells remain discernible, and if the detailed boundary data within a cell are not employed in the upscaling procedure, then the reservoir model will again be nonlocal. A number of researchers have recently begun developing nonlocal models of flow and transport (e.g. Koch & Brady, 1987, 1988; Cushman, 1991; Cushman & Ginn, 1993; Deng *et al.*, 1993; Neuman & Orr, 1993; Sahimi, 1993). We continue in that spirit here, and show by using first-order perturbation analysis (Gelhar & Axness, 1983) and in a Eulerian framework (Deng *et al.*, 1993), that chemicals undergoing non-equilibrium linear reactions with random forward and backward rate coefficients display a very complex nonlocal behavior; far more complex than others (see Brusseau & Rao, 1989) have presumed. A particular novelty of the present work is the appearance of more general nonlocality than in convolution Fickian dispersion. Specifically we will show existence of nonlocality in the dispersive flux, in an effective convective flux, and in sources and sinks.

2 GENERAL ANALYSIS

It is assumed on the local scale that the following equation for the concentration, C, in solution holds:

$$\frac{\partial C}{\partial t}+\frac{\partial S}{\partial t}=\nabla\cdot(\mathbf{d}\cdot\nabla C)-\nabla\cdot(\mathbf{V}C) \tag{1}$$

where S is the sorbed concentration defined as sorbed solute mass per formation solid volume, $\mathbf{d}$ is the local-scale dispersion tensor, and $\mathbf{V}$ is the locally homogeneous but macroscopically stochastic Darcy velocity. It is further assumed that the mean flow is constant in the x_1-direction so that $\bar{\mathbf{V}}=(V,0,0)$, and that the dispersion tensor is diagonal with entries d_1, d_2, and d_3, where d_1, d_2, and d_3 are the so-called Darcy-scale-longitudinal, transverse-horizontal and transverse-vertical dispersion coefficients, respectively. It is important to note that in this Eulerian framework we do not neglect local-scale dispersivities as is commonly done in the Lagrangian framework (see Dagan, 1989).

We further assume that the concentrations C and S are related by a rate equation of the form (Dagan & Cvetkovic, 1993)

$$\frac{\partial S}{\partial t}=K_r(K_dC-S)=K_fC-K_bS \tag{2}$$

where K_d is the usual partition coefficient, K_r is the reaction rate coefficient, $K_f=K_rK_d$ is known as the forward rate coefficient, and $K_b=K_r$ is the backward rate coefficient. In subsequent analysis it is assumed that $\ln K$ (K is hydraulic conductivity), K_f, and K_b are random variables.

In the usual fashion, decompose the forward and backward rate coefficients, velocity, and concentrations into means and fluctuations about the means:

$$K_f=\bar{K}_f+k_f \tag{3}$$

$$K_b=\overline{K}_b+k_b \tag{4}$$

$$V_i=\overline{V}_i+v_i \tag{5}$$

$$C=\overline{C}+c \tag{6}$$

and

$$S=\overline{S}+s \tag{7}$$

where $\overline{K}_f$, $\overline{K}_b$, and $\overline{V}_i$ are assumed constant.

Substituting (5)–(7) into (1), we obtain

$$\frac{\partial(\overline{C}+c)}{\partial t}+\frac{\partial(\overline{S}+s)}{\partial t}=d_i\frac{\partial^2(\overline{C}+c)}{\partial x_i^2}-\frac{\partial(\overline{V}_i+v_i)(\overline{C}+c)}{\partial x_i} \tag{8}$$

In deriving (8) we assume that the d_i are constant and that repeated indices imply summation.

Because we assume that the mean flow is constant and in the x_1-direction with $\overline{V}_1=V$, the mean equation corresponding to (8) is

$$\frac{\partial\overline{C}}{\partial t}+\frac{\partial\overline{S}}{\partial t}=d_i\frac{\partial^2\overline{C}}{\partial x_i^2}-V\frac{\partial\overline{C}}{\partial x_1}-\frac{\partial\overline{v_ic}}{\partial x_i} \tag{9}$$

and the mean removed equation is

$$\frac{\partial c}{\partial t}+\frac{\partial s}{\partial t}=d_i\frac{\partial^2 c}{\partial x_i^2}-V\frac{\partial c}{\partial x_1}-v_i\frac{\partial\overline{C}}{\partial x_i}-v_i\frac{\partial c}{\partial x_i}+\frac{\partial\overline{v_ic}}{\partial x_i} \tag{10}$$

In deriving (10) we have assumed that v_i is divergence-free, i.e.

$$\frac{\partial v_i}{\partial x_i}=0 \tag{11}$$

Substitute (3), (4), (6), and (7) into (2) to obtain

$$\frac{\partial(\overline{S}+s)}{\partial t}=(\overline{K}_f+k_f)(\overline{C}+c)-(\overline{K}_b+k_b)(\overline{S}+s) \tag{12}$$

The mean equation corresponding to (12) is

$$\frac{\partial\overline{S}}{\partial t}=\overline{K}_f\overline{C}+\overline{k_fc}-\overline{K}_b\overline{S}-\overline{k_bs} \tag{13}$$

and the mean-removed equation is

$$\frac{\partial s}{\partial t}=\overline{K}_fc+k_f\overline{C}-\overline{K}_bs-k_b\overline{S}+k_fc-\overline{k_fc}-k_bs+\overline{k_bs} \tag{14}$$

Equations (9), (10), (13), and (14) form a set of four equations which describe the mean and the fluctuating concentrations. Following the lead of Deng *et al.* (1993), we use spatial-Fourier and time-Laplace transforms to analyze and then solve these equations.

Define the time-Laplace transform L by

$$L[f(t)]=\int_0^\infty \exp[-\omega t]f(t)dt=\tilde{f}(\omega) \tag{15a}$$

and the Fourier spatial-transform F by

$$F[f(\mathbf{x})]=\int_{R^3}\exp[-i\mathbf{k}\cdot\mathbf{x}]f(\mathbf{x})d\mathbf{x}=\hat{f}(k) \tag{15b}$$

Recall some basic properties of Laplace and Fourier transforms:

$$L[df/dt]=\omega L[f]-f(0) \tag{16a}$$

$$L[f*_t g]=L[f]L[g] \tag{16b}$$

$$F[d^nf/dx^n]=(ik)^nF[f] \tag{16c}$$

$$F[f*_x g]=F[f]F[g] \tag{16d}$$

$$F[fg]=(2\pi)^{-N}F[f]*_{\mathbf{k}}F[g] \tag{16e}$$

where N is the dimensionality of the system. Here the asterisk indicates the convolution operator, and the subscript indicates the variable with respect to which it operates.

Apply (15) to (10) and (14), and use (16a) and (16c) to get

$$\omega\tilde{\hat{c}}+\omega\tilde{\hat{s}}=d_i(ik_i)^2\tilde{\hat{c}}-Vik_1\tilde{\hat{c}}-\left(v_i\frac{\partial\overline{C}}{\partial x_i}+v_i\frac{\partial c}{\partial x_i}-\frac{\partial\overline{v_ic}}{\partial x_i}\right)^{\tilde{\hat{}}} \tag{17}$$

$$\tilde{\hat{s}}=(\omega+K_b)^{-1}[\overline{K}_f\tilde{\hat{c}}+(k_f\overline{C}-k_b\overline{S}+k_fc-\overline{k_fc}-k_bs+\overline{k_bs})^{\tilde{\hat{}}}] \tag{18}$$

In deriving (17) and (18), we have assumed that $c(\mathbf{x},0)=0$ and $s(\mathbf{x},0)=0$, respectively. Substituting (18) into (17), we obtain

$$\tilde{\hat{c}}=-\tilde{\hat{B}}(\mathbf{k},\omega)\left[\left(v_i\frac{\partial\overline{C}}{\partial x_i}+v_i\frac{\partial c}{\partial x_i}-\frac{\partial\overline{v_ic}}{\partial x_i}\right)^{\tilde{\hat{}}}\right]-\tilde{\hat{G}}(\mathbf{k},\omega)[(k_f\overline{C}-k_b\overline{S}+k_fc$$

$$-\overline{k_fc}-k_bs+\overline{k_bs})^{\tilde{\hat{}}}] \tag{19}$$

where

$$\tilde{\hat{B}}(\mathbf{k},\omega)=\left[\omega\left(1+\frac{\overline{K}_f}{\omega+\overline{K}_b}\right)+d_ik_i^2+ik_1V\right]^{-1} \tag{20}$$

$$\tilde{\hat{G}}(\mathbf{k},\omega)=\frac{\omega}{\omega+\overline{K}_b}\tilde{\hat{B}}(\mathbf{k},\omega) \tag{21}$$

Taking inverse Laplace and Fourier transforms of (19), we find

$$c(\mathbf{x},t)=-\int_0^t\int_{R^3}B(\mathbf{x}-\mathbf{y},t-t')$$

$$\times\left[v_i(\mathbf{y})\frac{\partial\overline{C}(\mathbf{y},t')}{\partial y_i}+v_i(\mathbf{y})\frac{\partial c(\mathbf{y},t')}{\partial y_i}-\frac{\partial\overline{v_ic}}{\partial y_i}(\mathbf{y},t')\right]d\mathbf{y}dt'$$

$$-\int_0^t\int_{R^3}G(\mathbf{x}-\mathbf{y},t-t')[k_f(\mathbf{y})\overline{C}(\mathbf{y},t')-k_b(\mathbf{y})\overline{S}(\mathbf{y},t')$$

$$+k_f(\mathbf{y})\mathrm{c}(\mathbf{y},t')-\overline{k_fc}(\mathbf{y},t')-k_b(\mathbf{y})s(\mathbf{y},t')+\overline{k_bs}(\mathbf{y},t')]d\mathbf{y}dt' \tag{22}$$

Two types of correlation functions will be obtained in the subsequent analysis. One depends on space alone, and the other involves both space and time. We henceforth assume that correlation functions involving space alone are stationary, but that correlation functions involving both space and time are not stationary. Under these assumptions, after multiplying (22) by $v_j(\mathbf{x})$, taking expectations, and neglecting triplet correlations, we obtain

$$\overline{cv_j}(\mathbf{x},t)=-\int_0^t\int_{R^3}B(\mathbf{x}-\mathbf{y},t-t')\overline{v_iv_j}(\mathbf{x}-\mathbf{y})\frac{\partial\overline{C}}{\partial y_i}(\mathbf{y},t')d\mathbf{y}dt'$$
$$-\int_0^t\int_{R^3}G(\mathbf{x}-\mathbf{y},t-t')[\overline{v_jk_f}(\mathbf{x}-\mathbf{y})\overline{C}(\mathbf{y},t')$$
$$-\overline{v_jk_b}(\mathbf{x}-\mathbf{y})\overline{S}(\mathbf{y},t')]d\mathbf{y}dt' \quad (23)$$

In a similar fashion we find

$$\overline{k_fc}(\mathbf{x},t)=-\int_0^t\int_{R^3}B(\mathbf{x}-\mathbf{y},t-t')\overline{v_ik_f}(\mathbf{x}-\mathbf{y})\frac{\partial\overline{C}}{\partial y_i}(\mathbf{y},t')d\mathbf{y}dt'$$
$$-\int_0^t\int_{R^3}G(\mathbf{x}-\mathbf{y},t-t')[\overline{k_fk_f}(\mathbf{x}-\mathbf{y})\overline{C}(\mathbf{y},t')$$
$$-\overline{k_fk_b}(\mathbf{x}-\mathbf{y})\overline{S}(\mathbf{y},t')]d\mathbf{y}dt' \quad (24)$$

and

$$\overline{k_bc}(\mathbf{x},t)=-\int_0^t\int_{R^3}B(\mathbf{x}-\mathbf{y},t-t')\overline{v_ik_b}(\mathbf{x}-\mathbf{y})\frac{\partial\overline{C}}{\partial y_i}(\mathbf{y},t')d\mathbf{y}dt'$$
$$-\int_0^t\int_{R^3}G(\mathbf{x}-\mathbf{y},t-t')[\overline{k_bk_f}(\mathbf{x}-\mathbf{y})\overline{C}(\mathbf{y},t')$$
$$-\overline{k_bk_b}(\mathbf{x}-\mathbf{y})\overline{S}(\mathbf{y},t')]d\mathbf{y}dt' \quad (25)$$

Apply time-Laplace and spatial-Fourier transforms to (23)–(25), respectively, to obtain

$$\tilde{\hat{\overline{cv_j}}}=-\frac{1}{(2\pi)^3}\left\{\left[\tilde{\hat{G}}*_{\mathbf{k}}\hat{\overline{v_jk_f}}+(\tilde{\hat{B}}*_{\mathbf{k}}\hat{\overline{v_iv_j}})ik_i\right]\tilde{\overline{C}}-(\tilde{\hat{G}}*_{\mathbf{k}}\hat{\overline{v_jk_b}})\,\tilde{\overline{S}}\right\} \quad (26)$$

$$\tilde{\hat{\overline{k_fc}}}=-\frac{1}{(2\pi)^3}\left\{\left[\tilde{\hat{G}}*_{\mathbf{k}}\hat{\overline{k_fk_f}}+(\tilde{\hat{B}}*_{\mathbf{k}}\hat{\overline{v_ik_f}})ik_i\right]\tilde{\overline{C}}-(\tilde{\hat{G}}*_{\mathbf{k}}\hat{\overline{k_fk_b}})\tilde{\overline{S}}\right\} \quad (27)$$

and

$$\tilde{\hat{\overline{k_bc}}}=-\frac{1}{(2\pi)^3}\left\{\left[\tilde{\hat{G}}*_{\mathbf{k}}\hat{\overline{k_bk_f}}+(\tilde{\hat{B}}*_{\mathbf{k}}\hat{\overline{v_ik_b}})ik_i\right]\tilde{\overline{C}}-(\tilde{\hat{G}}*_{\mathbf{k}}\hat{\overline{k_bk_b}})\tilde{\overline{S}}\right\} \quad (28)$$

Apply Laplace and Fourier transforms to (13) and (14), respectively, to yield

$$\tilde{\overline{S}}=(\omega+\overline{K}_b)^{-1}\left[\hat{S}_0+\overline{K}_f\tilde{\overline{C}}+\tilde{\hat{\overline{k_fc}}}+\tilde{\hat{\overline{k_bs}}}\right] \quad (29)$$

and

$$\tilde{s}=(\omega+\overline{K}_b)^{-1}[\overline{K}_f\tilde{c}+\widetilde{(k_f\overline{C}-k_b\overline{S}+k_fc-\overline{k_fc}-k_bs+\overline{k_bs})}] \quad (30)$$

The inverse transforms of (29) and (30) are

$$\overline{S}(\mathbf{x},t)=e^{-\overline{K}_bt}S_0+\int_0^t e^{-\overline{K}_b(t-t')}[\overline{K}_f\overline{C}(\mathbf{x},t')+\overline{k_fc}(\mathbf{x},t')$$
$$-\overline{k_bs}(\mathbf{x},t')]dt' \quad (31)$$

and

$$s=\int_0^t e^{-\overline{K}_b(t-t')}[\overline{K}_fc(\mathbf{x},t')+k_f(\mathbf{x})\overline{C}(\mathbf{x},t')-k_b(\mathbf{x})\overline{S}(\mathbf{x},t')$$
$$+k_f(\mathbf{x})c(\mathbf{x},t')-\overline{k_fc}(\mathbf{x},t')-k_b(\mathbf{x})s(\mathbf{x},t')+\overline{k_bs}(\mathbf{x},t')]dt' \quad (32)$$

Multiply (32) by k_b, take ensemble averages, and neglect triplet correlations to yield

$$\overline{k_bs}(\mathbf{x},t)=\int_0^t e^{-\overline{K}_b(t-t')}[\overline{K}_f\overline{k_bc}(\mathbf{x},t')+\overline{k_bk_f}(0)\overline{C}(\mathbf{x},t')$$
$$-\overline{k_bk_b}(0)\overline{S}(\mathbf{x},t')]dt \quad (33)$$

Taking Laplace and Fourier transforms of (33), we obtain

$$\tilde{\hat{\overline{k_bs}}}=(\omega+\overline{K}_b)^{-1}\left[\overline{K}_f\tilde{\hat{\overline{k_bc}}}+\overline{k_bk_f}(0)\tilde{\overline{C}}-\overline{k_bk_b}(0)\tilde{\overline{S}}\right] \quad (34)$$

Insert (27) and (34) into (29) and rearrange terms to yield

$$\tilde{\overline{S}}=\tilde{I}_1^{-1}(\mathbf{k},\omega)\left[\hat{S}_0+\tilde{I}_2(\mathbf{k},\omega)\tilde{\overline{C}}\right] \quad (35)$$

where

$$\tilde{I}_1(\mathbf{k},\omega)=\omega+\overline{K}_b-\frac{\overline{k_bk_b}(0)}{\omega+\overline{K}_b}-\frac{1}{(2\pi)^3}(\tilde{\hat{G}}*_{\mathbf{k}}\hat{\overline{k_fk_b}})+\frac{1}{(2\pi)^3}\frac{\overline{K}_f}{\omega+\overline{K}_b}$$
$$(\tilde{\hat{G}}*_{\mathbf{k}}\hat{\overline{k_bk_b}}) \quad (36)$$

$$\tilde{I}_2(\mathbf{k},\omega)=\overline{K}_f-\frac{\overline{k_bk_f}(0)}{\omega+\overline{K}_b}-\frac{1}{(2\pi)^3}\left[\tilde{\hat{G}}*_{\mathbf{k}}\hat{\overline{k_fk_f}}+(\tilde{\hat{B}}*_{\mathbf{k}}\hat{\overline{v_ik_f}})ik_i\right]$$
$$+\frac{1}{(2\pi)^3}\frac{\overline{K}_f}{\omega+\overline{K}_b}\left[\tilde{\hat{G}}*_{\mathbf{k}}\hat{\overline{k_bk_f}}+(\tilde{\hat{B}}*_{\mathbf{k}}\hat{\overline{v_ik_b}})ik_i\right] \quad (37)$$

Insert (35) into (26) to obtain

$$\tilde{\hat{\overline{cv_j}}}=-\frac{1}{(2\pi)^3}\left[\tilde{\hat{G}}*_{\mathbf{k}}\hat{\overline{v_jk_f}}+(\tilde{\hat{B}}*_{\mathbf{k}}\hat{\overline{v_iv_j}})ik_i\right.$$
$$\left.-\frac{\left(\tilde{\hat{G}}*_{\mathbf{k}}\hat{\overline{v_jk_b}}\right)\tilde{I}_2(\mathbf{k},\omega)}{\tilde{I}_1(\mathbf{k},\omega)}\right]\tilde{\overline{C}}+\frac{1}{(2\pi)^3}\frac{\left(\tilde{\hat{G}}*_{\mathbf{k}}\hat{\overline{v_jk_b}}\right)}{\tilde{I}_1(\mathbf{k},\omega)}\hat{S}_0 \quad (38)$$

Apply time-Laplace and spatial-Fourier transforms to (9) to obtain

$$\omega\tilde{\overline{C}}-\hat{C}_0+\omega\tilde{\overline{S}}-\hat{S}_0=d_i(ik_i)^2\tilde{\overline{C}}-Vik_1\tilde{\overline{C}}-ik_i\tilde{\hat{\overline{v_ic}}} \quad (39)$$

Insert (35) and (38) into (39) and rearrange the terms to obtain

$$\bar{\tilde{C}}=\hat{\tilde{F}}^{-1}(\mathbf{k},\omega)\left\{\hat{C}_0+\left[1-\frac{1}{\hat{\tilde{I}}_1(\mathbf{k},\omega)}\left(\frac{ik_j}{(2\pi)^3}(\hat{\tilde{G}}*_{\mathbf{k}}\overset{\wedge}{\overline{v_jk_b}})+\omega\right)\right]\hat{S}_0\right\} \tag{40}$$

where

$$\hat{\tilde{F}}(\mathbf{k},\omega)=\omega+d_ik_i^2+ik_1V+[\omega\hat{\tilde{I}}_2(\mathbf{k},\omega)/\hat{\tilde{I}}_1(\mathbf{k},\omega)]$$
$$-\frac{ik_j}{(2\pi)^3}\left[\hat{\tilde{G}}*_{\mathbf{k}}\overset{\wedge}{\overline{v_jk_f}}+ik_i(\hat{\tilde{B}}*_{\mathbf{k}}\overset{\wedge}{\overline{v_iv_j}})-\frac{\hat{\tilde{I}}_2(\mathbf{k},\omega)}{\hat{\tilde{I}}_1(\mathbf{k},\omega)}(\hat{\tilde{G}}*_{\mathbf{k}}\overset{\wedge}{\overline{v_jk_b}})\right] \tag{41}$$

Once the parameters and correlation functions are given, (40) and (41) can be used to calculate the mean concentration via fast Fourier transform (FFT). In current hydrologic lab and field experiments, K_d and K_r are measured rather than K_f and K_b. Therefore, it may be advantageous to rewrite (40) and (41) in terms of K_d and K_r (see the Appendix to this chapter for details of this transformation).

3 SPECIAL CASES

DETERMINISTIC CONSTANT K_b (OR K_r)

If the backward rate coefficient, K_b (or K_r), is assumed to be a deterministic constant, which is the case discussed in Hu *et al.* (1995), then (36) and (37) become

$$\hat{\tilde{I}}_1(\mathbf{k},\omega)=\omega+K_b \tag{42}$$

$$\hat{\tilde{I}}_2(\mathbf{k},\omega)=\overline{K}_f-\frac{1}{(2\pi)^3}\left[\hat{\tilde{G}}*_{\mathbf{k}}\overset{\wedge}{\overline{k_fk_f}}+(\hat{\tilde{B}}*_{\mathbf{k}}\overset{\wedge}{\overline{v_ik_f}})ik_i\right] \tag{43}$$

where

$$\hat{\tilde{B}}(\mathbf{k},\omega)=\left[\omega\left(1+\frac{\overline{K}_f}{\omega+K_b}\right)+d_ik_i^2+ik_1V\right]^{-1} \tag{44}$$

and

$$\hat{\tilde{G}}(\mathbf{k},\omega)=\frac{\omega}{\omega+K_b}\hat{\tilde{B}}(\mathbf{k},\omega) \tag{45}$$

Equations (40) and (41) become

$$\bar{\tilde{C}}=\hat{\tilde{F}}^{-1}(\mathbf{k},\omega)\left\{\hat{C}_0+\frac{K_b}{\omega+K_b}\hat{S}_0\right\} \tag{46}$$

and

$$\hat{\tilde{F}}(\mathbf{k},\omega)=\hat{\tilde{B}}^{-1}(\mathbf{k},\omega)+\frac{k_ik_j}{(2\pi)^3}(\hat{\tilde{B}}*_{\mathbf{k}}\overset{\wedge}{\overline{v_iv_j}})-\frac{i2k_j}{(2\pi)^3}\frac{\omega}{\omega+K_b}$$
$$(\hat{\tilde{B}}*_{\mathbf{k}}\overset{\wedge}{\overline{v_jk_f}})-\frac{1}{(2\pi)^3}\left(\frac{\omega}{\omega+K_b}\right)^2(\hat{\tilde{B}}*_{\mathbf{k}}\overset{\wedge}{\overline{k_fk_f}}) \tag{47}$$

Under the assumption of constant K_b

$$k_f=K_rk_d \tag{48}$$

By substituting (48) into (47), it can be seen that (46) and (47) are identical to the expressions in Hu *et al.* (1995).

If both K_b and K_f are deterministic constants, then (47) becomes

$$\hat{\tilde{F}}(\mathbf{k},\omega)=\hat{\tilde{B}}^{-1}(\mathbf{k},\omega)+\frac{k_ik_j}{(2\pi)^3}(\hat{\tilde{B}}*_{\mathbf{k}}\overset{\wedge}{\overline{v_iv_j}}) \tag{49}$$

This latter result is identical to that obtained in Hu *et al.* (1995, Eq. (27)).

If both K_b and K_r are zero, as is the case for a conservative tracer, then (46) reduces to the Deng *et al.* (1993) result, i.e.

$$\bar{\tilde{C}}=\left\{\hat{\tilde{B}}^{-1}(\mathbf{k},\omega)+\frac{k_ik_j}{(2\pi)^3}(\hat{\tilde{B}}*_{\mathbf{k}}\overset{\wedge}{\overline{v_iv_j}})\right\}\hat{C}_0 \tag{50}$$

where

$$\hat{\tilde{B}}^{-1}(\mathbf{k},\omega)=\omega+ik_1V+k_i^2d_i \tag{51}$$

4 NONLOCALITY AND THE REAL-SPACE BALANCE LAW

The form of the real-space balance law for mean concentration suggests novel processes control mean concentration evolution. For illustrative purposes we present only the law with random K_d and deterministic constant K_r. This result contains all pertinent information on the structure of the more general balance law for K_r random, yet it offers a certain simplicity over the more general result. This balance law was first derived in Cushman *et al.* (1995) and subsequently discussed in detail in Hu *et al.* (1995). The governing law for mean concentration with deterministic and constant K_r is

$$\frac{\partial\overline{C}}{\partial t}+\left\{V\frac{\partial\overline{C}}{\partial x_1}-\left[\frac{\partial}{\partial x_j}\int_0^t\int_{R^3}G(\mathbf{x}-\mathbf{y},t-t')\overline{k_dv_j}(\mathbf{x}-\mathbf{y})\overline{C}(\mathbf{y},t')d\mathbf{y}dt'\right.\right.$$
$$+\int_0^t\left(\delta(t-t')-K_re^{-K_r(t-t')}\right)\int_0^{t'}\int_{R^3}B(\mathbf{x}-\mathbf{y},t'-t'')\overline{k_dv_j}(\mathbf{x}-\mathbf{y})$$
$$\left.\left.\times\frac{\partial\overline{C}(\mathbf{y},t'')}{\partial y_j}d\mathbf{y}dt''dt'\right]\right\}-\frac{\partial}{\partial x_j}\int_0^t\int_{R^3}D_{ij}(\mathbf{x}-\mathbf{y},t-t')\frac{\partial}{\partial y_i}\overline{C}(\mathbf{y},t')d\mathbf{y}dt'$$
$$=-K_r\left\{\overline{K}_d\overline{C}-e^{-K_rt}S_0-K_r\overline{K}_d\int_0^te^{-K_r(t-t')}\overline{C}(x,t')dt'\right.$$
$$-\int_0^t\left[\delta(t-t')-K_re^{-K_r(t-t')}\right]\int_0^{t'}\int_{R^3}G(\mathbf{x}-\mathbf{y},t'-t'')\overline{k_dk_d}(\mathbf{x}-\mathbf{y})$$
$$\left.\times\overline{C}(\mathbf{y},t'')d\mathbf{y}dt''dt'\right\} \tag{52}$$

where

$$D_{ij}(\mathbf{x}-\mathbf{y},t-t')=d_{ij}\delta(\mathbf{x}-\mathbf{y},t-t')+\overline{v_iv_j}(\mathbf{x}-\mathbf{y})B(\mathbf{x}-\mathbf{y},\ t-t') \tag{53}$$

$$\tilde{\hat{B}}(\mathbf{k},\omega)=\left[\omega\left(1+\frac{K_r\bar{K}_d}{\omega+K_r}\right)+d_ik_i^2+ik_1V\right]^{-1} \quad (54)$$

and

$$\tilde{\hat{G}}(\mathbf{k},\omega)=\frac{\omega K_r}{\omega+K_r}\tilde{\hat{B}}(\mathbf{k},\omega) \quad (55)$$

The various terms in (52) admit the following physical interpretation:

(i) The first term on the LHS represents the local change in mean concentration.

(ii) The second term on the LHS, in braces, { }, is a sum of 'actual' and 'effective' convective fluxes. The first term, in the braces, is the 'actual' convective flux, while the term in the brackets, [ ], is an effective nonlocal convective flux, which arises from the correlation between the fluctuating retardation coefficient and fluctuating velocity. If either K_d is deterministic or K_d is uncorrelated with v_i, then this latter flux is zero. Also if K_d and v_i have short-range correlation only, then this latter flux becomes 'local'.

(iii) The last term on the LHS is the effective nonlocal dispersive flux, which localizes if $\nabla\bar{C}$ varies slowly relative to $\mathbf{D}$.

(iv) The RHS represents 'actual' local sources without memory, and memory in 'effective' sources. The various terms on the RHS of (52) admit the following interpretations. The first term represents a local source or sink owing to the mean partition coefficient. Had K_d been nonrandom and with equilibrium sorption, this term would have been the sole source or sink of contamination. The second term also represents a local source/sink, but in this case it is due to the nonequilibrium nature of the sorbed phase. This portion of the source/sink decreases exponentially fast with time constant K_r^{-1}. The third term represents system memory in the source/sink; it is present even when K_d is nonrandom. The last term is associated with the randomness of K_d (it is zero for deterministic K_d).

5 CORRELATION FUNCTIONS

One needs to know $\overline{v_iv_j}$, $\overline{v_ik_f}$, $\overline{v_ik_b}$, $\overline{k_bk_f}$, $\overline{k_fk_f}$, and $\overline{k_bk_b}$ or their Fourier transforms, to calculate mean concentration, $\bar{C}(\mathbf{x},t)$. Let $\ln K=F+f$ with $\overline{\ln K}=F$. From Gelhar & Axness (1983), under suitable conditions, the Fourier transform of $\overline{v_iv_j}$ is

$$\widehat{\overline{v_iv_j}}(\mathbf{k})=\left(\frac{JK_g}{n}\right)^2\left[\delta_{i1}-\frac{k_1k_i}{k^2}\right]\left[\delta_{j1}-\frac{k_1k_j}{k^2}\right]\hat{\overline{ff}}(\mathbf{k}) \quad (56)$$

Here n is the assumed deterministic porosity, K_g is the geometric mean of the hydraulic conductivity, J is the mean hydraulic gradient in the x_1-direction, $k^2=k_1^2+k_2^2+k_3^2$, and $\hat{\overline{ff}}(\mathbf{k})$ is the Fourier transform of the covariance function for the log-fluctuating conductivity with variance σ_f^2 and integral scales l_i. To first order it can also be argued that

$$\widehat{\overline{v_ik_f}}(\mathbf{k})=\left(\frac{JK_g}{n}\right)\left[\delta_{i1}-\frac{k_1k_i}{k^2}\right]\widehat{\overline{fk_f}}(\mathbf{k}) \quad (57)$$

and

$$\widehat{\overline{v_ik_b}}(\mathbf{k})=\left(\frac{JK_g}{n}\right)\left[\delta_{i1}-\frac{k_1k_i}{k^2}\right]\widehat{\overline{fk_b}}(\mathbf{k}) \quad (58)$$

To the authors' knowledge, there are no experimental data on the correlations $\overline{fk_f}$ and $\overline{fk_b}$. Bellin *et al.* (1993), after reviewing experimental results, proposed three different correlations between the retardation factor R (or partition coefficient K_d) and log conductivity, $\ln K$: perfect positive correlation, perfect negative correlation, and no correlation. Hu *et al.* (1995) studied the effect of the three different correlation structures on reactive chemical transport under nonequilibrium sorption with random $\ln K$ and K_d, but deterministic constant K_r. Here we extend Hu *et al.*'s results by assuming that K_f and K_b may also be positively correlated, negatively correlated or uncorrelated with $\ln K$. Since K_f is directly related to K_d, and K_d was studied in Hu *et al.* (1995), we restrict our attention here to the relationship between $\ln K$ to K_b. For simplicity, we assume that $\ln K$ and K_f are negatively correlated:

$$k_f(\mathbf{x})=-\bar{K}_f f(\mathbf{x}) \quad (59)$$

So that

$$\widehat{\overline{v_ik_f}}(\mathbf{k})=-\frac{JK_g\bar{K}_f}{n}\left[\delta_{i1}-\frac{k_1k_i}{k^2}\right]\hat{\overline{ff}}(\mathbf{k}) \quad (60)$$

and

$$\widehat{\overline{k_fk_f}}(\mathbf{k})=-\bar{K}_f^2\hat{\overline{ff}}(\mathbf{k}) \quad (61)$$

We work with three correlation structures between K_b and $\ln K$:

Perfect positive correlation (Model A)

$$k_b(\mathbf{x})=\bar{K}_bf(\mathbf{x}) \quad (62a)$$

Perfect negative correlation (Model B)

$$k_b(\mathbf{x})=-\bar{K}_bf(\mathbf{x}) \quad (62b)$$

No correlation (Model C)

$$k_b(\mathbf{x})=\bar{K}_b\,w(\mathbf{x}) \quad (62c)$$

where $w(\mathbf{x})$ is a normally distributed random space function with zero mean, variance σ_w^2, and correlation function $\overline{ww}(\mathbf{r})=\sigma_w^2e^{-r/l_w}$, $r=|\mathbf{r}|$, and l_w is the integral scale of $w(\mathbf{x})$.

Models A and B give

$$\widehat{k_b k_b}(\mathbf{k}) = \overline{K}_b^2 \widehat{ff}(\mathbf{k}) \tag{63}$$

$$\widehat{v_i k_b}(\mathbf{k}) = \pm \frac{J K_g \overline{K}_b}{n} \left[\delta_{i1} - \frac{k_1 k_i}{k^2} \right] \widehat{ff}(\mathbf{k}) \tag{64}$$

with the plus sign for Model A and the minus sign for Model B

$$\widehat{k_b k_f}(\mathbf{k}) = \mp \overline{K}_b \overline{K}_f \widehat{ff}(\mathbf{k}) \tag{65}$$

with the minus sign for Model A and the plus sign for Model B.

Model C gives

$$\widehat{k_b k_b}(\mathbf{k}) = \overline{K}_b^2 \ \widehat{ww}(\mathbf{k}) \tag{66}$$

$$\widehat{v_i k_b}(\mathbf{k}) = 0 \tag{67}$$

and

$$\widehat{k_b k_f}(\mathbf{k}) = 0 \tag{68}$$

The correlation between k_b and k_f is associated with both chemical sorption and movement induced (physical) sorption. The former is most likely a negative correlation and the latter a positive correlation. The actual correlation structure is probably a combination. Lab and field experiments are needed to decide which correlation structures are most appropriate.

The final model we will study (Model D) takes K_b as a deterministic constant.

Once $\hat{C}_0$ is given, $\tilde{\hat{F}}$ and $\tilde{\overline{C}}$ can be calculated from (41), (42), and the above correlation functions. Following Deng *et al.* (1993) the FFT method is used to calculate the mean concentration field, $\overline{C}(\mathbf{x},t)$. Although the parameters K_f and K_b, not K_d and K_r, are used to calculate concentration and spatial moments in this work, the analysis developed in the Appendix, using parameters K_d and K_r, can also be used to do the same calculation.

It should be pointed out that the authors have used analytical methods to analyze the various spatial moments for reactive transport under kinetic sorption with constant K_d and K_r (Hu *et al.*, 1995). The problem considered here is far too complex to use a similar approach.

We take the initial concentration C_0 to be uniform over a rectangular prism $2a_1 \times \cdots \times 2a_N$ so that

$$\hat{C}_0(k) = C_m \prod_{i=1}^{N} \frac{2\sin(a_i k_i)}{k_i} \tag{69}$$

where C_m is the initial concentration and N is the dimension of the system. Since the sorption is nonequilibrium, and we assume the background concentration of the reactive chemical to be zero, the initial concentration in sorbed phase, S_0, is set to be zero. The computational procedure followed is similar to that presented in Deng *et al.* (1993).

6 DISCUSSION OF NUMERICAL RESULTS

Though our analysis is applicable in three dimensions, for the sake of simplicity and to expedite comparisons, we restrict our discussion to the two-dimensional case. We examine the variation of mean solution-phase concentration and spatial moments under various values of the mean backward and forward coefficients and several different correlation structures between v_i and k_b and between k_b and k_f. In this chapter our attention is focused on the role of randomness of K_b in transport. Mean concentration is examined by contour plotting and the calculation of various spatial moments. The first, second and third moments, as well as skewness, are given by

$$X_i = (1/M) \int_{R^2} n x_i \overline{C} d\mathbf{x} \tag{70}$$

$$X_{ii} = (1/M) \int_{R^2} n x_i^2 \overline{C} d\mathbf{x} - X_i^2 \tag{71}$$

$$X_{iii} = (1/M) \int_{R^2} n (x_i - X_i)^3 \overline{C} d\mathbf{x} \tag{72}$$

$$\text{skewness} = X_{iii} / X_{ii}^{3/2} \tag{73}$$

where

$$M = \int_{R^2} n \overline{C} d\mathbf{x} \tag{74}$$

Following Deng *et al.* (1993), the log conductivity structure is chosen to be isotropic-exponential

$$\overline{ff}(\mathbf{z}) = \sigma_f^2 \exp[-(z_1^2/l^2 + z_2^2/l^2)] \tag{75}$$

where $l = l_1 = l_2$ is correlation length. We choose $\sigma_f^2 = \sigma_W^2 = 0.2$ and $V = 1.0$ m/day. The local dispersivities are set to $d_1 = 0.05$ m²/day and $d_2 = 0.005$ m²/day. Also, for ease of comparison, we use dimensionless solution-phase concentration $C^* = \overline{C}/C_m$, dimensionless sorbed concentration $S^* = \dfrac{\overline{S}}{(\overline{K}_f/\overline{K}_b)\overline{C}}$, and, in some cases, dimensionless time $t' = tV/(1 + \overline{K}_f/\overline{K}_b)l = tV/Rl$.

Let us first study the effect that different correlations between k_f and k_b and between k_b and f have on the various spatial moments. We set $\overline{K}_b = 0.1$/day, $\overline{K}_f = 1.0$/day and let $l = l_w = 1.0$ m. Four different models between k_f and k_b and between k_b and f are used to calculate spatial moments. Fig. 2(a) shows that the various models have little influence on the

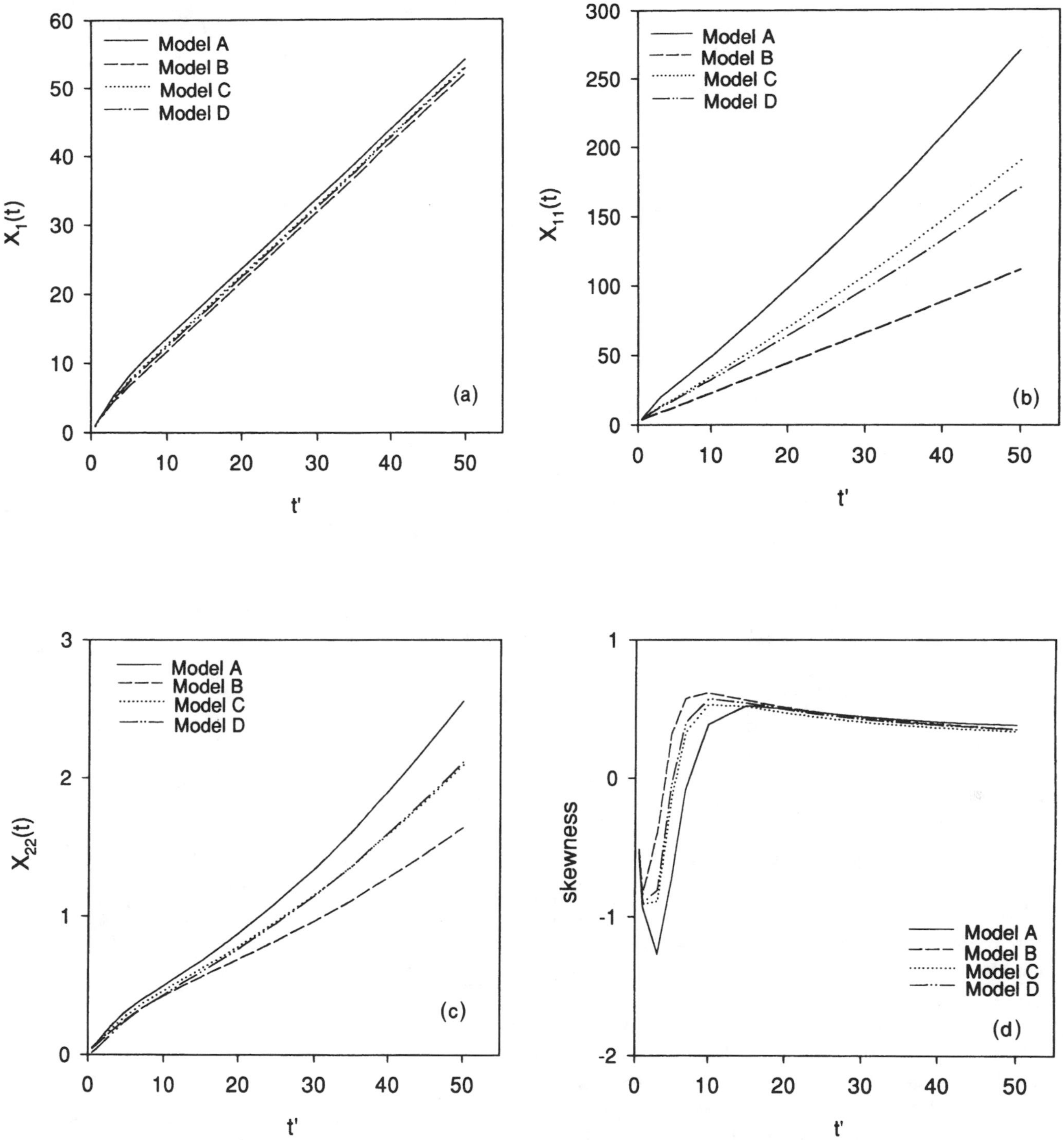

Fig. 2 Spatial moments for various models with $\overline{K}_f=1.0$/day, $\overline{K}_b=0.1$/day and $l=l_w=1.0$ m as functions of dimensionless time ($t'=tV/l(1+\overline{K}_f/\overline{K}_b)$): (a) first moment, (b) second longitudinal moment, (c) second transverse moment, and (d) skewness.

first moment. However, Figs. 2(b) and (c) show that the second moments are significantly affected by the choice of model. Relative to the uncorrelated model (Model C), positive correlation (Model A) increases the second longitudinal moment and negative correlation (Model B) decreases the second moment. The effect of the various correlation functions relating k_f to k_b and k_b to f on the second moment is opposite to the effect of similar correlation structures relating k_d to f (Hu *et al.*, 1995), where positive correlation decreases the second moment and negative correlation increases it. If K_b is a deterministic constant (Model D), and results are compared with those using Model C, then the second longitudinal moment decreases, but the transverse is unaffected. This phenomenon is similar to that observed in Hu *et al.* (1995), where the longitudinal second moment will decrease when K_d is a deterministic constant. Fig. 2(d) shows,

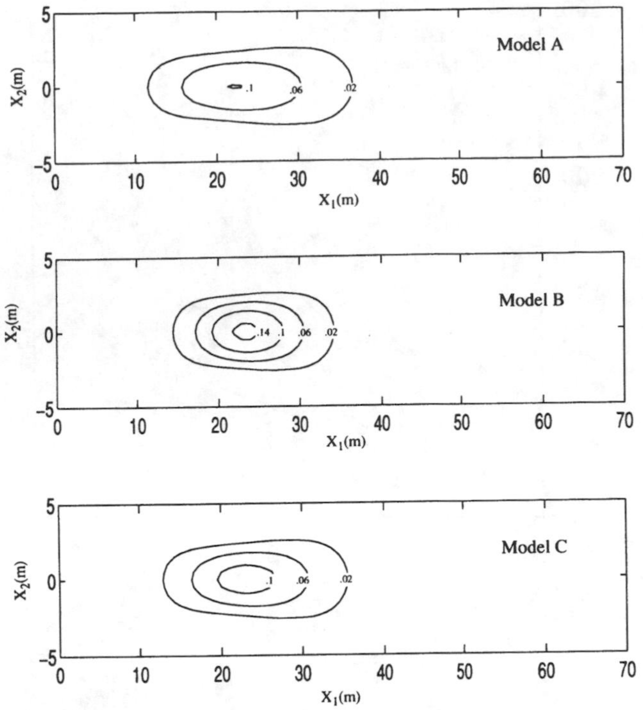

Fig. 3 Mean concentration contours for various models with $\overline{K}_f$=1.0/day, $\overline{K}_b$=0.1/day and l=l_w=1.0 m at t'=25.

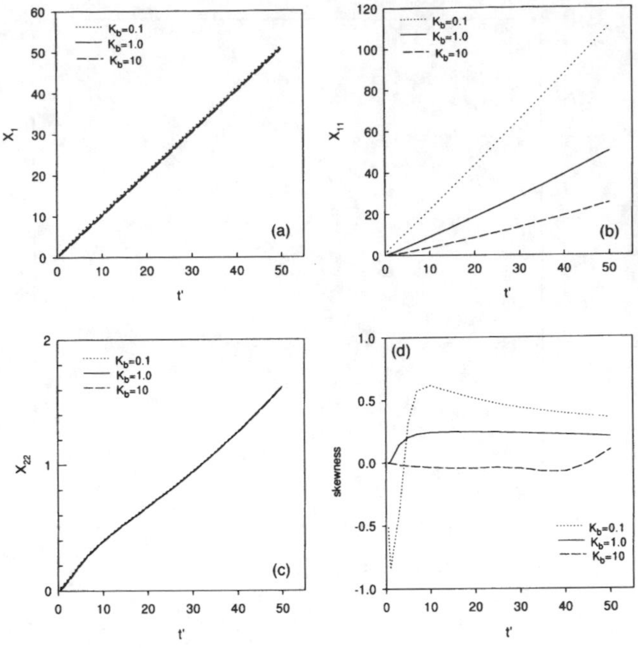

Fig. 4 Spatial moments for Model B with $\overline{K}_f$=1.0/day and l=l_w=1.0 m for various $\overline{K}_b$ as functions of dimensionless time (t'=$tV/l(1+\overline{K}_f/\overline{K}_b)$): (a) first moment, (b) second longitudinal moment, (c) second transverse moment, and (d) skewness.

at short times, that the skewness always increases rapidly from negative to positive with increasing t'. With increasing time there is little difference between the various models.

Fig. 3 shows the concentration distributions at t'=25 for the Models A, B and C (Model D's results are close to those of Model C). The parameters used in Fig. 3 are the same as those in Fig. 2. The mean maximum concentrations for the three models are similar. The dispersion of concentration for Model A is larger than that for Model C, which, in turn, is larger than that for Model B. All three concentration distributions are positively skewed. The results are consistent with the spatial moments of Fig. 2, as expected.

In Fig. 4 we investigate the influence of $\overline{K}_b$ on the various spatial moments. Here we set $\overline{K}_f$=1.0/day, l=l_w=1.0 m and the correlation structure is Model B. The different values of $\overline{K}_b$ are 0.1, 1.0 and 10/day. The results are given in terms of dimensionless time t'=$tV/(1+\overline{K}_f/\overline{K}_b)l$. Figs. 4(a) and (c) show that $\overline{K}_b$ does not influence the first moment and second transverse moment. From Fig. 4(b) it is apparent, however, that $\overline{K}_b$ has a significant effect on X_{11}, which increases with decreasing $\overline{K}_b$. The skewnesses shown in Fig. 4(d) are most interesting. When $\overline{K}_b$=0.1/day, the skewness initially decreases and reaches its minimum rapidly, then increases quickly to a maximum, and finally it gradually levels. The changes in skewness for $\overline{K}_b$=1.0/day and $\overline{K}_b$=10/day are much less dramatic.

Fig. 5 displays the role $\overline{K}_f$ plays in determining the various spatial moments. Here $\overline{K}_b$=0.1/day, l=l_w=1.0 m. Model B is

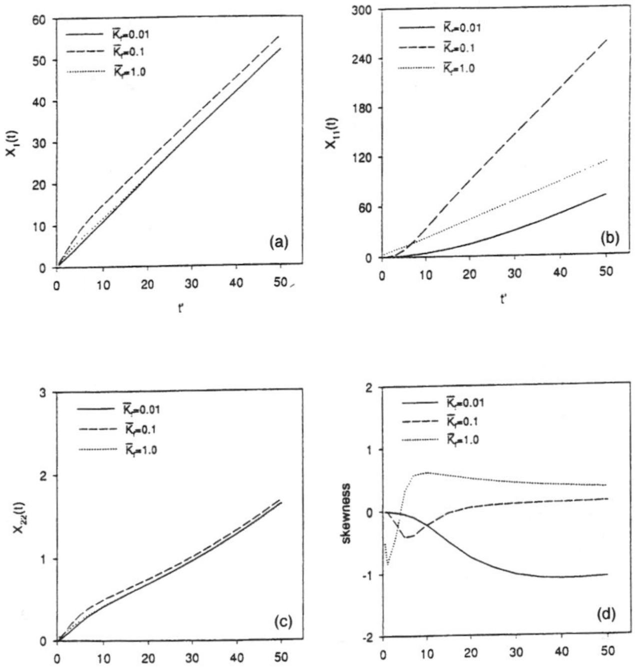

Fig. 5 Spatial moments for Model B with $\overline{K}_b$=0.1/day and l=l_w=1.0 m for various $\overline{K}_f$ as functions of dimensionless time (t'=$tV/l(1+\overline{K}_f/\overline{K}_b)$): (a) first moment, (b) second longitudinal moment, (c) second transverse moment, and (d) skewness.

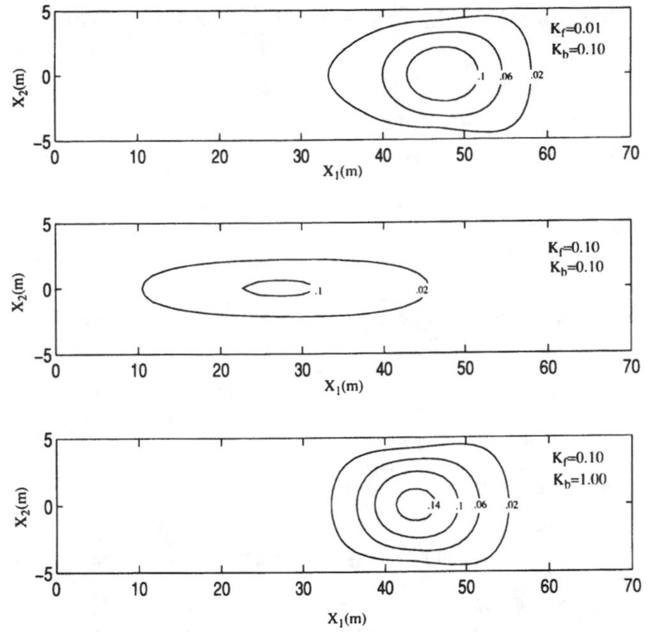

Fig. 6 Mean concentration contours for Model B and $l=l_w=1.0$ m with various $\overline{K}_f$ and $\overline{K}_b$ at $t=50$ days.

used. Three $\overline{K}_f$ values are used: 0.01, 0.1 and 1.0/day. Figs. 5(a) and (c) show that the first moment and second transverse moment for $\overline{K}_f=0.1$/day are larger than those for cases $\overline{K}_f=0.01$/day and $\overline{K}_f=1.0$/day. Fig. 5(b) shows that the second longitudinal moment for $\overline{K}_f=0.1$/day is larger than that for $\overline{K}_f=1.0$/day, which is larger than that of $\overline{K}_f=0.01$/day. The change in skewness (Fig. 5(d)) as a function of $\overline{K}_f$ is very complicated. The skewness for the case of $\overline{K}_f=1.0$/day decreases, reaches its minimum, and then increases dramatically. After reaching its maximum it decreases slowly. The skewness for the case of $\overline{K}_f=0.1$/day decreases initially, reaches a minimum, and then increases. For $\overline{K}_f=0.01$/day, the skewness continuously decreases as t' increases, but the rate slows with t'. At long times, the skewness increases with increasing $\overline{K}_f$.

Fig. 6 graphically illustrates the influence of various $\overline{K}_b$ and $\overline{K}_f$ on the mean concentration distribution.

Figs. 7 and 8 illustrate the effect that various correlation lengths have on the spatial moments and mean concentration (at $t=50$ days), for Model B with $\overline{K}_f=\overline{K}_b=1.0$/day. Correlation lengths are chosen as 0.2, 1.0 and 5.0 m. From Fig. 7 we see that the first moment changes little with l, but all the higher moments increase with l. This phenomenon is also clearly illustrated in the mean concentration contours in Fig. 8.

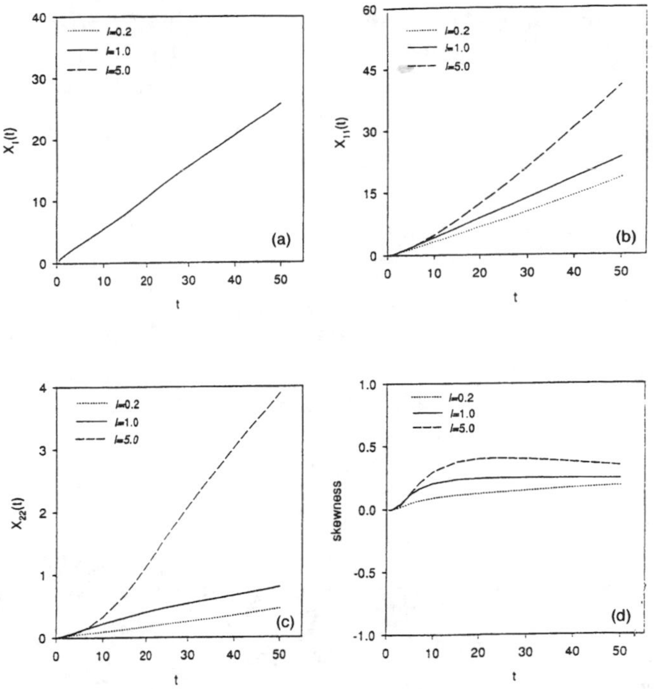

Fig. 7 Spatial moments for Model B with $\overline{K}_f=\overline{K}_b=1.0$/day for various l as functions of real time: (a) first moment, (b) second longitudinal moment, (c) second transverse moment, and (d) skewness.

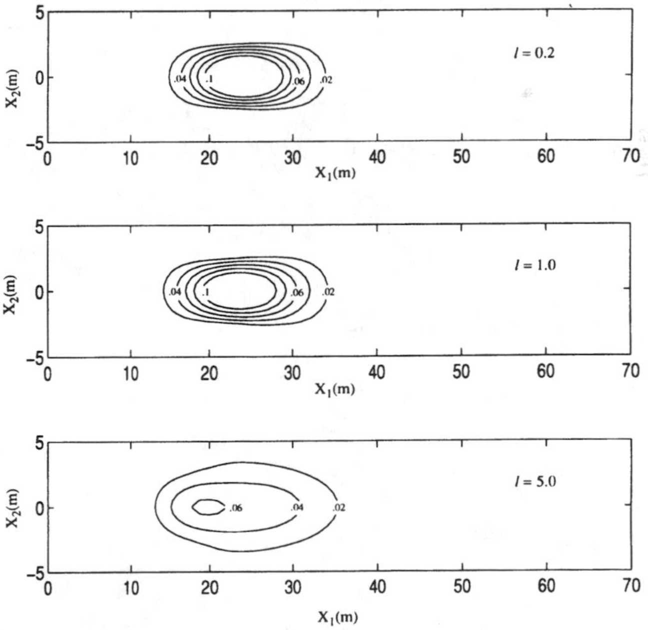

Fig. 8 Mean concentration contours of Model B with $\overline{K}_f=\overline{K}_b=1.0$/day for various l at real time $t=50$ days.

7 CONCLUSIONS

Using a first-order perturbation scheme in a Eulerian framework, a model for the mean concentration of a reactive chemical is developed. The chemical is allowed to sorb via a first-order nonequilibrium reaction with random forward and backward rate coefficients. The model is solved exactly in Fourier–Laplace space and numerically inverted to real space. Data required for the model include $\overline{ff}$, $\overline{fk_f}$, $\overline{fk_b}$, $\overline{k_bk_f}$,

$\overline{k_f k_f}$, $\overline{k_f k_b}$, C_0, and S_0. Because no data relating k_b, k_f, and f stochastically exist at present, the model suggests the need for a number of new experiments involving reactive chemicals.

ACKNOWLEDGMENTS

This research has been supported by the DOE/OHER Subsurface Science Program and the ARO Environmental Science Program under grants DEFG02–85–60310 and DAAL 03–90–G-0074, respectively.

APPENDIX

The relationship between the two sets of parameters is

$$\overline{K}_b = \overline{K}_r, \qquad k_b = k_r \tag{A1}$$

$$\overline{K}_f = \overline{K}_d \overline{K}_r + \overline{k_d k_r}(0) \tag{A2a}$$

and

$$k_f(\mathbf{x}) = \overline{K}_d k_r(\mathbf{x}) + \overline{K}_r k_d(\mathbf{x}) + k_r(\mathbf{x}) k_d(\mathbf{x}) - \overline{k_d k_r}(0) \tag{A2b}$$

Multiply (A2b) by $k_f(\mathbf{y})$, take ensemble averages, and neglect triplet correlations to yield

$$\overline{k_f k_b}(\mathbf{x}-\mathbf{y}) = \overline{K}_d \overline{k_r k_r}(\mathbf{x}-\mathbf{y}) + \overline{K}_r \overline{k_d k_r}(\mathbf{x}-\mathbf{y}) \tag{A3}$$

In deriving (A3) all correlation functions are assumed to be stationary. Take the spatial-Fourier transform of (A3) to obtain

$$\widehat{\overline{k_f k_b}}(\mathbf{k}) = \overline{K}_d \widehat{\overline{k_r k_r}}(\mathbf{k}) + \overline{K}_r \widehat{\overline{k_d k_r}}(\mathbf{k}) \tag{A4}$$

In the same way we can also obtain

$$\widehat{\overline{k_f k_f}}(\mathbf{k}) = \overline{K}_d^2 \widehat{\overline{k_r k_r}}(\mathbf{k}) + \overline{K}_r^2 \widehat{\overline{k_d k_d}}(\mathbf{k}) + 2\overline{K}_r \overline{K}_d \widehat{\overline{k_d k_r}}(\mathbf{k}) \tag{A5}$$

and

$$\widehat{\overline{k_f v_j}}(\mathbf{k}) = \overline{K}_d \widehat{\overline{k_r v_j}}(\mathbf{k}) + \overline{K}_r \widehat{\overline{k_r v_j}}(\mathbf{k}) \tag{A6}$$

Inserting (A1)–(A6) into (40) and (41), we obtain the mean concentration in terms K_d and K_r, where the expressions for $\tilde{\hat{I}}_1(\mathbf{k},\omega)$, $\tilde{\hat{I}}_2(\mathbf{k},\omega)$, $\tilde{\hat{B}}(\mathbf{k},\omega)$ and $\tilde{\hat{G}}(\mathbf{k},\omega)$ in the parameters K_d and K_r can be obtained by substituting (A1)–(A6) into (36), (37), (20), and (21), respectively.

REFERENCES

Bellin, A. A., Rinoldo, A., Bosma, W. J. P., Van Der Zee, S. E. A. T. M. & Rubin, Y. (1993). Linear equilibrium adsorbing solute transport in physically and chemically heterogeneous porous formations, 1. Analytical solution. *Water Resources Research*, 29(12), 4019–4030.

Brusseau, M. L. & Rao, P. R. C. (1989). Sorption nonideality during organic contaminant transport in porous media. *CRC Critical Reviews in Environmental Control*, 19(1), 33–99.

Cushman, J. H. (1991). Diffusion in fractal porous media. *Water Resources Research*, 27(4), 643–644.

Cushman, J. H. & Ginn, T. R. (1993). Non-local dispersion in porous media with continuously evolving scales of heterogeneity. *Transport in Porous Media*, 13(1), 123–138.

Cushman, J. H., Hu, X. & Ginn, T. R. (1994). Nonequilibrium statistical mechanics of preasymptotic dispersion. *Journal of Statistical Physics*, 75, 859–878.

Cushman, J. H., Deng, F.-W. & Hu, X. (1995). A nonlocal theory of reactive transport with physical and chemical heterogeneity: localization errors. *Water Resources Research*, 31(9), 2239–2255.

Dagan, G. (1989). *Flow and Transport in Porous Formations.* New York: Springer-Verlag.

Dagan, G. & Cvetkovic, V. (1993). Spatial moments of a kinetically sorbing solute plume in a heterogeneous aquifer. *Water Resources Research*, 29(2), 4053–4061.

Deng, F.-W., Cushman, J. H. & Delleur, J. W. (1993). A fast Fourier transform stochastic analysis of the contaminant transport problem. *Water Resources Research*, 29(9), 3241–3247.

Gelhar, L. W. (1993). *Stochastic Subsurface Hydrology.* Englewood Cliffs, N.J.: Prentice Hall.

Gelhar, L. W. & Axness, C. L. (1983). Three-dimensional stochastic analysis of macrodispersion in aquifers. *Water Resources Research*, 19(1), 161–190.

Hu, B. X., Deng, D.-W. & Cushman, J. H. (1995). Nonlocal reactive transport with physical and chemical heterogeneity: Linear nonequilibrium sorption with random K_d. *Water Resources Research*, 31(9), 2239–2252.

Koch, D. L. & Brady, J. F. (1987). A nonlocal description of advection diffusion with application to dispersion in porous media. *Chemical Engineering Science*, 42, 1377–1392.

Koch, D. L. & Brady, J. F. (1988). Anomalous diffusion in heterogeneous porous media. *Physics of Fluids*, 31, 965–973.

Neuman, S. P. & Orr, S. (1993). Prediction of steady state flow in nonuniform geologic media by conditional moments; exact nonlocal formalism, effective conductivities, and weak approximation. *Water Resources Research*, 29(3), 633–645.

Sahimi, M. (1993). Fractal and superdiffusion transport and hydrodynamic dispersion in heterogeneous porous media. *Transport in Porous Media*, 13(1), 3–40.

4 Perspectives on field-scale application of stochastic subsurface hydrology

LYNN W. GELHAR
Massachusetts Institute of Technology

ABSTRACT Recent advances in the application of stochastic methods to naturally heterogeneous aquifer systems are summarized in the context of field-scale contaminant transport descriptions. Key results on contaminant advection, spreading and dispersion, mixing and dilution, and retardation are illustrated by comparisons with field observations, focusing particularly on the Cape Cod glacial aquifer system where hydraulic characterization and contaminant transport observations are available over scales ranging from a few meters to several kilometers. Both theoretical results and field observations point to the importance of characterizing heterogeneity as fully three-dimensional and anisotropic, and of including unsteadiness in the flow description. Relative dispersion analysis indicates that larger plumes will have larger macrodispersivities; this plume-scale effect is demonstrated quantitatively for the Cape Cod site. Mixing and dilution are controlled by the interplay between local dispersion and small-scale variations in hydraulic conductivity, suggesting the need to characterize hydraulic conductivity variations at scales down to centimeters. The important influence of the heterogeneity of reactive transport properties in enhancing longitudinal dispersion points to the need for systematic measurements of variations in such chemical characteristics. Unresolved research areas critical to effective application of stochastic methods to field problems include efficient characterization of heterogeneity over a wide range of scales (centimeters to kilometers), large-scale controlled field experiments on reactive transport and multiphase flow, and theoretical treatments of multiphase flow and highly heterogeneous media.

1 INTRODUCTION

It is now widely acknowledged that natural aquifer materials are heterogeneous in terms of their flow properties such as hydraulic conductivity, and it is becoming more widely recognized that chemical and biological properties affecting transport of contaminants in aquifers also vary erratically in space. Applications of contaminant transport models, for example in hazardous waste site remediation or radioactive waste disposal, involve large spatial scales (100 m to 10 km) and long time scales (10 to 10 000 years), making direct measurements of the transport properties at pertinent field scales unfeasible. It is equally impractical to map out the detailed heterogeneity of aquifers at these pertinent application scales. Over the last decade or so, the problem of treating the large-scale behavior of heterogeneous aquifer systems has been extensively explored theoretically using stochastic descriptions for the relatively small-scale heterogeneity of flow and transport properties.

The overall goals of these stochastic approaches can be put into two broad categories, the first of which deals with the scientific elements of the issue; that being to understand and quantify the dominant large-scale flow and transport processes. The second overall goal is to develop field-scale predictions addressing site-specific applications. In this applied context, we seek aggregated large-scale descriptions of the pertinent processes, and at the same time we need to quantify the local error introduced in such predictions as a result of the heterogeneity in the system. This stochastic framework can provide a systematic basis for designing data collection programs, indicating, from the nature of the large-scale aggregated behavior, the factors which are important as opposed to those which may be scientifically interesting but largely irrelevant to the large-scale behavior of such systems. Many of the theoretical developments in this new field of stochastic subsurface hydrology are summarized in Dagan (1989) and Gelhar (1993).

The goal here is to review some of the key applied results in the field of stochastic subsurface hydrology in the context of contaminant transport problems, which can be thought of in terms of the usual one-dimensional transport equation

$$\frac{\partial c}{\partial t}+V\frac{\partial c}{\partial x}=AV\frac{\partial^2 c}{\partial x^2}-\kappa\left(c,\frac{\partial c}{\partial t},\ldots\right)c \tag{1}$$

This equation represents, through the advection term, the bulk transport of a dissolved contaminant of concentration c at some aggregated average velocity V, a spreading and mixing effect through the dispersion term characterized by a longitudinal dispersivity A, and a possible reaction or decay term represented here by a rate constant κ, which may itself depend on concentration and time rate of change of concentration.

The purpose here is to summarize some of the key stochastic results which address the behavior of the aggregated transport parameters appearing in this traditional advection–dispersion-reaction equation. The focus is on observed field behavior as the ultimate test of the viability of any of the theoretical results.

We will explore the applicability of some of the key theoretical results in the context of large-scale field observations which demonstrate the actual performance of stochastic methods and highlight some of the unresolved issues of importance, particularly in an applied context. Consequently, the emphasis is on the interplay between theoretical developments and field observations rather than on methodological issues relating to analytical or numerical approaches. Only selected theoretical developments relating directly to the applied issues are cited; several of the topics discussed here are the subject of detailed review articles in this volume. This chapter will be developed in the context of field experiments and observations demonstrating the efficacy of stochastic methods in deriving large-scale transport parameters.

2 SOME FIELD OBSERVATIONS

The famous Borden experiment (Freyberg, 1986; Sudicky, 1986), in a relatively homogeneous glaciolacustrine fine to medium sand, demonstrated that measurements of the variability of hydraulic conductivity can be used to develop independent predictions of the longitudinal macrodispersivity

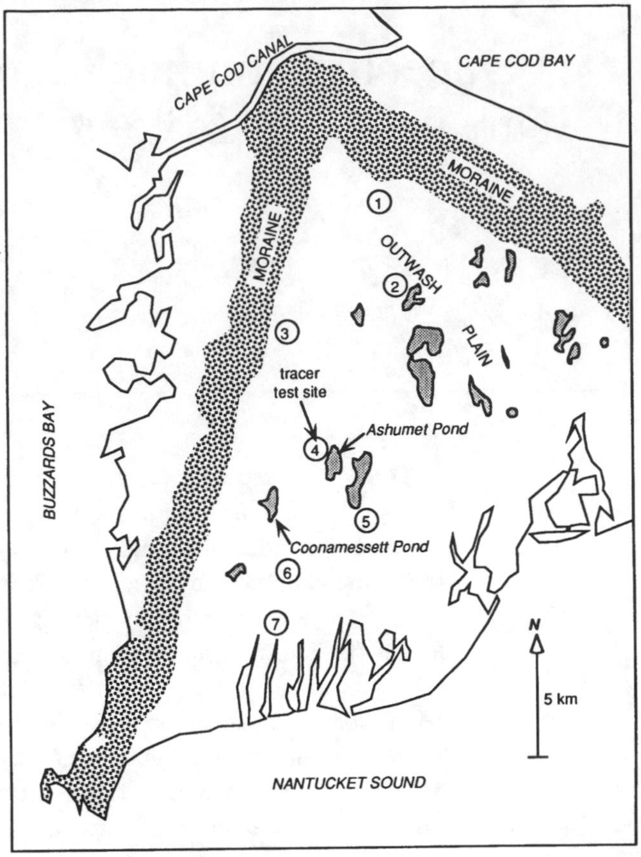

Fig. 1 Western Cape Cod regional hydrogeologic setting and locations of geologic facies mapping sites (circled numbers).

which are in agreement with observations from a large-scale tracer test at the site. At a second major field site near Columbus, Mississippi (Adams & Gelhar, 1992; Boggs *et al.*, 1992; Rehfeldt, Boggs & Gelhar, 1992) in a very heterogeneous fluvial sand and gravel aquifer exhibiting large-scale trends in hydraulic conductivity, it was shown that measurements of the variability of hydraulic conductivity used in conjunction with stochastic theory provide a plausible prediction of the longitudinal macrodispersivity independently derived from large-scale tracer experiments. A third major field site in relatively homogeneous glacial outwash sand and gravel on Cape Cod (Garabedian *et al.*, 1991; LeBlanc *et al.*, 1991) will be explored in greater detail here, particularly focusing on the much larger scale associated with several contamination plumes which have been encountered in this aquifer.

The regional geologic setting for the Cape Cod aquifer system is illustrated in Fig. 1, which shows the location of the two moraines and the extensive outwash plain which constitutes the primary aquifer in the area. Also shown are several lakes or ponds, which represent the intersection of the water table with the land surface, and the site of the Cape Cod tracer test. The relationship between the geologic conditions of the outwash plain and the hydraulic properties of the

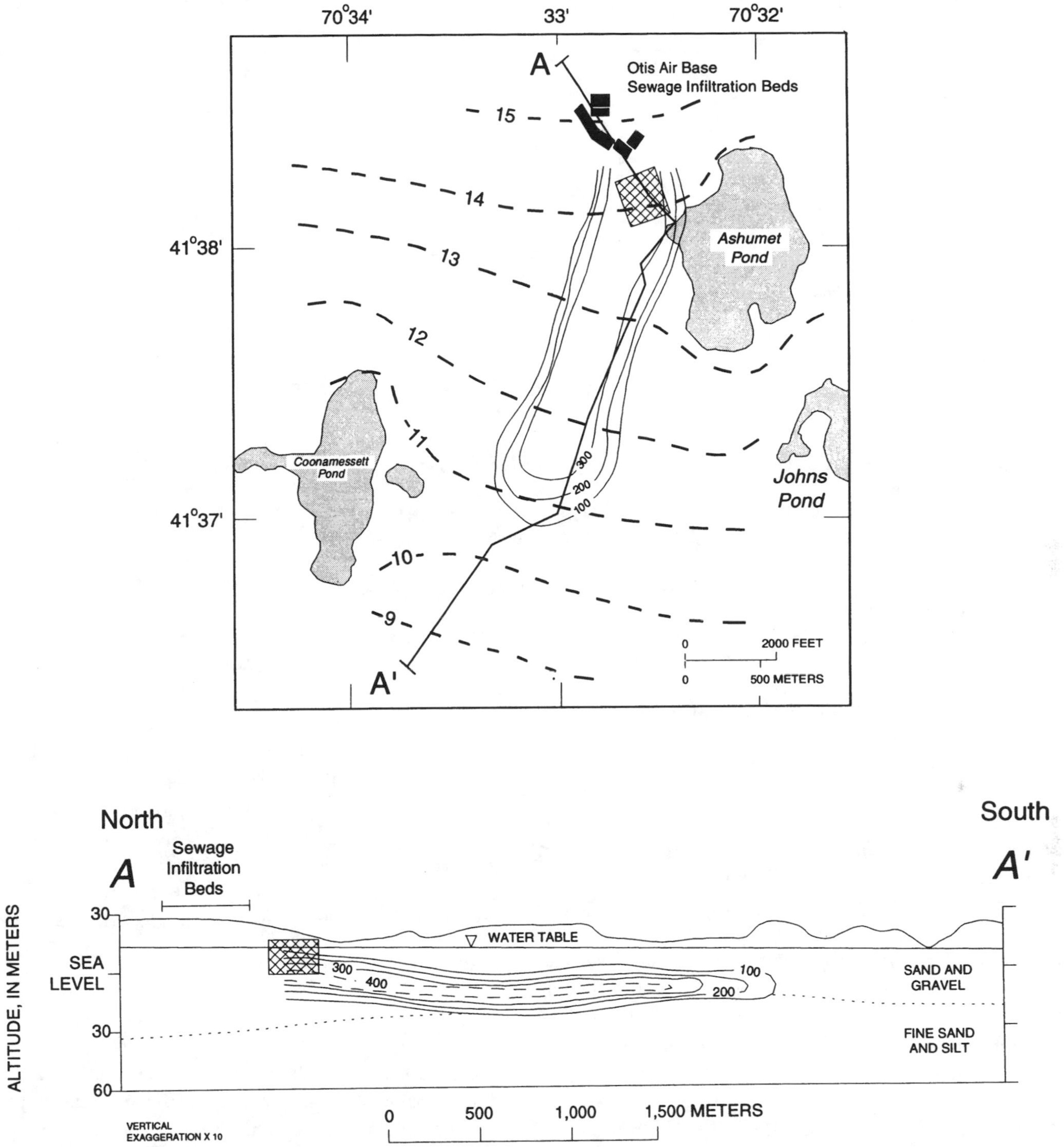

Fig. 2 Cape Cod sewage plume plan (top: solid contours show boron concentration in micrograms per liter; dashed contours are water levels in meters) and vertical section (bottom: contours of boron concentration in micrograms per liter) with the location of the tracer test site (cross-hatched box). Based on LeBlanc *et al.* (1991).

aquifer characterized by its mean and variability has been explored extensively by Thompson (1994). Based on geologic facies mapping information from the numbered sediment exposure sites identified in Fig. 1, systematic trends in the mean and variance of the log conductivity have been identified as being related to the distance from the moraine on the north (Thompson & Gelhar, 1993). Fig. 2 shows one of the extensive contamination plumes on Cape Cod, in this case associated with the sewage disposal operation for the Otis Air Base. The Cape Cod tracer test site is located just above the sewage plume, which is limited to a relatively narrow horizon in the vertical, as shown in Fig. 2.

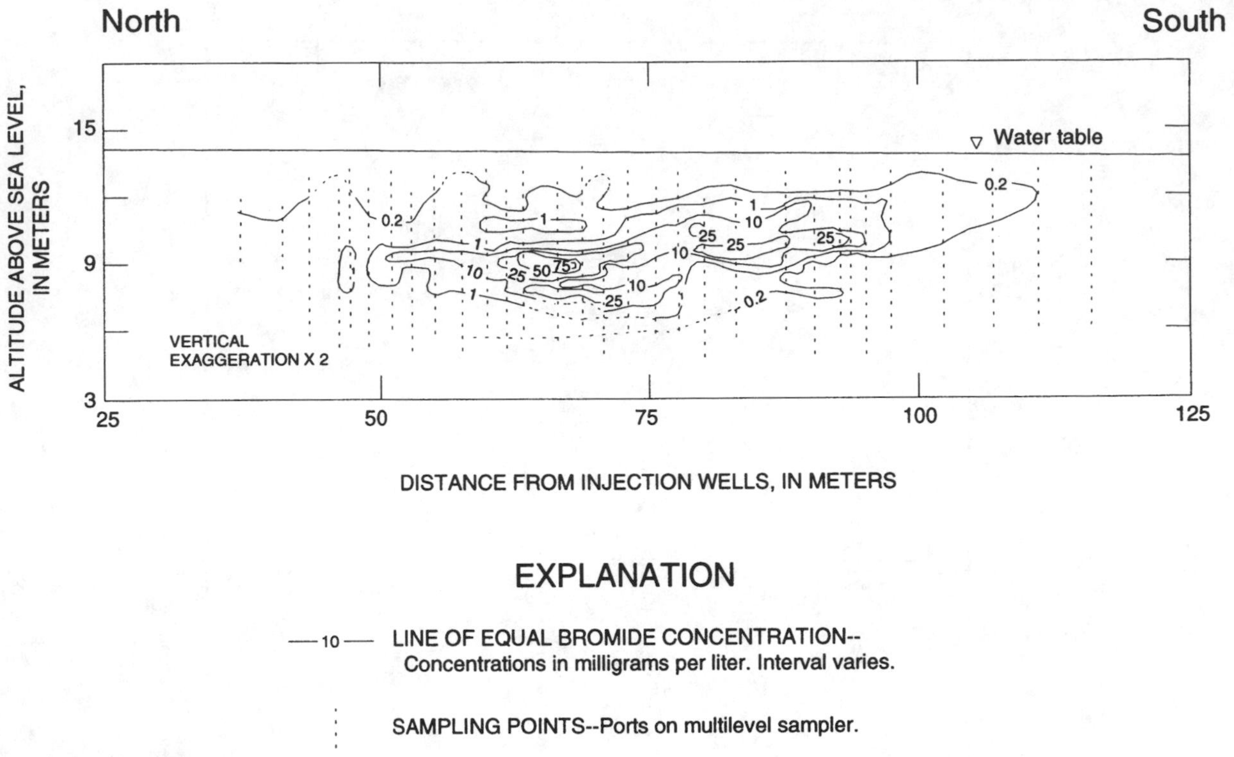

Fig. 3 Vertical longitudinal section through the tracer plume at 173 days; based on LeBlanc *et al.* (1991).

Fig. 3 shows a longitudinal vertical cross-section through the bromide tracer cloud produced 173 days into the tracer test. The initial concentration of bromide was 640 mg/l, so that we see about a ten-fold decrease in the maximum concentration reflecting the dilution effect of mixing in the aquifer. It is also seen that the plume is very irregular in shape, with tongues and lobes of high concentration reflecting the complex velocity field associated with the heterogeneity of hydraulic conductivity. Note that the vertical scale in Fig. 3 is exaggerated by a factor of 2; the plume has spread very extensively in the longitudinal direction, but shows only limited vertical spreading. Some quantitative features of the tracer experiment are explored in the following sections.

3 SOLUTE ADVECTION RESULTS

The results of the bromide tracer experiment on Cape Cod are summarized, in Fig. 4, in terms of the first and second spatial moments of the solute distribution as it evolves in time. Even though the details of the tracer distribution are very complex (Fig. 3), the aggregated spatial moments show systematic variations in time. From the first spatial moment in Fig. 4(a), the velocity of the center of mass of the tracer cloud is easily evaluated to be 0.42 m/day. The implied horizontal effective hydraulic conductivity is evaluated from the Darcy equation,

$$K_{11}=nV/J$$

where n is the porosity and J is the mean hydraulic gradient. The calculated effective conductivities using this approach for the Borden (Sudicky, 1986) and Cape Cod (Hess, Wolf & Celia, 1992) sites are shown in Fig. 5(a), which also includes the theoretical result of Gelhar and Axness (1983) based on a conjecture extending the first-order results to large variance of the natural logarithm of hydraulic conductivity, σ_f^2:

$$K_{11}/K_G=\exp[((1/2)-g_{11})\sigma_f^2]=\gamma$$

$$g_{11}=\frac{1}{2(\rho^2-1)}\left[\frac{\rho^2}{(\rho^2-1)^{1/2}}\tan^{-1}[(\rho^2-1)^{1/2}]-1\right];\ \rho=\lambda_1/\lambda_3>1 \quad (2)$$

Here it is assumed that the mean flow is essentially horizontal in the plane of bedding; λ_1 is the correlation scale in the plane of bedding and λ_3 is the vertical correlation scale. Statistical isotropy in the horizontal plane is assumed. We see that, even for these rather homogeneous aquifers, the observed effective conductivity is significantly higher than the geometric mean K_G, as would be predicted by the theoretical result. The observations generally fall within the range indicated by the theory, but do not show strict agreement in terms of the λ_1/λ_3 ratio. These differences could easily be explained, however, by uncertainties in the estimated porosity, which is difficult to evaluate directly in the field.

In Fig. 5(b) some results for the effective conductivity determined from three-dimensional numerical simulations

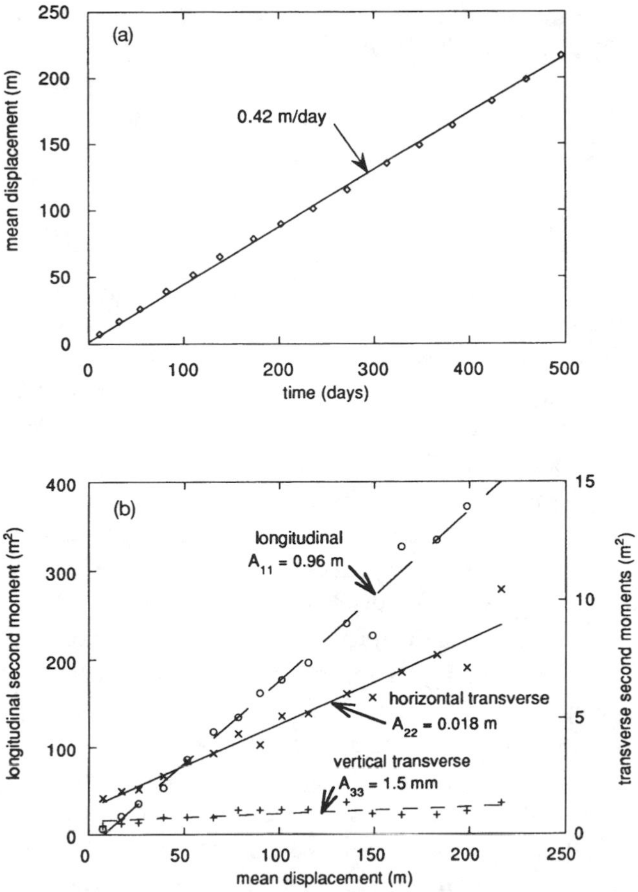

Fig. 4 Evolution of the first and second spatial moments of tracer cloud for the Cape Cod tracer test.

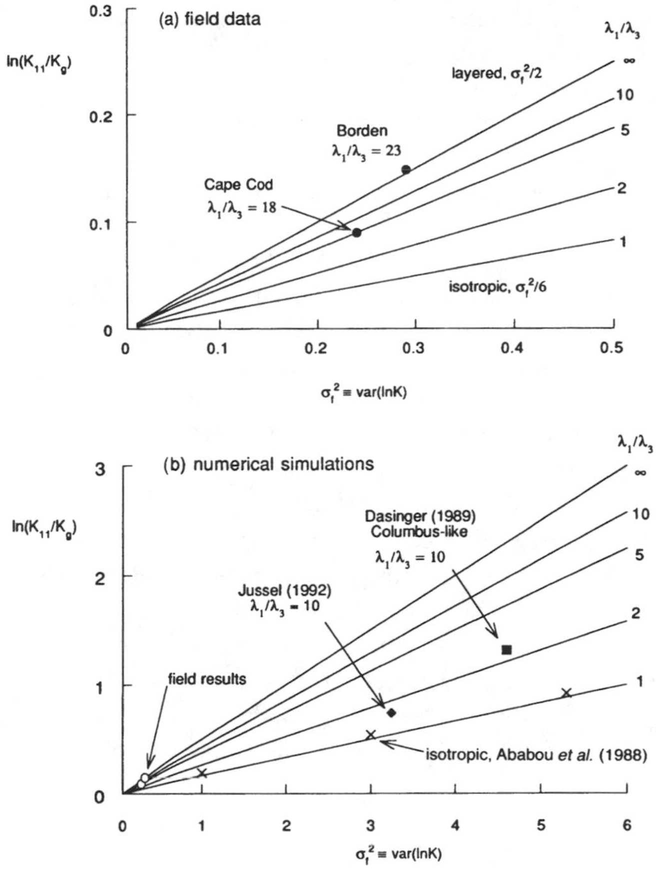

Fig. 5 Effective conductivities based on (a) field data, (b) numerical simulations compared with theory, eq. (2).

are compared with (2). The isotropic simulation results of Ababou, Gelhar & McLaughlin (1988) are seen to follow quite closely the theoretical result of (2) ($g_{11}=1/3$)

$$K_{11}/K_g=\exp(\sigma_f^2/6) \tag{2a}$$

even for strongly heterogeneous materials. Other numerical simulations also confirm this isotropic relationship for σ_f^2 up to 7 (Dykaar & Kitanidis, 1992; Neuman & Orr, 1993). In the case of an anisotropic, very heterogeneous, Columbus-like system (Dasinger, 1989; Dasinger & Gelhar, 1991), the theory is seen to predict a significantly higher effective conductivity than is found from the simulations. The anisotropic simulations of Jussel (1992) also show a lower effective conductivity than predicted by (2). For the isotropic case (Dagan, 1993), it has been shown that the perturbation analysis, extended to second order in σ_f^2, is in agreement with the second-order expansion of the exponential conjecture of (2a). A similar second-order analysis for the anisotropic case (Indelman & Abramovich, 1994) finds a second-order result which is not in agreement with that corresponding to (2), but unfortunately this result cannot apply to very heterogeneous materials. For example, if one attempts to calculate the effective conductivity for the Columbus-like condition (Fig. 5(b)) from the second-order expression, one finds that the effective conductivity would be negative. The field observations of Fig. 5(a) are also included in Fig. 5(b) for reference. Burr, Sudicky & Naff (1994) report on numerical simulations of a Borden-like system which fall within this same range in terms of effective conductivity.

Recent theoretical analyses have considered the influence of large-scale flow nonuniformities on the effective hydraulic conductivity. For steady radial mean flow to a well in a three-dimensionally heterogeneous aquifer with a statistically anisotropic ln$K$ covariance, Naff (1991) finds that the effective conductivity is reduced relative to that for a uniform mean flow. This is in contrast to the two-dimensional numerical simulation results of Smith & Freeze (1979), which find that effective conductivity in a converging flow configuration is greater than the geometric mean, the effective conductivity of the corresponding uniform mean flow. Nonstationary analyses of the influence of a linear longitudinal trend in the mean ln$K$ (Rubin & Seong, 1994; Indelman & Rubin, 1995; Li & McLaughlin, 1995) indicate that the effective conductivity is decreased relative to that for a constant mean conductivity, but these results were developed only in the practically uninteresting two-dimensional case. A direct comparison of the theoretical results with the field experiment at the Columbus site is not meaningful because of the complex large-scale nonuniformity in the flow field and the hydraulic gradient.

The above results indicate that the effective conductivity, and consequently the bulk translation of a solute plume, are well predicted in the case of relatively homogeneous materials such as those at the Borden and Cape Cod sites, but there remain major uncertainties in the case of very heterogeneous anisotropic systems. There is a need for carefully designed three-dimensional numerical experiments which address this issue. This is a particularly challenging problem because boundary effects can become much more important in anisotropic systems. Ababou *et al.* (1988) find that the head covariance function shows large persistence in the vertical direction perpendicular to bedding even when the vertical correlation scale is much smaller than the horizontal correlation scales. Consequently, the anisotropic systems will be much more sensitive to boundary conditions. There is also a need for carefully designed field experiments in very heterogeneous systems. Theoretical methods of incorporating the influences of more general large-scale flow nonuniformities in realistic three-dimensionally anisotropic systems are also needed.

4 SOLUTE SPREADING RESULTS

The spreading of solute at the Cape Cod site, as reflected in Fig. 3, can be characterized in terms of a longitudinal macrodispersivity. A crude quantitative characterization of a dispersivity in this case can be developed from a simple visual inspection of this cross-section. From the standard Gaussian solution for a pulse, we know that the width of the mixed zone, characterized by the region where the concentration exceeds e^{-1} times the maximum concentration (35 m from Fig. 3), is four times the square root of the product of the longitudinal dispersivity and the mean displacement,

$$4\sqrt{A_{11}\bar{x}}=35 \text{ m}$$

and with a mean displacement of about 70 m we find that the implied longitudinal dispersivity is about 1 m. It is important to recognize that this modest dispersivity corresponds to very substantial longitudinal spreading of the solute, which at the point of injection was concentrated in a small region a few meters in horizontal extent. A more refined evaluation of the longitudinal macrodispersivity can be found from the evolution of the longitudinal second moment as shown in Fig. 4(b); this leads to a dispersivity of 0.96 m, in reasonable agreement with our crude estimate based on visual inspection of the plume. An independent prediction of the longitudinal dispersivity at the Cape Cod site was developed by measuring the variability of hydraulic conductivity using the borehole flow meter technique (Hess, 1989; Rehfeldt *et al.*, 1989; Hess *et al.*, 1992) and using the resulting estimates of the statistical parameters characterizing the hydraulic conductivity in the theoretical results of Gelhar & Axness (1983):

$$A_{11}=\sigma_f^2\lambda_1/\gamma^2 \qquad (3)$$

Here the flow factor γ is given by (2). A typical result for the hydraulic conductivity determined from the borehole flow meter technique is shown in Fig. 6. Sixteen flow meter wells, located roughly along a 25 m long transect, adjacent to, but not within, the path of the actual tracer cloud, were used to evaluate the variability of the hydraulic conductivity. The corresponding predictions of the longitudinal dispersivity ranged from 0.7 to 1.0 m, depending on how one fits the estimated variograms. Considering the inherent uncertainty in evaluating the statistical parameters, this is considered to be in good agreement with the longitudinal dispersivity of 0.96 m determined from the tracer test. The implicit assumption is that the hydraulic conductivity field is stationary, so that sampling this small region adjacent to the plume will represent the variability of hydraulic conductivity over the entire area traversed by the plume. The positive results indicate that this is a reasonable assumption, and, in addition, we can conclude that the classical linearized theory (Gelhar & Axness, 1983; Dagan, 1988) does represent the dominant mechanism of longitudinal spreading in this experiment. Similar findings for the Borden (Sudicky, 1986; Rajaram & Gelhar, 1991) and Columbus (Adams & Gelhar, 1992) sites support the viability of the classical theories for the treatment of longitudinal spreading, although the results for the more heterogeneous Columbus site are less definitive because of incomplete sampling of the plume. Recent results for a very heterogeneous fluvial sand and gravel site in Germany (Ptak & Teutsch, 1994; Schad & Teutsch, 1994) are of interest in view of the non-Gaussian behavior of the observed log conductivity, which was characterized via indicator variograms. Based on the agreement of dispersivity predictions using second moment characterization in the classical Gaussian theory and the results of tracer tests, it was suggested that non-Gaussian refinements may not be necessary even for this very heterogeneous site.

The appropriate theoretical description of the longitudinal dispersivity for very heterogeneous media ($\sigma_f^2>1$) is not well established. For the ideal isotropic case, the numerical simulations of Jussel, Stauffer & Dracos (1990) and Tompson and Gelhar (1990) indicate agreement with (3) for $\sigma_f^2>3$, whereas those of Chin & Wang (1992) indicate agreement with (3) using $\gamma=1$. Corrsin's conjecture, an approximation which partially takes into account nonlinear effects in the transport analysis (Dagan, 1988), does not seem to capture some important effects; numerical results (Chin &

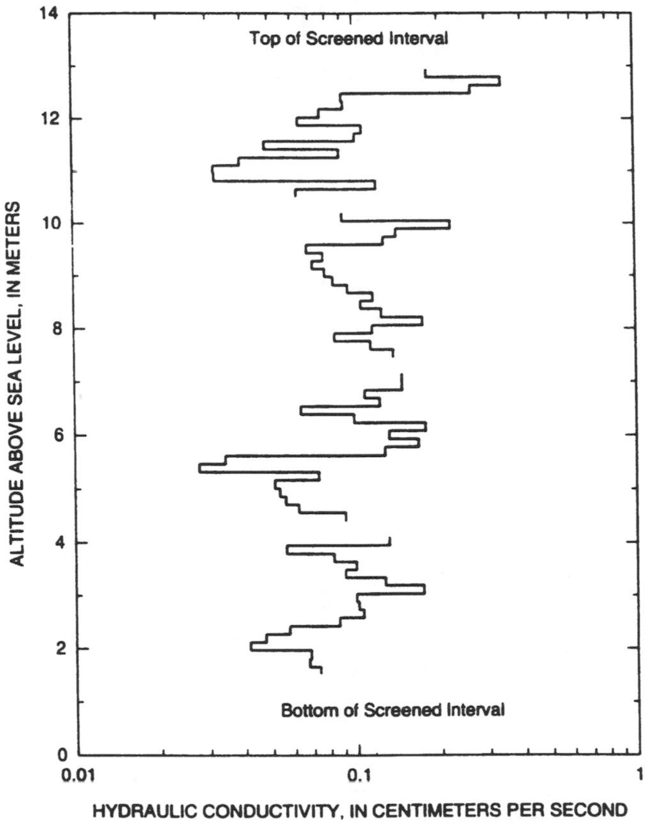

Fig. 6 Typical vertical variation of hydraulic conductivity at the Cape Cod tracer test site determined from a borehole flow meter log; based on Hess (1989).

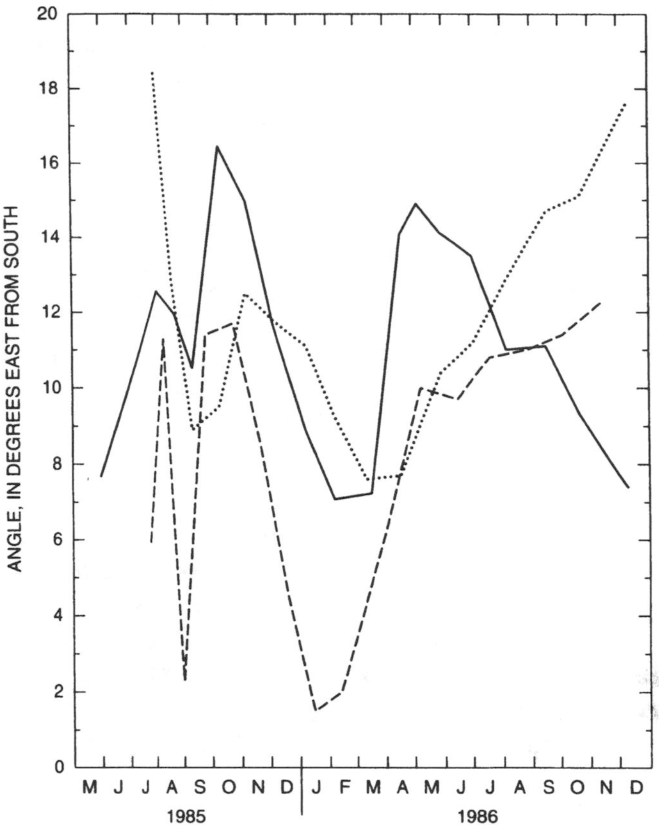

Fig. 7 Unsteady flow effects at the Cape Cod tracer test site: direction of the hydraulic gradient (solid line), the direction of displacement of the center of mass of the tracer cloud (dashed line), and orientation of the axis of the plume (dotted line). Based on Garabedian *et al.* (1991).

Wang, 1992, fig. 13) indicate that Corrsin's conjecture, as implemented by Neuman & Zhang (1990) for small displacement, does not yield improved dispersivity predictions for more heterogeneous aquifers. For the practically important case of anisotropic media, some simulations (Dasinger, 1989; Dasinger & Gelhar, 1991) produce longitudinal dispersivities with $\gamma>1$ in (3) whereas others (Jussel, Stauffer & Dracos, 1994) indicate agreement using $\gamma=1$.

In the case of transverse spreading, the classical steady flow theory clearly does not explain the observed horizontal transverse spreading. Fig. 7 illustrates observations of the hydraulic gradient, and the plume displacement and orientation, as they vary with time during the Cape Cod tracer test. The fluctuations in the direction of the displacement of the center of mass of the plume are seen largely to follow that of the hydraulic gradient, though there is a small but significant difference between the direction of mean displacement and the direction of the applied hydraulic gradient. This implies that the bulk hydraulic behavior of the aquifer is anisotropic in the horizontal plane, and that the assumption of statistical isotropy within the horizontal plane is not strictly correct. Such horizontal anisotropy is certainly plausible from a geologic point of view, but because the sampling network used to determine the hydraulic variability of the aquifer was largely along a transect, it is not possible to detect this anisotropy directly. Fig. 7 shows that the orientation of the plume axis is generally at a somewhat larger angle than the direction of the movement of the center of mass of the plume. In essence, the axis of the plume is deflected more toward the direction of the hydraulic gradient, and the principle axis of the dispersivity tensor is not oriented with the direction of the mean flow. This tendency of the axis of the plume to turn toward the direction of the mean hydraulic gradient in a horizontally anisotropic system is the same as that seen in the Gelhar & Axness (1983) analysis of a three-dimensional, horizontally anisotropic system in the case of steady flow.

Analysis of the effects of the fluctuation in the direction of the hydraulic gradient on transverse dispersion (Rehfeldt, 1988; Rehfeldt & Gelhar, 1992) predicts that the transverse horizontal dispersivity will be significantly enhanced as a result of fluctuations in the direction of the hydraulic gradient. The addition to the macrodispersivity is described by

$$A^*_{22}=\sigma^2_\phi\lambda_\phi V/\gamma^2 \qquad (4)$$

where σ^2_ϕ is the variance of the angle of the hydraulic gradient (in radians), λ_ϕ is the temporal correlation scale of the fluctuations in the angle of the hydraulic gradient, and γ is the flow factor given in (2). Using the parameters estimated from the observed variation in direction of the hydraulic gradient at the Cape Cod site, it is found that the unsteady relationship in (4) predicts the proper order of magnitude, about 0.02 m, for the transverse dispersivity in the horizontal direction. Using recent measurements of the fluctuation in the hydraulic gradient at the Borden site, Farrell *et al.* (1994) have clearly shown that the magnitude of the transverse dispersion at that site is determined by the unsteady fluctuations in the direction of the hydraulic gradient. In fact, the larger transverse horizontal dispersivity at the Borden site (0.04 m) is explained by the larger degree of variation in the direction of the hydraulic gradient at that site in comparison with the Cape Cod site.

Many efforts to interpret the dispersion process at these field sites have adopted the convenient assumption that the spatial variation in the flow field can be represented as two-dimensional depth-averaged quantities (Freyberg, 1986; Dagan, 1987; Barry, Coves & Sposito, 1988; Neuman & Zhang, 1990; Graham & McLaughlin, 1991; Woodbury & Sudicky, 1991; Deng, Cushman & Delleur, 1993), taking advantage of the greater initial transverse spreading reflected in the two-dimensional solution (Dagan, 1984). It is now clear that the apparent agreement between the two-dimensional theory and the field observations is coincidental; the actual transverse spreading process is clearly controlled by the degree of temporal fluctuation in the hydraulic gradient.

This point is further reinforced by the very interesting recent large-scale field investigation of a contamination plume near Regina in Canada (Van der Kamp *et al.*, 1994). This site, identified as the Condie aquifer, is unique in terms of the unusually steady flow that occurs in the sand and gravel aquifer because the aquifer is isolated from near surface seasonal changes by a thick till layer; fluctuations of water level in this aquifer are at most a few centimeters. An 8 km long contamination plume has been carefully monitored three-dimensionally in this aquifer. Interestingly, it is observed that the plume is extremely narrow horizontally, and even after 8 km of travel it has not mixed vertically over the 6 m thickness of the aquifer. Fig. 8 shows the resulting horizontal and vertical transverse dispersivities estimated by Van der Kamp *et al.* (1994), in comparison with the dispersivities from previous field studies as summarized by Gelhar, Welty & Rehfeldt (1992). The unusually low values of the transverse dispersivities at these very large scales further illustrate the point that transverse dispersion is controlled largely by fluctuations in the direction of the hydraulic gradient under field conditions.

Recognition of the dominant role of unsteady flow in determining the transverse dispersion characteristics of aquifers is of major practical significance, and it suggests that more theoretical work should incorporate this influence. For problems which are dominated by transverse dispersion, as is the case for contamination plumes involving a continuous release of contaminant from a source, this is a particularly important consideration. Recognizing this control of transverse dispersion by temporal fluctuations in the direction of the hydraulic gradient suggests that greater emphasis should be placed on measurements characterizing the temporal variation in the hydraulic gradient. Such measurements of hydraulic gradient are relatively simple to make and should routinely be included as part of site characterization programs.

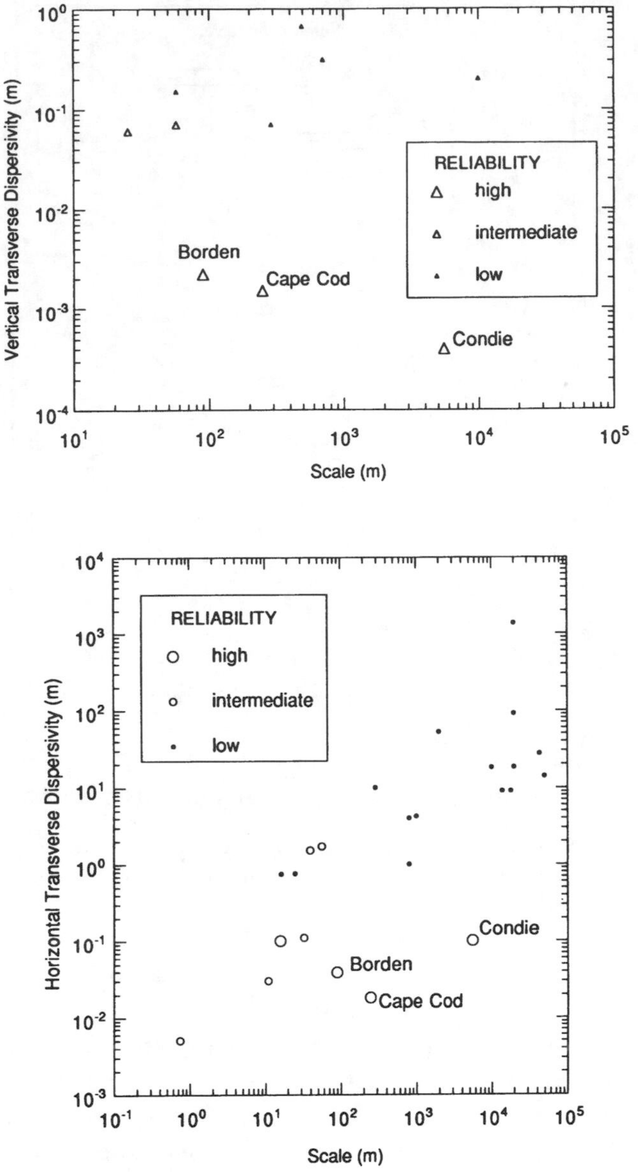

Fig. 8 Summary of field data on vertical and horizontal transverse dispersivities including the Condie site.

Fig. 9 shows the horizontal extent of several large-scale contamination plumes that have been observed at the Cape Cod site. In the context of the above discussion, one may

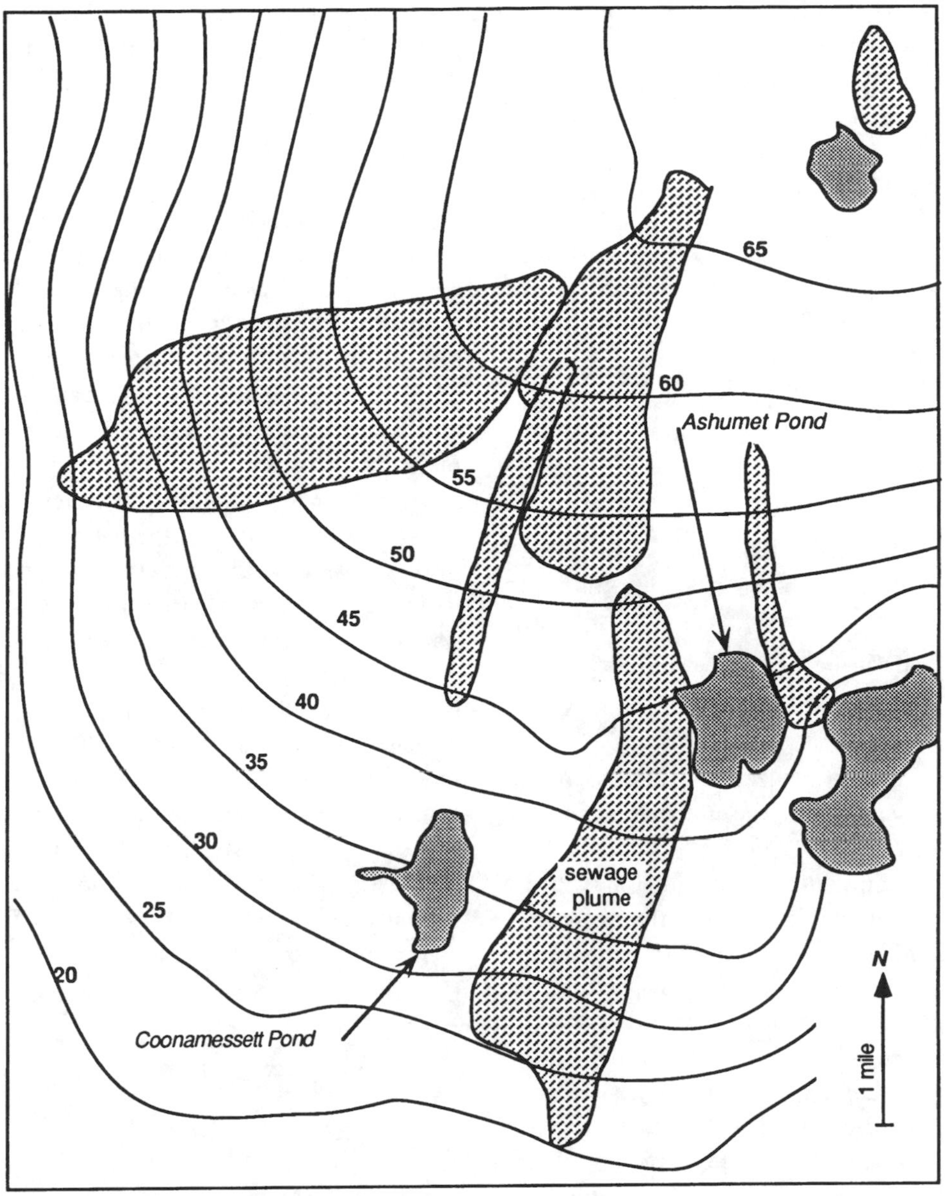

Fig. 9 Western Cape Cod contaminant plumes and water table elevations in feet above sea level.

speculate whether the major differences in shapes of the different plumes are to some degree associated with differences in the variations in flow regime in the aquifer. As can be seen from the head pattern, there is also probably an important influence of diverging flow patterns. The very broad plume migrating to the west seems to be influenced by the diverging flow pattern in that area. Theoretical treatments of dispersive transport will need to take into account large-scale flow nonuniformities and unsteady flow effects if they are to be relevant to the kinds of contamination situations reflected in Fig. 9. Fig. 10 shows the classical summary of longitudinal dispersivity data (Gelhar *et al.*, 1992), with the additional longitudinal dispersivity determined by Van der Kamp *et al.* (1994) for the Condie aquifer. In the following section we explore some interpretations of the scale dependence reflected in these observations.

5 DILUTION AND MIXING EFFECTS

In applying stochastic theory to the dispersal of contaminants in individual aquifers, it is important to recognize that the classical theory describes the average behavior of plumes in an ensemble of aquifers all having the specified statistical properties. When the individual plumes associated with a given realization are relatively small compared with the scale of heterogeneity involved in the aquifer, then the position of the center of mass of the individual plumes may be quite

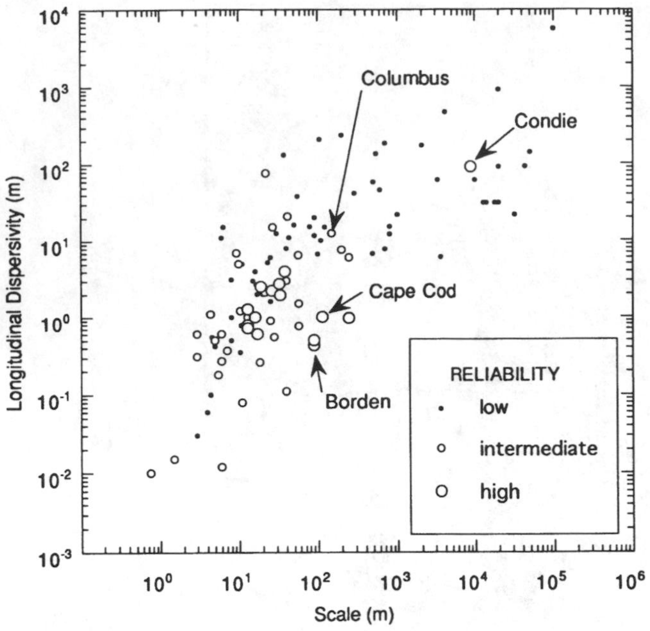

Fig. 10 Summary of field data on longitudinal dispersivity including the Condie site.

different among the different realizations. Consequently, an ensemble second moment around the ensemble mean center of mass reflects the contaminant spreading around the center of mass plus the uncertainty in the location of the center of mass of an individual plume. The classical theory treats the behavior of the ensemble mean concentration and moments thereof. The second moments, which represent the spreading effect, will consequently tend to overestimate the degree of spreading that occurs in an individual aquifer. The classical analysis, viewed in a Lagrangian context, considers the behavior of a single particle moving in a random velocity field (Taylor, 1921; Dagan, 1982); these classical results will tend to overestimate the degree of spreading and dilution in a finite-sized plume.

In order to treat the influence of the size of the contamination plume and approximate the behavior of a individual plume in a specific aquifer, it is necessary to consider the spreading of the solute relative to the center of mass of each realization making up the ensemble. This so-called 'relative dispersion analysis' was first developed by Richardson (1926) in the context of turbulent diffusion. Essentially it considers the simultaneous motion of two particles in a random velocity field, so that one can treat the behavior relative to the center of mass. Recently this approach has been applied in subsurface hydrology for three-dimensionally statistically anisotropic systems of applied interest (Rajaram 1991; Rajaram & Gelhar 1993b, 1995), as well as for more idealized one- and two-dimensional flow systems (Dagan, 1990, 1991, 1994; Rajaram & Gelhar, 1993a). A key result obtained by Rajaram & Gelhar (1993b) expresses a relative macrodispersivity tensor, which incorporates the influence of plume size, in terms of a three-dimensional wave number domain integral as follows:

$$A_{ij}^{(r)}=\frac{1}{2}\frac{d\Sigma_{ij}(\bar{x})}{d\bar{x}}=\iiint_k \frac{\sin k_1\bar{x}}{V^2k_1}S_{ij}(\mathbf{k})\left[1-e^{-k_ik_l\Sigma_{il}(\bar{x})}\right]d\mathbf{k} \qquad (5)$$

Here Σ_{ij} is the second moment of the solute mass distribution relative to the realization center of mass, $\bar{x}$ is the mean displacement in the direction of the flow with mean velocity of magnitude V, and S_{ij} is the spectrum of the components of the velocity fluctuations. The final portion of the integrand in (5), in square brackets, is essentially a low wave number filter which is related to the size of the plume as reflected in the second moment term. At large time, as the plume becomes very large, the dispersivity approaches the classical ensemble result, but when the plume is smaller the relative dispersivity is reduced. Physically, (5) represents the fact that, when the plume is small, it is dispersed by only the shorter-wavelength or higher-wave-number fluctuations in the velocity field, but as the plume grows it is also dispersed by larger-wavelength, lower-wave-number fluctuations.

Generally (5) is a system of differential equations that must be solved numerically to determine the second moment tensor and consequently the relative macrodispersivity, but it is possible to derive some simple analytical approximations for large displacement that offer more direct insight into the influence of plume scale. For a medium with finite log conductivity correlation scales (λ_1,λ_2,λ_3), it is found that (Rajaram & Gelhar, 1993b):

$$A_{11}^{(r)}=\frac{\sigma_f^2\lambda_1}{\gamma^2}\left(1-\frac{\lambda_2\lambda_3}{\sqrt{\Sigma_{22}\Sigma_{33}}}\right)$$

$$\bar{x}>>\lambda_1,\ \sqrt{\Sigma_{22}}>>\lambda_2,\ \sqrt{\Sigma_{33}}>>\lambda_3 \qquad (6)$$

This result illustrates that the longitudinal macrodispersivity is reduced relative to the ensemble result in (3) by a factor proportional to the product of the correlation scale relative to the transverse plume dimension in the two transverse directions. The reduction in dispersivity for the Borden tracer experiment is about 25 per cent and for the Cape Cod experiment around 15 per cent (Rajaram & Gelhar, 1993b).

The effects of plume scale become particularly dramatic in the case of heterogeneity with variability over a wide range of scales, as in the case of fractional Gaussian noise or self-similar media. Several analyses of such cases (Philip, 1986; Koch & Brady, 1988; Neuman, 1990; Glimm & Sharp, 1991; Kemblowski & Wen, 1993) have evaluated ensemble macrodispersivities for such models, and find that the macrodispersivity tends to grow as a power of the mean displacement. When the influence of the plume size is included according to (5), much smaller macrodispersivities are predicted, and a simple power law dependence on displacement is not found (Rajaram & Gelhar, 1995). Dagan (1994) also notes similar differences in the case of a two-dimensional heterogeneous field, but these results are less relevant to the field because of the important role played by the vertical correlation scale. As illustrated by (6), because of the highly stratified nature of many natural materials, λ_3 will typically be quite small compared with the vertical thickness of the plume, and this leads to dispersive behavior which is much closer to the ensemble result. For this reason it is very important to include the real three-dimensional velocity variations in the treatment of the effects of plume scale on macrodispersion.

The effects of multiple scales of variability in hydraulic conductivity in a field context will be illustrated by considering data on the variability in hydraulic conductivity over the scale of the sewage plume on Cape Cod (see Fig. 2). Fig. 11 shows the location of some 350 monitoring wells in the area of the sewage plume where slug tests were performed to evaluate the hydraulic conductivity variation in this area (Springer, 1991). Fig. 12 shows a north–south vertical section into which all of the three-dimensionally distributed observations of hydraulic conductivity have been projected. It is seen that the hydraulic conductivity varies over four orders of magnitude in this region; aside from a few systematically low values deep in the aquifer, it is difficult to discern visually any systematic trends. An anisotropic Gaussian spatial filter was used to identify a large-scale trend in log conductivity as illustrated in Fig. 13. Variogram analysis was then applied to the residuals; a horizontal variogram incorporating the slug test data discussed here plus the borehole flow meter information at the tracer test site (Hess *et al.*, 1992) is shown in Fig. 14. The variogram corresponding to a two-scale exponential covariance with the variances and correlation scales as indicated was used to represent the variogram. Also shown in Fig. 14 for reference is a power law variogram corresponding to a self-similar representation (the dashed line).

This two-scale characterization of the log conductivity variability along with a similar two-scale structure in the vertical were used in the framework of the relative dispersion treatment of (5) to develop an illustration of the influence of initial source size on the macrodispersivity for the Cape Cod area represented by these data. Two different source sizes were used in the calculations corresponding to the sewage plume and the tracer test. The results in Fig. 15, from Rajaram & Gelhar (1995), show that the source size has a strong influence on the magnitude of the longitudinal dispersivity. It is also seen that the ensemble dispersivity A_{11} is about three times as large as the relative dispersivity for the large source. The relative dispersivity is more appropriate for describing the longitudinal spreading in the sewage plume. In the case of the small source, the dispersivity grows slowly up

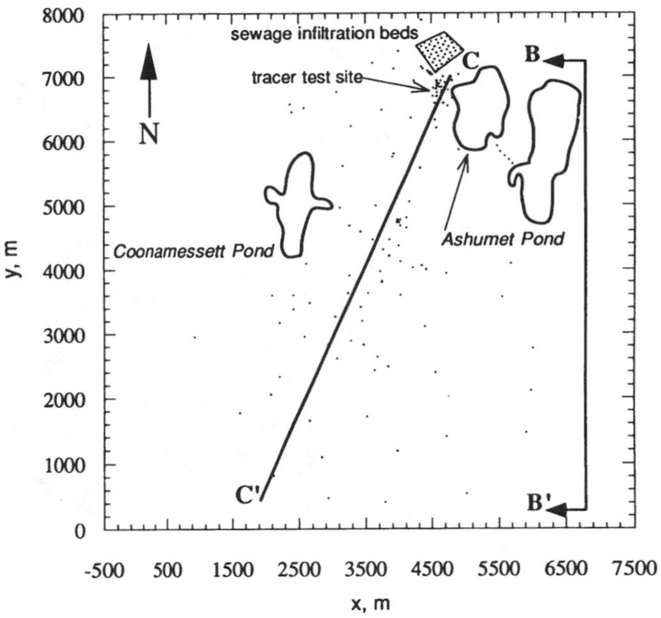

Fig. 11 Cape Cod sewage plume monitoring well locations where slug tests were done.

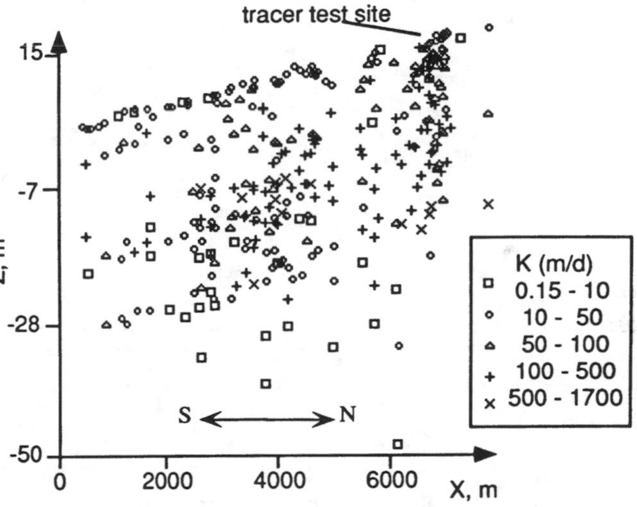

Fig. 12 Slug test hydraulic conductivities projected onto a north–south vertical section.

to a few hundred meters, producing a value similar to that observed, but then begins to increase more rapidly as the influence of the larger-scale heterogeneity is affecting the tracer plume. These calculations illustrate the important influence of plume size on macrodispersivity and suggest that summary data, such as those given in Fig. 10, will be importantly influenced by an additional parameter, the initial plume size. Larger macrodispersivities are to be expected when the initial plume size is large.

It is important to draw the distinction between contaminant spreading, as reflected in a macrodispersion coefficient representing the rate of increase of the second moment of the solute distribution, and the actual degree of mixing that

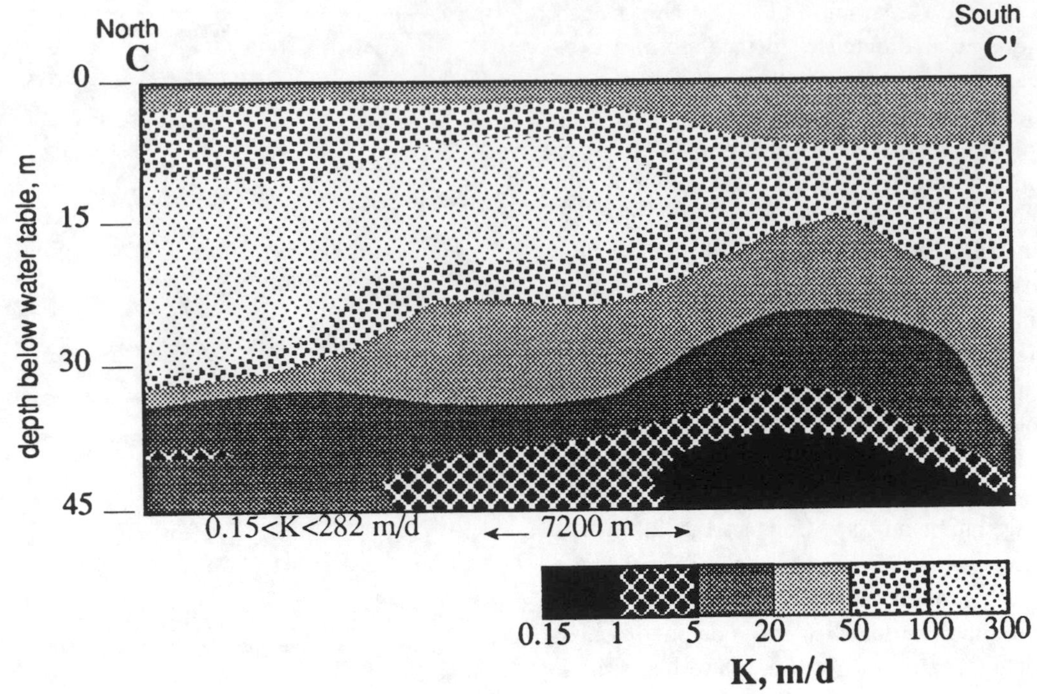

Fig. 13 Large-scale features of the hydraulic conductivity data illustrated in a longitudinal vertical section showing the filtered ln*K* field.

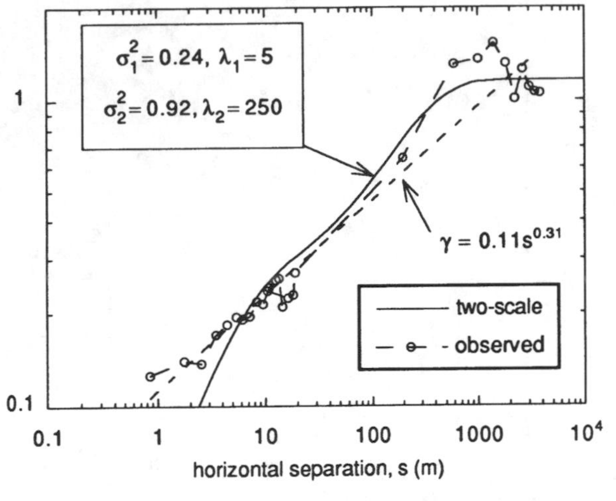

Fig. 14 Observed horizontal variogram of log conductivity from slug tests and borehole flow meter data represented by a two-scale model.

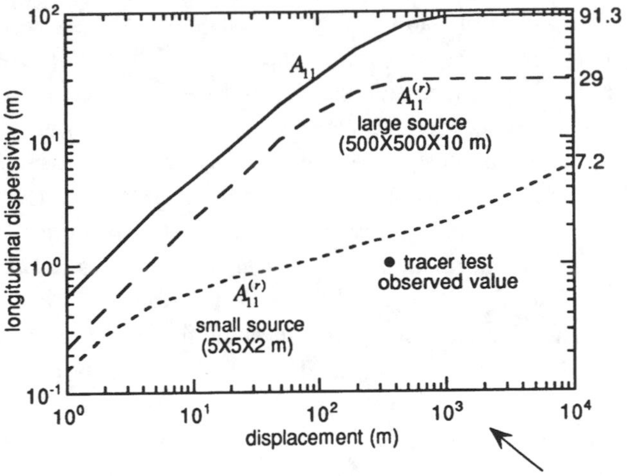

Fig. 15 Relative longitudinal dispersivity for Cape Cod with two different initial source sizes.

takes place in the aquifer. This important distinction is illustrated by the two contrasting situations sketched in Fig. 16. The upper part of the figure represents the hypothetical situation in which there is no local dispersion or diffusion affecting the transport process. In that situation, the tracer blob, initially at a constant concentration c_0, is advected by the complex heterogeneous velocity field to produce the much more irregular distribution of solute at time t_1. As a result of the variations in velocity, the plume has spread longitudinally but, because there is no local mixing, the concentration in the area within the plume remains at c_0. The longitudinal section below illustrates the concentration distribution that would occur in this hypothetical nondispersive case; the concentration is either c_0 or zero as shown. If one then thinks of repeating this kind of experiment many times for different realizations of a random medium, the average over all of these realizations will, after large enough displacement, appear as the Gaussian dashed line, the ensemble mean. On the other hand, when we consider the influence of local dispersion, as indicated in the lower sketch, the concentration distribution shows irregular fluctuations around the ensemble mean concentration, and the maximum concentration is now less than c_0 due to the important influence of local dispersion in producing mixing

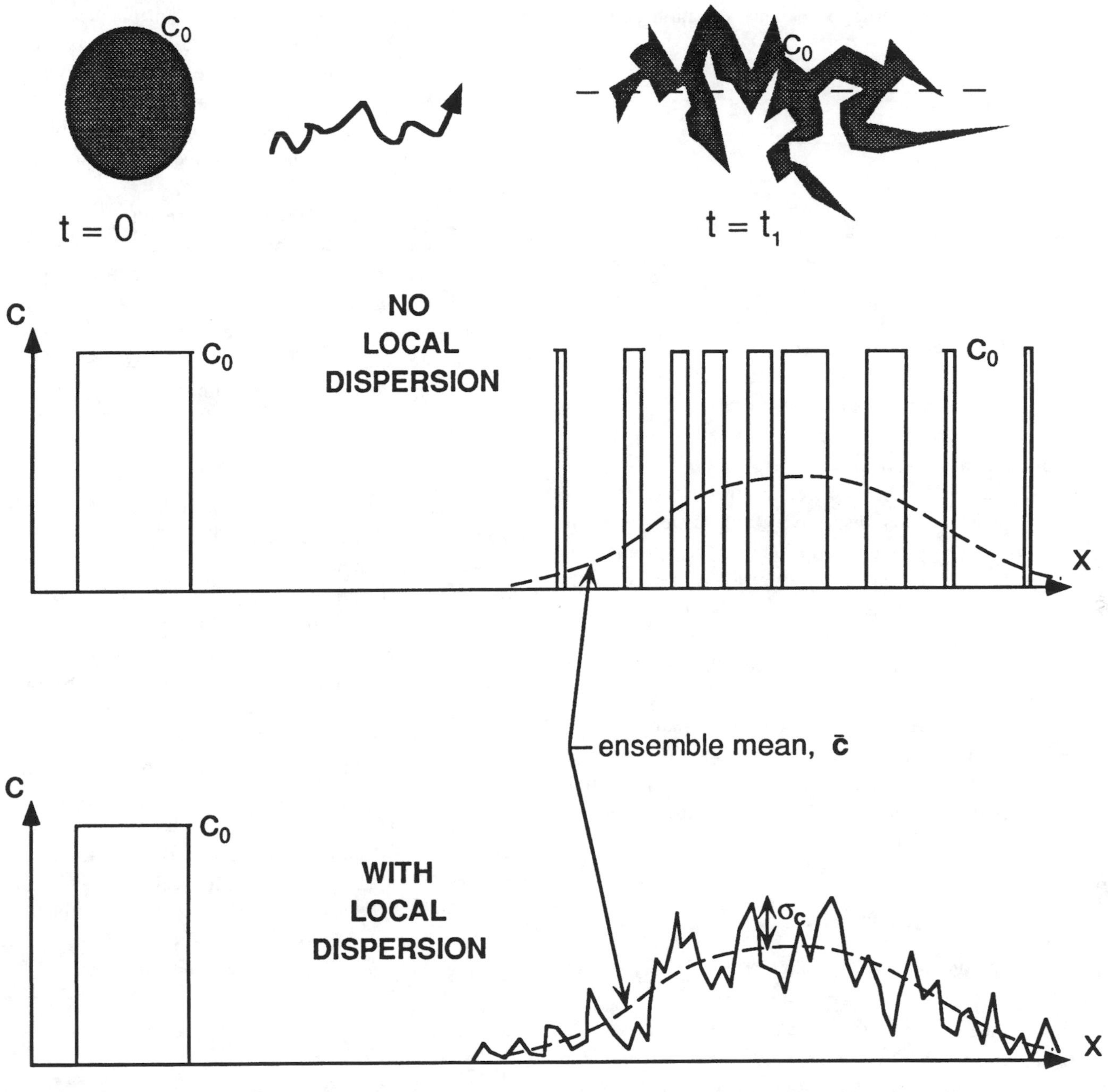

Fig. 16 Contrasting behavior of concentration fluctuations without (top) and with (bottom) local dispersion.

and thereby decreasing the magnitude of the maximum concentration. Here the magnitude of the concentration fluctuation is characterized by the concentration standard deviation σ_c, that is, the square root of the concentration variance.

From these elementary considerations, it is clear that any realistic model representing the concentration variance as a measure of the fluctuations in concentration, and hence the degree of dilution or decrease in maximum concentration, must incorporate the influence of local dispersion. However, some of the previous analyses of this issue (Dagan, 1982, 1990; Kabala & Sposito, 1994; Zhang & Neuman, 1995) have emphasized the behavior of the concentration variance in the nondispersive case. In the nondispersive case, there is a simple general solution for the case of an initially constant concentration based on a bimodal distribution of concentration. However, a formal analysis is hardly needed in this case, since, as is evident from the upper half of Fig. 16, the maximum concentration is always c_0 and there is no dilution in the aquifer. Only if one introduces some spatial averaging due to sampling or extraction effects will the concentration appear to decrease.

The behavior of the concentration variance under the influence of local dispersion can be understood by considering the concentration variance balance equation approximated in the form

$$\frac{\partial \sigma_c^2}{\partial t}+V\frac{\partial \sigma_c^2}{\partial x_1}-V(A_{ij}+\alpha_{ij})\frac{\partial^2 \sigma_c^2}{\partial x_i \partial x_j} \cong 2VA_{ij}\frac{\partial \bar{c}}{\partial x_i}\frac{\partial \bar{c}}{\partial x_j}-2V\alpha_{ij}\overline{\frac{\partial c'}{\partial x_i}\frac{\partial c'}{\partial x_j}} \qquad (7)$$

This form of the balance equation, as developed by Kapoor & Gelhar (1994), includes the effect of the local dispersivity, α_{ij}, and is valid for relatively large displacement where the macrodispersivity A_{ij} can be regarded as constant. The left-hand side of this equation represents the transport effect for variance, and on the right are the source/sink terms. The first term on the right represents the variance production as a result of mean concentration gradients interacting with the macrodispersivity, and the second term reflects the dissipation produced by gradients in concentration fluctuations interacting with the local dispersivity. This dissipation term is a key element in the concentrations variance problem in that it leads to a decrease in concentration variance with increasing time, indicating that the ensemble mean becomes a useful approximation of the actual concentration.

The product of the derivatives of the concentration fluctuations in this dissipation term is expressed as the concentration variance divided by the square of the concentration microscale. Kapoor & Gelhar (1994) developed an approximate treatment showing that the concentration microscale is proportional to the microscale of log conductivity which is defined by

$$\delta_f^2=\frac{\int_{-\infty}^{\infty} S_{ff}(k)dk}{\int_{-\infty}^{\infty} k^2 S_{ff}(k)dk}; \quad f=\ln K$$

The microscale is, in a sense, a measure of the smallest scale of variability in the hydraulic conductivity fluctuations. Using this approach, the dissipation term in (7) is proportional to

$$\frac{2V\alpha}{\delta_f^2}\sigma_c^2$$

which shows that the dissipation term acts as a first-order decay term in which the decay rate is proportional to the local dispersivity and inversely proportional to the square of the ln$K$ microscale. Note that the microscale for the widely used exponential covariance is zero. This implies an unrealistic infinite dissipation rate if the exponential covariance function is used. An exponential covariance function is not physically consistent in this situation, and strictly is not mathematically consistent because it does not represent a differentiable random field. Essentially the exponential covariance produces a field which has significant fluctuations at arbitrarily fine scales. This is not physically reasonable because the continuum equations for flow in a porous medium apply only above a certain averaging volume, the representative elementary volume. Analysis of millimeter-scale permeability data from a sandstone demonstrates that a finite microscale occurs (Kapoor, 1993).

Kapoor & Gelhar (1994) developed an explicit solution of (7) and showed that, after large displacement, the left-hand side of (7) involving the transport terms is practically negligible, so that the essential balance is between the production and dissipation terms. Under these circumstances, we find, away from the point of maximum mean concentration, the following simple local variance relationship:

$$\sigma_c^2 \propto \frac{A_{ij}\delta_f^2}{\alpha}\frac{\partial \bar{c}}{\partial x_i}\frac{\partial \bar{c}}{\partial x_j} \qquad (8)$$

This form is similar to the local variance relationship found by Vomvoris & Gelhar (1990) under the assumption of a stationary concentration field. Their result shows the concentration variance to be proportional to the square of the mean concentration gradient and inversely proportional to the local dispersivity, but dependence of the microscale is not included. Numerical solutions for the concentration variance also show the general behavior indicated by (8); Graham & McLaughlin (1989) find that the concentration variance is large in regions of steep mean concentration gradient.

The application of the concentration variance balance, (7), was illustrated by Kapoor & Gelhar (1994) for the Cape Cod tracer test. Fig. 17 shows a result of that calculation with the assumption that the microscale is 0.6 times the correlation scale. Here the plume volume is the square root of the product of the three principal spatial second moments of the

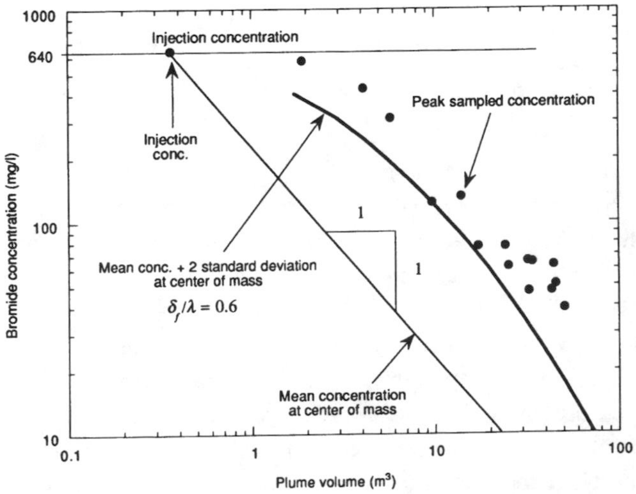

Fig. 17 Peak sampled bromide concentrations for the Cape Cod tracer test; based on Kapoor & Gelhar (1994).

concentration distribution; plume volume increases with time. The mean concentration plus two standard deviations is seen to represent reasonably well the maximum sampled concentration .

The above analyses clearly show that local dispersivity must be included in any physically meaningful treatment of the concentration variance, and that the rate of dissipation of concentration fluctuations is importantly controlled by the finest scale of variation in the log conductivity. The reason for the importance of the fine-scale variation is that the resulting small-scale variations in the velocity field create sharp concentration gradients and essentially a very large surface area over which local dispersion can act effectively to dissipate concentration fluctuations. These results emphasize the features that will need to be included in any realistic treatment of the uncertainty of predictions of solute concentration. The results also point to the need for measurements of the variations in hydraulic conductivity at smaller scales than have customarily been investigated in order to evaluate the microscale. These results also strongly suggest that any efforts to simulate transport numerically with the goal of describing the variations in concentration must involve fine-scale resolution down to the microscale. The common practice of employing an exponential covariance function in numerical simulations is inconsistent and is implicitly introducing a microscale that is on the order of the grid spacing in a numerical model. Consequently, the uncertainty predicted from such simulations is going to be dependent on the grid resolution in the simulations, as can be recognized from (8). Kitanidis (1994) has proposed a dilution index to characterize the degree of dilution in a plume. This technique has been applied to evaluate the dilution characteristic for the Borden and Cape Cod plumes (Thierrin & Kitanidis, 1994).

Considerations of mixing and dilution within solute plumes are particularly important in problems of reactive contaminant transport involving multiple species where the local *in situ* concentration can have an important influence on the reactions. These features are discussed in the following section.

6 REACTIVE AND COUPLED TRANSPORT

Many applications in subsurface contaminant transport involve solutes which undergo chemical and/or biological transformations within the aquifer. The processes that control these reaction phenomena are almost certain to vary in space, as do the physical properties like hydraulic conductivity. The question of how such variations in chemical properties affect contaminant transport and how we can develop effective parameters describing field-scale behavior needs to be explored.

The case of simple linear equilibrium sorption is an elementary example which demonstrates the role of chemical heterogeneity in contaminant transport. At the local scale, the transport of a linearly sorbing species is described by

$$\left(1+\frac{\rho K_d}{n}\right)\frac{\partial c}{\partial t}+\frac{q_i}{n}\frac{\partial c}{\partial x_i}=D_{ij}\frac{\partial^2 c}{\partial x_i \partial x_j} \quad (9)$$

where n is the porosity (assumed constant), ρ is the bulk density of the aquifer, q_i is the local specific discharge, and K_d is the distribution coefficient, which is regarded to vary in space as characterized by a stationary random field.

The stochastic solution for the mean concentration distribution, taking into account the variations in the specific discharge as a result of variations in hydraulic conductivity as well as the variations in K_d, is found in the form

$$\frac{\partial \bar{c}}{\partial t}+\frac{q}{nR_e}\frac{\partial \bar{c}}{\partial x_i}=A_{ij}\frac{q}{nR_e}\frac{\partial^2 \bar{c}}{\partial x_i \partial x_j} \quad (10)$$

where x_1 is in the direction of the mean specific discharge of magnitude q and A_{ij} is the macrodispersivity tensor. The effective retardation coefficient, R_e, is found to be of the form

$$R_e=1+\frac{\rho E[K_d]}{n} \quad (11)$$

where the bulk density has been regarded as a constant. The important result here is that the effective retardation factor is found simply by using the arithmetic mean of the distribution coefficient.

The longitudinal macrodispersivity, A_{11}, is affected by the variability of K_d as is shown in the following:

$$A_{11}=\frac{\sigma_f^2\lambda_1}{\gamma^2}\left[1-\frac{\gamma\rho\beta}{nR_e}\right]^2+\frac{\rho^2\sigma_\eta^2\lambda_\eta}{n^2R_e^2} \quad (12)$$

This result is based on a model of the variability of K_d which allows for partial correlation with the variations in log conductivity, as shown in Fig. 18. The distribution coefficient has a correlated part reflected by the slope β and an uncorrelated residual η, which is assumed to be independent of $\ln K$. If the correlation between K_d and $\ln K$ is negative, as in Fig. 18 and as would be expected since fine-grained materials of low permeability have larger surface area and are likely to be more reactive, the longitudinal macrodispersivity will be increased relative to that of a nonsorbed solute. The contribution of the uncorrelated residual will always produce an increase in the longitudinal dispersivity. With a plausible degree of variability in K_d and a modest negative correlation of K_d with $\ln K$, the longitudinal macrodispersivity can increase by an order of magnitude. It is easily shown that the transverse dispersivity for the heterogeneously sorbing species is not affected by the variability of K_d (Garabedian, Gelhar & Celia, 1988).

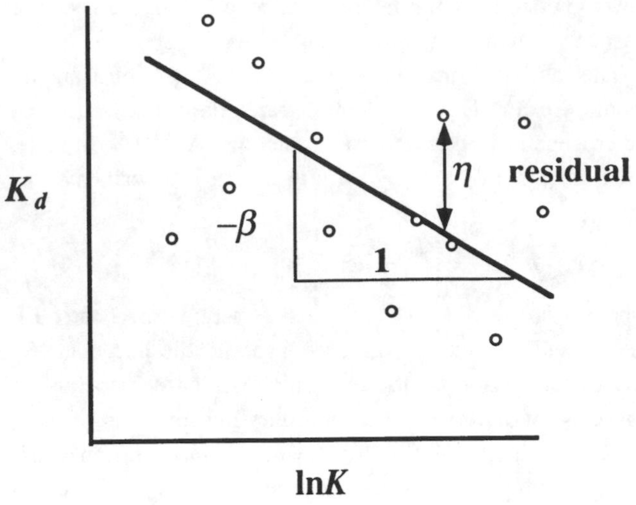

Fig. 18 Sorption coefficient related to log conductivity.

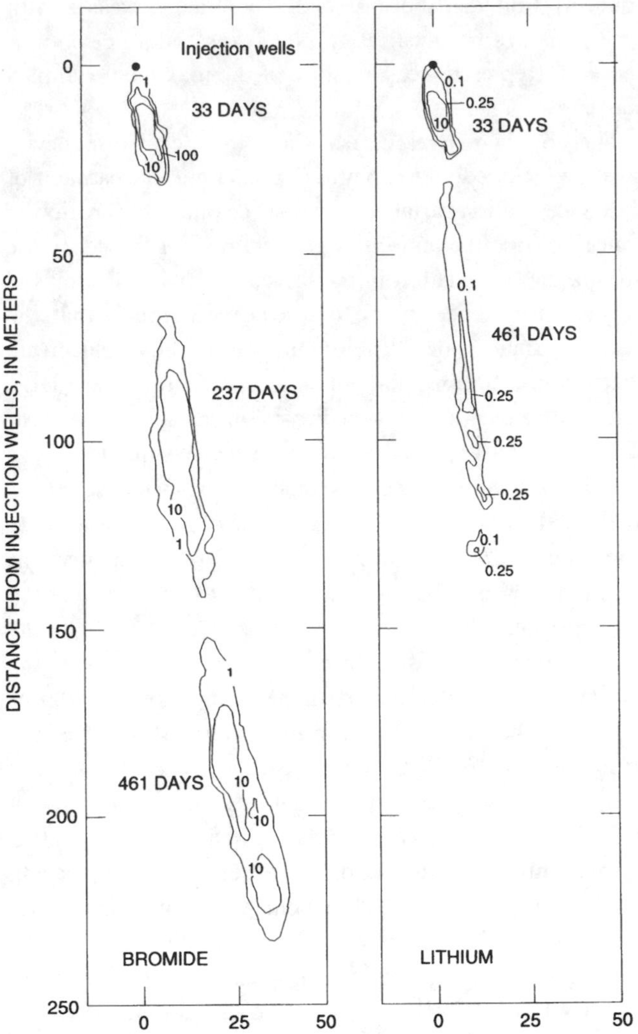

Fig. 19 Contrasting longitudinal spreading of nonsorbing (bromide) and sorbing (lithium) plumes in the Cape Cod tracer test (concentrations in milligrams per liter); based on LeBlanc *et al.* (1991).

The lithium tracer results of the Cape Cod tracer test exhibit features which are qualitatively similar to the enhancement effect of (12), but because of the complications of a nonlinear sorption isotherm, direct quantitative comparisons have not been possible. Comparing the bromide and lithium plumes in Fig. 19, it is clear that, at the same mean displacement, the lithium plume is much more elongated, indicating an enhanced longitudinal macrodispersivity. Some laboratory studies (Robin *et al.*, 1991; Foster-Reid, 1994) have demonstrated a modest correlation between K_d and ln$K$, but field tracer tests quantitatively demonstrating the predicted macrodispersivity enhancement of (12) are not available.

The longitudinal dispersivity enhancement effect in (12) is very important because it suggests that different macrodispersivities should apply to different transported species. This important effect has been recognized for over a decade. I first discussed this effect at a research workshop sponsored by the International Association for Water Pollution Research and Control in Denmark in 1984, and the results were formally presented a year later (Garabedian & Gelhar, 1985). Subsequently there have been numerous theoretical studies with various enhancements which confirm the basic behavior predicted by (12), but little has been done to bring these results to bear on practical problems. It is still common practice to assume that the same dispersivity applies to all transported species, whether reactive or not, and chemical research still proceeds with the implicit assumption that laboratory experimental results can be used directly to characterize field-scale reactive processes. One reason that so little has been done in applying the theoretical results for heterogeneous reactive transport seems to be the usual reluctance of practitioners in one field (environmental chemistry) to adopt new results that originate in another field (subsurface hydrology). The process of adopting such results in practice is further stymied by the natural tendency to be skeptical about new results, particularly when it appears that the researchers in the field are in disagreement about methodology. The fact is that there is substantial agreement about the essential theoretical results for reactive heterogeneous transport.

Recent theoretical developments (Kabala & Sposito, 1991; Miralles-Wilhelm, 1993; Miralles-Wilhelm & Gelhar, 1996) have shown that locally heterogeneous linear equilibrium sorption can produce a significant macrokinetic effect in the field-scale sorption process. This kinetic effect can dominate field-scale behavior so that much of the effort to explore detailed kinetic effects on a laboratory scale may not be useful for describing field processes. The importance of the enhanced macrodispersivity effect in an application of contaminant transport modeling for a low level radioactive waste site (Talbott & Gelhar, 1994) shows dramatic effects of the increased longitudinal dispersivity due to heterogeneous sorption. Concentration predictions for species undergoing first-order decay can be several orders of magnitude higher when the enhanced dispersivity is used.

An analysis of a three-component multispecies transport problem describing oxygen-limited biodegradation under conditions of heterogeneous flow and reaction parameters (Miralles-Wilhelm, 1993; Miralles-Wilhelm, Kapoor & Gelhar, 1994) shows that macrodispersivities for the different species can be very different, and that field-scale rate constants in the case of spatially variable rate processes can be significantly lower than the mean. Analyses of coupled multicomponent reactive systems of this type show, through the influence on the effective parameters, which local-scale reactive transport parameters will need to be characterized in terms of their variability in order to predict field-scale behavior. The theoretical results thereby form a rational basis for setting priorities for data collection.

Analysis of the transport of components which alter the density and/or viscosity of the fluid must incorporate the coupling between the flow description and the transport equation. Eulerian-based coupled analyses show that the longitudinal dispersivity can be significantly altered by density and/or viscosity differences (Welty & Gelhar, 1991, 1992). The classical Lagrangian approach, being restricted to passive solutes for which the velocity field is independently prescribed, is not applicable to this important class of problems.

7 SOME RESEARCH CHALLENGES

The theoretical results and field observations outlined here point to a number of important research areas which will need to be explored in order to address realistically field-scale behavior and to bring stochastic methods more into the arena of real problem solving. This is not intended to be an exhaustive list of potential problem areas, but rather focuses on key areas which are likely to make the greatest difference in field applications.

The issue of characterization of aquifer heterogeneity probably represents the greatest limitation in what can be done in site-specific field situations. There is a need to characterize heterogeneity over a wide range of scales from millimeters up to kilometers. Smaller-scale heterogeneity is sensibly characterized stochastically but, based on our understanding of the controls on the concentration variance, it will be necessary to develop some experience characterizing microscale variations, say down to the scale of centimeters or less. At a specific site, large-scale heterogeneity is most plausibly characterized deterministically, including the conditioning effect of head observations through inverse procedures. Large-scale characterization should also include temporal variations in flow and head conditions because of the important influence of temporal variability on transverse dispersion. It is important to make use more effectively of geologic information in characterizing aquifer heterogeneity, and research also should focus on improvements of measurement techniques which can reliably and economically determine heterogeneity of hydraulic properties. Characterization work particularly should focus on systematic measurements of chemical and biological heterogeneity which are carried out in parallel with hydraulic characterization. In chemical and biological characterization, there is a need to develop simple screening techniques which will allow reliable determination of the variation of parameters for large numbers of samples, rather than to focus on extremely detailed investigations of a small number of arbitrarily selected samples. There is also a need to improve statistical inference techniques, particularly dealing with fully three-dimensional heterogeneity and multiscale variability. The usual Gaussian description implied by the use of a covariance function seems to be adequate for many of the problems we face. When are non-Gaussian models needed?

A second important research area is that of large-scale controlled field experiments, say at scales of 100 m or more. Particularly important are carefully designed experiments involving reactive solutes including problems of *in situ* biodegradation. Field experiments on multiphase flow and transport involving, for example, the flow of dense nonaqueous phase liquids, the transport of compounds in solution which are dissolving from the nonaqueous liquids, and the behavior of gases injected below the water table. There is also a need for large-scale unsaturated solute transport experiments to evaluate results on macrodispersion in this setting. Such field experimentation is central to advancing our basic knowledge and applied capabilities because it provides a basis for evaluating theories and a site-specific setting in which characterization techniques can be evaluated. Dealing with a specific site forces one to address the real problems of characterization of heterogeneity at a pragmatic level.

Although theoretical analyses have clearly developed well beyond the needs in characterization and field experimentation, there is still a need for some further enhancements of theory. For example, there is no comprehensive theory for even nonreactive solute transport in very heterogeneous media. It seems unlikely that an approach of simply extending a perturbation scheme to higher-order terms will be of much benefit in this difficult problem; an entirely new approach will probably have to be found. Theoretical work is definitely needed in the area of multiphase flow and transport as a prerequisite to intelligent design of meaningful field experiments in this area. Theoretical work on the effects of large-scale trends on dispersive mixing is also needed. All of the theoretical developments should focus on the realistic case of three-dimensionally heterogeneous statistically anisotropic media with nonzero local dispersion. The common two-dimensional nondispersive representation is largely irrelevant to real applications. I know of no field situations where it is plausible to describe the heterogeneity and consequently the variation in the flow field as two-dimensional. Such conceptualizations should be recognized for what they are: assumptions of convenience to facilitate the analytical or numerical treatment of the problem.

Numerical simulation techniques should focus on adequately resolving the wide range of scale variation that is now understood to be important in the problem of macrodispersion and evaluation of concentration fluctuations. There is a need for carefully designed fully three-dimensional statistically anisotropic simulations of flow and transport to evaluate theories for this important practical situation. One can easily envision large problems involving billions to trillions of nodes, in which case it seems likely that new hardware configurations and algorithms will be needed.

The field of stochastic subsurface hydrology has seen major advances during the last decade but has matured to the point that much of the current research emphasizes theoretical refinements of practically insignificant, but conveniently solvable, problems. While researchers debate theoretical minutia, billions of dollars are being spent – some say wasted – trying to clean up serious contamination and to protect aquifers from future degradation. The important advances in the next decade will be those which more broadly focus on integrating appropriate methodological developments with the realities of field observations at specific sites, thereby bringing the powerful stochastic framework more directly to bear on real subsurface environmental problems.

REFERENCES

Ababou, R., Gelhar, L. W. & McLaughlin, D. (1988). *Three-Dimensional Flow in Random Porous Media*. Parsons Laboratory Report 318, R88–08. Cambridge: Massachusetts Institute of Technology.

Adams, E. E. & Gelhar, L. W. (1992). Field study of dispersion in a heterogeneous aquifer: 2. Spatial moments analysis. *Water Resources Research*, 28(12), 3293–3308.

Barry, D. A., Coves, J. & Sposito, G. (1988). On the Dagan model of solute transport in groundwater: Application to the Borden site. *Water Resources Research*, 24(10), 1735–1747.

Boggs, J. M., Young, S. C., Beard, L. M., Gelhar, L. W., Rehfeldt, K. R. & Adams, E. E. (1992). Field study of dispersion in a heterogeneous aquifer, 1, Overview and site description. *Water Resources Research*, 28(12), 3281–3292.

Burr, D. T., Sudicky, E. A. & Naff, R. L. (1994). Nonreactive and reactive solute transport in three-dimensional heterogeneous porous media – mean displacement, plume spreading, and uncertainty. *Water Resources Research*, 30(3), 791–815.

Chin, D. A. & Wang, T. Z. (1992). An investigation of the validity of first-order stochastic dispersion theories in isotropic porous media. *Water Resources Research*, 28(6), 1531–1542.

Dagan, G. (1982). Stochastic modeling of groundwater flow by unconditional and conditional probabilities, 2, The solute transport. *Water Resources Research*, 18(4), 835–848.

Dagan, G. (1984). Solute transport in heterogeneous porous formations. *Journal of Fluid Mechanics*, 145, 151–177.

Dagan, G. (1987). Theory of solute transport by groundwater. *Annual Reviews in Fluid Mechanics*, 19, 183–215.

Dagan, G. (1988). Time-dependent macrodispersion for solute transport in anisotropic heterogeneous aquifers. *Water Resources Research*, 24(9), 1491–1500.

Dagan, G. (1989). *Flow and Transport in Porous Formations*. New York: Springer-Verlag.

Dagan, G. (1990). Transport in heterogeneous formations: Spatial moments, ergodicity, and effective dispersion. *Water Resources Research*, 26, 1281–1290.

Dagan, G. (1991). Disperson of a passive solute in non-ergodic transport by steady velocity fields in heterogeneous formations. *Journal of Fluid Mechanics*, 233, 197–210.

Dagan, G. (1993). Higher-order correction of effective conductivity of heterogeneous formations of lognormal conductivity. *Transport in Porous Media*, 12, 279–290.

Dagan, G. (1994). The significance of heterogeneity of evolving scales to transport in porous formations. *Water Resources Research*, 30(12), 3327–3336.

Dasinger, A. (1989). *Large-Scale Simulations of Groundwater Flow and Solute Transport in an Anisotropic Heterogeneous Aquifer*. M.S. thesis, Department of Civil Engineering. Cambridge: Massachusetts Institute of Technology.

Dasinger, A. & Gelhar, L. W. (1991). Simulation of flow and transport in an anisotropic, three-dimensionally heterogeneous aquifer. American Geophysical Union, *EOS*, 72(44), 211.

Deng, F.-W., Cushman, J. H. & Delleur, J. W. (1993). A fast Fourier transform stochastic analysis of the contaminant transport problem. *Water Resources Research*, 29(9) 3241–3247.

Dykaar, B. B. & Kitanidis, P. K. (1992). Determination of the effective hydraulic conductivity for heterogeneous porous media using a numerical spectral approach 2. Results. *Water Resources Research*, 28(4), 1167–1178.

Farrell, D. A., Woodbury, A. D., Sudicky, E. A. & Rivett, M. O. (1994). Stochastic and deterministic analysis of dispersion in unsteady flow at the Borden Tracer-Test site, Ontario, Canada. *Journal of Contaminant Hydrology*, 15(3), 159–186.

Foster-Reid, G. (1994). *Variability of Hydraulic Conductivity and Sorption in a Heterogeneous Aquifer*, M.S. thesis, Department of Civil Engineering. Cambridge: Massachusetts Institute of Technology.

Freyberg, D. L. (1986). A natural gradient experiment on solute transport in a sand aquifer: 2. Spatial moments and the advection and dispersion of nonreactive tracers. *Water Resources Research*, 22(13), 2031–2046.

Garabedian, S. P. & Gelhar, L. W. (1985). Effect of correlation between distribution coefficients and hydraulic conductivity on the macrodispersivity of non-conservative solutes. American Geophysical Union, *EOS*, 66(46), 903.

Garabedian, S. P., Gelhar, L. W. & Celia, M. A. (1988). *Large-Scale Dispersive Transport in Aquifers: Field Experiments and Reactive Transport Theory*. Parsons Laboratory Report 315. Cambridge: Massachusetts Institute of Technology.

Garabedian, S. P., LeBlanc, D. R., Gelhar, L. W. & Celia, M. A. (1991). Large-scale natural-gradient tracer test in sand and gravel, Cape Cod, Massachusetts: 2, Analysis of tracer moments for a nonreactive tracer. *Water Resources Research*, 27(5), 911–924.

Gelhar, L. W. (1993). *Stochastic Subsurface Hydrology*. Englewood Cliffs, NJ: Prentice Hall.

Gelhar, L. W. & Axness, C. L. (1983). Three-dimensional stochastic analysis of macrodispersion in aquifers. *Water Resources Research*, 19(1), 161–180.

Gelhar, L. W., Welty, C. & Rehfeldt, K. R. (1992). A critical review of data on field-scale dispersion in aquifers. *Water Resources Research*, 28(7), 1955–1974.

Glimm, J. & Sharp, D. H. (1991). A random field model for anomalous diffusion in heterogeneous porous media. *Journal of Statistical Physics*, 62, 415–424.

Graham, W. & McLaughlin, D. (1989). Stochastic analysis of nonstationary subsurface solute transport: 1, Unconditional moments. *Water Resources Research*, 25(2), 215–232.

Graham, W. D. & and McLaughlin, D. (1991). A stochastic model of solute transport in groundwater: Application to the Borden, Ontario tracer test. *Water Resources Research*, 27(6), 1345–1360.

Hess, K. M. (1989). Use of a borehole flowmeter to determine spatial heterogeneity of hydraulic conductivity and macrodispersion in a sand and gravel aquifer, Cape Cod, Massachusetts. In *Proceedings of the Conference on New Field Techniques for Quantifying the Physical and Chemical Properties of Heterogeneous Aquifers*, eds. F. J. Molz, J. G. Melville & O. Guven. Dublin, Ohio: National Water Well Association, 497–508.

Hess, K. M., Wolf, S. H. & Celia, M. A. (1992). Large-scale natural gradient tracer test in sand and gravel: 3. Hydraulic conductivity variability and calculated macrodispersivities. *Water Resources Research*, 28(8), 2011–2027.

Indelman, P. & Abramovich, B. (1994). A higher-order approximation to effective conductivity in media of anisotropic random structure. *Water Resources Research*, 30(6), 1857–1864.

Indelman, P. & Rubin, Y. (1995). Flow in heterogeneous media displaying a linear trend in the log conductivity. *Water Resources Research*, 31(5), 1257–1265.

Jussel, P. (1992). Modellierung des Transports gelöster Stoffe in inhomogenen Grundwasserleitern. In *Institut für Hydromechanik und Wasserwirtschaft, R29–92*. Zurich: Swiss Federal Institute of Technology (ETH).

Jussel, P., Stauffer, F. & Dracos, T. (1990). Three-dimensional simulation of solute transport in inhomogeneous fluvial gravel deposits using stochastic concepts. In *Proceedings of the 8th International Conference on Computational Methods in Water Resources*, eds. G. Gambolati, A. Rinaldo, C. A. Brebbia, W. G. Gray & G. F. Pinder. Berlin: Springer-Verlag.

Jussel, P., Stauffer, F. & Dracos, T. (1994). Transport modelling in heterogeneous aquifers: 2. Three-dimensional transport model and stochastic numerical tracer experiments. *Water Resources Research*, 30(6), 1819–1831.

Kabala, Z. J. & Sposito, G. (1991). A stochastic model of reactive transport with time-varying velocity in a heterogeneous aquifer. *Water Resources Research*, 27(6), 1819–1831.

Kabala, Z. J. & Sposito, G. (1994). Statistical moments of reactive solute concentration in a heterogeneous aquifer. *Water Resources Research*, 30(3), 759–768.

Kapoor, V. (1993). *Macrodispersion and Concentration Fluctuations in Three-Dimensionally Heterogeneous Aquifers*. ScD Thesis, Department of Civil and Environmental Engineering. Cambridge: Massachusetts Institute of Technology.

Kapoor, V. & Gelhar, L. W. (1994). Transport in three-dimensionally heterogeneous aquifers: 2. Concentration variance for a finite size impulse input. *Water Resources Research*, 30(6), 1789–1801.

Kemblowski, M. W. & Wen, J.-C. (1993). Contaminant spreading in stratified soils with fractal permeability distribution. *Water Resources Research*, 29(2), 419–426.

Kitanidis, P. K. (1994). The concept of the dilution index. *Water Resources Research*, 30(7), 2011–2026.

Koch, D. L. & Brady, J. F. (1988). Anomalous diffusion in heterogeneous porous media. *Physics of Fluids*, 31(5), 965–973.

LeBlanc, D. R., Garabedian, S. P., Hess, K. M., Gelhar, L. W., Quadri, R. D., Stollenwerk, K. G. & Wood, W. W. (1991). Large-scale natural gradient tracer test in sand and gravel, Cape Cod, Massachusetts: 1, Experimental design and observed tracer movement. *Water Resources Research*, 27(5), 895–910.

Li, S.-G. & McLaughlin, D. (1995). Using the nonstationary spectral method to analyze flow through heterogeneous trending media. *Water Resources Research*, 31(3), 541–551.

Miralles-Wilhelm, F. (1993). *Stochastic Analysis of Sorption and Biodegradation in Three-Dimensionally Heterogeneous Aquifers*. Ph.D. Thesis, Department of Civil and Environmental Engineering. Cambridge: Massachusetts Institute of Technology.

Miralles-Wilhelm, F. & Gelhar, L. W. (1996). Stochastic analysis of sorption macrokinetics in heterogeneous aquifers. *Water Resources Research* (in press).

Miralles-Wilhelm, F., Kapoor, V. & Gelhar, L. W. (1994). *Modeling Oxygen-Transport Limited Biodegradation in Three-dimensionally Heterogeneous Aquifers*. Final Report under Contract API GW-14-360-8. Washington, D.C.: American Petroleum Institute.

Naff, R. L. (1991). Radial flow in heterogeneous porous media – an analysis of specific discharge. *Water Resources Research*, 27 (3), 307–316.

Neuman, S. P. (1990). Universal scaling of hydraulic conductivities and dispersivities in geologic media. *Water Resources Research*, 26(8), 1749–1758.

Neuman, S. P. & Orr, S. (1993). Prediction of steady state flow in nonuniform geologic media by conditional moments – exact nonlocal formalism, effective conductivities, and weak approximation. *Water Resources Research*, 29(2), 341–364.

Neuman, S. P. & Zhang, Y.-V. (1990). A quasi-linear theory of nonfickian and Fickian subsurface dispersion. 1: Theoretical analysis with application to isotropic media. *Water Resources Research*, 26(5), 887–902.

Philip, J. R. (1986). Issues in flow and transport in heterogeneous porous media. *Transport in Porous Media*, 1(4), 319–338.

Ptak, T. & Teutsch, G. (1994). Forced and natural gradient tracer tests in a highly heterogeneous porous aquifer – instrumentation and measurements. *Journal of Hydrology*, 159, 79–104.

Rajaram, H. (1991). *Scale-Dependent Dispersion in Heterogeneous Porous Media*. Ph.D. Thesis, Department of Civil Engineering. Cambridge: Massachusetts Institute of Technology.

Rajaram, H. & Gelhar, L. W. (1991). Three-dimensional spatial moments analysis of the Borden tracer test. *Water Resources Research*, 27(6), 1239–1251.

Rajaram, H. & Gelhar, L. W. (1993a). Plume scale dependent dispersion in heterogeneous aquifers: 1, Lagrangian analysis in a stratified aquifer. *Water Resources Research*, 29(9), 3249–3260.

Rajaram, H. & Gelhar, L. W. (1993b). Plume scale dependent dispersion in heterogeneous aquifers: 2, Eulerian analysis and three-dimensional aquifers. *Water Resources Research*, 29(9), 3261–3276.

Rajaram, H. & Gelhar, L. W. (1995). Plume scale dependent dispersion in aquifers with a wide range of scales of heterogeneity. *Water Resources Research*, 31(10), 2469–2482.

Rehfeldt, K. R. (1988). *Prediction of Macrodispersivity in Heterogeneous Aquifers*. Ph.D. Thesis, Department of Civil Engineering. Cambridge: Massachusetts Institute of Technology.

Rehfeldt, K. R., Boggs, J. M. & Gelhar, L. W. (1992). Field study of dispersion in a heterogeneous aquifer: 3. Geostatistical analysis of hydraulic conductivity. *Water Resources Research*, 28(12), 3309–3324.

Rehfeldt, K. R. & Gelhar, L. W. (1992). Stochastic analysis of dispersion in unsteady flow in heterogeneous aquifers. *Water Resources Research*, 28(8), 2085–2099.

Rehfeldt, K. R., Hufschmied, P., Gelhar, L. W. & Schaefer, M. E. (1989). *Measuring Hydraulic Conductivity with the Borehole Flowmeter*, EPRI EN-6511, Research Project 2485-5, Topical Report. Palo Alto, CA: Electric Power Research Institute.

Richardson. L. F. (1926). Atmospheric diffusion shown on a distance-neighbor graph. *Proceedings of the Royal Society*, Series A, 110, 709–737.

Robin, M. J. L., Sudicky, E. A., Gillham, R. W. & Kachanoski, R. G. (1991). Spatial variability of strontium distribution coefficients and their correlation with hydraulic conductivity in the Canadian forces base Borden aquifer. *Water Resources Research*, 27(10), 2619–2632.

Rubin, Y. & Seong, K. (1994). Investigation of flow and transport in certain cases of nonstationary conductivity fields. *Water Resources Research*, 30(11), 2901–2911.

Schad, H. & Teutsch, G. (1994). Effects of the investigation scale on pumping test results in heterogeneous porous aquifers. *Journal of Hydrology*, 159, 61–77.

Smith, L. & Freeze, R. A. (1979). Stochastic analysis of steady state groundwater flow in a bounded domain, 2, Two-dimensional simulations. *Water Resources Research*, 15(6), 1543–1559.

Springer, R. K. (1991). *Application of an Improved Slug Test Analysis to the Large-Scale Characterization of Heterogeneity in a Cape Cod Aquifer*. MS Thesis, Department of Civil Engineering. Cambridge: Massachusetts Institute of Technology.

Sudicky, E. A. (1986). A natural gradient experiment on solute transport in a sand aquifer: Spatial variability of hydraulic conductivity and its role in the dispersion process. *Water Resources Research*, 22(13), 2069–2082.

Talbott, M. E. & Gelhar, L. W. (1994). *Performance Assessment of a Hypothetical Low-Level Waste Facility: Groundwater Flow and Transport Simulation*. U.S. Nuclear Regulatory Commission Report NUREG/CR-6114, Vol. 3. Washington, D.C.

Taylor, G. I. (1921). Diffusion by continuous movements. *Proceedings of the London Mathematical Society*, 2(20), 196–211.

Thierrin, J. & Kitanidis, P. K. (1994). Solute dilution at the Borden and Cape Cod groundwater tracer tests. *Water Resources Research*, 30(11), 2883–2890.

Thompson, K. D. (1994). *The Stochastic Characterization of Glacial Aquifers Using Geologic Information*. Ph.D. Thesis, Department of Civil Engineering. Cambridge: Massachusetts Institute of Technology.

Thompson, K. D. & L. W. Gelhar (1993). A generalized stochastic characterization for outwash, 1993 Fall Meeting of the American Geophysical Union, *EOS*, 74(43), 282.

Tompson, A. F. B. & Gelhar, L. W. (1990). Numerical simulation of solute transport in three-dimensional, randomly heterogeneous porous media. *Water Resources Research*, 26(10), 2541–2562.

Van der Kamp, G., Luba, L. D., Cherry, J. A. & Maathuis, H. (1994). Field study of a long and very narrow contaminant plume. *Ground Water*, 32(6), 1008–1016.

Vomvoris, E. G. & Gelhar, L. W. (1990). Stochastic analysis of the concentration variability in a three-dimensional, heterogeneous aquifer. *Water Resources Research*, 26(10), 2591–2602.

Welty, C. & Gelhar, L. W. (1991). Stochastic analysis of the effects of fluid density and viscosity variability on macrodispersion in heterogeneous porous media. *Water Resources Research*, 27(8), 2061–2075.

Welty, C. & Gelhar, L. W. (1992). Simulation of large-scale transport of variable density and viscosity fluids using a stochastic mean model. *Water Resources Research*, 28(3), 815–827.

Woodbury, A. D. & Sudicky, E. A. (1991). The geostatistical characteristics of the Borden aquifer. *Water Resources Research*, 27(4), 533–546.

Zhang, D. & Neuman, S. P. (1995). Eulerian-Lagrangian analysis of transport conditioned on hydraulic data, 1. Analytical-numerical approach. *Water Resources Research*, 31(1), 39–52.

V

Fractured rocks and unsaturated soils

1 Component characterization: an approach to fracture hydrogeology

JANE C. S. LONG
Lawrence Berkeley National Laboratory

CHRISTINE DOUGHTY
Lawrence Berkeley National Laboratory

AKHIL DATTA-GUPTA
Lawrence Berkeley National Laboratory and Texas A&M University

KEVIN HESTIR
Lawrence Berkeley National Laboratory and Utah State University

DON VASCO
Lawrence Berkeley National Laboratory

1 INTRODUCTION

Within the last few decades, the problems in hydrogeology and reservoir engineering have changed, from simply being concerned with how much fluid could be extracted or injected, to more complex problems such as contaminant transport and multi-fluid extraction schemes. Engineers began to realize that field-scale observations of these complex phenomena could not be predicted from laboratory measurements. A prime example was that lab-scale measurements of dispersivity were too small to account for the amount of contaminant spreading actually observed in the field. Scientists came to the conclusion that this amount of spreading was due to the heterogeneous nature of the medium and was really advective in nature. The large-scale dispersion was actually the tortuosity of flow as fluids moved through the more permeable parts of the medium and around the less permeable parts. Work began to define the nature of geologic heterogeneity and its effect on flow and transport.

This chapter examines the special nature of *fracture* heterogeneity and the implications of this nature for system characterization and the development of predictive models for hydrologic behavior. Examples are drawn from work at various sites in different rock types. The philosophy expounded in this chapter is primarily a rationalist approach. That is, the rock type and the history determine the nature of the hydrology. Observations should illuminate this nature and identify the components of the system which affect flow. This conceptual understanding of the system components should then guide further characterization efforts, placing emphasis on measurements made judiciously (rather than randomly or regularly) and based on understanding of the role of the various components of the system. We often take this rational approach in porous media by identifying components such as units which are aquifers or aquitards, but in fractured rock it is less common to classify and understand the types of components that exist. The science of fracture geology and the ability to use geophysics to image rock have advanced significantly in recent years, and these provide a basis for improving our ability to identify fracture system components and change the way in which we approach fracture hydrology.

This chapter also looks at new inverse techniques developed for dealing with the special characteristics of fractured rock in an efficient manner. These models are based on representing the fracture network as an equivalent lattice of conductors, which is a simple way of defining systems that are not well interconnected. We give a number of examples, emphasizing those where the results of the inversion are used to make a prediction that has subsequently been checked through further measurement.

2 SPECIAL CHARACTERISTICS OF FRACTURE HETEROGENEITY

Fractures in rocks are one type of heterogeneity that is particularly difficult to deal with for four fundamental reasons. The first is that fracture systems are distinguished from most porous systems in that the contrasts in permeability are more extreme and localized. Fractures can be highly permeable conduits imbedded in a matrix rock that is practically impermeable. The second reason is that a fracture in a rock is the equivalent of a pore in a porous medium. Fracture 'pores' are large compared with boreholes and measurement devices. The third reason is that, in a given system, many, or even nearly all, observable fractures, even those that appear to be interconnected, can be non-conductive and play no significant role in flow. The fourth is that fracture systems are created through chaotic processes, which can result in features of many different scales that are related to each other in a complex manner.

3 CONNECTIVITY

High contrasts in permeability between the fractures and the matrix mean that flow in a fracture network depends strongly on the interconnection or 'connectivity' of the fractures. Where the matrix permeability is low, a fracture will not have a significant role in hydrologic behavior unless it is connected to a network of permeable fractures. The classic evidence for imperfect connectivity is in a well field where a nearby observation well does not respond to the pumping well while a well farther away does.

In contrast, connectivity is not an issue in many porous systems. In sand deposits, small changes in porosity, resulting from slightly different depositional conditions, provide contrasts in permeability that channel transport through the medium. These slight changes vary relatively smoothly in space; flow is able to move from any part of the medium to any other part. That is, connectivity is not an issue. Likewise, connectivity is less of an issue where the matrix permeability is comparable to the fracture permeability. However, at the other end of the spectrum, flow in a fractured granite is confined to the fractures and cannot move significantly through the adjacent matrix rock. So, if the fractures are not interconnected, fluid does not move. Of course there are porous systems that also have the characteristic of high contrast, such as sand lenses in clay. The characterization and analysis of such porous systems may be very analogous to that for fractures.

The connectivity of a fracture network has a strong influence over whether or not a rock can be treated as an equivalent continuum. Connectivity is a critical phenomenon, in that, if one starts with an intact rock and adds fractures randomly one at a time, the resulting clusters of interconnected fractures will suddenly become infinite in size at a critical limit (the percolation limit) and thus will be permeable on the 'infinite' scale. Near this critical limit, further increases in the number of fractures will increase the permeability exponentially. Well above the critical limit, the addition of more fractures will cause a linear increase in permeability.

A network of equally conductive fractures which is above the critical percolation limit will behave like a continuum if it is observed on a large enough scale. Numerical studies of homogeneous, two-dimensional Poisson networks of equally conductive fractures have been constructed to examine how this equivalent continuum behavior develops as a function of scale and a function of connectivity. These percolation studies for simple fracture systems have some elegant results. Hestir & Long (1990) have shown that the mean permeability of such systems can be given by:

$$K_c/K_s = \zeta_c(\zeta_c - 4)/(\zeta_c^2 - 4)$$

where K_c is the conductivity of the network normalized by 'Snow's conductivity', K_s. K_s is the conductivity of the related network which has the same fracture frequency (i.e. same number of fractures intersecting a line sample) but where all the fractures are infinite in length; ζ is the connectivity defined as the average number of fracture intersections per fracture. The subscript c refers to the fracture network minus the shorter fractures up to the cutoff, $\ell = c$. c is defined by removing fractures one by one from the network starting with the shortest one. c is the length of the longest fracture that can be removed without changing the conductivity of the network. ζ_c is given by:

$$\zeta_c = \lambda_A \langle \ell_c \rangle^2 H(\theta)$$

where λ_A is the number of fracture centers per unit area; the term $\langle \ell_c \rangle$ is the mean of the fracture lengths greater than c. $H(\theta)$ is given by:

$$H(\theta) = \int_0^{\pi} \int_0^{\pi} \sin|\theta_0 - \theta| g(\theta_0) d\theta d\theta_0$$

where $g(\theta)$ is the orientation distribution.

Hestir & Long (1990) also estimated the minimum size of the network that would exhibit equivalent continuum behavior (the REV), which was given by :

$$\mathrm{REV} = \frac{a\langle \ell_c \rangle}{\xi_c} \{P(\zeta_c) - P_{crit}(\zeta_c)\}^{4/3}$$

where a is a dimensionless constant (Hestir & Long (1990) found $a = 15$ to work well). $P(\zeta)$ is the percentage of elements present in a square lattice model with the equivalent degree of connectivity:

$P(\zeta)=\zeta/(\zeta+2)$

P_{crit} is the critical percolation threshold.

This work provides some insight, but in a practical sense it is not usual to find a homogeneously fractured rock. Most rocks appear to be heterogeneously fractured. In fact, the larger the scale you look at, the larger are the features that can be identified. Consequently, the REV defined assuming the system was homogeneous would always be a minimum.

4 FRACTURES ARE VERY LARGE PORES

The design and interpretation of borehole tests in fractured media are complicated by the fact that the response in the well may be from a single fracture. In a porous medium, this would be equivalent to measuring the response in a single 'pore'. Variability of the response on this scale is very high. An even more dramatic possibility is to get no response at all because the borehole has just missed intersecting a nearby conductive fracture or has hit a non-conductive part of the fracture. In a tight rock, a 'miss is as good as a mile', in that you cannot tell whether you have just missed intersecting the conductive feature, or whether such a feature just does not exist. The net result is that there can be a large variation in the measurements, and any particular measurement may not be representative.

In order to treat the medium as an equivalent continuum, one could consider increasing the size of the wellbore, for example by using a very large diameter hole, a long interval, or a tunnel. Theoretically, this measurement might well be on the scale of the REV. In this case, you can get an average response representing the equivalent continuum behavior on that scale. However, such large measurements can also create short circuits in the fracture flow pattern, which means that the system you are trying to measure is changed by the act of making the measurement. The result is that it is extremely difficult to design a test that gives results that apply generally to any other flow situation.

5 FEW FRACTURES ARE HYDROLOGICALLY IMPORTANT

To be hydrologically significant, a fracture must both be open and connected. A fracture can be impermeable because it never opened, or because it is healed or filled or closed by high normal stresses. A fracture need not be impermeable to be hydraulically insignificant. It only needs to be hydraulically isolated. Sometimes, only the intersections between fractures are conductive and the fractures themselves play little role in flow.

A number of field studies have been conducted which demonstrate that few of the observable fractures at the site, even those that are apparently interconnected, play a significant role in flow. An important example is the Validation Drift at the Stripa Mine in Sweden (Olsson, 1992). Here, all the fractures in a 50 m long drift through a granitic rock were mapped and the inflow to the drift was monitored in 1 m^2 intervals (Fig. 1). The center of the drift intersected a major fracture zone called the H-zone. The distribution of inflow to the drift was very skewed. Approximately 95% of the water came into the drift through the H-zone and approximately 85% came in through a single fracture in that zone. Thousands of other fractures were seen in the drift walls, but they played a much less significant role in the flow of fluids. Clearly, it is the existence of the fracture zone and the internal morphology of this zone that controls the majority of the flow. It can be surmised that most of the fractures outside of such zones are either impermeable or not hydraulically connected, or both.

In the Stripa case, there are fractures of almost every orientation, but the orientation of the fracture that conducted almost all the flow was perpendicular to the maximum principal compressive stress. In this case, stress is evidently not a strong determiner of which fractures are open and conductive. In other cases, it can be shown that the significant fractures are those that are parallel to the maximum compressive stress. For example, at Ekofisk in the North Sea, permeability anisotropy aligns with the direction of the maximum compressive stress, even though this direction is not parallel to the orientation of the densest fracture set

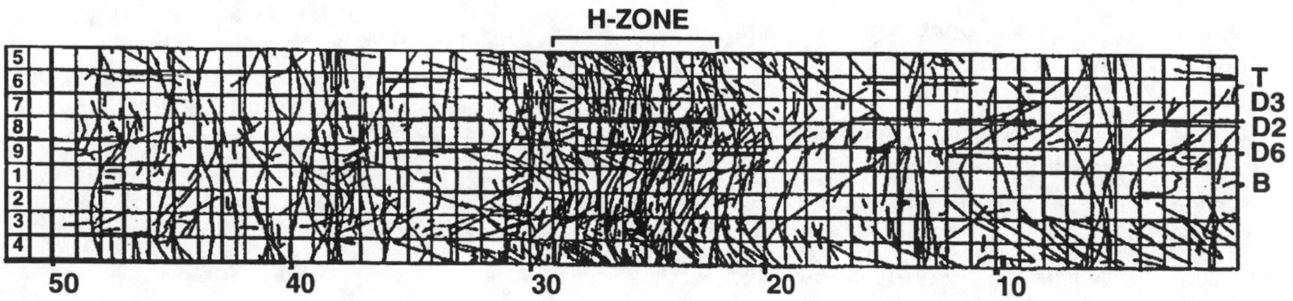

Fig. 1 'Unrolled' fracture map of the Validation Drift, viewed from the outside.

(Teufel, Rhett & Farrell, 1991). One can surmise that most of the fractures perpendicular to the maximum principal stress are not open and are not hydrologically important. Similarly, where the direction of the maximum compressive stress in a given stratum is different in different parts of a reservoir, and this direction correlates with the direction of maximum permeability, it is evident that stress controls which fractures are significant for flow.

In both of these cases, it is not possible to understand the flow properties of the rock simply by understanding the fracture patterns. In the Stripa case, the fracture statistics would lead you to believe that flow should be ubiquitous. In the North Sea example, fracture statistics would give the wrong direction for permeability anisotropy. It is only the interconnection of the open fractures that determines the flow behavior, and it is not always immediately obvious which these are.

6 COMPLEX FRACTURE GENESIS

There are three major factors that determine fracture hydrology: the mechanical properties of the rock, the stress history, and the subsequent geochemical alteration. The fracturing of rock is a recursive phenomena. The rock itself is heterogeneous, containing flaws and materials of different mechanical properties. These heterogeneities serve to concentrate stress; stress concentrations induce failure in the form of fractures. The fractures themselves cause significant perturbations in the stress field, and these perturbations influence the next iteration of fracturing. Figs. 2(a) and (b) show an example of stress levels perturbed by two en-echelon fractures. Fig. 2(c) shows an example of a real fracture pattern that can be explained by stress perturbations similar to that shown in Fig. 2(b) (from Olmacher & Aydin, 1995). Further complications occur because stress directions can change and major geologic processes affect the formation of fractures over eons.

Unfortunately, the generation of fractures is not all that is required to understand hydrology. After the fractures form, they may be filled, partially dissolved, or materials inside the fractures may be redistributed. These processes may in turn interact with the stress state. For example, Moore & Vrolijk (1992) describe fluid circulation in an accretionary prism. Hot, mineral-rich fluids upwell through fractures, depositing mineral fillings which decrease the permeability of fractures. The resulting decrease in permeability increases the pressure, which causes the fractures to open and conduct more fluid, which leaves more precipitate, etc. The evidence for this behavior is in 'book quartz' fracture fillings which have layer after layer of precipitate representing each episode of opening and subsequent filling. Diagenetic alteration can

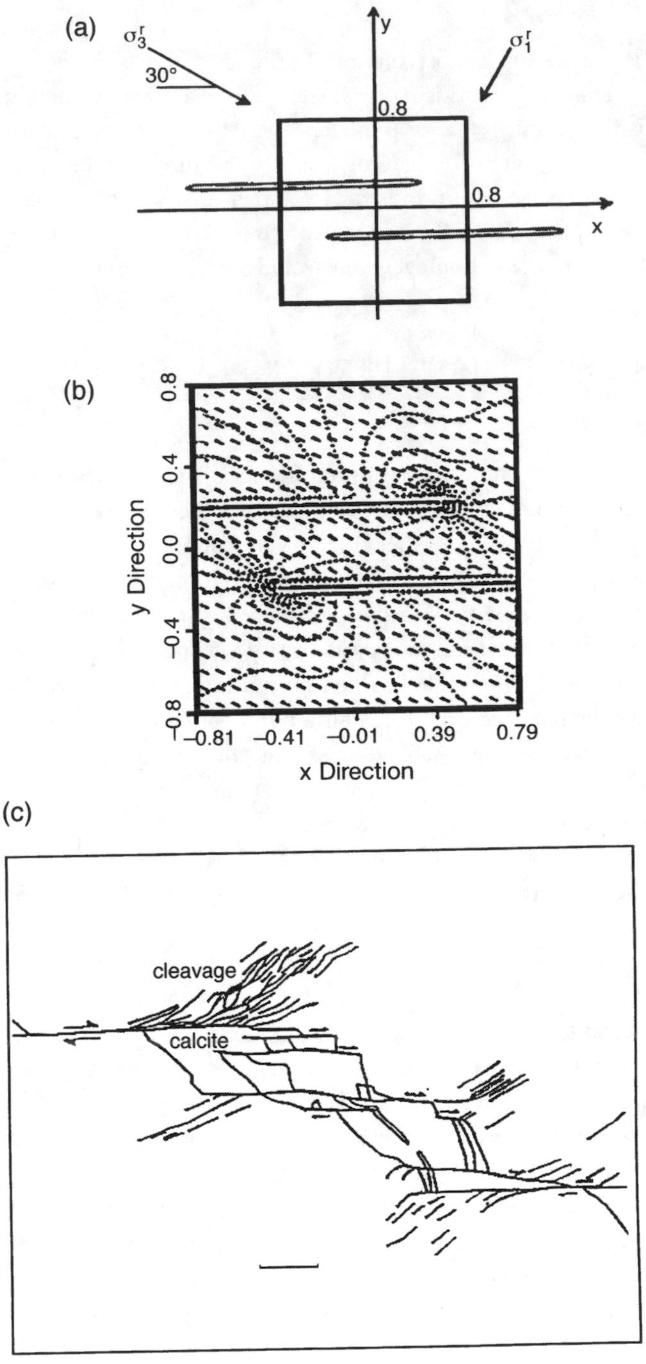

Fig. 2 (a) Loading configuration. (b) Stress distribution around two en-echelon mode-II fractures in an elastic plate subjected to biaxial compressive principal stresses. Contours indicate the relative magnitude of the greatest principal stress, and tick marks indicate their orientation. (c) An array of en-echelon faults and the associated structures. Calcite filling at the extensional steps and pressure solution cleavages at the contractional quadrants correspond to the stress fields shown in (b). The location and the orientation of various associated structures are consistent with the stress field shown in (b).

also cause mineral bridges to form across a fracture, resulting in the fracture being highly permeable and insensitive to increases in effective stress (Dyke, 1992). The effect of geochemistry on fracture systems is very complicated and can result in stable, metastable, or unstable hydrologic conditions. These processes mean that any understanding of fracture hydrology based only on fracture pattern is questionable.

7 THE IMPLICATION OF THESE DIFFICULTIES ON CHARACTERIZATION AND ANALYSIS

What are the implications of these four types of special problems on the approaches used to characterize and model fracture systems? The usual approaches followed fall generally within two classifications: stochastic permeability approaches and stochastic pattern approaches. Although both of these approaches are useful, they also have significant inadequacies.

Stochastic permeability approaches were developed for porous media systems. Well tests are used to infer permeability values, which are assumed to apply on the scale of the measurement itself. Then, the distribution of permeability in space is described through a statistical analysis of the inferred values of permeability. This approach essentially ignores the *cause* of the permeability and focuses on the inferred *value* of permeability. The difficulty here is that what you measure in the test is a flow rate for a given perturbation and one must then assume a model as the basis for interpreting the test data. The assumed model is usually very simple, not the same as the fracture network, and almost always homogeneous. The vagaries of the real fracture geometry can mean that what you measure has little or nothing to do with the conditions that control flow and transport behavior under conditions different from the test. In other words, permeability is not an absolute parameter which is a fundamental property of the medium. Actually, this is another way of saying that REV concepts do not apply. Prediction of transport in these systems requires that we have more understanding of the cause of the permeability, i.e. the fracture system geometry. Another more practical problem with this approach is that the space of the model is filled with porous blocks that are each assigned a permeability. The fact that some regions are not connected to others can be represented in such a model, but doing this can require very fine discretization that can be a very inefficient way to go about modeling the system.

Dissatisfaction with the stochastic permeability approach led to the discrete fracture network approach, which is similar to that which has been used in rock mechanics to understand the mechanical behavior of a fractured rock mass. In this approach, one tries to recreate the fracture network statistics and to use this information to predict the behavior of the system. First applications of this method used fracture statistics such as fracture density, length, orientation, and conductance, and created stochastic models with discrete fractures that had the same statistical properties (e.g. Long *et al.*, 1982; Dershowitz, 1984). These models were attractive because they can represent poorly connected fracture systems very efficiently. However, these models completely misrepresent the spatial relationships between the fractures, i.e. the fracture pattern. Further work has tried to do more to represent higher level aspects of fracture pattern, such as clustering (e.g. Billaux *et al.*, 1989; Lee, Veneziano & Einstein, 1990), but it is difficult to believe that overall fracture statistics are important in predicting flow in cases like the Stripa Validation Drift experiment, where so very few of the fractures actually conduct fluid.

8 THE COMPONENT APPROACH

Where do we go from here? The classical approach to hydrogeology includes identifying components of the system that are significant features. We call out each significant hydrogeologic unit. We even assign these units names and recognize what role they play in the hydrologic system (aquifer, aquitard, etc.) We try to explain the relationships between these units and characterize the variability within them. This is a fundamentally rational and sensible approach. It incorporates a geologic understanding and accounts for the genesis of the system. It subdivides the system into functional components and, most importantly, it provides a context for understanding the overall system behavior.

The same philosophical approach is appropriate for fracture systems. There are components in a fracture system that are identifiable and play a specific role in the hydrology. Granted, these components can be harder to perceive than calling out a limestone aquifer between two shale aquitards, but in many cases hydrogeologic units for fracture systems can be described, located and characterized.

One of the promising approaches towards identifying hydrogeologic units in fracture systems is to infer these from the way fractures are formed. Fractures are often divided into categories reflecting their origin. In geology, fractures are generally distinguished by whether there has been displacement along the fracture (shears) or not (joints). However, to get to the point where we understand hydrology, we have to go beyond the origin of single features to the evolution of fracture patterns.

It is possible to surmise that the fracture patterns resulting from a recursive process would form attractors and possibly

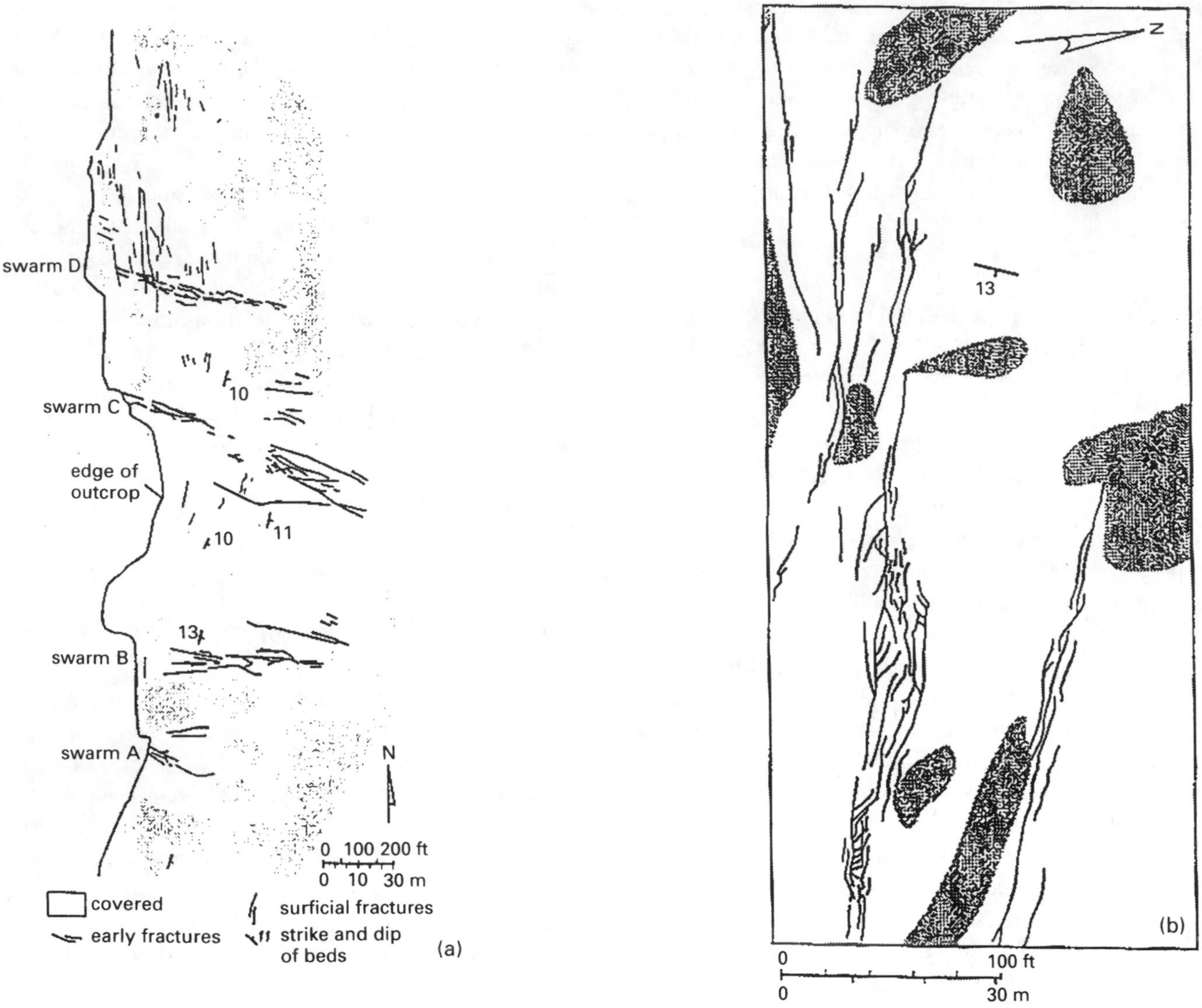

Fig. 3 (a) Map showing patterns of fracture swarms in sandstone (Laubach, 1991). (b) Details of swarm D in the map (Laubach, 1992).

fractals. Certainly, it has been shown that fracture patterns have some fractal properties (e.g. Barton, 1993). What is definitely true is that fractures form recognizable patterns. Recent work has illuminated some of these patterns in a way that can provide insight about the hydrologic behavior. Two salient examples are the development of joint clusters and the growth of faults in brittle rocks.

Laubach (1991, 1992) has described joint systems in sandstones (Fig. 3, for example). In these systems the joints are not uniformly distributed; rather, they are clustered. Olson (1990) has developed a model of joint genesis that helps to explain these types of patterns. In this elastic model based on fracture mechanics, the iterative growth of joints is calculated, taking into account joint interaction. The result is that longer fractures tend to grow at the expense of shorter ones. Further, the extent of clustering is related to the rate of fracture formation. Faster joint propagation results in a greater degree of clustering. These clusters represent the areas of the rock that are more permeable. Questions remain about how these clusters interconnect, especially in the vertical dimension.

Faults in brittle rocks often exhibit step-overs (e.g. Aydin & Nur, 1985; Martel, Pollard & Segall, 1988). These step-overs are either regions of compression or extension depending on the sense of the step with respect to the sense of movement on the fault system. For example, these steps can form when a fault develops on a pre-existing joint system (Fig. 4, taken from Martel, 1990). The regions of extension are regions where the rock is being pulled apart and joints are developing. These joints are often open and very conductive in the direction along the fault perpendicular to the sense of movement along the fault. In fact, it is common to find these features filled with precipitate in rhomboidal-shaped veins as a record of old flow systems. The hydrologic role of the com-

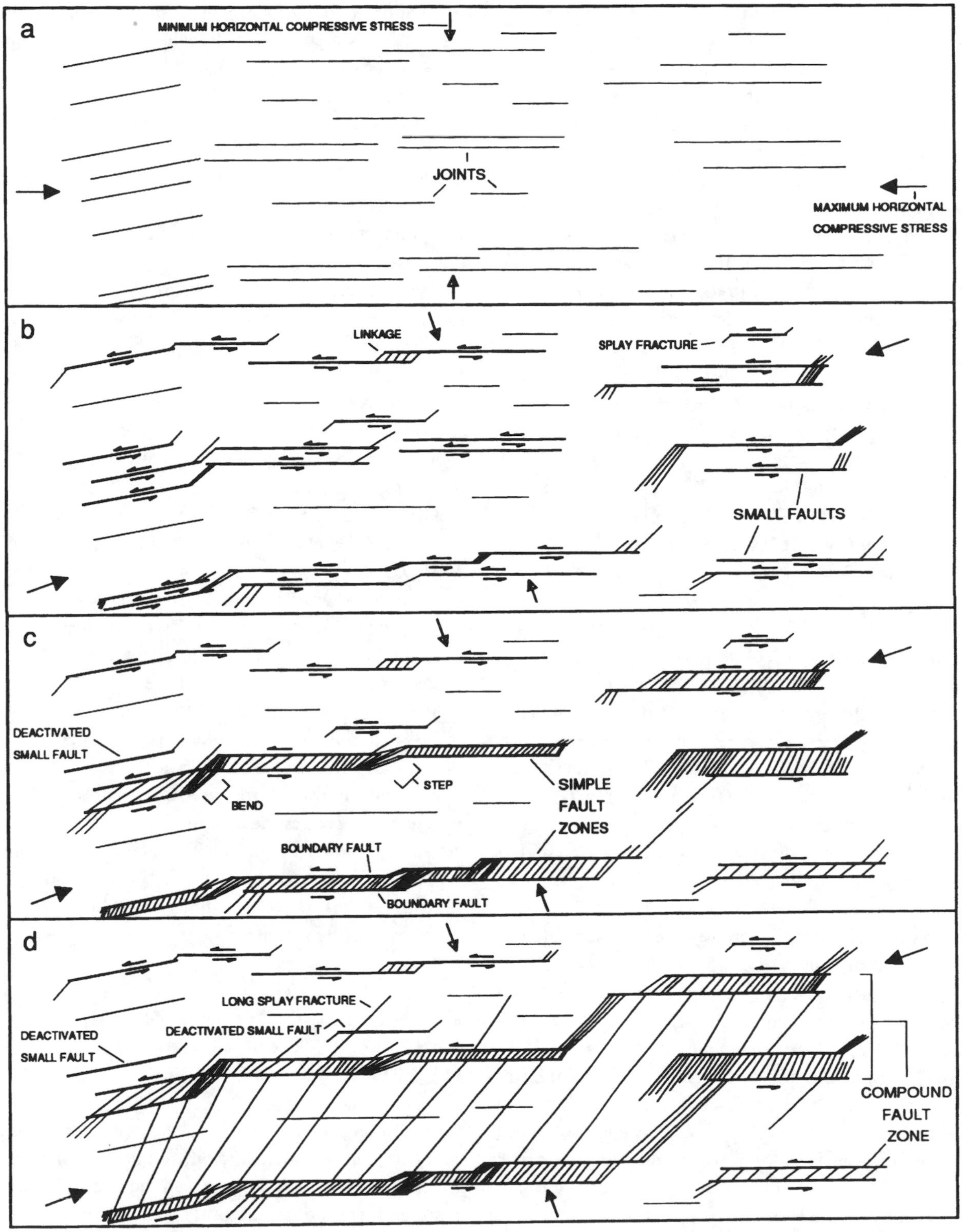

Fig. 4 A schematic illustration of the development from (a) to (d) of a strike-slip fault system and fault zones of different magnitudes in the granitic rocks of Sierra Nevada, California. From Martel (1990).

pressed steps is not well understood. It is possible that such regions extend perpendicular to the compression, which would increase permeability parallel to the compression, or that they fail along shears that may decrease permeability. These fault patterns are known to occur, but it remains to completely understand the hydrologic implications.

These are just two of many possible examples. The key idea is that some of these patterns appear often enough to be recognized. These patterns could be looked at critically with respect to their likely hydrology. Fracture patterns are usually studied in two dimensions, either the plane of the exposure (geologically) or the plane of the maximum and minimum principal stresses (mechanically). To understand the hydrology, we will have to study these systems in three dimensions. Most of these patterns will have enhanced permeability in the direction of the intermediate principal stress, which ironically is perpendicular to the plane that is usually studied in a two-dimensional mechanical model. In other words, the interesting hydrology may be taking place in the plane perpendicular to the interesting mechanics. As a result, the way in which open fractures develop and interconnect is inherently a three-dimensional problem. This problem will depend on the stress history, the material properties, and the anisotropy of the rock material.

9 LOCATING HYDROGEOLOGIC COMPONENTS – ADVANCES IN GEOPHYSICS

Once we obtain an understanding of the types of features we are looking for, it is very useful to have geophysical methods for finding these features in the rock we wish to characterize. For example, in the Stripa case, the hydrology is dominated by fracture zones. These are faults that have probably moved in more than one direction over time. Within these faults there are some highly permeable fractures that carry most of the flow. The pertinent geophysical problem is to find the fracture zones. For the sandstones, the problem is to find the fracture clusters, often called 'sweet spots' by petroleum geophysicists.

A variety of methods have recently been developed to look for fracture features underground. New seismic and radar tomography methods see highly fractured regions as low velocity, high attenuation anomalies. Shear waves hold great promise for detecting fracture features because the shear wave oscillating in the direction perpendicular to the fractures will be slowed and attenuated much more than the shear wave oscillating parallel to the fractures. Any one method or any one way of using a method may not be successful. It is best if a suite of methods can be calibrated against a known feature of the type that is of interest. For example, at the Grimsel Rock Laboratory in Switzerland, seismic tomography was performed to image a fracture zone from two parallel tunnels. Certain frequencies gave a much better image of the fracture zone than others (Majer *et al.*, 1990). In the Stripa case, many single-hole and cross-hole methods were applied to the same rock. A weighted parameter was developed called the Fracture Zone Index (FZI; Olsson, 1992), which was derived from the eigenvector of the correlation matrix between these measurements. The FZI was extremely useful for dividing the rock into fracture zones and 'good rock'.

From a hydrologic point of view, some of the most exciting geophysical methods are the difference methods. In this approach, the flow system is perturbed in such a way that the perturbation also changes the geophysical properties of the medium. By subtracting the image before and after the perturbation, one obtains a picture of the perturbation in the flow system. Some excellent examples are available.

The first example is from Stripa (Olsson, 1992), where radar tomography was performed before and during the injection of saline water into the H-zone of the Validation Drift experiment. The saline water increases the electrical conductivity, which increases the attenuation of radar waves. Images produced in this way allowed the visualization of the saline migration (Fig. 5). The saline water changed the electrical properties, which changed the radar response.

The second example is from the Conoco Borehole Test Facility (Majer *et al.*, 1990; Datta-Gupta, Vasco & Long, 1994). Cross-hole seismic measurements were made in a five-spot pattern where the wells go through a fractured limestone formation (Fig. 6). The well GW-2 is hydraulically connected to well GW-5, but well GW-3 in the middle is not well connected to either GW-2 or -5. The question was whether that connection between GW-2 and -5 was to the north or south of GW-3. Seismic cross-hole measurements were taken between all well pairs before and after the injection of air into GW- 5 and withdrawal from GW-2. Air introduced in a fracture greatly increases the compressibility of the fracture and consequently increases the ability of the fracture to attenuate seismic waves. Fig. 6 shows the resulting increase in attenuation to the north of GW-3, indicating that the air was following a fracture in that location.

10 HYDROLOGIC AND TRACER TESTING OF FRACTURE HYDROGEOLOGIC UNITS

The advantage of knowing what the important components of a fracture system are and where they are in space is that the hydrologic testing program can then be targeted at these features. If we know a fracture zone is the important feature, then a test sequence can be designed to develop an understanding of the hydrologic properties of the zone. Boreholes can be designed to pierce the zone, and the zone can be packed off in each borehole to isolate the responses. Communication between the fracture zone and other zones or parts of the fracture system can similarly be tested.

If the dominant feature is a fracture cluster, one may be able to design tests that determine the interconnectivity in

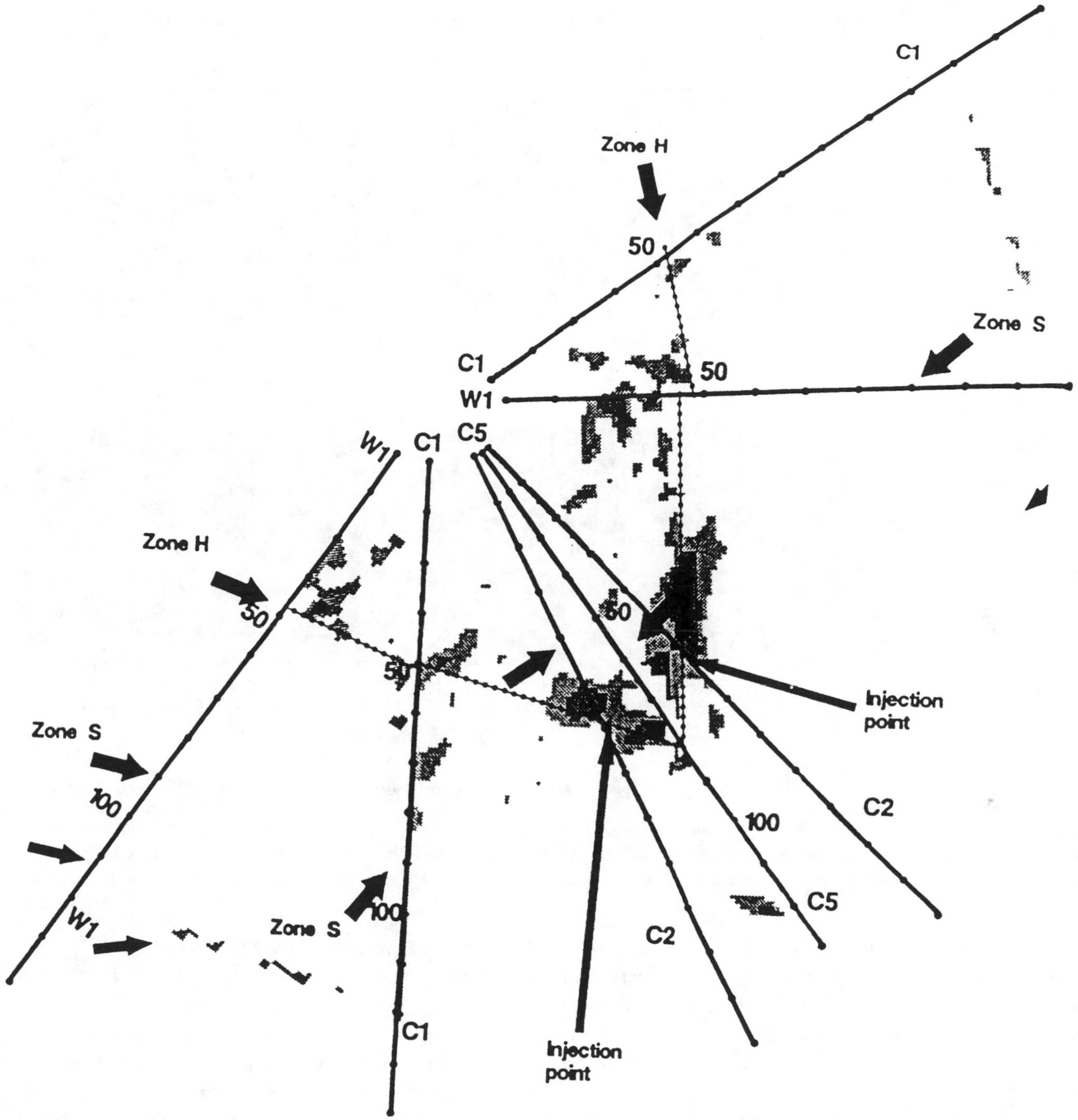

Fig. 5 Composite of difference tomograms showing the distribution of tracer in the first radar/saline tracer experiment approximately 290 hours after start of injection. (After Olsson, 1992.)

the cluster or the communication between one cluster and another. If you have determined that steps in a fault zone are the important features, then a test can be designed to measure the properties of these steps. The important point is that the testing is not purely random or regular. It is rational and focused on features that play an important role in the hydrology. Within these features, the details may be less important and these can be successfully represented stochastically.

11 INVERSE METHODS

Given knowledge of the major components of the fracture system, a good way to go about developing a representation of the heterogeneity within these features is through inverse models. Inverse models have been developed in order to address the problem of representing heterogeneity or complex parameters in such a way that the models are self-consistent with the data collected in the field. The inverse

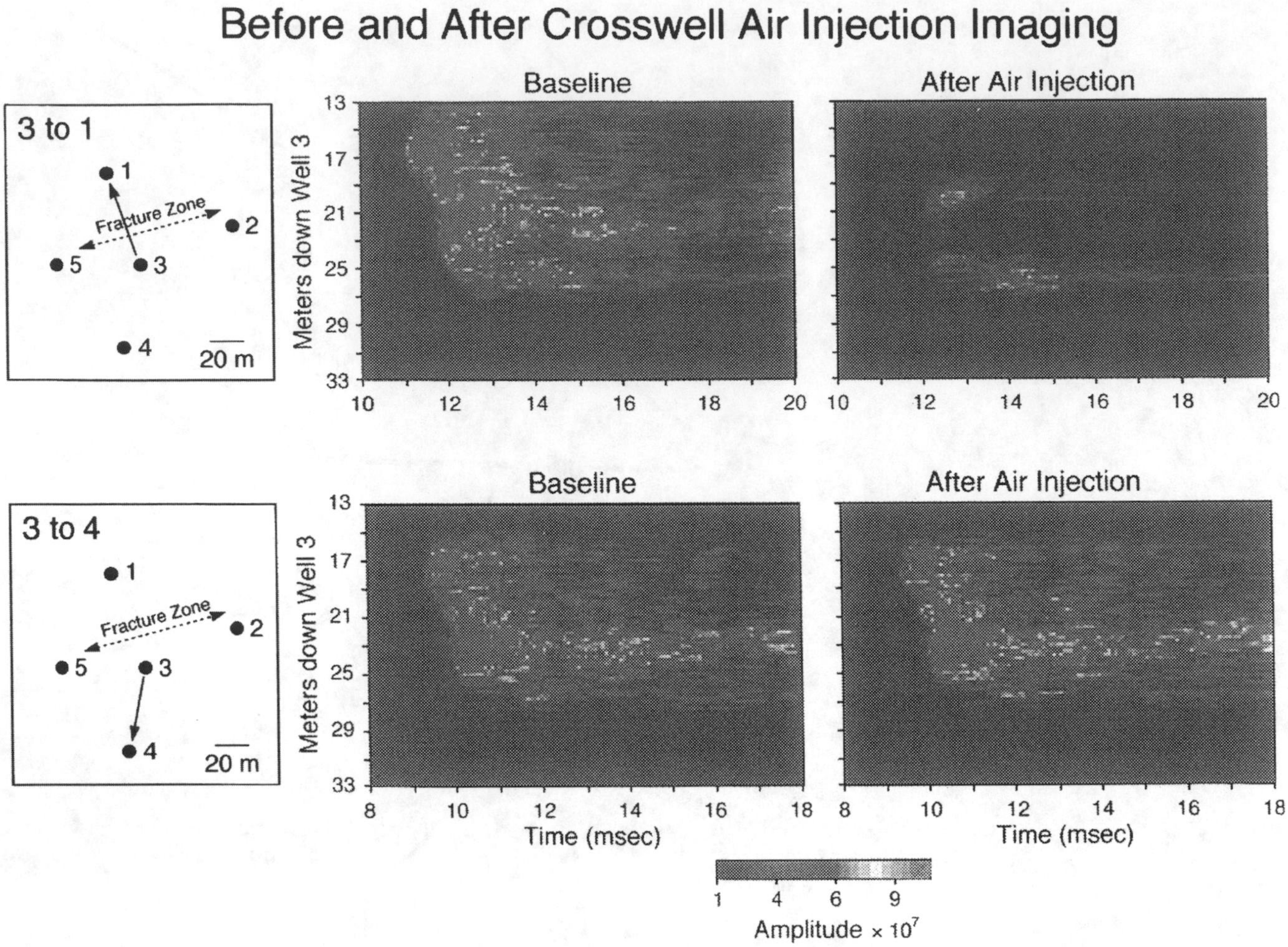

Fig. 6 Cross-well seismic signals between well pair GW-3 and GW-4, and well pair GW-3 and GW-1, before and after air was injected into well GW-5. Seismic energy was greatly attenuated between wells GW-3 and GW-1 after air injection, whereas it was unchanged between wells GW-3 and GW-4, suggesting that the air-filled fracture lies to the north of well GW-3. This is consistent with the well-test inversions.

approach creates a model by finding a specific model or series of models which reproduce the observed data, and integrates data interpretation and model building into one process. The hope is that models which can at least reproduce the data are more likely to make good predictions of similar phenomena.

One of the benefits of a component approach is that it often allows for a two-dimensional or quasi-two-dimensional approximation. For example, if the dominant features are fracture zones, it may be possible to model the system as a lattice of one-dimensional conductors lying on the two-dimensional plane of the fracture zone. If there is more than one zone, these can be represented by intersecting two-dimensional lattices. Such a model was developed for the Stripa site (Long *et al.*, 1992) and is shown in Fig. 7. In this way the detail in the model matches the dominant features of the hydrologic system.

12 SIMULATED ANNEALING ON A LATTICE

One of the special needs of fracture system modeling is to represent the connectivity of the system. This means that we are asking the inverse routine to look for structure in the model. The more classic inverse approach is to define the structure of the medium *a priori* and ask the inverse model to look for the best parameters for the prescribed elements of the structure. In some fracture networks, the connectivity is the dominant phenomenon and the appropriate model becomes a lattice of 'on' or 'off' conductors. In this case, the parameter set under investigation is a disjoint set, e.g. a vector consisting of zeros and ones. An optimization technique that works well for disjoint sets is simulated annealing (Long *et al.*, 1992). Simulated annealing has been used successfully to find lattice configurations that match the hydrologic data. We call these models 'equivalent discontinuum models'.

The simulated annealing process consists of changing an element of the lattice, say turning an element off that was previously on, or vice versa. Then we recalculate the model response and compare it with the data. If the response is closer to the data, accept the change in the model. If not, then accept the change with a probability that is inversely proportional to the decrease in model performance. As annealing proceeds, we decrease the constant of proportionality, allowing the model to make large changes at the beginning of the process in order to jump out of local minima but converge towards the end to a stable solution. Fig. 8 gives a theoretical example. Fig. 8(a) shows a possible fracture configuration that is tested by wells indicated on the figure. We start by representing the system as a regular lattice (Fig. 8(b)). Then annealing is applied to the lattice to find a configuration (Fig. 8(c)) that matches the well test data derived from the fractures in Fig. 8(a).

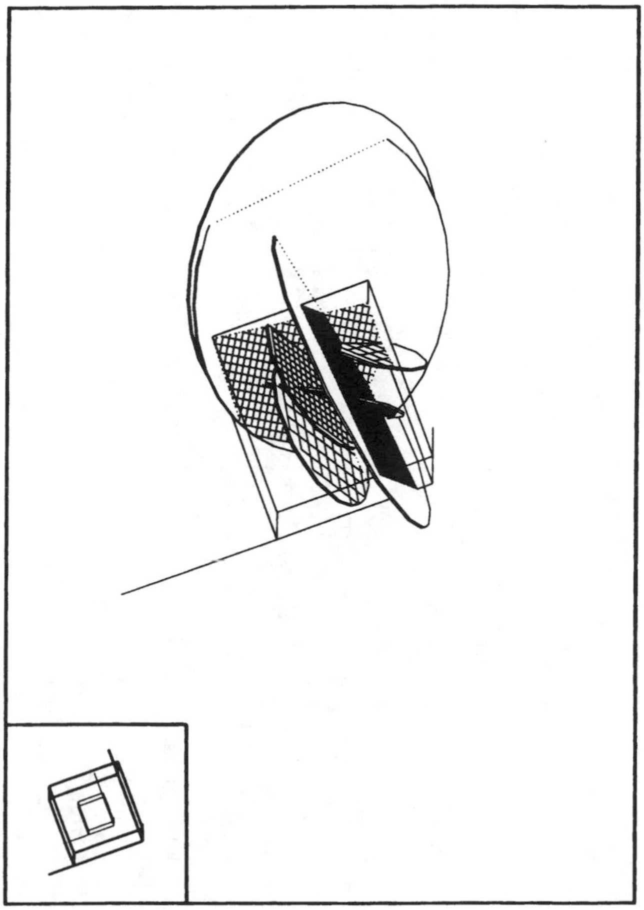

Fig. 7 The three-dimensional zone mode for the Stripa site. Each fracture zone was represented by a lattice of conductors lying in the plane of the fracture zone.

An example of the application of lattice annealing is to the wells at the Conoco Borehole Test Facility mentioned above (Datta-Gupta *et al.*, 1994). Boreholes at this site extend through a fracture limestone unit bounded on top and bottom by low permeability shale units. Two well tests were conducted at the site, one pumping GW-2 and one pumping GW-5. In each case the other four wells were observation wells. The salient feature of these tests is that GW-3, -4, and -1 do not respond very much to either pumping test. However, GW-2 and -5 respond to each other almost immediately and strongly. Apparently, there is a fracture connecting GW-2 and -5 that does not intersect GW-3. Fig. 9 shows three different annealing solutions for these well tests using three different types of lattices. Each of these, in fact all of the hundreds of inversions that were done, show a hydraulic connection between GW-2 and GW-5 to the north of GW-3.

13 ENSEMBLE ANALYSIS

One of the issues with this style of inversion is that it results in a non-unique solution. We view this as realistic because there are rarely enough data to determine any hydrologic system uniquely. The only way a unique solution can be

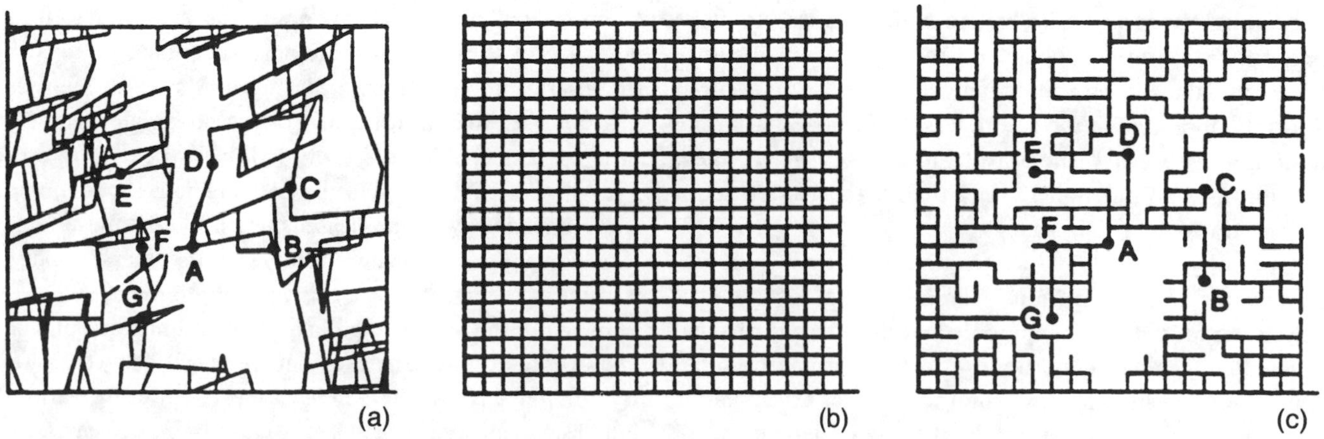

Fig. 8 (a) A synthetic fracture network where we have generated well test data by pumping hole A with monitoring in holes B–G; (b) template model; and (c) the pattern of conductors resulting from annealing.

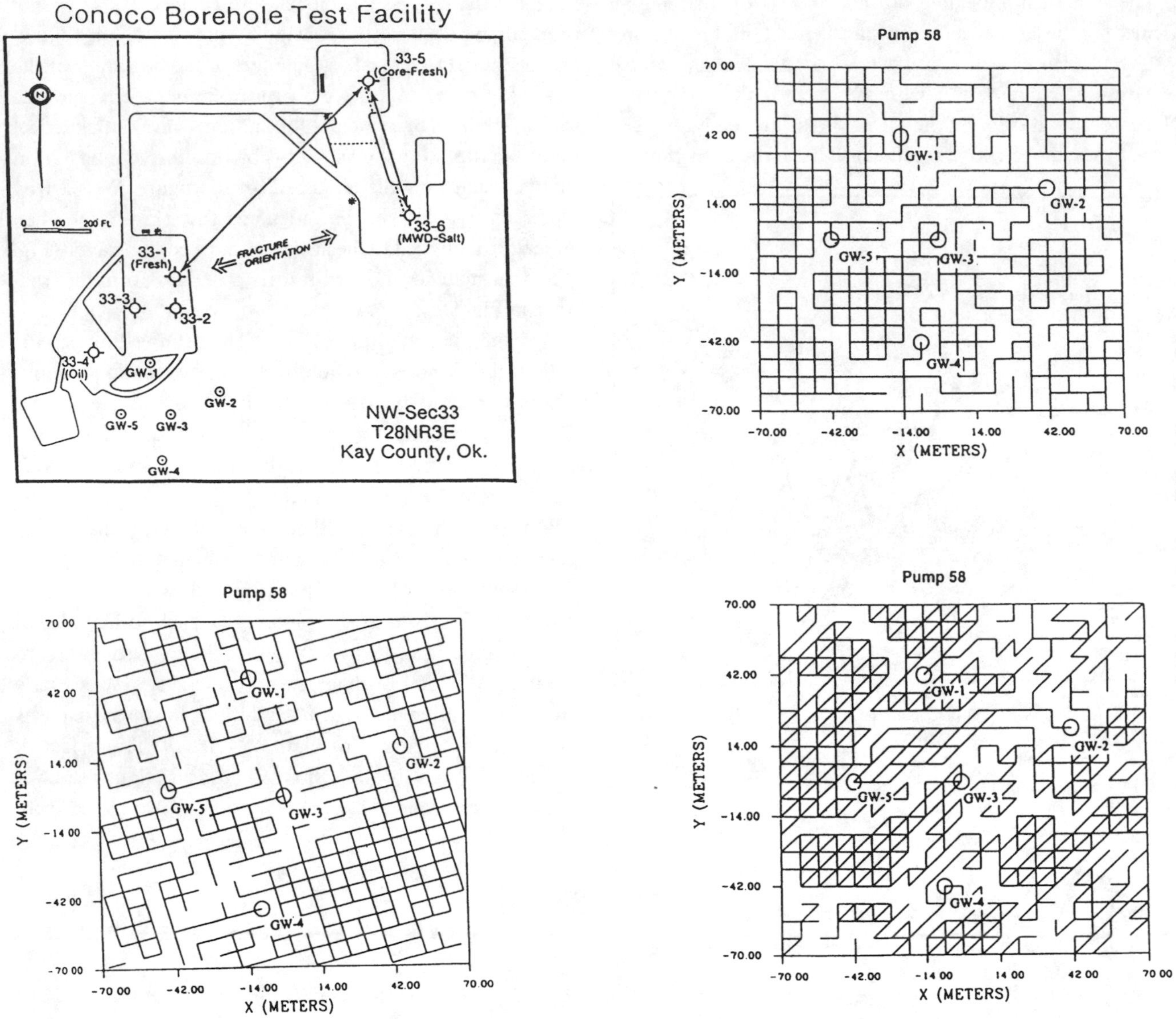

Fig. 9 Lattice annealing inversions of well test data taken at the Conoco Borehole Test Facility in Oklahoma (upper left). These results were obtained by finding lattices that could reproduce well test data taken in Boreholes GW-1 to GW-5. All models show a connection between GW-2 and GW-5 to the north of GW-3.

accomplished mathematically is to make some assumptions about the structure of the heterogeneity. The more assumptions we make, the fewer parameters that need to be resolved and the more unique the solution is. If we make the wrong assumptions, we have no way to test the impact of these assumptions on the model, other than making a new set of assumptions and trying again.

One can reasonably reduce the non-uniqueness by conditioning the solution on non-hydrologic data, such as geophysical information or an understanding of the process that created the geologic conditions. This research approach has important bearing on providing the basis for extrapolating from regions where there are hydrologic data and, for example, geophysical information, to regions where there may only be, for example, geophysical information.

Since simulated annealing inversions are unique, it is advisable to study the distribution of possible solutions to the inverse problem. Simulated annealing used in multiple inversions samples a distribution of approximately equally probable models. This series of models can be used to estimate that distribution, find the best predictive model, and study uncertainty in those predictions. Unlike the unique inversion approach, the uncertainty is more general because it is not linked to as many *a priori* assumptions.

In lieu of the unique approach, the non-unique approach allows us to explore structures that are suggested by the data. By running an inversion many times over using various initial structures, an ensemble of solutions can be generated. In these solutions, one can look for common features and have some confidence that these features are robustly indicated by the data. For example, in the Conoco case study, we performed many inversions of the well test data. In each and every case, the connection between wells GW-2 and GW-5 ran north of GW-3. We tried different types of lattices and different orientations for the lattice. Every case was the same. The existence of a fracture connection north of GW-3 was substantiated with geophysics as described above (see Fig. 6). In this case we begin to feel confident that the well test data are indicative of a connection to the north of GW-3. However, because connectivity is the key issue, care must be taken in interpreting ensemble data. When the mean of the ensemble conductances is presented, the information about the nature of interconnection is lost.

14 ITERATED FUNCTION SYSTEMS INVERSION

A variation of lattice annealing is to generate the model parameters considered during the inversion process using an iterated function system (IFS) (Doughty *et al.*, 1994). The IFS's we use are composed of two to four affine transforms. An affine transform is a function which may rotate, reflect, deform, contract, and translate a set of points. When multiple transforms are applied iteratively, a set of points with a fractal geometry, known as an attractor, results. The attractor can be mapped to a distribution of transmissivity or storativity in a mathematical model. We then use the mathematical model to simulate the well test and compare the calculated heads with those observed during the test. If the match is not satisfactory, the parameters of the IFS are varied, and the simulation is repeated. This procedure continues until the IFS creates an attractor which yields a hydrologic property distribution that produces a good match to observed hydrologic data.

The advantage of using an IFS generator is two-fold. First, the IFS can create a very complicated pattern of conductances with a few parameters. In fact, IFS's were first developed to store graphical information in a computer. The complex graphics take large amounts of computer storage if information about each pixel is to be kept. By representing the information with IFS, the information is reduced to a much smaller number of parameters. Reproducing the figure requires computer time to operate the iteration rather than storage space. In our case, we wish to represent information about heterogeneity in a small number of parameters. The second advantage is that the fractal nature of the medium is naturally represented by the iterative process of creating the heterogeneity field. The resulting IFS may be a very natural way to represent scale effects.

Fig. 10 shows an example of an IFS used to generate a Sierpinski's gasket. There are 18 parameters needed to generate this fractal. Fig. 10 shows what happens to the fractal as the parameters are gradually changed. Each of the points in this fractal can be used to increment or decrement the conductance of the nearest lattice element. We can then optimize on the IFS parameters to match the observed data.

This method was applied to Stripa data from the H-zone (Olsson, 1992; Doughty *et al.*, 1994). The H-zone was identified with an extensive geophysical characterization. This zone dominated the hydrology of the Validation Drift experiment as mentioned above. A number of boreholes intersected this zone, and an interference test, called the C1–2 test, was conducted in these boreholes. Observations were made in packed-off intervals in the H-zone. Another series of boreholes, the D-holes, were drilled through the H-zone. These were meant to mimic the behavior of a drift from the hydrologic point of view by placing six parallel boreholes within a ring. The inflow to these boreholes was predicted based on a model derived by inverting the C1–2 interference test. No data from the D-holes were included in the inversion of the C1–2 test.

Three different inversions are shown in Fig. 11: S1, S2, and S3. S1 shows the case when the IFS is used to increment and decrement the conductance distribution. S2 shows the case where the IFS is used to increment the conductance, representing a case where the fracture is basically closed with some open areas. S3 shows the resulting permeability distribution for the case that the IFS was used to decrement the permeability of nearby lattice elements. This would correspond to a basically open fracture that is sealed in some places. In each case the model was used to predict the measured inflow to the six D-holes, which was 0.21 cm^3/s. All the models did quite well, but the increment plus decrement model did the best and the decrement model did the worst.

15 RESOLUTION ANALYSIS

Another important part of inversion analysis is knowing what it is that the data are capable of resolving and on what scale. To that end, Vasco, Datta-Gupta & Long (1997) have developed an application of geophysical resolution analysis to hydrologic data. Resolution is a measure of how well the spatial variation of subsurface properties can be determined with the data we have available. Vasco *et al.* (1997) derived a resolution matrix, R_{ij}, based on the inversion of hydrologic well and tracer test data. The diagonal elements of the matrix, R_{ii}, correspond to each element of the model. The values of R_{ii} vary from 0 to 1, where 1 means that the value of the parameter of interest, say permeability k_i, can be resolved perfectly. Zero means the parameter k_i cannot be resolved at all. Any off-diagonal element R_{ij} of the matrix is an averaging coefficient; R_{ij} shows how much the value of any

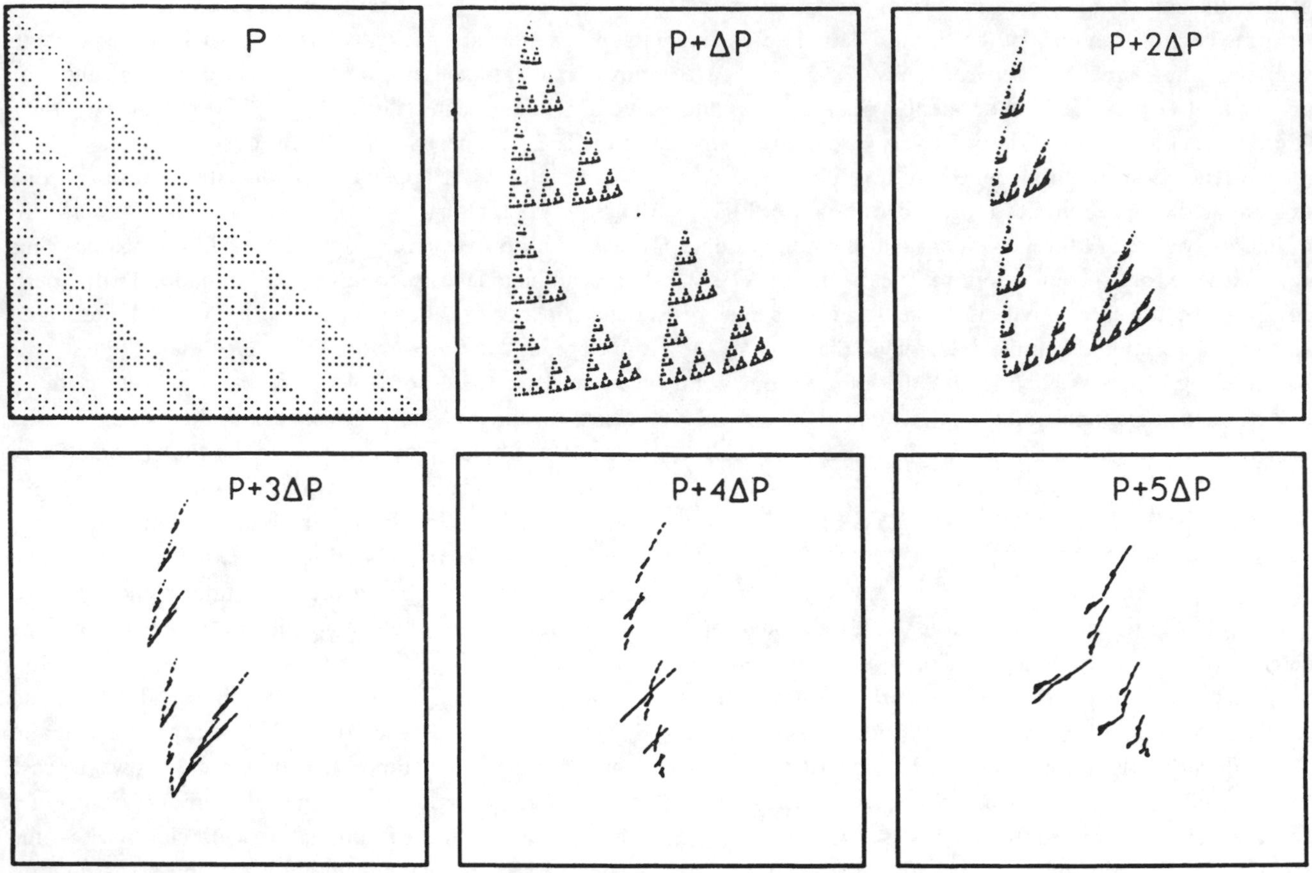

Fig. 10 The variation in an attractor as the parameters of the IFS change.

other parameter value, k_j, influences the resolution of k_i. Thus the rows, or averaging kernels, of the matrix give an indication of the amount of smearing in the inversion of the data.

Given a model of the system, the resolution matrix is derived from the sensitivity matrix,

$$G_{ij}=\partial C_i/\partial k_j$$

where C_i is any observed data point, such as a particular concentration value in a tracer test. G_{ij} gives the sensitivity of any data point to any parameter. The sensitivity matrix can be calculated in a numerical model for flow and transport by perturbing each parameter k_i by a small amount and noting the consequent change in C_i. This operation is a normal step in a conjugate gradient inversion scheme. The change that would be observed in the data as a result of any change in model parameters, assuming a linear relationship between model parameters and data, is given by:

$$dC_i=G_{ij}dk_j$$

Current work is focused on extending this methodology beyond the linearized relationships. We would like to be able to invert this equation to estimate the change in parameters that is indicated by a change in the data. However, G_{ij} cannot be inverted exactly because it is singular: there are usually more model parameters then there are data to determine them. Thus G_{ij} is not a square matrix. By using singular value decomposition, Vasco *et al.* (1997) derived a generalized inverse of G_{ij} called G_{ij}^{-g}. The resolution matrix R_{ij} is then defined as:

$$R_{ij}=G_{ij}^{-g}G_{ij}$$

If the data were sufficient to determine all the parameters perfectly, then $G_{ij}^{-g}G_{ij}$ would be equal to the identity matrix. Essentially, $G_{ij}^{-g}G_{ij}$ relates the true parameter field to the estimated parameter field. A key aspect of this method is the need to linearize the problem so that an inverse of G_{ij} can be found.

Resolution analysis can be used in two ways. First, one can propose a series of field tests and examine the resolution that they are likely to provide, assuming some distribution of permeability. Fig. 12 shows an example application of this method to such a synthetic problem in a heterogeneous fracture zone. In this case, a series of interference tests was proposed in the well field, using each well in turn as the pumping well and all the other wells as observation wells. Then a similar series of convergent tracer tests was proposed, with each well in turn used as the sink. Vasco *et al.* (1997) used

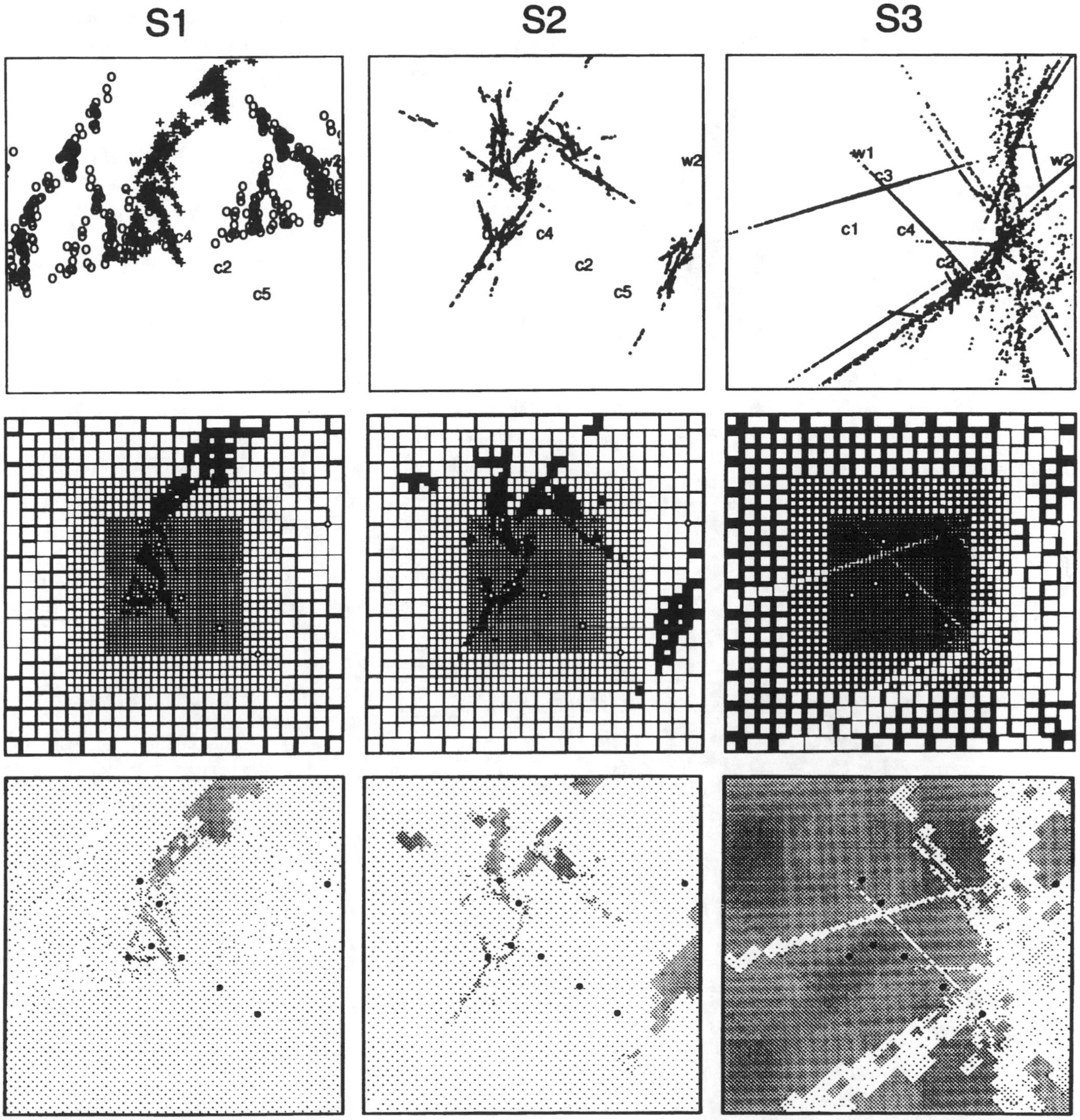

Fig. 11 IFS inversion of an interference test in a fracture zone (H-zone) at Stripa. The top row shows three different attractors from three different IFS schemes. The middle row shows the resulting numerical grids, where higher conductances are shown by thicker lines. The bottom row is the average distribution of conductance that results from the attractor, where lighter tones indicate lower conductance. The left column, S1, is for the case that the IFS increases and decreases permeability; the middle column, S2, is for incrementing; and the right column, S3, is for decrementing. The measured flow into the D-holes was 0.21±0.6 cc/s. The flows predicted by S1, S2 and S3 were 0.22 cc/s, 0.13 cc/s and 0.38 cc/s, respectively.

resolution analysis to determine how well these test data could be used to resolve the permeability distribution in the fracture zone. In this case the inversion was accomplished using a flow and transport model based on a semi-analytical model for tracer flow (Datta-Gupta & King, 1995). The figure shows the diagonal values of R_{ij}, i.e. the resolution, for the interference tests, the tracer tests, and both tests combined. This example illustrates the dramatic theoretical increase in resolution that results from tracer test data as opposed to interference test data. Of course, in reality, tracer tests are much harder to perform than interference tests. Problems which dramatically affect tracer test results, such as

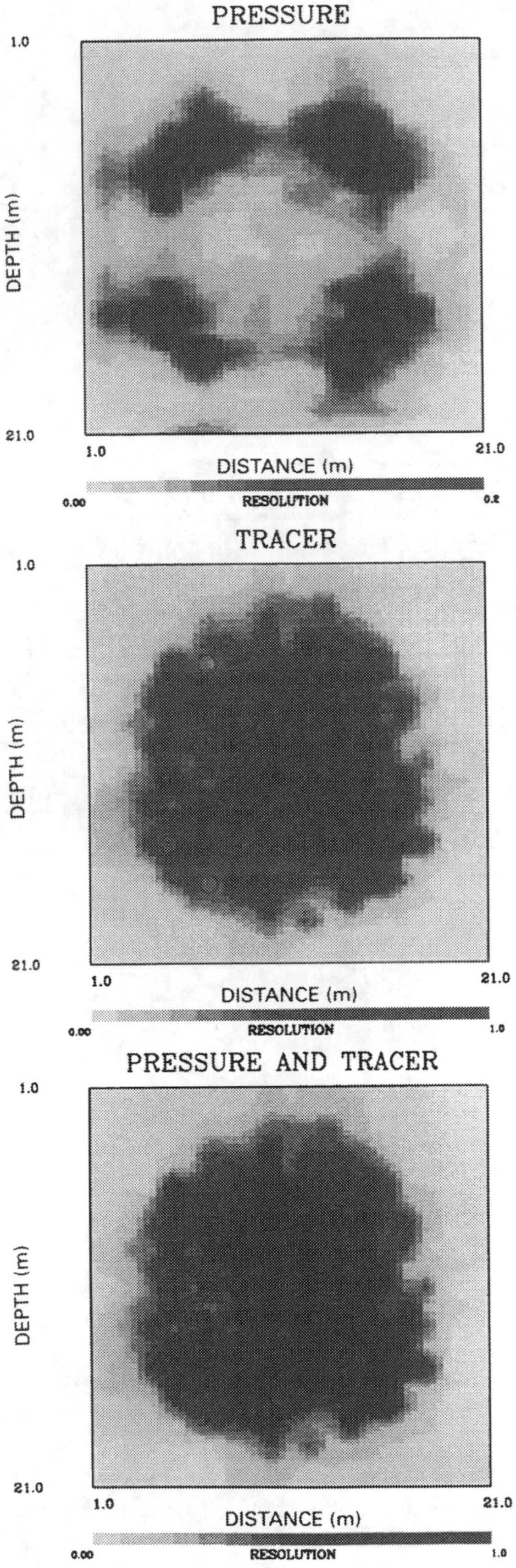

Fig. 12 Model parameter resolution corresponding to a synthetic experiment. The synthetic tracer and pressure values are computed with respect to a specified set of permeability variations. The model parameter resolution is shown for three data sets: transient pressure data only, tracer data only, and transient pressure and tracer data.

not knowing the boundary conditions or not recovering the tracer, are not yet included in the resolution analysis.

The second way to use resolution analysis is to perform an actual test and invert the data. Then, use the inverted model to examine the resolution that was provided by the data. In this way, resolution analysis is used to see how effective the test plan was in determining the parameters of interest and how much confidence one might have in the results of the inversion.

16 CONCLUSIONS

This chapter has tried to set the context for a rational approach to the characterization and modeling of flow and transport in fractured rock. The fundamental point is that we need to look at fracture systems in terms of defining hydrogeologic units and then characterize these units. Then the units themselves form the context for modeling the system. Because of the vagaries of hydraulic connectivity in fracture systems, we recommend the use of inverse models which do not require the prior interpretation of field data. Use of inverse models eliminates the need for an intermediate conceptual model, which is only used to interpret the data and may result in a 'parameter' that has little relevance to any flow system but the test itself.

Many of the points made in this chapter are based on limited experience. Only a few applications of each technique exist. As more real sites are characterized and more predictions are made with models and tested against real behavior, we will gain more confidence that the approaches we use are efficient and appropriate. The maturing of fracture hydrogeology will require such testing facilities in a number of rock types and fracture settings that are relevant to critical problems.

ACKNOWLEDGMENTS

Funding for this work came from a variety of sources. We would like to thank the following from the US Department of Energy: Bill Luth of Basic Energy Sciences/Energy Research; Bob Lemmon of Bartlesville Operations Office/Fossil Energy; Royal Watts of Morgantown Energy Research/Fossil Energy; and Bob Levich of the Yucca Mountain Project/Office of Civilian Radioactive Waste Management. In addition, much of our work was done in collaboration with other institutions. Previously published work done in collaboration with Bill Rizer, Pete D'Onfro, and John Queen of Conoco is reviewed here, and we take another opportunity to thank them. Ed Witterholt of British Petroleum, the Swedish Nuclear Fuel and Waste Management Company (SKB), Olle Olsson of Conterra, and the Swiss National Cooperative for Disposal of Radioactive Waste (NAGRA) also have played a significant part in this research.

REFERENCES

Aydin, A. & Nur, A. (1985). The types and role of stepovers in strike-slip tectonics. *Society of Economic Paleontologists and Mineralogists*, Special Publication, 37, 35–45.

Barton, C. C. (1993). Fractal analysis of the scaling and spatial clustering of fractures. In *Fractals and their Use in Earth Sciences*, eds. C. C. Barton & P. R. La Pointe. Geological Society of America. New York: Plenum Publishing Co.

Billaux, D., Chiles, J. P., Hestir, K. & Long, J. C. S. (1989). Three-dimensional statistical modeling of a fractured rock mass – an example from the Fanay-Augeres Mine. *International Journal of Rock Mechanics and Mineral Sciences and Geomechanics*, Abstract, 26, 281–299.

Datta-Gupta, A. & King, M. J. (1995). A semianalytical approach to tracer flow modeling in heterogeneous permeable media. *Advances in Water Resource*, 18, 9–24.

Datta-Gupta, A., Vasco, D. W. & Long, J. C. S. (1994). *Detailed Characterization of a Fractured Limestone Formation Using Stochastic Inverse Approaches*. SPE/DOE 9th Symposium on Improved Oil Recovery, Tulsa, Oklahoma, SPE 27744.

Dershowitz, W. S. (1984). *Rock Joint Systems*. Ph.D. thesis, Cambridge, MA: Massachusetts Institute of Technology.

Doughty, C., Long, J. C. S., Hestir, K. & Benson, S. M. (1994). Hydrologic characterization of heterogeneous geologic media with an inverse method based on iterated function systems. *Water Resources Research*, 30(6), 1721–1745.

Dyke, X. X. (1992). How sensitive is natural fracture permeability at depth to variation in effective stress? In *Proceedings of the Fractured and Jointed Rockmasses International ISRM Symposium*, Rotterdam. A. A. Balkema.

Hestir, K. & Long, J. C. S. (1990). Analytical expressions for the permeability of random two-dimensional poisson fracture networks based on regular lattice percolation and equivalent media theories. *Journal of Geophysical Research*, 95 (B13), pp. 21, 565–71, 581.

Laubach, S. E. (1991). *Fracture Patterns in Low-Permeability Sandstone Gas Reservoir Rocks in the Rocky Mountain Region.* Proceedings, Joint SPE Rocky Mountain Regional Meeting/Low-Permeability Reservoir Symposium, SPE Paper 21853, pp. 503–510.

Laubach, S. E. (1992). Fracture networks in selected Cretaceous sandstones of the Green River and San Juan basins, Wyoming, New Mexico, and Colorado. In *Geological Studies Relevant to Horizontal Drilling: Examples from Western North America*, eds. J. W. Schmoker, E. B. Coalson & C.A. Brown. Denver: Rocky Mountain Association of Petroleum Geologists, pp. 61–73.

Lee, J. S., Veneziano, D. & Einstein, H. H. (1990). *Hierarchical Fracture Trace Model.* 31st U.S. Symposium on Rock Mechanics, Golden Co., Balkema.

Long, J. C. S., Remer, J. S., Wilson, C. R. & Witherspoon, P. A. (1982). Porous media equivalents for networks of discontinuous fractures. *Water Resources Research*, 18(3), 645–658.

Long, J. C. S., Mauldon, A., Nelson, K., Martel, S., Fuller, P. & Karasaki, K. (1992). Prediction of flow and drawdown for the site characterization and validation site in the Stripa Mine. Stripa Project Technical Report 92–05. Stockholm, Sweden: SKB.

Majer, E. L., Myer, L. R., Peterson, J. E., Karasaki, K., Long, J. C. S., Martel, S. J., Blümling, P. & Vomvoris, S. (1990). Joint seismic hydrogeological and geomechanical investigations of a fracture zone in the Grimsel Rock Laboratory. *Swi -46, NAGRA*, Wettingen, Switzerland.

Majer, E. L. *et al.* (1995). Fracture detection using cross well and single well surveys. *Geophysics* (in press).

Martel, S. J. (1990). Formation of compound strike-slip fault zones, Mount Abbot quadrangle, California. *Journal of Structural Geology*, 12, 869–882.

Martel, S. J., Pollard, D. D. & Segall, P. (1988). Development of simple strike-slip fault zones in granitic rock, Mount Abbot quadrangle, Sierra Nevada, California. *Geologists Society of American Bulletin*, 99, 1451–1465.

Moore, J. C. & Vrolijk, P. (1992). Fluids in accretionary prisms. *Reviews of Geophysics*, 30, 113–135.

Olmacher, G. & Aydin, A. (1995). Progressive deformation in the Bays Mountain Syncline, Kingsport, TN. *American Journal of Science*, 295, 943–987.

Olson, J. (1990). *Fracture Mechanics Analysis of Joints and Veins*. Ph.D. dissertation. California: Stanford University, 174 pp.

Olsson, O., ed. (1992). *Site Characterization and Validation – Final Report*. Stripa Project Technical Report 92–22. Stockholm, Sweden: SKB, p. 364.

Teufel, L. W., Rhett, D. W. & Farrell, H. E. (1991). Effect of reservoir depletion and pore pressure drawdown on in-situ stress and deformation in Ekofisk Field, North Sea. *Proceedings of the 32nd U.S. Rock Mechanics Symposium*, Rotterdam. A. A. Balkema, pp. 63–72.

Vasco, D. W., Datta-Gupta, A. & Long J. C. S. (1997). Resolution and uncertainty in hydrologic characterization. *Water Resources Research* (in press).

2 Stochastic analysis of solute transport in partially saturated heterogeneous soils

DAVID RUSSO

The Volcani Center, Bet Dagan

ABSTRACT The problem of solute transport through the unsaturated zone of heterogeneous porous formations is discussed, focusing on mechanistic, stochastic vadose-zone transport models based on two different approaches to modeling flow in these formations, namely the independent vertical columns and the stochastic continuum approaches. It is shown here that the resulting transport models differ in their predictions with respect to (i) how solute spreading evolves with time; and (ii) how water saturation of the formation affects solute spreading. Predictions of transport models based on the first approach are restricted to solute spreading in the longitudinal direction only. Predictions of transport models based on the second approach are in qualitative agreement with the results of numerical simulations of flow and transport in partially saturated, heterogeneous formations, and are consistent with results obtained by the theory of transport by groundwater flow. The need for additional field-scale characterizations of the spatial variability of the soil properties relevant to vadose-zone transport, and for additional, carefully designed field-scale transport experiments and simulations, to validate further existing stochastic transport models, is emphasized.

1 INTRODUCTION

Quantitative field-scale descriptions of chemical transport in the unsaturated (vadose) zone are essential for improving the basic understanding of the transport process in near-surface geological environments, and for providing predictive tools that, in turn, will be used to predict the future spread of pollutants in these environments. Our interest is focused on transport occurring on the field (formation) scale. One of the distinctive features of a natural formation at this scale is the spatial heterogeneity of its properties that affect transport. This spatial heterogeneity is generally irregular; it occurs on a scale beyond the scope of laboratory samples and has a distinct effect on the spatial distribution of solutes, which results from transport through the formation.

To illustrate the type of heterogeneity one may encounter on the field scale in the vadose zone, we present in Fig. 1 measured soil properties from Bet Dagan, Israel (Russo & Bouton, 1992). This figure shows contour lines of log-transformations of two formation properties which characterize the unsaturated conductivity, i.e. the saturated conductivity, K_s, and the reciprocal of the macroscopic capillary length scale, α. The irregular, ostensibly erratic, spatial variation of both properties is evident in this figure. It is clear that such heterogeneity in the formation properties must affect flow and transport at the field scale. The complexity and irregular distribution of solutes one might encounter in the vadose zone at the field scale is illustrated in Fig. 2. This figure displays contour lines of measured chloride concentrations in the unsaturated zone, obtained from a tracer experiment conducted under natural soil, vegetative, and climatic conditions at Creux de Chippis, Switzerland (Schulin *et al.*, 1987). Clearly, the chloride concentrations show considerable variability. A fundamental question raised by the experimental evidence displayed in Figs. 1 and 2 is that of how to develop predictive models that incorporate the impact of field-scale spatial variability of soil properties on vadose-zone flow and transport. In the following, a few advances in this area will be presented and analyzed.

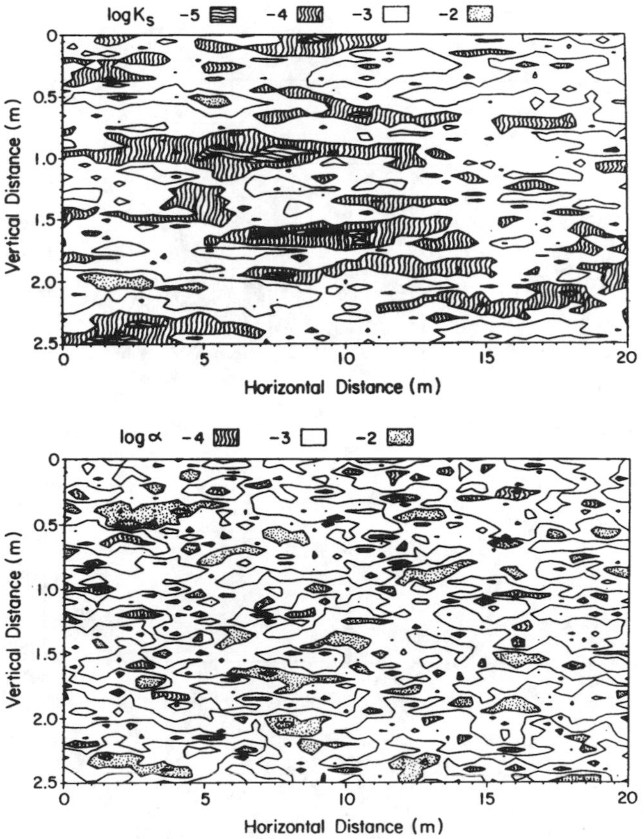

Fig. 1 Distribution of logK_s (K_s in cm/min) and of logα (α in cm^{-1}) in the upper 2.5 m of the vertical cross-section at Bet Dagan site (from Russo & Bouton, 1992). Vertical exaggeration 2.1×.

2 MODELING OF TRANSPORT IN PARTIALLY SATURATED, HETEROGENEOUS POROUS FORMATIONS

The uncertainty in soil properties affecting transport (due to their inherent erratic nature and paucity of measurements) generally precludes the use of the traditional, deterministic modeling approach for the prediction of flow and transport at the field scale. In an alternative approach, uncertainty is set in a mathematical framework by modeling the relevant formation properties (e.g. parameters which characterize the hydraulic conductivity and water retention functions) as random space functions (RSFs). As a consequence, the flow and transport equations are of a stochastic nature, and the dependent variables (e.g. fluid pressures, solute concentrations) are also RSFs. The aim of the stochastic approach, therefore, is to evaluate the statistical moments of variables of interest, given the statistical moments of the formation properties. In general, this is a formidable task, and it is common to regard as stationary each of the formation properties, as well as the various flow-controlled attributes, $p(\mathbf{x})$, where $\mathbf{x}$ is the spatial coordinate vector. They are characterized at second-order by a spatially invariant mean, $P=\langle p(\mathbf{x})\rangle$, and a covariance, $C_{pp}(\mathbf{x}',\mathbf{x}'')=\langle p'(\mathbf{x}')p'(\mathbf{x}'')\rangle$, in which $p'(\mathbf{x})=p(\mathbf{x})-P$, that depends on the separation vector, $\xi=\mathbf{x}'-\mathbf{x}''$, and not on $\mathbf{x}'$ and $\mathbf{x}''$ individually.

Vadose-zone transport models may be classified as either mechanistic or nonmechanistic. Mechanistic is taken here to imply that the model incorporates the most fundamental mechanisms of the process, as understood at present (e.g. describing local flow and transport by Richards' equation and by the convection-dispersion equation (CDE), respectively). Nonmechanistic implies that the model disregards the internal physical mechanisms which contribute to the transport process, and concentrates on the relationships between an input function and an output response function. In this chapter we will focus on mechanistic transport models, and, furthermore, we will concentrate on the transport of conservative, nonreactive, nonvolatile solutes in the absence of plant roots. It should be emphasized, however, that the transport models discussed in this chapter can serve as a basis for analyzing the behavior of more complex solutes.

The starting point is the considerable uncertainty in point values of the solute concentration, c, that is attributed to the complex heterogeneity of the soil hydraulic properties. For practical purposes (including the assessment of transport model formulation) therefore, larger-scale integrated measures of the solute transport, such as the spatial moments of the distribution of the point values of c, are most appropriate inasmuch as they have a lesser degree of uncertainty than the point values. In turn, these moments are related to the statistical moments of the solute particle displacement probability density function (PDF).

Quantification of solute transport in terms of the aforementioned integrated entities may be accomplished in a two-stage approach: the first stage involves relating the statistical moments of the PDF of the particle displacement to the statistical moments of the velocity PDF, while the second stage involves relating the statistical moments of the velocity PDF to those of the properties of the heterogeneous porous formation. The former stage, which is independent of whether the flow is saturated or unsaturated, may be accomplished in a general Lagrangian framework, as was done by Dagan (1984) for transport by groundwater flow.

Using a general Lagrangian description of the motion of a solute particle, its trajectory, $\mathbf{X}(t)$, is the solution of the kinematic equation $d\mathbf{X}/dt=\mathbf{V}(\mathbf{X})$, with $\mathbf{X}=\mathbf{a}$ for $t=0$, where $\mathbf{V}=\mathbf{U}+\mathbf{u}$ is the Eulerian velocity, with $\mathbf{U}=\langle\mathbf{V}\rangle$ being its expected value (assumed to be constant and uniform in time and space) and $\mathbf{u}$ being its fluctuation. Neglecting pore-scale dispersion, and for $\mathbf{a}$ within a release volume Ω_0, the trajectory $\mathbf{X}=(X_1,X_2,X_3)$ at time t is given (Dagan, 1989) by

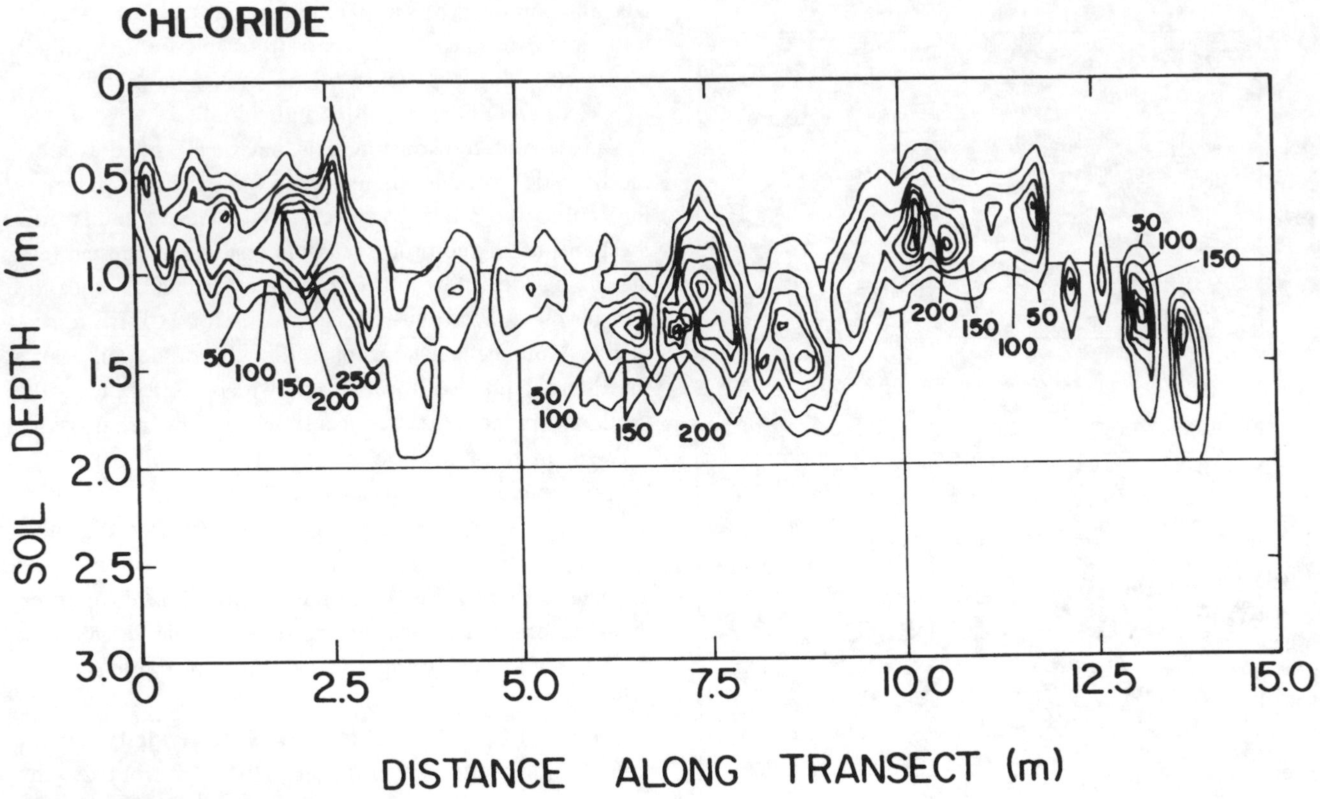

Fig. 2 Distribution of observed chloride concentration (in μg/g dry soil) in unsaturated soil at Creux de Chippis site (from Schulin *et al.*, 1987). Vertical exaggeration 2.5×.

$$\mathbf{X}(t;\mathbf{a})=\mathbf{a}+\mathbf{U}t+\int_0^t \mathbf{u}[\mathbf{X}(t';\mathbf{a})]dt' \tag{1}$$

For fixed **a**, assuming ergodic conditions, Lagrangian and Eulerian stationarity and homogeneity, and assuming that the streamlines of the advecting solute particles do not deviate significantly from the mean flow direction, Dagan (1982, 1984) approximated the actual trajectory of the particle path by its mean path. Consequently the particle displacement mean, $\langle \mathbf{X}(t;\mathbf{a})\rangle$, and, by a first-order approximation in the velocity variance, the particle displacement covariance, $X_{ij}(t;\mathbf{a})=\langle X_i'(t;\mathbf{a})X_j'(t;\mathbf{a})\rangle$ $(i,j=1,2,3)$ at time t, where $\mathbf{X}'=\mathbf{X}-\langle\mathbf{X}\rangle$ is the fluctuation, are given by:

$$\langle \mathbf{X}(t)\rangle=\mathbf{U}t \tag{2a}$$

$$X_{ij}(t)=2\int_0^t (t-\tau)u_{ij}(\mathbf{U}\tau)d\tau \tag{2b}$$

where $u_{ij}=u_{ij}(\xi)$ is the velocity covariance tensor and $\mathbf{a}=(a_1,a_2,a_3)$ is suppressed for simplicity of notation. For a multivariate normal (MVN) Eulerian velocity field, the solute particle displacement, whose PDF is completely defined by (2a,b), is also MVN.

It should be emphasized that (2b) is restricted to flow regimes associated with relatively large Peclet numbers UI_{y1}/D_L and $UI_{y2}/I_{y1}D_T$ (Dagan, 1988), where D_L and D_T are the longitudinal and transverse components of the pore-scale dispersion tensor, respectively. From a physical point of view, the ratios $X_{ii}/2t$ may be regarded as the apparent dispersion coefficients which would lead to the same X_{ij} as the actual time-dependent ones, in the solution of the convection-dispersion equation with constant coefficients.

The second stage, which relates the statistical moments of the velocity PDF to those of properties of the heterogeneous porous formation, depends on whether the flow is saturated or unsaturated. Under unsaturated flow conditions, quantification of these relationships is rather difficult inasmuch as the relevant flow parameters – hydraulic conductivity, K, and water capacity, C – depend both on formation properties (e.g. K_s and α) and on dependent flow variables (water saturation, Θ, or capillary pressure head, ψ) in a highly nonlinear fashion. Analysis of unsaturated flow is further complicated by the fact that the flow in the vadose zone may be subject to time-dependent processes that occur at the soil–atmosphere interface. Accomplishment of the second stage, therefore, requires several simplifying assumptions regarding both the structure of the formation heterogeneity and the flow regime.

3 THE INDEPENDENT VERTICAL COLUMNS APPROACH

Stochastic, vadose-zone transport models which are based on the independent vertical columns approach are either mechanistic (e.g. Dagan & Bresler, 1979; Bresler & Dagan, 1981; Destouni & Cvetkovic, 1989, 1991) or nonmechanistic (e.g. Jury, 1982). Both groups of models assume that the heterogeneous inputs are stationary RSFs, and ergodic over the region of interest, so that the ensemble moments can be derived from sample spatial statistics. In addition, it is assumed that steady water flow takes place under imposed recharge rate, R. Both groups of models treat the heterogeneous soil as though it were composed of a series of isolated stream tubes (vertically homogeneous, independent soil columns) with different velocities. In the mechanistic approach, the velocity in each vertical soil column is related to the soil properties by using the one-dimensional Darcy's law, assuming a deterministic constant unit head gradient. It is further assumed that the local relationships between conductivity, K, and water content, θ, are isotropic and described by the expression (Brooks & Corey, 1964)

$$K(\theta,\mathbf{x})=K_s(\mathbf{x})[(\theta-\theta_{ir})/(\theta_s-\theta_{ir})]^{1/\beta} \tag{3}$$

The irreducible water content, θ_{ir}, is taken as zero, and both saturated water content, θ_s, and β, a parameter related to the soil pore-size distribution, are viewed as constant and deterministic, whereas K_s, the saturated conductivity, is assumed to be proportional to the square of a scaling factor, δ, which is viewed as a stationary RSF in the horizontal plane only. Because of the assumption of independent, homogeneous vertical soil columns, the correlation scale of the input formation property, $\lambda=\log\delta$, is infinite in the vertical direction, and the variance, σ^2_λ, describes soil property variations among the soil columns, regardless of the position of the columns in the horizontal plane.

From (3), if $\log\Theta$ is normally distributed, $\log K(\Theta)$, being a linear combination of $\log\delta$ and $\beta^{-1}\log\Theta$, is also normally distributed, and its PDF is characterized completely by the mean, $Y(\Theta)=2\Lambda+\log K^*_s+\beta^{-1}W$, and the variance, $\sigma^2_y(\Theta)=4\sigma^2_\lambda+\beta^{-2}\sigma^2_w+2\beta^{-1}\sigma^2_{\lambda w}$. Here $\Theta=\theta/\theta_s$ is water saturation, W and w are the mean and the fluctuation of $\log\Theta$, Λ and λ are the mean and the residual of $\log\delta$, σ^2_w is the variance of w, and $\sigma^2_{\lambda w}$ is the cross-variance between λ and w.

Under the aforementioned assumptions, the transverse components of velocity, u_2 and u_3, vanish and the longitudinal component of velocity, $u_1=u_1(\lambda,R,K^*_s,\beta,\theta_s)$, is given by

$$u_1=\frac{R\exp(2\lambda\beta)}{(R/K^*_s)^\beta\theta_s} \quad \text{for} \quad R<K^*_s\exp(2\lambda) \tag{4a}$$

$$u_1=\frac{K^*_s\exp(2\lambda)}{\theta_s} \quad \text{for} \quad R\geq K^*_s\exp(2\lambda) \tag{4b}$$

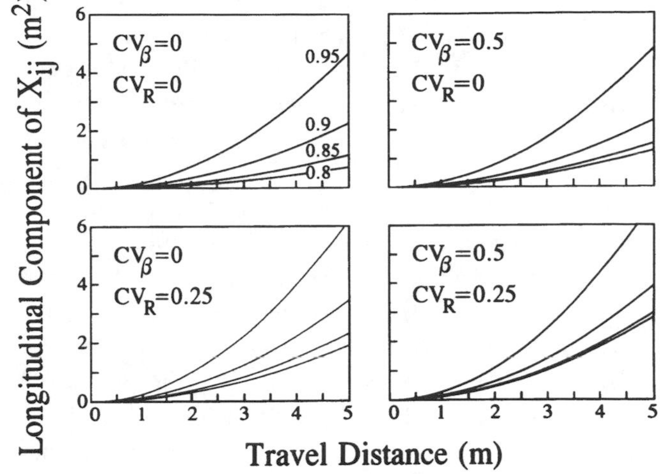

Fig. 3 Longitudinal component of the displacement covariance tensor (5b) as a function of travel distance, calculated for various values of mean water saturation (the numbers labeling the curves), using $\sigma^2_\lambda=0.25$, $\beta=0.14$ and $\theta_s=0.4$.

where $K^*_s=(\int(K_s)^{1/2}f(K_s)dK_s)^2=\langle K_s\rangle\exp(\sigma^2_\lambda)$.

Because of the neglect of the variability in the vertical direction, the longitudinal component of $u_{ij}(\xi)$ in the direction of the mean flow, $u_{11}(\xi_1)=\langle u'_1(x'_1)u'_1(x'_1+\xi_1)\rangle$ is independent of the separation distance, $\xi_1=U_1t$, i.e. $u_{11}(\xi_1)$ has an infinite correlation scale. Inasmuch as $u_2=u_3=0$, the transverse and the off-diagonal components of the particle displacement covariance tensor, X_{ij} $i,j=2,3$, are zero, while the first two moments of the PDF of its longitudinal component, X_{11}, are obtained from the first two moments of the PDF of u_1. They are given by

$$\langle X_1(t)\rangle=t\int_{-\infty}^{\infty}\int_{-\infty}^{\infty} u_1f(\lambda)f(R)d\lambda\, dR \tag{5a}$$

and

$$X_{11}(t)=t^2\int_{-\infty}^{\infty}\int_{-\infty}^{\infty} [u_1-\langle u_1\rangle]^2f(\lambda)f(R)d\lambda\, dR \tag{5b}$$

where $f(\lambda)$ and $f(R)$ are the PDFs of $\lambda=\log\delta$ and of the recharge rate, R, respectively, viewed as independent variates. For normally distributed λ and uniformly distributed R, closed-form expressions for the cumulative distribution of u_1 have been obtained previously and published elsewhere (Dagan & Bresler, 1979). In general, u_1 is a highly skewed variate. At the small limit of the recharge rate variance, $\sigma^2_R\to0$, $\hat{R}$ is deterministic and spatially uniform and u_1 has a bimodal distribution.

The time-dependent longitudinal component of the particle displacement covariance tensor, $X_{11}(t)$ (5b), is illustrated graphically in Fig. 3 for selected values of mean water saturation, $\langle\Theta\rangle$, associated with different values of the ratio $\langle R\rangle/K^*_s$ and the coefficient of variation of the recharge rate,

$CV_R=\sigma_R/\langle R\rangle$. For a given $\langle\Theta\rangle$ and given statistics of logδ, $X_{11}(t)$ increases with CV_R. When the soil parameter β is assumed to be a stationary RSF (e.g. Destouni, 1993), $X_{11}(t)$ will increase also with CV_β. Generally, the dependence of $X_{11}(t)$ on CV_R and CV_β becomes stronger with decreasing $\langle\Theta\rangle$.

For a given $\langle\Theta\rangle$ and given statistics of logδ, R and β, the transport model based on the independent vertical columns approach predicts a purely convection-dominated transport process in which X_{11} is proportional to the square of the travel time. In other words, the longitudinal component of the ensemble macrodispersion tensor, D_{11}, grows linearly with solute residence time as the dispersion process develops. The same result can be obtained by using a nonmechanistic approach (e.g. the transfer function model of Jury (1982)) if an assumption is made that convective solute movement dominates over lateral mixing (Jury, Sposito & White, 1986).

4 THE STOCHASTIC CONTINUUM APPROACH

The stochastic continuum approach (Yeh, Gelhar & Gutjahr, 1985a,b) considered an unbounded flow domain of a partially saturated, heterogeneous soil with a three-dimensional, statistically anisotropic structure. The steady-state flow obeys Darcy's law locally for unsaturated flow, and may be visualized as being generated under a predetermined capillary pressure head imposed on the soil surface, corresponding to a mean water saturation, Θ. Although the mean flow is considered to be vertically unidirectional, because of the three-dimensional variability of the formation properties, perturbations of the capillary pressure head gradient and the velocity fields are three-dimensional. It is assumed further that the local relationships between conductivity, K, and the capillary pressure head, ψ, are isotropic, described by the expression (Gardner, 1958)

$$K(\psi,\mathbf{x})=K_s(x)\exp[-\alpha(\mathbf{x})\psi] \qquad (6)$$

where the relevant soil properties, logK_s and α, are viewed as either cross-correlated or independent, multivariate normal (MVN), second-order stationary RSFs, ergodic over the region of interest. They are characterized by means F and A, respectively, and by three-dimensional, statistically anisotropic covariances with finite correlation scales, I_{fi} and I_{ai}, $i=1,2,3$, respectively.

Equation (6) implies that, for a given mean capillary pressure head $H(\Theta)$, the log-unsaturated conductivity, being a linear combination of logK_s and α, is also MVN. Its PDF is characterized by the mean $Y(\Theta)=F-AH(\Theta)$ and the covariance $C_{yy}(\xi;\Theta)$ given by

$$C_{yy}(\xi;\Theta)=C_{ff}(\xi)+H^2(\Theta)C_{aa}(\xi)-2H(\Theta)C_{fa}(\xi)$$
$$+A^2C_{hh}(\xi;\Theta)-2AC_{fh}(\xi;\Theta)+2AH(\Theta)C_{ah}(\xi;\Theta) \qquad (7)$$

where F and A are the mean values of logK_s and α, respectively, $C_{ff}(\xi)$, $C_{aa}(\xi)$ and $C_{hh}(\xi;\Theta)$ are the covariances of logK_s, α and the capillary pressure head, respectively, and $C_{fa}(\xi;\Theta)$, $C_{fh}(\xi;\Theta)$ and $C_{ah}(\xi;\Theta)$ are the cross-covariances between logK_s and α, logK_s and ψ, and α and ψ, respectively.

In view of the relatively small variability in water saturation as compared with the variability in conductivity, and in order to simplify further the analysis, it is assumed that, for a given mean capillary pressure head H, the water saturation Θ is a deterministic constant, given by (Russo, 1988)

$$\Theta(H)=\left[\exp\left(-\frac{1}{2}AH\right)\left(1+\frac{1}{2}AH\right)\right]^{2/(m+2)} \qquad (8)$$

where $m>-2$ is a parameter that accounts for the dependence of tortuosity and correlation between pores at two different cross-sections of the porous formation, on water saturation.

By employing the aforementioned assumptions and by linearization of Darcy's law for unsaturated flow and for a given water saturation, the first-order approximation of the velocity perturbation is given by (Russo, 1995b)

$$u_i(\Theta)=(K_g(\Theta)/n\Theta)\{J_i[f-aH(\Theta)-Ah(\Theta)]+[\partial h(\Theta)/\partial x_i]\} \qquad (9)$$

Here J_i is the mean head gradient vector, $K_g(\Theta)=\exp[F-AH(\Theta)]$ is the geometric mean conductivity, n is porosity (considered as a deterministic constant), f, a and $h(\Theta)$ are the fluctuations of logK_s, α and $\psi(\Theta)$, respectively, and $i=1,2,3$.

Inasmuch as for a given water saturation Θ, log$K(\Theta)$ is MVN, and $\psi(\Theta)$, through the linearization of the flow equation, is a linear function of log$K(\Theta)$, it follows that, for a given Θ, $\psi(\Theta)$ is also MVN. Then by (9), for a given Θ, so is the velocity, characterized completely by the mean vector $(i=1,2,3)$

$$U_i(\Theta)=K_g(\Theta)J_i/n\Theta \qquad (10)$$

and the covariance tensor, $u_{ij}(\xi)$ $(i,j=1,2,3)$

$$u_{ij}(\xi;\Theta)=\left[\frac{K_g(\Theta)}{n(\Theta)}\right]^2\left[J_iJ_jC_{yy}(\xi;\Theta)-J_i\frac{\partial C_{yh}(\xi;\Theta)}{\partial\xi_j}\right.$$
$$\left.+J_j\frac{\partial C_{yh}(-\xi;\Theta)}{\partial\xi_i}+\frac{\partial^2C_{hh}(\xi;\Theta)}{\partial\xi_i\xi_i}\right] \qquad (11)$$

Here $C_{yh}(\xi;\Theta)$ is the cross-covariance between log$K$ and ψ, and it is given by

$$C_{yh}(\xi;\Theta)=C_{fh}(\xi)-H(\Theta)C_{ah}(\xi)-AC_{hh}(\xi) \qquad (12)$$

For heterogeneous porous formations of axisymmetric anisotropy (i.e. $I_{yv}=I_{y1}$ and $I_{yh}=I_{y2}=I_{y3}$) with mean flow in the vertical direction, perpendicular to the formation bedding, expressions for the components of the velocity covariance tensor (11) were obtained for the general case in which the separation vector is inclined to the mean gradient vector **J** (Russo, 1995b). Expressions for the components of the displacement covariance tensor, $X_{ij}(t;\Theta)$, for mean flow in an arbitrary direction with respect to the principal axes of the formation heterogeneity were obtained by using (2b) and integrating the respective components of the resultant velocity covariance tensor, $u_{ij}(\xi;\Theta)$, along a line inclined to the mean gradient, **J** (Russo, 1995b).

If **J** coincides with the longitudinal axis of $C_{yy}(\xi;\Theta)$ (whose correlation scale is I_{yv}), and assuming that the correlation scales of $\log K_s$ and α are identical, the principal components of X_{ij}, X_{ii}, $i=1,2,3$, are given by

$$X_{ii}(t;\Theta)=\sigma_y^2(\Theta)I_{yv}^2\int_{-1}^{+1} C_i\left[\frac{b_2}{[(d^2+r^2e^{-2})^2-r^2e^{-4}g^2]^2}\right]dr \qquad (13a)$$

Here $d=\sqrt{(1-r^2)}$, $C_1=2(1-r^2)^2$, $C_2=C_3=r^2(1-r^2)e^{-2}$, b_2 is given by

$$b_2=\frac{\exp(|r|Ut)(d^2+r^2e^{-2}-|r|e^{-2}g)^2-(d^2+r^2e^{-2})^2(1+|r|Ut)}{r^2\exp(|r|Ut)}$$

$$-\frac{r^2e^{-4}g^2(1-|r|Ut)}{r^2\exp(|r|Ut)}+\frac{2|r|e^{-2}g(d^2+r^2e^{-2})}{r^2\exp\left[\frac{r^2e^{-2}gUt}{d^2+r^2e^{-2}}\right]} \qquad (13b)$$

$e=I_{yv}/I_{yh}$, $g=(2J_1-1)AI_{yv}$ and $\sigma_Y^2(\Theta)=C_{yy}(0;\Theta)$, the log-conductivity variance, is given by (7) (for $\xi=0$).

The time-dependent $X_{ii}(t;\Theta)$ $(i=1,2,3)$ in (13) are illustrated graphically in Fig. 4, for selected values of Θ. They are consistent with those derived for mean flow in the horizontal direction, parallel to the bedding of saturated formations (Dagan, 1984, 1988). For a given Θ and given statistics of $\log K_s$, the components of X_{ij} increase with increasing $CV_\alpha=\sqrt{\sigma_a^2/A}$. The effect of CV_α on X_{ij} is generally larger with decreasing Θ. For a given Θ and given statistics of $\log K_s$ and α, for unidirectional vertical mean flow (i.e. **J** coincides with the vertical axis, x_1), the transport model based on the stochastic continuum approach predicts a continuous transition from a convection-dominated transport process (for which the principal components of the particle displacement covariance tensor, X_{ii}, $i=1,2,3$, are proportional to the square of the travel time), to a convection-dispersion transport process (for which X_{11} increases linearly with travel time, while the transverse components, X_{22} and X_{33}, approach constant asymptotic values), characterized by an asymptotic, constant macrodispersivity tensor, $D_{ij}(i,j=1,2,3)$, with D_{11} being its only nonzero component.

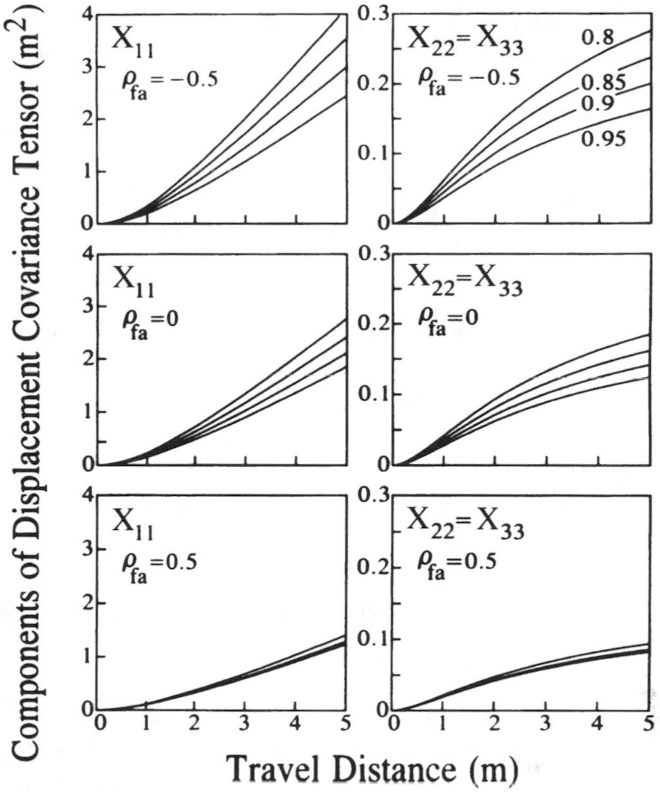

Fig. 4 Principal components, X_{ii}, $i=1, 2, 3$ of the displacement covariance tensor (13), as a function of travel distance, calculated for various values of mean water saturation (the numbers labeling the curves) and ρ_{fa}, using $\sigma_f^2=1$, $\sigma_a^2/A^2=0.25$, $A=0.4$ m^{-1}, $I_{yv}=0.25$ m and $e=I_{yv}/I_{yh}=0.25$.

5 EFFECT OF WATER SATURATION ON SOLUTE SPREADING

In the independent vertical columns approach, for given statistics of $\log\delta$, the variability in velocity is controlled by the magnitude of the recharge rate, R, relative to the magnitude of saturated conductivity, K_s. For given statistics of R, when $\langle R\rangle$ is larger than $\langle K_s\rangle$, the variability in the velocity is mainly controlled by the variability in K_s. As the ratio $\langle R\rangle/\langle K_s\rangle$ decreases (and mean water saturation decreases), however, the variability in the velocity decreases rapidly, because, for an increasing portion of the flow domain, only Θ contributes to the variability in velocity, and the variability in Θ (which slightly increases with CV_β, when β is considered as a stationary RSF) is negligibly small as compared with $\log K_s$. For given statistics of $\log\delta$, R and β, therefore, the transport model based on the independent vertical columns approach predicts an increase in solute spreading with increasing water saturation (Fig. 3).

In the stochastic continuum approach, for given statistics of $\log K_s$ and α, the variability in velocity is controlled by water saturation owing to its effect on the variability in unsaturated hydraulic conductivity, K, and the variability in capillary pressure head gradients, $\partial\psi/\partial x_i$, $i=1,2,3$. The effect of Θ on the variability in velocity depends on the cross-correlation coefficient ρ_{fa} between $\log K_s$ and α (Russo, 1995a). When $\rho_{fa}\leq 0$, $u_{ij}(\xi;\Theta)$ is a monotonic decreasing function of the water saturation Θ. For $\rho_{fa}>0$, however, $u_{ij}(\xi;\Theta)$ is a nonmonotonic function of Θ which increases with decreasing Θ when $\Theta<\Theta_c(H_c)$, where H_c is given (Russo, 1995a) by $H_c=\rho_{fa}/A\sqrt{(\sigma_a^2/\sigma_f^2)}$, while the converse is true when Θ is greater than Θ_c. For given statistics of $\log K_s$ and α, therefore, the transport model based on the stochastic continuum approach generally predicts an increase in solute spreading with decreasing water saturation (Fig. 4). However, when $\log K_s$ and α are positively correlated ($\rho_{fa}>0$), and $\Theta>\Theta(H_c)$, solute spreading increases with increasing water saturation.

6 ASYMPTOTIC MACRODISPERSION COEFFICIENTS

Asymptotic constant macrodispersion coefficients, D_{ij} ($i,j=1,2,3$), determine the mean dispersive flux after the center of mass of the solute plume has traveled a few tens of log-conductivity correlation scales and the transport approaches a Fickian behavior. First-order approximations of D_{ij} ($i,j=1,2,3$) for mean flow in an arbitrary direction with respect to the principal axes of the formation heterogeneity can be derived by means of the transport model based on the stochastic continuum approach (Russo, 1995b).

Principal and off-diagonal components of the macrodispersivity coefficients, $D_{ij}(\Theta)/U$ ($i,j=1,2,3$), are displayed in Figs. 5 and 6, respectively, as functions of the angle β' between **J** and the vertical axis x_1 and for selected values of $A=\langle\alpha\rangle$ and water saturation, Θ. Furthermore, the angle between the projection of **J** on the horizontal x_2x_3 plane and the x_2 axis is $\alpha'=\pi/4$. The results shown in Figs. 5 and 6 suggest that, under unsaturated flow, for given statistics of the formation properties, macrodispersion will diminish in formations of coarser texture (larger $A=\langle\alpha\rangle$) and will increase with decreasing water saturation, Θ. Furthermore, for given A and Θ, the longitudinal component of the displacement covariance tensor decreases and its transverse components increase when **J** is more inclined to the longitudinal axis of the formation heterogeneity.

Nonzero off-diagonal components of D_{ij} (Fig. 6) imply that the principal axes associated with the principal components of D_{ij} are deflected in the same direction (positive D_{12} and/or D_{13}), or in a direction opposite (negative D_{12} and/or D_{13}) to that of **J** relative to the principal axes of $C_{yy'}(\xi;\Theta)$. Consequently, for a given Θ, when **J** does not coincide with the principal axes of the heterogeneous formation, soil materials of coarser texture (larger A) will diminish the deflection of the principal axes of the macrodispersion tensor, D_{ij}, in the same direction as that of **J**, and may enhance their deflection in a direction opposite to that of **J**, relative to the principal axes of the formation heterogeneity.

The effect of Θ on D_{ij} is explained by its impact on the variability in unsaturated hydraulic conductivity and capillary pressure head gradients and, concurrently, on the variability in velocity, as discussed earlier. The effect of the soil texture (expressed in terms of A) on D_{ij} is explained as followed by (Russo, 1995a,b); an increase in $A=\langle\alpha\rangle$ expresses a transition from a fine-textured soil material, associated with significant

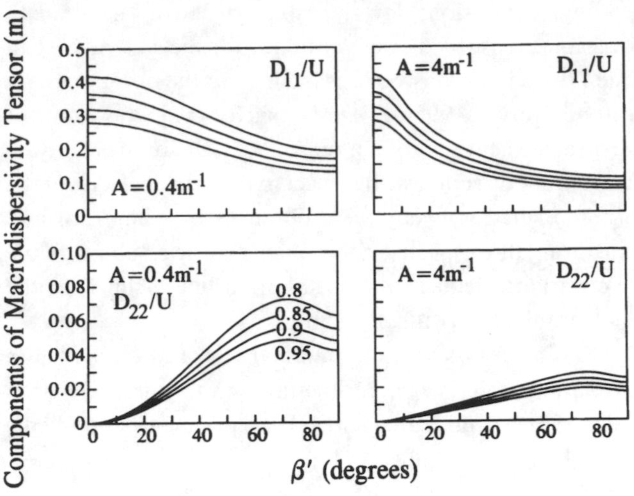

Fig. 5 Principal components of the asymptotic macrodispersion tensor, $D_{jj'}$, as a function of the angle β', calculated for the angle $\alpha'=\pi/4$ and various values of mean water saturation (the numbers labeling the curves) and A, using $\sigma_f^2=1$, $\sigma_a^2/A^2=0.25$, $\rho_{fa}=0$, $I_{yv}=0.25$ m and $e=I_{yv}/I_{yh}=0.25$.

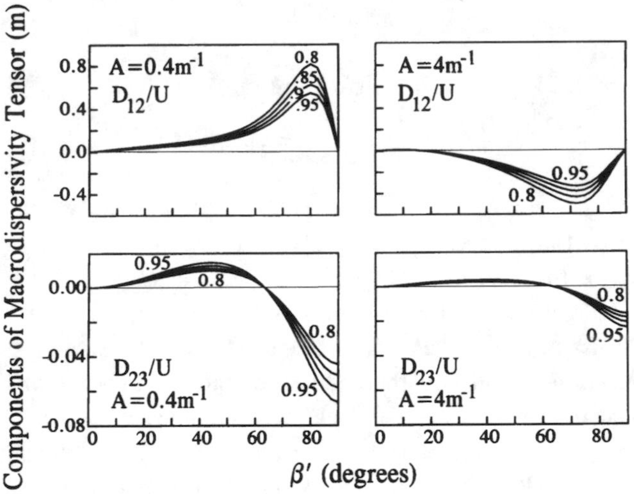

Fig. 6 Off-diagonal components of the asymptotic macrodispersion tensor, $D_{jj'}$, as a function of the angle β', calculated for the angle $\alpha'=\pi/4$ and various values of mean water saturation (the numbers labeling the curves) and A, using $\sigma_f^2=1$, $\sigma_a^2/A^2=0.25$, $\rho_{fa}=0$, $I_{yv}=0.25$ m and $e=I_{yv}/I_{yh}=0.25$.

capillary forces, to a coarse-textured soil material, associated with negligible capillary forces. Therefore, as A increases, the capillary pressure head gradients fluctuations, $\partial h/\partial x_i$ (i=1,2,3), vanish and the flow becomes gravity dominated. Furthermore, for a given Θ, the magnitude of $C_{y'y'}(0;\Theta)$ decreases while the persistence of $C_{yy}(\xi;\Theta)$ increases with increasing A. Inasmuch as the first term of (9), which dominates u_1, decreases with A at a rate slower than that of $\partial h/\partial x_i$ (i=1,2,3) the transverse components of u_{ij} and D_{ij} are more sensitive to changes in A as compared with the longitudinal components of u_{ij} and D_{ij}. The discrepancy between the sensitivity of u_1 and u_i (u_i=2,3) to $A=\langle\alpha\rangle$, however, diminishes with decreasing e.

7 SIMULATION OF SOLUTE TRANSPORT

Simulation is a powerful tool that can be regarded as a 'numerical experiment', which, in turn, can provide much more detailed information (that is unaffected by measurement errors or inadequate sampling) about the effects of spatial variability in soil properties on the response of the flow system than that which is attainable in practice from field investigations. Furthermore, results of transport simulations may be used to test the applicability of current stochastic vadose-zone transport models to more realistic field situations.

The transport of a conservative, nonreactive solute in a vertical cross-section of a hypothetical, partially saturated, heterogeneous soil, under various flow regimes, was simulated recently by Russo, Zaidel & Laufer (1994). It was assumed that water flow and solute transport are described locally by Richards' equation and by the CDE, respectively, and that, locally, the constitutive relationships for unsaturated flow are given by the model of van Genuchten (1980). Parameters of this model were taken as realizations of RSFs characterized by anisotropic, axisymmetric, exponential covariances. The simulations were performed by combining a statistical generation method for producing realizations of the heterogeneous formation properties with sufficient resolution, combined with an efficient numerical method for solving the partial differential equations governing flow and transport. Different upper boundary conditions for water flow were considered in a series of simulations. It should be emphasized that the simulated transport processes were based on single realizations of the input formation properties and the output flow-attributed variables. Since the size of the solute inlet zone extended over 17 correlation scales of the input formation properties in the transverse horizontal direction, it was assumed that ergodicity requirements were at least partially satisfied. For more details see Russo *et al.* (1994).

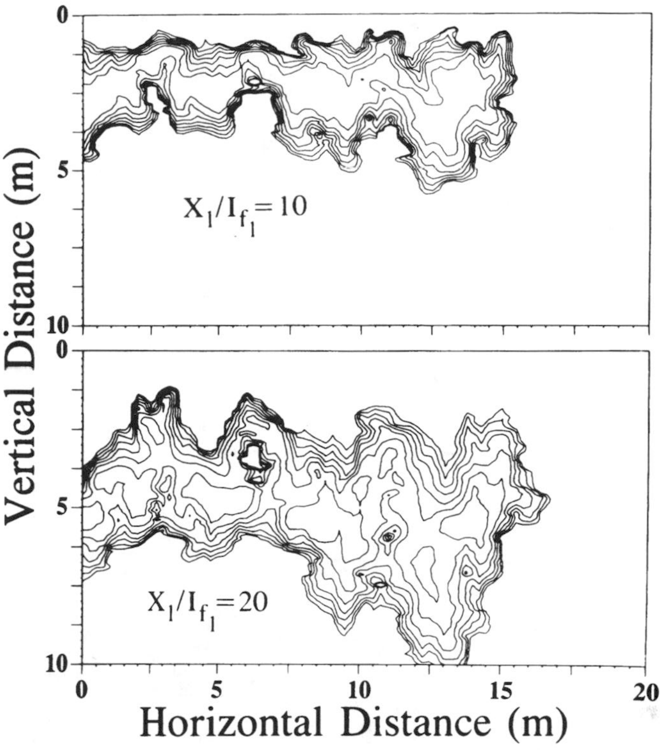

Fig. 7 Contours of the simulated point concentrations under intermediate infiltration (from Russo *et al.*, 1994). Results are depicted for Δt=3 days and two different scaled travel distances, $X_1'=X_1/I_{f1}$. The exterior contour is $0.5\langle c(\mathbf{x};\, t)\rangle$ and contour spacing is $0.10\langle c(\mathbf{x};\, t)\rangle$. Vertical exaggeration 1.15×.

Simulated solute spreading is demonstrated in Fig. 7. This figure displays contour lines of the simulated solute concentrations in the vertical cross-section of the heterogeneous soil, obtained under periodic influx characterized by a constant time interval between successive water applications, Δt=3 days, for two different scaled travel distances, $X_1'=X_1/I_{f1}$. The simulated results depicted in Fig. 7 clearly demonstrate the considerable spatial variability in the solute concentration due to the complex spatial heterogeneity of the soil hydraulic properties, in agreement with field data (see Fig. 2 data from Schulin *et al.*, 1987).

The principal components of the dimensionless displacement covariance tensor, $X_{ii}'(t)=X_{ii}(t)/\sigma_f^2 I_{f1}$ (i=1,2) calculated from the spatial moments of the distribution of the simulated solute concentrations under both quasi-steady-state and transient, nonmonotonic flow regimes (depicted in figures 3 and 4 in Russo *et al.*, 1994, respectively) are depicted in Figs. 8 and 9 as functions of dimensionless travel time, $t'=Ut/I_{f1}$. Under both quasi-steady-state (Fig. 8) and transient, nonmonotonic (Fig. 9) flow regimes, the lower mean water saturation (established either by decreasing the capillary pressure head at the entry zone or by increasing the time interval, Δt, between successive water applications, respectively) is shown to increase the simulated solute spreading, in

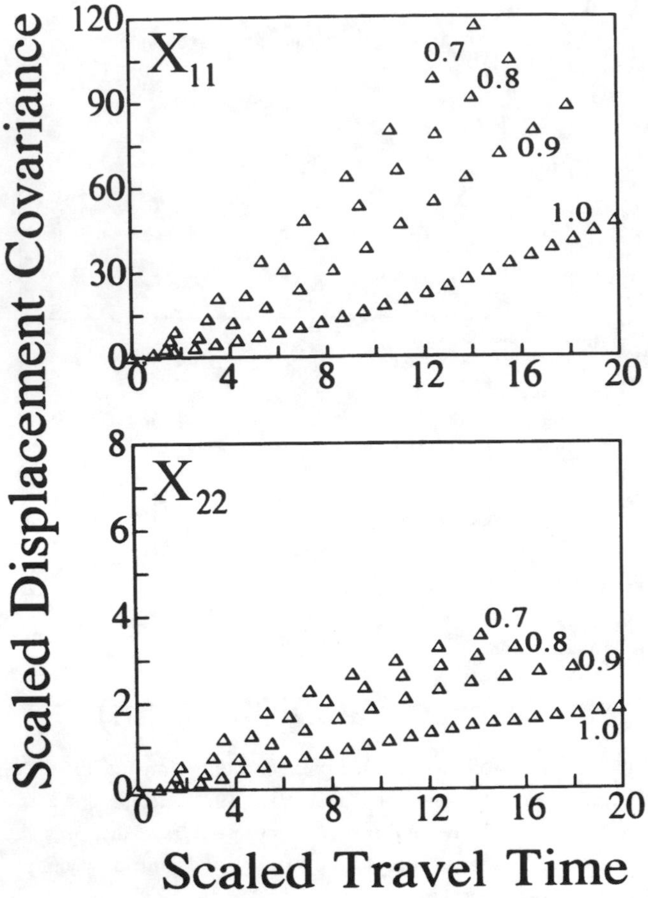

Fig. 8 Scaled longitudinal ($X'_{11}=X_{11}/I^2_{f1}\sigma^2_f$) and transverse ($X'_{22}=X_{22}/I^2_{f1}\sigma^2_f$) components of the displacement covariance tensor as a function of scaled travel time, $t'=tU_1/I_{f1}$, calculated from the spatial moments of the distribution of the simulated point concentrations under continuous infiltration (Russo *et al.*, 1994), associated with various water saturations, Θ_0, denoted by the numbers labeling the curves.

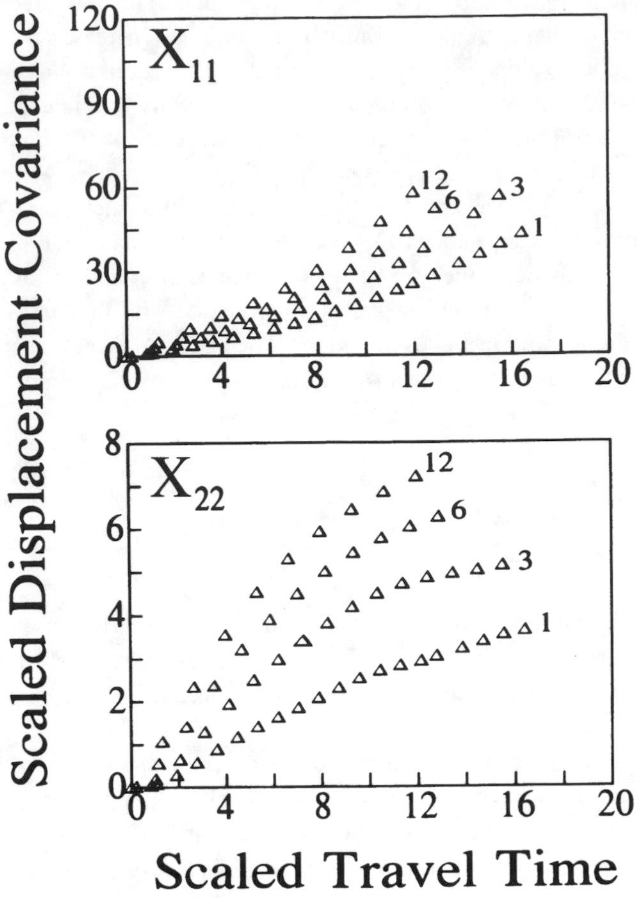

Fig. 9 Scaled longitudinal ($X'_{11}=X_{11}/I^2_{f1}\sigma^2_f$) and transverse ($X'_{22}=X_{22}/I^2_{f1}\sigma^2_f$) components of the displacement covariance tensor as a function of scaled travel time, $t'=tU_1/I_{f1}$, calculated from the spatial moments of the distribution of the simulated point concentrations under intermediate infiltration (Russo *et al.*, 1994), associated with various time intervals between successive water applications, Δt, denoted by the numbers labeling the curves, in days.

agreement with the predictions based on the stochastic continuum approach (Russo, 1993, 1995a,b).

The simulation results for the transient, nonmonotonic flows caused by periodic influx at the entry zone (Fig. 9) reveal that longitudinal spreading of the solute body is restricted and that its transverse spreading is enhanced with increasing Δt. This finding is explained by the fact that, during the redistribution periods, the flow regime is relaxed from the essentially unidirectional mean flow (i.e. $J_2/J_1 \rightarrow 0$) imposed by the uniform influx during the infiltration periods. Consequently, during the former periods, the unsaturated conductivity and capillary pressure head gradients and, concurrently, the velocity, are adjusted to the spatial heterogeneity in the formation properties, creating a multidimensional flow pattern, in which J_2/J_1 increases with increasing Δt (see table 3 in Russo *et al.*, 1994). The increase in transverse spreading and the decrease in longitudinal spreading with increasing J_2/J_1 are in agreement with the predictions based on the stochastic continuum approach (Russo, 1995b).

The simulation study of Russo *et al.* (1994) demonstrated the impact of the upper boundary conditions for the flow on both flow and transport through partially saturated and heterogeneous formations. Under conditions of a predetermined capillary pressure head, ψ_0, imposed on the entry zone throughout the flow event, the distribution of velocity is controlled mainly by the distribution of the unsaturated conductivity, which, in turn, is controlled by the spatial distribution of the formation properties and the water saturation corresponding to ψ_0. These conditions are consistent with the unbounded flow domain considered in the stochastic continuum approach.

Under conditions of steady-state influx, R, imposed on the entry zone, and if R is small compared with $\langle K_s \rangle$, the capillary pressure head gradients must be adjusted to compensate for the variability in $K(\Theta)$ as the flow proceeds, in order to enable the constraint imposed by R to be met. Consequently, the variability in the flow velocity is considerably restrained. These conditions prevail in the simulation study of Tseng & Jury (1994), and are consistent with the upper boundary conditions considered in the independent vertical columns approach. These latter boundary conditions, however, should be distinguished from conditions of a periodic influx imposed on the entry zone. During redistribution periods between successive water applications, the flow regime is relaxed from the restraint imposed by the uniform influx which prevails during the infiltration periods and is adjusted to the spatial heterogeneity in the formation properties. Consequently, the variability in unsaturated conductivity and pressure head gradients and, concurrently, in velocity, increases with increasing time intervals between successive water applications (Russo *et al.*, 1994).

8 SUMMARY AND CONCLUDING REMARKS

This chapter discusses the problem of solute transport through partially saturated, heterogeneous porous formations, addressing vadose-zone transport models based on different models of flow through heterogeneous formations. The starting point is the considerable spatial variability in point values of the solute concentration, attributed to the complex spatial heterogeneity of the soil hydraulic properties. This has been observed in field experiments (e.g. Schulin *et al.*, 1987; Butters, Jury & Ernst, 1989; Ellsworth *et al.*, 1991) and demonstrated in simulations (e.g. Russo, 1991; Russo *et al.*, 1994) of transport in heterogeneous soils. The emphasis in this chapter, therefore, is on prediction of integrated measures of the transport, which are subject to a much lesser degree of uncertainty than are the point values.

The main conclusions of this chapter are summarized below:

(1) Transport models which are based on the independent vertical columns approach (e.g. Dagan & Bresler, 1979; Destouni & Cvetkovic, 1989; Destouni, 1993) are restricted to solute spreading in the longitudinal direction only. For a given mean water saturation, Θ, they predict a purely convection-dominated transport process for which the longitudinal component of the macrodispersion tensor, D_{11}, increases linearly with travel time; for a given travel time, they predict an increase in solute spreading with increasing Θ.

(2) For a given Θ, transport models which are based on the stochastic continuum approach (e.g. Russo, 1993, 1995a,b) predict a continuous transition from a convection-dominated transport process to a convection-dispersion transport process characterized by an asymptotic, constant macrodispersivity tensor, D_{ij} $(i,j=1,2,3)$, the magnitude of whose components depends on the direction of the mean gradient vector relative to the principal axes of the formation heterogeneity. For a given travel time, they predict an increase in solute spreading with decreasing Θ.

(3) Results of the latter transport models are in qualitative agreement with results of numerical simulations of flow and transport in a partially saturated, heterogeneous formation (Russo *et al.*, 1994); for a given mean, Θ, they are consistent with results obtained by the theory of transport by groundwater flow (e.g. Gelhar & Axness, 1983; Dagan, 1984, 1988).

(4) Analyses of flow and transport based on the stochastic continuum approach demonstrate the importance of the soil parameter α. For given statistics of $\log K_s$ and given water saturation, transport in unsaturated flow will approach asymptotic Fickian behavior more slowly as $\mathbf{I}_a$ and/or $\rho_{fa}(\rho_{fa}>0)$ increase(s), while the magnitude of the components of the asymptotic macrodispersion tensor, D_{ij}, will decrease with decreasing σ_a^2 and $\mathbf{I}_a$ and increasing $\rho_{fa}(\rho_{fa}>0)$ and $A=\langle\alpha\rangle$. Larger A (coarser soil material) will also diminish the deflection of the principal axes associated with the principal components of D_{ij} in the same direction as that of $\mathbf{J}$, and may enhance their deflection in a direction opposite to that of $\mathbf{J}$, relative to the principal axes of the formation heterogeneity, when $\mathbf{J}$ does not coincide with the latter.

Unfortunately, detailed experimental information on the spatial behavior of the relevant formation properties (in particular the soil parameter α) is limited. Furthermore, controlled, field-scale, transport experiments, which could be used to validate existing stochastic transport models, are also limited in number. There is an immediate need, therefore, for additional field studies to characterize the spatial variability in soil properties relevant to vadose-zone transport, and for controlled, field-scale, transport experiments (and simulations) further to validate existing stochastic transport models. Inasmuch as stochastic models of solute transport predict the ensemble-average spread of a solute body, while field experiments provide information on the spread of a single solute body for a particular, site-specific application (i.e. a single realization from an ensemble of many plausible realizations), care should be taken in the design of transport experiments.

Finally, we believe that the analyses presented in this chapter will improve our understanding of field-scale solute spread in heterogeneous, partially saturated porous formations and will serve as a basis for the subsequent development of field-scale solute transport models for more complex, real-world scenarios.

ACKNOWLEDGMENTS

This is contribution no. 1526–E, 1994 series, from the Agricultural Research Organization, The Volcani Center, Bet Dagan, Israel. The author is grateful to Mr Asher Laufer for his technical assistance during this study.

REFERENCES

Bresler, E. & Dagan, G. (1981). Convective and pore scale dispersive solute transport in unsaturated heterogeneous fields. *Water Resources Research*, 17, 1683–1689,

Brooks, R. H. & Corey, A. T. (1964). Hydraulic properties of porous media. *Hydrol. Pap.* 3, Fort Collins, CO: Colorado State University.

Butters, G. L., Jury, W. A. & Ernst, F. F. (1989). Field scale transport of bromide in an unsaturated soil. 1. Experimental methodology and results. *Water Resources Research*, 25(7), 1575–1581.

Dagan, G. (1982). Stochastic modeling of groundwater flow by unconditional and conditional probabilities. 2. The solute transport. *Water Resources Research*, 18(4), 835–848.

Dagan, G. (1984). Solute transport in heterogeneous porous formations. *Journal of Fluid Mechanics*, 145, 151–177.

Dagan, G. (1988). Time-dependent macrodispersion for solute transport in anisotropic heterogeneous aquifers. *Water Resources Research*, 24(9), 1491–1500.

Dagan, G. (1989). *Flow and Transport in Porous Formations*. New York: Springer-Verlag.

Dagan, G. & Bresler, E. (1979). Solute transport in unsaturated heterogeneous soil at field scale, 1. Theory. *Soil Science Society of America Journal*, 43, 461–467.

Destouni, G. (1993). Stochastic modeling of solute flux in the unsaturated zone at the field scale. *Journal of Hydrology*, 143, 45–61.

Destouni, G. & Cvetkovic, V. (1989). The effect of heterogeneity on large scale solute transport in the unsaturated zone. *Nordic Hydrology*, 20, 43–52.

Destouni, G. & Cvetkovic, V. (1991). Field scale mass arrival of sorptive solute into the groundwater. *Water Resources Research*, 27(6), 1315–1325.

Ellsworth, T. R., Jury, W. A., Ernst, F. F. & Shouse, P. J. (1991). A three-dimensional field study of solute transport through unsaturated layered porous media, 1. Methodology, mass recovery and mean transport. *Water Resources Research*, 27(5), 951–965.

Gardner, W. R. (1958). Some steady state solutions of unsaturated moisture flow equations with application to evaporation from a water table. *Soil Science*, 85, 228–232.

Gelhar, L. W. & Axness, C. (1983). Three-dimensional stochastic analysis of macrodispersion in aquifers. *Water Resources Research*, 19(1), 161–190.

Jury, W. A. (1982). Simulation of solute transport using a transfer function model. *Water Resources Research*, 18(2), 363–368.

Jury, W. A., Sposito, G. & White, R. E. (1986). The transfer function model of solute transport through soil. I. Fundamental concepts. *Water Resources Research*, 22(2), 243–247.

Russo, D. (1988). Determining soil hydraulic properties by parameter estimation: On the selection of a model for the hydraulic properties. *Water Resources Research*, 24(3), 453–459.

Russo, D. (1991). Stochastic analysis of vadose-zone solute transport in a vertical cross section of heterogeneous soil during nonsteady water flow. *Water Resources Research*, 27(3), 267–283.

Russo, D. (1993). Stochastic modeling of macrodispersion for solute transport in a heterogeneous unsaturated porous formation. *Water Resources Research*, 29(2), 383–397.

Russo, D. (1995a). On the velocity covariance and transport modeling in heterogeneous anisotropic porous formations II. Unsaturated flow. *Water Resources Research*, 31(1), 139–145.

Russo, D. (1995b). Stochastic analysis of the velocity covariance and the displacement covariance tensors in partially saturated heterogeneous anisotropic porous formations. *Water Resources Research*, 31(7), 1647–1658.

Russo, D. & Bouton, M. (1992). Statistical analysis of spatial variability in unsaturated flow parameters. *Water Resources Research*, 28(7), 1911–1925.

Russo, D., Zaidel, J. & Laufer, A. (1994). Stochastic analysis of solute transport in partially saturated heterogeneous soil: I. Numerical experiments. *Water Resources Research*, 30(3), 769–779.

Schulin, R., van Genuchten, M. Th., Fluhler, H. & Ferlin, P. (1987). An experimental study of solute transport in a stony field soil. *Water Resources Research*, 23(9), 1785–1794.

Tseng, P. H. & Jury, W. A. (1994). Comparison of transfer function and deterministic modeling of area-average solute transport in heterogeneous field. *Water Resources Research*, 30(7), 2051–2064.

van Genuchten, M. Th. (1980). A closed-form equation for predicting the hydraulic conductivity of unsaturated soils. *Soil Science Society of America Journal*, 44, 892–898.

Yeh, T.-C., Gelhar, L. W. & Gutjahr, A. L. (1985a). Stochastic analysis of unsaturated flow in heterogeneous soils, 1. Statistically isotropic media. *Water Resources Research*, 21(4), 447–456.

Yeh, T.-C., Gelhar, L. W. & Gutjahr, A. L. (1985b). Stochastic analysis of unsaturated flow in heterogeneous soils, 2. Statistically anisotropic media with variable α. *Water Resources Research*, 21(4), 457–464.

3 Field-scale modeling of immiscible organic chemical spills

JACK C. PARKER

Environmental Systems & Technologies, Inc.

1 INTRODUCTION

Problems involving the flow of multiple fluid phases in porous media arise in a number of scientific and engineering disciplines. In electrical engineering, packed bed reactors are widely used to facilitate various homogeneous (single-phase) as well as heterogeneous (multiphase) reactions. In certain processes, coexisting liquid and gas phases occur which call for the application of multiphase flow models to facilitate design optimization. Historically, the greatest impetus for development of models for multiphase flow has come from petroleum engineering, induced by the lure of more efficient oil and gas recovery from hydrocarbon reservoirs. Early investigations tackled the problem of two-phase flow in gas–oil and oil–water systems, facilitating subsequent development of models for three-phase flow of gas, oil, and water in reservoir rocks.

Also interested in flow in geologic media, but operating at shallower depths, hydrologists, agricultural engineers, and soil physicists have for many years been concerned with fluid movement in air–water porous media systems prompted by motives ranging from groundwater resource utilization to optimizing water use by agricultural crops. More recently, environmental concerns have led to an increased interest in the movement of fluids in the so-called vadose or unsaturated zone interposed between the atmosphere and the groundwater (saturated zone).

Many potential groundwater contaminants are introduced at or near the soil surface via atmospheric deposition, spills, leakage from underground tanks, subsurface waste disposal, etc. Soluble components may migrate in the aqueous phase through the vadose zone to groundwater. Volatile components may move by mass flow and diffusion in the gas phase. Furthermore, a wide class of environmental contaminants consists of organic compounds of low water solubility which can occur as a separate nonaqueous phase liquid (NAPL) in the soil. Such liquids include many widely used industrial solvents and automobile and jet fuels, which unfortunately often enter the ground via surface spills or leaks from underground storage tanks. As in the general petroleum reservoir engineering problem, analyses of such systems require consideration of three mobile fluid phases: water, air, and organic liquid.

In this chapter, we will review the fundamentals of modeling multiphase flow and transport in porous media, discuss the constitutive relations required to solve the governing equations, discuss limitations and areas of uncertainty that exist in the present state of knowledge, and present a practical modeling approach that can be used for certain types of field problems.

2 BASICS OF MULTIPHASE FLOW

2.1 Generalized Darcy equation

Fluid flow in porous media normally occurs at velocities that are below the threshold for turbulent flow. Under such conditions, flow is generally described satisfactorily by a generalized form of the equation first set out for the case of single-phase flow by the French engineer Darcy in the nineteenth century. In its generalized form for multiphase systems (e.g. Aziz & Setari, 1979; Greenkorn, 1983), Darcy's law has the form

$$q_{pi}=\frac{k_{rp}k_{ij}}{\eta_p}\left[\frac{\partial P_p}{\partial x_j}+\rho_p g e_j\right] \tag{1}$$

where i and j are direction indices (i,j=1,2,3), with repeated values indicating summation in tensor notation, x_i (or x_j) is the ith (or jth) Cartesian coordinate, q_{pi} is the volume flux density or Darcy velocity of fluid phase p in the i direction [L^3,L^{-1},T^{-1}], k_{ij} is the intrinsic permeability tensor of the porous medium [L^2], k_{rp} is the relative permeability to phase p [L^0] which varies from 0 when no p-phase is present to 1 when the medium is saturated with p-phase, η_p is the absolute viscosity of phase p, ρ_p is the density of phase p [LM^{-3}], g is gravitational acceleration [LT^{-2}], and $e_j = \partial z / \partial x_j$ is the j component of a unit gravitational vector, where z is elevation.

In petroleum reservoir engineering, (1) is the commonly employed form of Darcy's law. In groundwater hydrology it is more common to utilize fluid heads, rather than pressures, yielding the form

$$q_{pi} = -(K_{rp} k_{sp_{ij}}) \left[\frac{\partial h_p}{\partial x_j} + \rho_{rp} e_j \right] \tag{2a}$$

in which

$$K_{sp_{ij}} = \mathrm{k}_{p_{ij}} \rho_R g / \eta_p \tag{2b}$$

$$\rho_{rp} = \rho_p / \rho_R \tag{2c}$$

$$h_p = P_p / \rho_R g \tag{2d}$$

where $K_{sp_{ij}}$ is the saturated conductivity of phase p [LT^{-1}], h_p is the fluid pressure head [L], and ρ_{rp} is the ratio of the fluid density to that of the reference fluid of density ρ_R. If the reference density is taken to be that of water at standard conditions for all phases, then h_p represents the water height equivalent pressure of phase p.

Equations (1) and (2) are empirical in nature and invoke several implicit assumptions. A fundamental assumption is that flow of the p-phase is not directly affected by pressure gradients in other phases. This, it may be noted, is not absolutely true, but requires that slippage zones at phase interfaces are thin relative to the total film thicknesses of the phases. The assumption may be increasingly prone to breakdown in fine-grained porous media and at low fluid saturations. However, since phase permeabilities ($K_{pij} = K_{rp} K_{ij}$) will become extremely low under such circumstances, this should be a largely moot issue from a practical point of view.

Another critical assumption is the validity of the concept of intrinsic permeability, which purports to separate fluid-dependent and porous medium-dependent effects on fluid flow. The assumption that intrinsic permeability is a unique (tensorial) characteristic of the porous medium is reasonably justifiable for rigid granular soils which do not exhibit swelling or consolidation in response to interaction with fluids. For fine-grained materials the assumption is again prone to break down, as such materials may exhibit order of magnitude differences in 'intrinsic' permeability when saturated with different fluids (Acar *et al.*, 1985). Furthermore, the notion that the tensorial nature of fluid conduction can be relegated to k_{ij} while relative permeability remains a scalar has little to justify itself beyond convenience. Recent evidence in fact indicates that relative permeability may have a tensorial nature that varies with fluid saturation (Bear, Braester & Menier, 1987; Mantoglou & Gelhar, 1987). This is unfortunate because it adds considerable complication to an already difficult problem. Future studies will be required to resolve the importance of these effects and to develop practical methods for accommodating them if necessary.

2.2 Continuity equations

To model an N-phase system, N-phase continuity equations are needed to stipulate phase mass conservation constraints. To avoid the mathematical complications of translating coordinate systems, as well as the troublesome practical problem of how to relate porous media volume changes to fluid pressures, we shall assume the porous medium itself to be incompressible. This is not a mandatory assumption, and a substantial body of literature exists which addresses the problem of matrix compressibility. However, such investigations are peripheral to the aim of this chapter, so we will circumvent them. For an incompressible porous medium, the fluid-phase continuity relations are of the form

$$\phi \frac{\partial(\rho_p S_w)}{\partial t} = -\frac{\partial q_{pi}}{\partial x_i} + \gamma_p \tag{3}$$

where ϕ is the medium porosity, S_p is the fraction of the pore space containing phase p (phase saturation), ρ_p is the phase density, q_{pi} is the Darcy velocity of the p-phase, and γ_p [$ML^{-3}T^{-1}$] is a source–sink term due to a transfer of mass between phases.

Substituting Darcy's equation of q_p yields

$$\phi = \frac{\partial(\rho_p S_p)}{\partial t} = \frac{\partial}{\partial x_i} \left[\rho_p K_{pij} \left(\frac{\partial h_p}{\partial x_j} + \rho_{rp} e_j \right) \right] + \gamma_p \tag{4}$$

Fluid compressibility can generally be described adequately by a linear relation of the form

$$\rho_p = \rho_p^{ref} + a_p h_p \tag{5}$$

where a_p is the p-phase compressibility and ρ_p^{ref} is the reference at $h_p = 0$. For gas phase flow, assuming ideal gas behavior yields a good approximation for a_p as

$$a_p = \rho_R g M_p / RT \tag{6}$$

where M_p is the average molecular weight of the phase [M mol^{-1}], R is the ideal gas constant, and T is absolute temperature. For liquids near atmospheric pressure, $a_p \rightarrow 0$ and ρ_p may be factored from (4) with negligible error if fluid composition effects on phase density may also be disregarded. In compositional models, consideration may be given to temporal variations in phase density associated with changes in phase composition caused by mass transfer between phases and subsequent transport by convection and diffusion. Such effects will be considered later in conjunction with the discussion of component transport.

Considering now the specific case of a three-fluid phase porous media system with water (w), organic liquid (o), and air (a), (4) becomes the system of phase conservation equations given by

$$\phi\frac{\partial(\rho_w S_w)}{\partial t}=\frac{\partial}{\partial x_i}\left[\rho_w K_{wij}\left(\frac{\partial h_w}{\partial x_i}+\rho_{rw}e_j\right)\right]+\gamma_w \tag{7a}$$

$$\phi\frac{\partial(\rho_o S_o)}{\partial t}=\frac{\partial}{\partial x_i}\left[\rho_o K_{oij}\left(\frac{\partial h_o}{\partial x_i}+\rho_{ro}e_j\right)\right]+\gamma_0 \tag{7b}$$

$$\phi\frac{\partial(\rho_a S_a)}{\partial t}=\frac{\partial}{\partial x_i}\left[\rho_a K_{aij}\left(\frac{\partial h_a}{\partial x_i}+\rho_{ra}e_j\right)\right]+\gamma_a \tag{7c}$$

Since, in general, $S_p=S_p(h_w,h_o,h_a)$, the left-hand sides of eqs. (7) may be expanded in terms of pressure head time derivatives of all phases, and a system of coupled partial differential equations results. Coupling arises because of the implicit occurrence of $\partial h_p/\partial t$ terms for all phases in each equation, as well as because of the interdependence of k_{rp} and $\partial S_p/\partial h_q$ terms ($p,q=a,o,w$) on phase pressures, as will be discussed shortly. If mass transfer between phases is disregarded, the γ_p terms in eqs. (7) may be dropped. Otherwise an additional set of equations must be introduced to describe component transport in each phase, as will be discussed in a later section.

Under certain circumstances, the system of flow equations given by (7) may be simplified. In the complete absence of one or more fluid phases from the model system, the pertinent equations may be dropped entirely. For example, in the absence of an organic liquid phase, (7b) would be eliminated from the system of equations, and no $\partial h_o/\partial t$ terms would occur on expansion of saturation derivatives for (7a) and (7c). Even in the presence of a gas phase, elimination of the gas equation is sometimes justified. Under natural conditions in the zone of aeration above a water table aquifer (the 'vadose zone'), air pressure is generally controlled at the soil–atmosphere boundary by atmospheric pressure, which normally varies over a small range. Furthermore, gas phase conductivities at a given phase saturation will be markedly greater than those of the liquids because of low gas viscosities. For example, the ratio of air to water viscosity at 20 °C is about 0.018. The combination of these two factors tends to make gas phase pressure gradients in the vadose zone small under natural conditions, if the gas phase remains continuous with the atmosphere. That is, the gas phase pressure will remain nearly constant at atmospheric pressure. Coupling between the water and the gas phase flow equations thus becomes small, and gas phase flow need not be considered at all to model liquid phase flow. Elimination of the gas flow equation assuming $\partial h_a/\partial t\rightarrow 0$ and $\partial h_a/\partial x_i\rightarrow 0$ is often referred to as Richards' approximation and is the basis for conventional analyses of two-phase air-water flow in the vadose zone.

Assuming the validity of the intrinsic permeability hypothesis, coefficients required to model multiphase flow include phase viscosities (η_p); reference state densities ρ_r^{ref} and

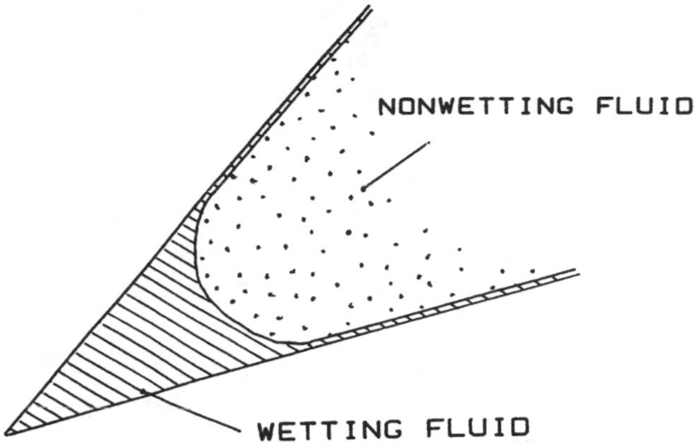

Fig. 1 Idealized pore cross-section with two fluids.

compressibilities (a_p); porous medium porosity (ϕ); intrinsic permeability (k) or saturated conductivity of a single phase (K_{sp}); and functional relations between fluid saturations (S_p), pressure heads (h_p), and relative permeabilities (k_{rp}). We turn next to the characterization of the latter functional relationships.

2.3 Two-phase saturation–capillary pressure relations

When two fluids coexist in a porous medium, one fluid will generally have preferential wettability for the solid phase, and will occupy the smaller voids, while the less wetting fluid is consigned to larger voids. At the pore-scale level, fluid–fluid interfaces exist and across them a pressure difference occurs whose magnitude depends on the interface curvature (Fig. 1). This pressure difference is termed the capillary pressure defined by

$$P_{nw}=P_n-P_w \tag{8}$$

where P_n is the pressure in the nonwetting phase, and P_w is the one in the wetting phase. Since the nonwetting phase pressure must, by definition, exceed that in the wetting phase, P_{nw} must be positive. (For mixed wettability systems, the terms 'wetting' and 'nonwetting' phases become ambiguous and negative capillary pressures may occur, regardless of how the pairs are defined.)

We may similarly define the capillary pressure head as

$$h_{nw}=h_n-h_w \tag{9}$$

where we shall assume that both heads are defined with respect to the same reference fluid, which we select to be water $\rho_R=\rho_w$.

If fluid pressures are controlled purely by capillary pressure, Laplace's equation of capillarity holds:

$$R=2\sigma_{nw}/P_{nw} \tag{10}$$

where R is the mean radius of curvature of the fluid–fluid interfaces, and σ_{nw} is the interfacial tension between the two fluids. As liquid film thicknesses on the solid grains approach molecular dimensions, forces other than capillary forces may substantially affect fluid pressures, so that (10) will not be accurate. However, except for fine-grained soils, this condition will only occur at very low moisture content, which is not often of great practical interest.

Equation (10) indicates that, as capillary pressure gradually increases, the wetting phase will be progressively displaced from larger voids by the nonwetting phase. Consider the behavior of a soil core that is initially fully saturated with water. We place the core in contact with a fine water-saturated porous plate to which a suction is applied in gradual increments. That is, we gradually make the water pressure more negative relative to the gas phase, which is kept at an atmospheric pressure (or equivalently, we increase the gas pressure in the core while keeping the water at the exit at atmospheric pressure). In either case, the air–water capillary pressure increases. At a certain point, the air–water interfaces on the upper boundary achieve a radius of curvature that is smaller than the largest pore open to the boundary, and air enters the core. As the capillary pressure is increased further, the radius of the curvature interface decreases, and more air progressively enters the system. If we increment the suction slowly enough to attain equilibrium at each step, then an equilibrium relationship between capillary pressure and fluid saturation will be obtained.

Since we are now considering systems with only two fluids, clearly

$$S_{nw}(h_{nw})=1-S_w(h_{nw}) \tag{11}$$

and the saturation–capillary pressure relations of the system are fully defined.

The foregoing discussion implies that saturation–capillary pressure relationships reflect an underlying pore size distribution of the porous medium. The Laplace equation suggests that the two-phase saturation versus capillary pressure relations for a given porous medium will depend on the interfacial tension for the specific fluid pair involved. In particular, if we assume that $S_w(R)$ represents a pore size distribution of the porous medium, then (10) implies that saturation–capillary pressure relations may be scaled by the interfacial tension (e.g. Corey, 1986). The scaling relationship may be written in the form

$$\bar{S}_w(\beta_{nw}h_{nw})=S^*(h) \tag{12}$$

where $\bar{S}_w=(S_w-S_m)/(1-S_m)$ is the effective wetting phase saturation, S_m is an apparent minimum wetting phase saturation, $S^*(h^*)$ is the effective wetting fluid saturation versus capillary head function for a reference two-fluid phase system, $S_w(h_{nw})$ is the wetting fluid saturation versus capillary head relation for the same porous medium with arbitrary fluids n and w, and β_{nw} is a fluid dependent scaling factor defined approximately by

$$\beta_{nw}=\sigma^*/\sigma_{nw} \tag{13}$$

where σ^* is the interfacial tension of the reference fluid pair, and σ_{nw} is the interfacial tension between fluids n and w. The choice of reference system is arbitrary, but a natural choice in groundwater investigations is the air–water fluid pair.

An example of unscaled and scaled saturation–capillary pressure relations for air–oil, oil–water, and air–water fluid pairs in a sandy porous medium is illustrated in Fig. 2. The scaling factors in (12) are fitted by a nonlinear regression procedure after assuming saturation–capillary pressure relations to be described by the empirical parametric model of van Genuchten (1980):

$$S^*=[1+(\alpha h)^n]^{-m} \quad (h^*>0) \tag{14a}$$

$$S^*=1 \quad (h^*<0) \tag{14b}$$

where α and n are van Genuchten model parameters, and $m=1-(1/n)$. A comparison of scaling factors determined by directly fitting (12) and (14) to measured two-phase saturation–capillary pressure data versus factors estimated from interfacial tension data using (13) for four organic fluids is given in Table 1. Scaling factors predicted by (13) appear generally to be accurate to within 10–20% for these systems.

Although simple single-valued parametric models such as (14) are frequently used to describe saturation–capillary pressure relations in porous media, real behavior is unfortunately rather more complicated. Actual measurements of saturation–capillary pressure relations exhibit hysteresis (Fig. 3), with higher capillary pressures observed at given saturations during intervals of decreasing wetting phase saturation (drainage) than during increasing wetting phase saturation (imbibition). Hysteretic effects may be attributed to irregularities in pore geometry, contact angle hysteresis, nonwetting fluid entrapment, and other phenomena. If nonwetting phase entrapment occurs during wetting phase imbibition, nonclosure of the saturation–capillary pressure function will arise. The magnitude of nonwetting fluid entrapment will depend on the fluid pair properties as well as on the saturation at which reversal from drainage to imbibition occurs, as demonstrated in various experimental studies (e.g. Land, 1968; Lenhard *et al.*, 1991).

Models have been developed to describe hysteresis in two-phase saturation–capillary pressure relations (e.g. Havercamp & Parlange, 1983; Kool & Parker, 1987, Parker & Kaluarachchi, 1991) which involve only slightly greater parametric complexity than monotonic models. It has been found that for air–water problems controlled by flux-type boundary conditions, hysteresis may have relatively minor effects on transient water flow (Kaluarachchi & Parker, 1987).

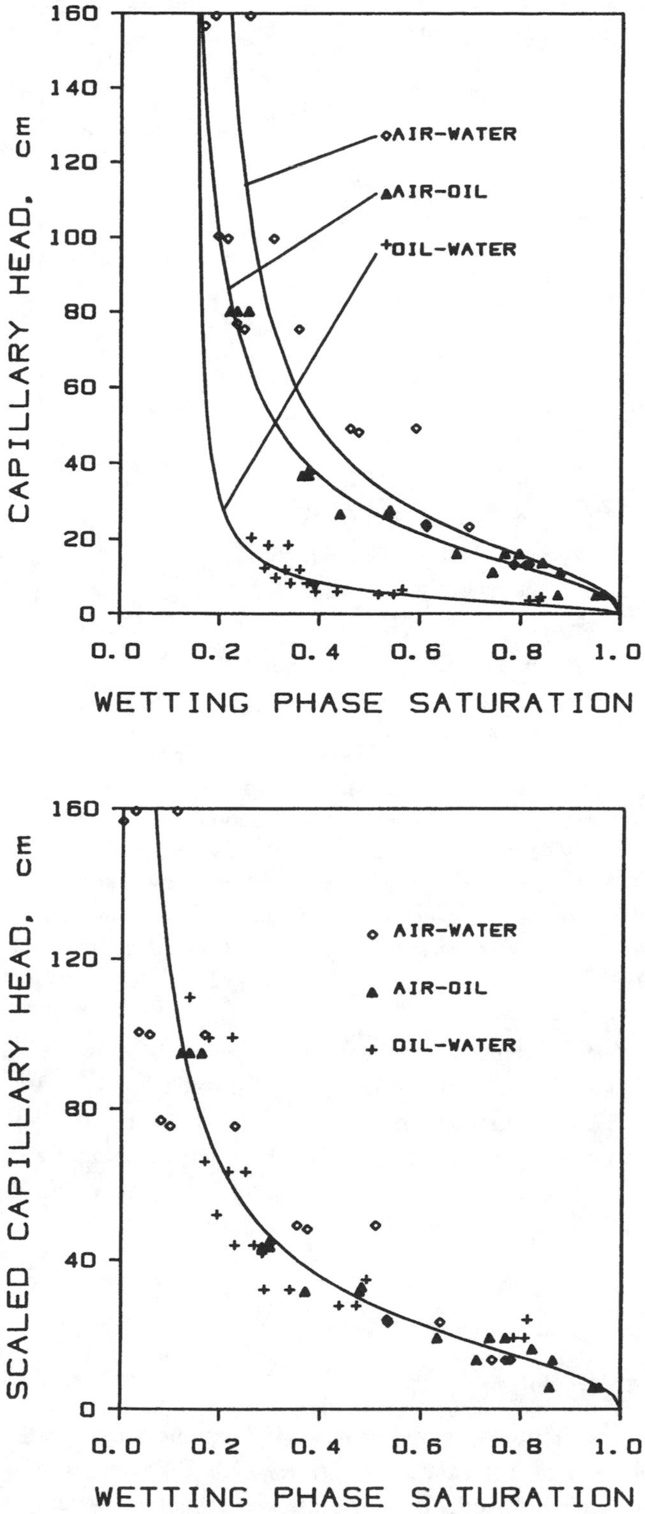

Fig. 2 Wetting fluid saturation versus capillary head for air–water, air–benzyl alcohol and benzyl alcohol–water fluid pairs in the same porous media: (a) unscaled, and (b) scaled.

Furthermore, parameter uncertainty due to measurement errors and other factors associated with model calibration may often overwhelm the effects of hysteresis (Kool & Parker, 1988). Thus for two-phase systems in which the wetting phase

Table 1. *Comparison of measured and predicted scaling factors for two-phase saturation–capillary pressure relations.*

	β_{ao}		β_{aw}	
Fluid	*Measured*	*Predicted*	*Measured*	*Predicted*
Benzene	2.18	1.94	1.85	2.12
o-Xylene	2.11	2.12	1.91	2.27
p-Cymene	1.90	1.97	2.11	2.14
Benzyl alcohol	1.22	1.26	5.59	5.97

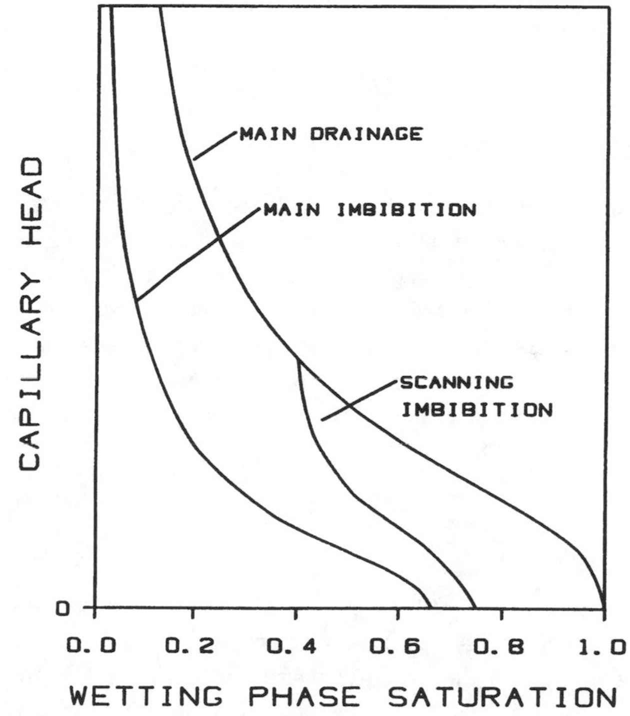

Fig. 3 Hysteresis in two-phase saturation–capillary pressure relations.

is of primary importance, there seems to be some justification for disregarding hysteresis, although certain conditions may arise where this is not so. For problems in which the behavior of the nonwetting phase is of primary concern (e.g. hydrocarbon in an air–hydrocarbon–water system), these conclusions may be different, as will be discussed shortly.

2.4 Three-phase saturation–capillary pressure relations

Descriptions of three-phase saturation–capillary–pressure relations are conventionally predicated on the assumption that wettability follows the well defined sequence of water (w) oil (o), then air (a), in order from wetting to nonwetting. Thus, water is assumed to occupy the pore space in immediate contact with the solids, oil occupies pore space with immediate contact with water, and air occupies the remaining space in contact with oil (Fig. 4).

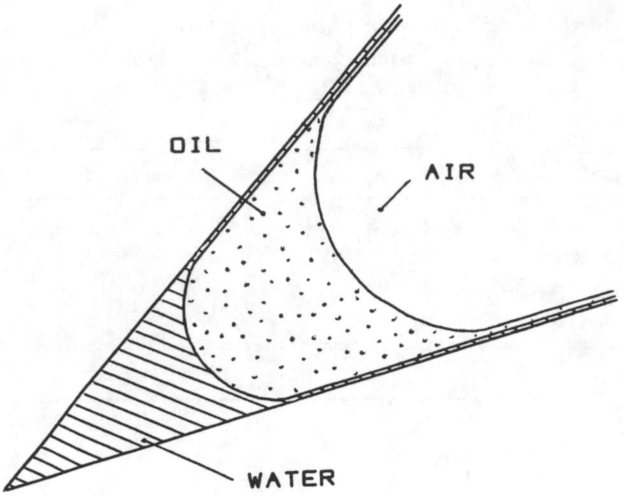

Fig. 4 Idealized pore cross-section with three fluids.

Accordingly, water saturation should depend only on the pressure difference between water and oil phases, while the total liquid saturation ($S_t = S_w + S_o$) should depend only on the pressure difference between air and oil (Leverett, 1941). Invoking the scaling arguments introduced in describing two-phase saturation–capillary pressure relations, we arrive at the three-phase scaling relations (Parker, Lenhard & Kuppasamy, 1987)

$$\bar{S}_w(\beta_{ow} h_{ow}) = S^*(h)^* \quad (15a)$$

$$\bar{S}_t(\beta_{ao} h_{ao}) = S^*(h)^* \quad (15b)$$

where $\bar{S}_w = (S_w - S_m)/(1 - S_m)$ is the effective water saturation, $\bar{S}_t = (S_w + S_o - S_m)/(1 - S_m)$ is the effective total liquid saturation, h_{ow} and h_{ao} are oil–water and air–oil capillary heads, respectively, β_{ow} and β_{ao} are fluid pair dependent scaling coefficients, and $S^*(h^*)$ is a scaled saturation–capillary head function defined here by

$$S^*(h^*) \equiv \bar{S}_w^{prist}(h_{aw}) \quad (15c)$$

where $\bar{S}_w^{prist}$ denotes the effective saturation of water in the pristine air–water system, and h_{aw} is the air–water capillary head. The capillary heads are defined by

$$h_{ao} = h_a - h_o \quad (16a)$$

$$h_{ow} = h_o - h_w \quad (16b)$$

$$h_{aw} = h_a - h_w \quad (16c)$$

The scaling factors β_{ao} and β_{ow} may be approximated by

$$\beta_{ao} = \sigma_{aw}/\sigma_{ao} \quad (17a)$$

$$\beta_{ow} = \sigma_{aw}/\sigma_{ow} \quad (17b)$$

where σ_{aw} is the surface tension of uncontaminated water, σ_{ao} is the surface tension of the organic liquid, and σ_{ow} is the interfacial tension between oil and water (Schiegg, 1983; Lenhard & Parker, 1987a, 1988; Kaluarachchi & Parker, 1989a). Contamination of the air–water system by an infinitesimal volume of oil such that (15a) and (15b) hold and $S_o \to 0$ indicates that the form of the contaminated air–water–saturation–capillary pressure relation is

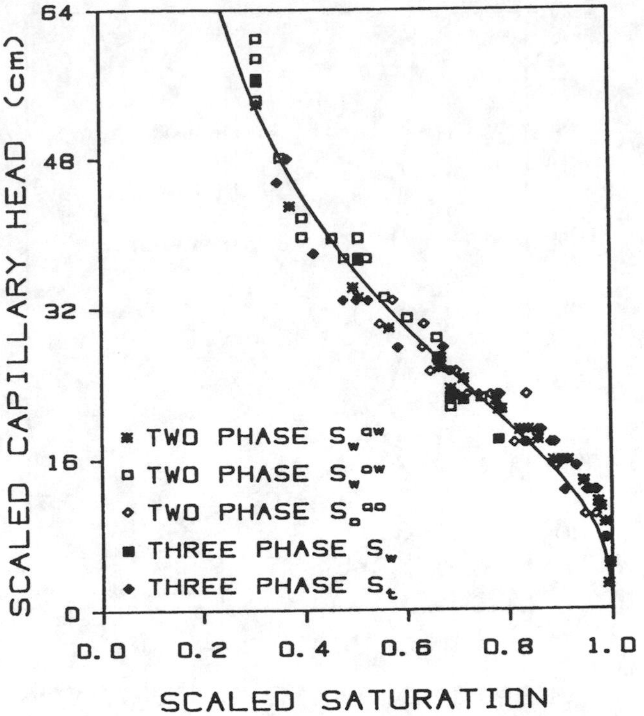

Fig. 5 Two-phase water saturation vs. air–water capillary head (asterisks), two-phase oil saturation vs. air–oil capillary head (open diamonds), two-phase water saturation vs. oil–water capillary head (open squares), three-phase water saturation vs. oil–water capillary head (closed squares), and three-phase total liquid saturation vs. air–oil capillary head (closed diamonds) for monotonic drainage for a sandy soil (Lenhard & Parker, 1988).

$$S_w^{contamin}(\beta'_{aw} h_{aw}) = S_w^{prist}(h_{aw}) \quad (18a)$$

where

$$\beta'_{aw} = \sigma_{aw}/\sigma'_{aw} \quad (18b)$$

in which σ'_{aw} is the surface tension of organic contaminated water. Fig. 5 illustrates a direct test of the relations given by (15) with (14) for $S^*(h^*)$ for a case with monotonic water and total liquid drainage saturation paths given by Lenhard & Parker (1988).

To consider hysteresis in three-fluid phase systems, Parker & Lenhard (1987) introduce the notion of apparent saturation (see also Kaluarachchi & Parker, 1989b). Disregarding the existence of an 'irreducible' wetting phase saturation for our present purposes, apparent water and total liquid saturations may be defined by

$$S_w^{app}=S_w+S_{ot}+S_{atw} \quad (19a)$$

$$S_t^{app}=S_w+S_o+S_{atw}+S_{ato} \quad (19b)$$

where S_w is water saturation, S_o is oil saturation, S_{ot} is the saturation of oil occluded within the water phase, S_{atw} is the saturation of air occluded in water, and S_{ato} is the saturation of air occluded within the oil phase. Oil entrapment by advancing oil–water interfaces and air entrapment by advancing air–oil interfaces (or air–water interfaces in the absence of oil) are assumed to be linearly related to the respective apparent saturations and a maximum trapped saturation inferred from two-phase saturation–capillary pressure relations. Apparent water and total liquid saturations are assumed to scale in the fashion described by (15) to yield a hysteretic $S^*(h^*)$ function which exhibits closure at $h^*\rightarrow 0$ and ∞. Scanning curves in the $S^*(h^*)$ function are computed by an empirical procedure which forces closed loops at all reversal points.

Experimental methodology for determining three-phase capillary pressure relations has been presented by Lenhard & Parker (1988), and a comparison of predicted and observed results for nonmonotonic saturation paths has been presented by Lenhard *et al.* (1991). Effects of hysteresis on well oil thickness during periods of fluctuating water tables have been reported by Kemblowski & Chiang (1990) and Parker *et al.* (1994).

2.5 Vertical equilibrium fluid distributions

To clarify the significance of three-phase saturation–capillary pressure relations, it is instructive to consider the problem of vertical equilibrium fluid distributions in a three-phase system. In particular, we consider a scenario involving a spill of lighter-than-water NAPL (LNAPL), which, after seeping into the ground, spreads laterally at the water table. After a period of time, vertical redistribution of hydrocarbon essentially ceases, and pressure distributions in the vertical direction at any areal point will be nearly hydrostatic. If observation wells with water and oil permeable screens are installed in the aquifer, an oil lens will be observed floating on the water at some depth. A common practical problem is how to interpret these data from observation wells in terms of the amount of hydrocarbon in the soil. To resolve this problem, we seek first to evaluate the vertical fluid pressure distribution from observation well fluid levels, and then, via the three-phase saturation–capillary pressure relations, to determine the vertical distribution of water, oil, and air saturations.

It is possible to characterize the vertical fluid pressure distributions in terms of various fluid 'table' elevations. Observation well fluid levels may be described by an air–oil

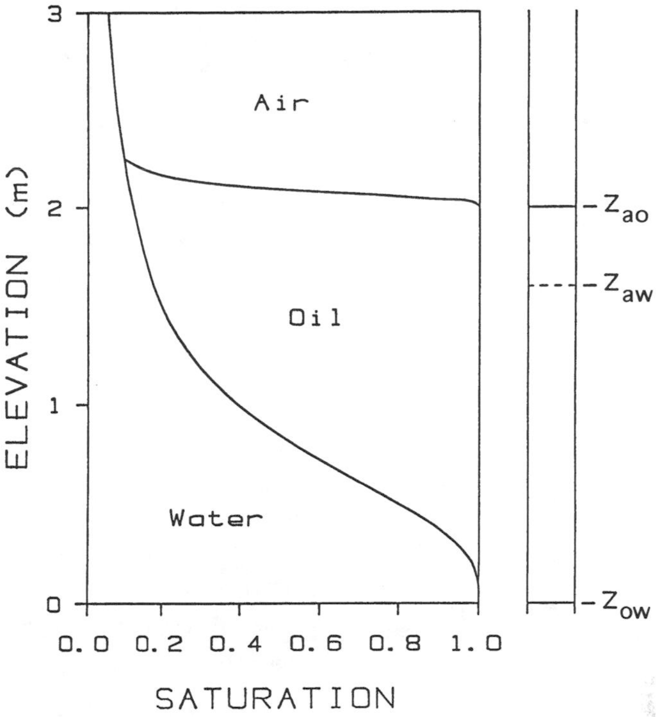

Fig. 6 Water and total liquid saturation versus elevation under vertically hydrostatic conditions for a sandy soil.

table elevation, z_{ao}, at which location the oil pressure is atmospheric, and an oil–water table elevation, z_{ow}, at which water and oil pressures are equal (Fig. 6). We may also theoretically define an air–water table elevation, z_{aw}, at which water and air pressure are equal. To determine the vertical pressure distribution, it is useful first to introduce the concept of piezometric heads for water and oil phases as

$$\Psi_w=h_w+z \quad (20a)$$

$$\Psi_o=h_o+\rho_{ro}z \quad (20b)$$

where Ψ_w and Ψ_o are water and oil piezometric heads, and z is elevation above an arbitrary datum. Note that piezometric head gradients are the driving force for the flow equations. Therefore, vertical equilibrium requires that $\partial\Psi_w/\partial z=0$ and $\partial\Psi_o/\partial z=0$; hence, if air pressure is uniform,

$$\Psi_w=z_{aw} \quad (21a)$$

$$\Psi_o=z_{ao} \quad (21b)$$

and

$$h_w=z_{aw}-z \quad (22a)$$

$$h_o=\rho_{ro}(z_{ao}-z) \quad (22b)$$

From eqs. (22) and the definitions of z_{aw}, z_{ao}, and z_{ow}, we observe that

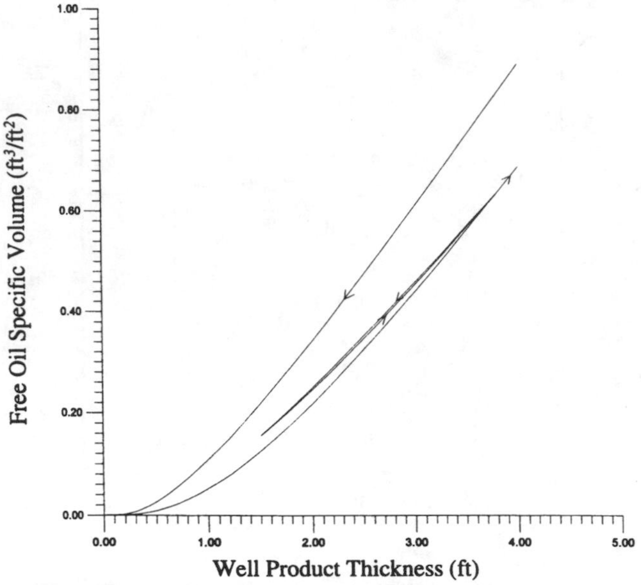

Fig. 7 Free oil specific volume versus well product thickness for a sandy soil for oil imbibition and drainage conditions.

$$z_{ow}=z_{aw}-\rho_{ro}H_o \quad (23a)$$

where $H_o=z_{ao}-z_{ow}$ is the well product thickness (Fig. 6). The various table elevations are thus related by

$$z_{ow}=(z_{aw}-\rho_{ro}z_{ao})/(1-\rho_{ro}) \quad (23b)$$

such that stipulation of any two of the three table elevations completely defines the three-phase static vertical head distributions. From (22) and (23), the vertical distributions of oil–water capillary head, h_{ow}, and air–oil capillary head, h_{ao}, are prescribed by the fluid table elevations as

$$h_{ao}=\rho_{ro}(z-z_{ao}) \quad (24a)$$

$$h_{ow}=(1-\rho_{ro})(z-z_{ow}) \quad (24b)$$

Given relations for $S_w(h_{ow})$ and $S_t(h_{ao})$, vertical distributions of water and free oil saturation in the system in equilibrium with the prescribed table elevations may be easily calculated. A typical case is shown in Fig. 6. Note that the well oil thickness, $H_o=z_{ao}-z_{ow}$, represents the oil thickness which would be observed in a well in equilibrium with the soil. The actual oil volume per unit area, which we term oil specific volume, will always be significantly less than H_o. The free oil specific volume V_{of}, may be computed from the free oil saturation distribution as

$$V_{of}=\int_{z_l}^{z_u}\phi S_o dz \quad (25)$$

where z_l and z_u are the lower and upper elevations with oil, relatively. Free oil specific volume versus well product thickness for a sandy soil is shown in Fig. 7. A comparison of calculated and field-measured oil saturation distributions has been presented by Huntley & Hawk (1992).

2.6 Relative permeability relations

Darcy's law is a macroscopic equation which represents multiple fluids in porous media as overlapping continua occupying the same physical space. In the process of passing from the pore scale to the macroscopic Darcy scale, we intentionally blur out the details of the smaller-scale behavior by introducing permeability coefficients. The form of the coefficients can, in principle, be related to the pore-scale behavior. Theoretical analyses predicated on such approaches are useful in providing insight into the physical nature of flow behavior and in developing physically based procedures for estimating macroscopic behavior.

The most rigorous analyses of this type involve the application of network models to investigate multiphase flow behavior. The basic approach is to construct a theoretical model of the porous medium at the pore scale described by a two- or three-dimensional network of nodes and interconnecting channels. Each node is characterized by a 'pore' radius, the distribution of which obeys some theoretical model, for example a log-normal probability distribution. The interconnecting channels also are characterized by some distribution of sizes which may be dependent on the sizes of the pores which they connect. The geometry of the network itself, i.e. the manner in which 'pores' and 'channels' are interconnected, is sometimes also controlled by a stochastic algorithm. For a given set of statistical parameters, which describe a specific porous medium, a theoretical network analog may be generated and used to investigate the macroscopic system behavior. For example, subject to incremental changes in the capillary pressure applied on network model boundaries, the drainage and imbibition of fluids may be simulated to determine macroscopic saturation–capillary pressure relations, or steady flow regimes may be imposed and pore-scale simulations of flow through the network may be used to compute apparent phase permeabilities at different phase saturations and for different saturation paths. A review of network modeling has been given by Dullien (1979).

Network models are very instructive for investigating the details of multiphase flow behavior at the pore scale. A practical drawback is that the statistical parameters which govern network models are not readily determined. In addition, the complexity of network models precludes their use in deriving analytic functional expressions for permeability–saturation relations. An alternative approach, which circumvents the latter difficulties at the expense of sacrificing some rigor in describing the pore-scale behavior, is predicated upon the conceptualization of a porous medium as a bundle of capillary channels. The channels are regarded as one-dimensional tubes with lengths greater than the macroscopic streamline length due to tortuosity. The channels exhibit a distribution of diameters which is the same on any cross-section.

According to the Hagan–Poiselle equation for flow in a single straight capillary tube, flow velocity is proportional to the square of the tube radius. For a system comprising a distribution of capillary tubes, the relative permeability to wetting phase may be regarded as a function of the mean radius of wetting phase saturated pores:

$$k_{rw}(S_w)=\tau_w \bar{R}_w(S_w)^2/\bar{R}_w(1)^2 \qquad (26)$$

where $\bar{R}_w(S_w)$ is the mean radius of pores filled with wetting phase at a saturation of S_w and τ_w is a tortuosity coefficient. Various theoretical analyses indicate that τ_w has the form

$$\tau_w=\bar{S}_w^b \qquad (27)$$

where $\bar{S}_w$ is effective water saturation (see eqs. (14)) and b is a tortuosity exponent. The wetting phase will occupy the smallest pores if local equilibrium occurs within the averaging volume implicit in the definition of the macroscopic scale. The Laplace equation indicates an inverse relationship between pore radius and capillary pressure, as discussed previously. Thus, the cumulative pore size distribution may be represented by $S^*(\lambda/h^*)$, where λ is an aggregate constant from the Laplace equation. Since the factor λ may be cancelled from the ratio in (26), the expression for k_{rw} may be written as

$$k_{rw}=\bar{S}_w^b\left[\frac{\left(\int_0^{\bar{S}_w}\frac{1}{h^*(S^*)}dS^*\right)}{\left(\int_0^1\frac{1}{h^*(S^*)}dS^*\right)}\right]^2 \qquad (28)$$

Note that the limits of the integral in the numerator from 0 to $\bar{S}_w$ imply that water occupies the smallest pores in the system. Based on analyses of experimental results for a large number of soils, Mualem (1976) concluded that $b=1/2$.

Similar reasoning may be employed to obtain expressions for oil and gas phase relative permeabilities assuming air is the least wetting phase and oil is the phase of intermediate wettability. The expressions take the form

$$k_{ro}=(\bar{S}_t-\bar{S}_w)^b\left[\frac{\left(\int_{\bar{S}_w}^{\bar{S}_t}\frac{1}{h^*(S^*)}dS^*\right)}{\left(\int_0^1\frac{1}{h^*(S^*)}dS^*\right)}\right]^2 \qquad (29)$$

$$k_{ra}=\bar{S}_a^b\left[\frac{\left(\int_{\bar{S}_t}^1\frac{1}{h^*(S^*)}dS^*\right)}{\left(\int_0^1\frac{1}{h^*(S^*)}dS^*\right)}\right]^2 \qquad (30)$$

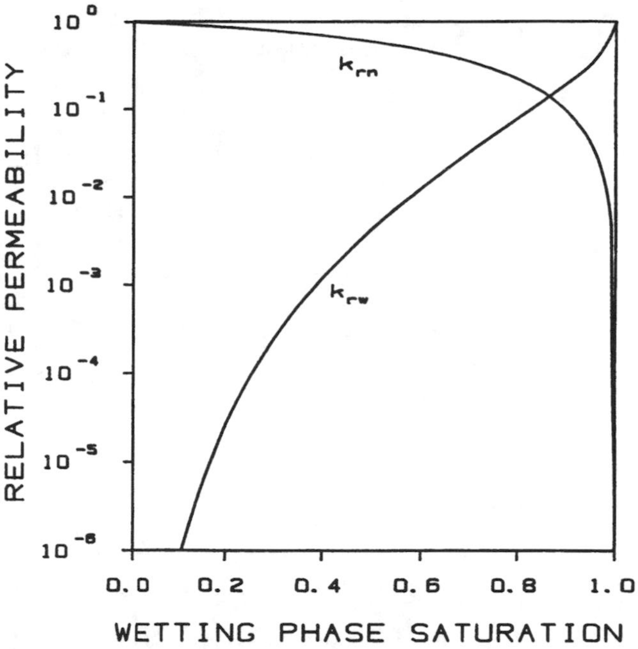

Fig. 8 Typical two-phase relative permeability–saturation relations.

where $\bar{S}_t=(S_w+S_o-S_m)/(1-S_m)$, and $\bar{S}_a=S_a/(1-S_m)$.

If the saturation–capillary pressure function is of a form amenable to exact integration in the foregoing equations, closed form expressions for relative permeabilities may be derived. In the case of the nonhysteretic van Genuchten model, (eqs. (14), the result is (Parker *et al.*, 1987)

$$k_{rw}=\bar{S}_w^{1/2}[1-(1-\bar{S}_w^{1/m})^m]^2 \qquad (31a)$$

$$k_{ro}=(\bar{S}_t-\bar{S}_w)^{1/2}\{(1-\bar{S}_w^{1/m})^m-(1-\bar{S}_t^{1/m})^m\}^2 \qquad (31b)$$

$$k_{ra}=(1-\bar{S}_t)^{1/2}(1-\bar{S}_t^{1/m})^{2m} \qquad (31c)$$

where $m=1-(1/n)$ and n is the van Genuchten parameter. Note that k_{rw} is found to be a function of S_w only, k_{ra} is a function of S_t (or S_a) only, while k_{ro} depends on both S_w and S_t. The basic forms of k_{rw} and $k_{rm}=k_{ra}$ or k_{ro} for the simple case of a two-fluid phase system are illustrated in Fig. 8. An exponential reduction in k_{rw} as wetting phase saturation decreases from unity is observed along with an exponential increase in k_{ra} as nonwetting phase displaces wetting phase from the largest pores in the system.

Variants of the foregoing simple theoretical model for three-phase permeabilities have been reported in the literature using somewhat different integral expressions in lieu of (28) to (30) for k_{rp} and employing different analytical forms for $S^*(h^*)$. Other widely used methods for estimating three-phase relative permeabilities are based on empirical predictions from two-phase relative permeability measurements (Stone, 1970, 1973). Baker (1988) has evaluated a number of three-phase relations by comparison with data in the litera-

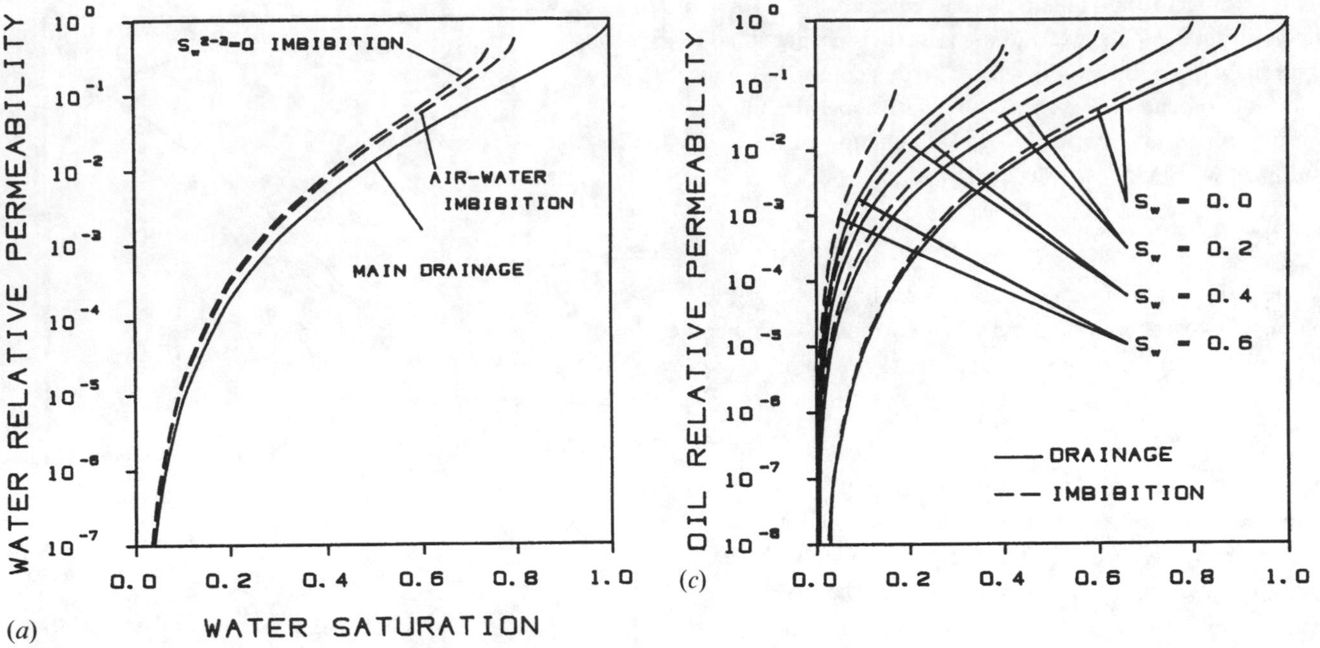

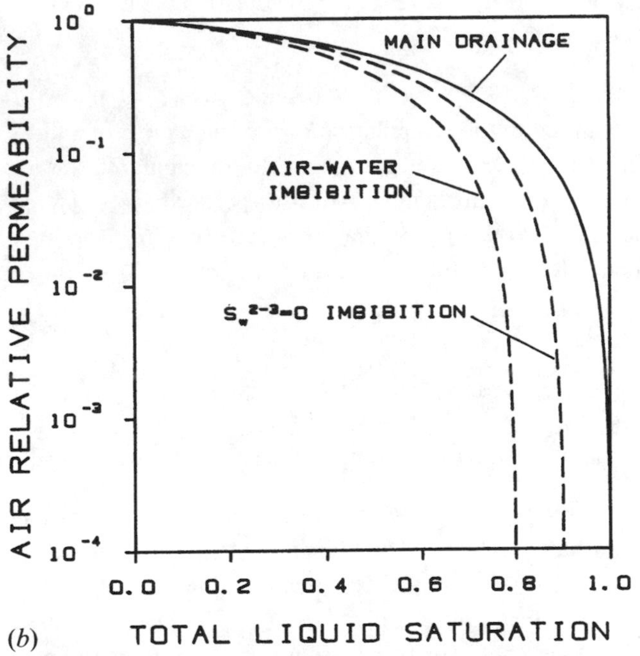

Fig. 9 Hysteresis in three-phase relative permeability relations predicted by the model of Lenhard & Parker (1987b) for a sandy soil: (a) water relative permeability in air–water system for main drainage and main imbibition, and main imbibition in air–oil–water system with free oil near zero; (b) air relative permeability in air–water system for main drainage and main imbibition, and main imbibition in air–oil–water system with free oil near zero; and (c) oil relative permeability for main drainage and main imbibition with various fixed water saturations.

ture. The accuracy of the different methods varied with the data set used for comparison, and no single method stood out as universally superior.

The derivation of eqs. (31) disregards hysteresis that may occur in saturation–capillary pressure relations. Accounting for such effects leads to nonunique relative permeability functions. Lenhard & Parker (1987b) and Lenhard, Parker & Kaluarachchi (1989) have presented a general theoretical model to predict three-phase relative permeabilities via modifications of (28) to (30) which correct for effects of nonwetting fluid entrapment. Predicted permeabilities for a sandy soil are shown in Fig. 9 for main drainage paths and for imbibition paths in two-phase air–water systems and in three-phase systems with specified water saturations at the moment oil is introduced ($S_w^{2\to 3}$).

3 BASICS OF MULTIPHASE TRANSPORT

In the preceding sections, we have considered the bulk convective flow of water, NAPL, and air phases in a porous medium. In certain cases, bulk fluid may be of principal concern (for example to model hydraulic recovery of separate phase hydrocarbon spreading on a groundwater table). However, fluids present in groundwater systems are virtually always heterogeneous, composed of a variety of individual chemical constituents. Of particular concern from the standpoint of groundwater contamination is the behavior of components of the organic liquid phase that may exhibit finite solubility in the aqueous phase and volatility to the gas phase. Within each phase, component transport may occur due to bulk phase convection as well as to diffusion and mechanical dispersion effects. The analysis of component transport in multiphase systems will be described in this section.

3.1 Mass flux equations

Chemical constituents that are present in aqueous, NAPL, and gas phases may move in the subsurface by various mechanisms. At the pore scale, chemical transport occurs due to convection in mobile phases and to diffusion due to the occurrence of concentration gradients within individual phases. However, we are not really interested in pore-scale descriptions since practicality demands a coarser characterization in geologic media. At the macroscopic scale, phase convection is represented by the Darcy velocity, and the corresponding convective flux of species α in phase p is given by

$$J^{conv}_{\alpha pi} = c_{\alpha p} q_{pi} \tag{32}$$

where $J^{conv}_{\alpha pi}$ is the convective mass flux density of constituent α in phase p in the i direction [$ML^{-2}T^{-1}$], q_{pi} is the Darcy velocity of the p-phase in the i direction [LT^{-1}], and $c_{\alpha p}$ is the concentration of α in phase p [ML^{-3}]. For purely convective transport, (32) implies solute 'particles' moving at a mean velocity equal to the Darcy velocity divided by the fraction of gross soil area perpendicular to flow occupied by the phase.

Diffusion is a process driven by random thermal motion of molecules which tends to equalize concentrations within nonhomogeneous phases. In a single-phase system with no porous medium present, Fick's law of diffusion has the well known form

$$J^{diff}_{\alpha pi} = -D^{o}_{\alpha p} \partial c_{\alpha p} / \partial x_i \tag{33}$$

where $J^{diff}_{\alpha pi}$ is the diffusive mass flux density of species α in phase p in the i direction [$ML^{-2}T^{-1}$], and $D^{o}_{\alpha p}$ is the molecular diffusion coefficient of α in the p-phase [L^2T^{-1}]. According to the Stokes–Einstein equation, single-phase liquid diffusion coefficients may be related to molecular size and phase viscosity by

$$D^{o}_{\alpha p} = RT/6\eta_p r_\alpha \tag{34}$$

where R is the gas constant, T is absolute temperature, η_p is the p-phase viscosity, and r_a is the molecular radius of species α (Lyman, Reehl & Rosenblatt, 1982).

Within a porous medium, the diffusive flux at a given concentration gradient will be less than that given by (33), because diffusion in the p-phase in the porous medium occurs only within the fraction of the pore space occupied by the p-phase and because diffusion paths are more circuitous due to the presence of other phases. Thus, within the porous medium, (33) is modified to

$$J^{diff}_{\alpha pi} = -\phi S_p D^{diff}_{\alpha p} \partial c_{\alpha p} / \partial x_i \tag{35}$$

in which $D^{diff}_{\alpha p}$ is an effective diffusion coefficient given by

$$D^{diff}_{\alpha p} = \tau_p D^{0}_{\alpha p} \tag{36}$$

where τ_p is a p-phase 'tortuosity' coefficient. A theoretical analysis by Millington & Quirk (1959) suggests the latter to be of the form

$$\tau_p = \phi^{1/3} S_p^{7/3} \tag{37}$$

assuming an isotropic porous medium with respect to diffusion. Little information is actually available to justify the assumption of diffusional isotropy in porous media, and it is assumed out of convenience that effects of anisotropic diffusion would in most cases be of minor importance in field-scale problems due to the dominance of mechanical dispersion over molecular diffusion. In the process of averaging from the pore scale to the macroscopic scale, effects of smaller-scale variations in fluid velocities on macroscopic transport are lost from the mean convection equation (32). If the scale of observation is sufficiently large compared with the scale of heterogeneity in the flow field, the dispersive process may asymptotically appear to be diffusive. That is, the process of mechanical dispersion may, under certain constraints, obey a diffusion-type equation commonly written in the form

$$J^{hyd}_{\alpha pi} = -\phi S_p D^{hyd}_{pij} \partial c_{\alpha p} / \partial x_j \tag{38}$$

where $J^{hyd}_{\alpha pi}$ is the dispersive mass flux density α in phase p in the i direction [$ML^{-2}T^{-1}$], and D^{hyd}_{pij} is a dispersion tensor in the p-phase [L^2T^{-1}].

In a statistically homogeneous and isotropic porous medium at a scale of observation such that (38) is obeyed, the asymptotic dispersion tensor may be shown theoretically to have the form (Bear, 1972)

$$\phi S_p D^{hyd}_{pij} = \lambda_T \bar{q}_p \delta_{ij} + (\lambda_L - \lambda_T) |q_{pi} q_{pj}| / \bar{q}_p \tag{39}$$

where λ_L and λ_T are the longitudinal and transverse dispersivities [L], q_{pi} and q_{pj} are p-phase Darcy velocities in the i and j directions, $q_p = |\Sigma q^2_{pi}|^{1/2}$ is the absolute magnitude of the p-phase velocity, and δ_{ij} is Kronecker's delta, which has the property $\delta_{ij}=1$ if $i=j$ and $\delta_{ij}=0$ if $i \neq j$. For nonisotropic porous media, the form of the dispersion tensor becomes much more complicated and the number of independent parameters becomes so great that calibration is impractical. Therefore, dispersion is almost always assumed to be isotropic, although this may not be strictly so.

The physical significance of (38) is more readily apparent if we consider the special case in which the coordinate system is oriented with the flow field in the x_1 direction such that $q_{p1} = \bar{q}_p$ and $q_{p2} = q_{p3} = 0$. Then the dispersion tensor takes on the simple form

$$[\phi S_p D^{hyd}_{pij}] = \begin{bmatrix} q_p\lambda_L & 0 & 0 \\ 0 & q_p\lambda_T & 0 \\ 0 & 0 & q_p\lambda_T \end{bmatrix} \tag{40}$$

Thus, the dispersion coefficient is $q_p\lambda_L/\phi S_p$ in the direction of water flow and $q_p\lambda_T/\phi S_p$ transverse to the direction of flow.

Dispersivity in geologic media is a very elusive quantity due to the approximate nature of (38). Since heterogeneities occur at many scales ranging from the pore scale to the regional geologic scale, asymptotic dispersion regimes may develop over a certain range of solute travel distances, only to be disturbed as larger-scale heterogeneities are encountered. Thus, dispersivity is often observed to exhibit apparent scale dependence. Longitudinal dispersivities typically are found to range from 0.01 to 0.1 of the mean travel distance for distances less than 1 km and to diminish to somewhat smaller fractions of travel distance beyond this (Gelhar, 1986). Transverse dispersivity is typically in the range of 0.1 to 0.3 times the longitudinal dispersivity although factors ranging from 0.003 to above 1.0 have been reported.

3.2 Combined mass flux equations

Since mechanical dispersion is assumed, asymptotically at least, to have the same form as Fick's law, the equations for hydrodynamic dispersion and diffusion may be combined as

$$J^{dh}_{\alpha pi} = -\phi S_p D_{\alpha pij} \partial c_{\alpha p} / \partial x_j \tag{41}$$

where $D_{\alpha pij}$ is an aggregate coefficient given by

$$D_{\alpha pij} = \tau_p D^0_{\alpha p} + D^{hyd}_{pij} \tag{42}$$

Since $\tau_p D^0_{\alpha p}$ is on the order of only 1 cm^2 day^{-1} for liquid phases, the second term will generally predominate for liquid phase transport except at short travel distances and for low fluid velocities. Diffusion may be more important in gas phase transport because $D^0_{\alpha p}$ is larger.

The total mass flux density of component α in phase p, $J_{\alpha pi}$, may now be defined by adding fluxes due to all mechanisms as

$$J_{\alpha pi} = J^{conv}_{\alpha pi} + J^{diff}_{\alpha pi} + J^{hyd}_{\alpha pi} \tag{43}$$

which, upon substituting (32), (33) and (38), gives

$$J^{hyd}_{\alpha pi} = c_{\alpha p} q_{pi} - \phi S_p D_{\alpha pij} \partial c_{\alpha p} / \partial x_j \tag{44}$$

which is the desired equation for total mass flux of α in phase p.

3.3 Continuity equations for transport

Assuming no chemical or biochemical transformations of component α within the p-phase, mass conservation of species α in phase p requires that

$$\phi \frac{\partial (c_{\alpha p} S_p)}{\partial t} + \frac{\partial J_{\alpha pi}}{\partial x_i} - R_{\alpha p} = 0 \tag{45}$$

where $R_{\alpha p}$ is the net mass transfer rate per porous medium volume of species α into (positive) or out of (negative) the p-phase. Incorporating (44) into (45) yields the combined transport equation

$$\phi \frac{\partial (c_{\alpha p} S_p)}{\partial t} + \frac{\partial (c_{\alpha p} q_{pi})}{\partial x_i} - \frac{\partial}{\partial x_i}\left[\phi S_p D_{\alpha pij} \frac{\partial c_{\alpha p}}{\partial x_j}\right] - R_{\alpha p} = 0 \tag{46}$$

Expanding the first and second terms in (46), employing the bulk p-phase continuity equation (3), and assuming density derivative terms to be of second-order importance yields the desired form of the transport equation

$$\phi S_p \frac{\partial c_{\alpha p}}{\partial t} + q_{pi} \frac{\partial c_{\alpha p}}{\partial x_i} - \frac{\partial}{\partial x_i}\left(\phi S_p D_{\alpha pij} \frac{\partial c_{\alpha p}}{\partial x_j}\right) - R_{\alpha p} + \frac{c_{\alpha p} \gamma_p}{\rho_p} = 0 \tag{47}$$

where the total phase mass transfer rate, γ_p, it may be noted, is related to the component mass transfer rates by

$$\gamma_p = \sum_{\alpha} R_{\alpha p} \tag{48}$$

For the three-fluid phase system, (47) constitutes a system of three equations which we may write for the water phase ($p=w$), the organic liquid phase ($p=o$), and the gas phase ($p=a$) as

$$\phi S_a \frac{\partial c_{\alpha a}}{\partial t} + q_{ai} \frac{\partial c_{\alpha a}}{\partial x_i} - \frac{\partial}{\partial x_i}\left(\phi S_a D_{\alpha aij} \frac{\partial c_{\alpha a}}{\partial x_j}\right) - R_{\alpha a} + \frac{c_{\alpha a} \gamma_a}{\rho_a} = 0 \tag{49a}$$

$$\phi S_o \frac{\partial c_{\alpha o}}{\partial t} + q_{oi} \frac{\partial c_{\alpha o}}{\partial x_i} - \frac{\partial}{\partial x_i}\left(\phi S_o D_{\alpha oij} \frac{\partial c_{\alpha o}}{\partial x_j}\right) - R_{\alpha o} + \frac{c_{\alpha o} \gamma_o}{\rho_o} = 0 \tag{49b}$$

$$\phi S_w \frac{\partial c_{\alpha w}}{\partial t} + q_{wi} \frac{\partial c_{\alpha w}}{\partial x_i} - \frac{\partial}{\partial x_i}\left(\phi S_w D_{\alpha wij} \frac{\partial c_{\alpha w}}{\partial x_j}\right) - R_{\alpha w} + \frac{c_{\alpha w} \gamma_w}{\rho_w} = 0 \tag{49c}$$

If adsorption of α by the solid phase can occur, a fourth continuity equation is required, which may be written as

$$\partial c_{\alpha s} / \partial t = R_{\alpha s} \tag{49d}$$

where $c_{\alpha s}$ is the solid phase concentration expressed as mass of adsorbed α per porous medium volume [ML^{-3}].

3.4 Phase-summed equation for local equilibrium transport

Coupling between the phase transport equations arises because of the interphase transfer terms. Explicit consideration of interphase transfer kinetics may often be justifiably avoided by assuming phase transfer to be equilibrium controlled. That is, local thermodynamic considerations fix the relationship between phase concentrations. We consider here the case of linear partitioning and the thermodynamic relations

$$c_{\alpha o} = \Gamma_{\alpha o} c_{\alpha w} \tag{50a}$$

$$c_{\alpha a} = \Gamma_{\alpha a} c_{\alpha w} \tag{50b}$$

$$c_{\alpha s} = \Gamma_{\alpha s} c_{\alpha w} \tag{50c}$$

where $\Gamma_{\alpha o}$ is the equilibrium partition coefficient for species α between water and organic liquid (Raoult's constant), $\Gamma_{\alpha a}$

is the equilibrium partition coefficient for species α between water and air (dimensionless Henry's constant), and $\Gamma_{\alpha s}$ is the equilibrium partition coefficient between water and solid phases.

Using the equilibrium relations, we may rewrite the phase transport equations in terms of a single concentration. For water-wet systems it is logical to retain the water phase concentration since water will always be present in the system. Using (50) to eliminate oil, gas, and solid phase concentrations from (49) and summing the equations, noting that

$$R_{\alpha w}+R_{\alpha o}+R_{\alpha a}+R_{\alpha s}=0 \quad (51)$$

leads to the phase-summed transport equation for species α as

$$\phi^*\frac{\partial c_{\alpha w}}{\partial t}+q_{\alpha i}{}^*\frac{\partial c_{\alpha w}}{\partial x_i}-\frac{\partial}{\partial x_i}\left(D_{\alpha ij}{}^*\frac{\partial c_{\alpha w}}{\partial x_j}\right)-\mu^*c_{\alpha w}=0 \quad (52a)$$

where

$$\phi^*=\phi S_w+\phi S_o\Gamma_o+\phi S_a\Gamma_{\alpha a}+\Gamma_{\alpha s} \quad (52b)$$

$$D_{\alpha ij}{}^*=\phi S_w D_{\alpha wij}+\phi S_o D_{\alpha oij}\Gamma_{\alpha o}+\phi S_a D_{\alpha aij}\Gamma_{\alpha a} \quad (52c)$$

$$q_{\alpha i}{}^*=q_{wi}+q_{oi}\Gamma_{\alpha o}+q_{ai}\Gamma_{\alpha a} \quad (52d)$$

$$\mu_\alpha{}^*=\frac{\gamma_w}{\rho_w}+\frac{\gamma_o\Gamma_{\alpha o}}{\rho_o}+\frac{\gamma_a\Gamma_{\alpha a}}{\rho_a} \quad (52e)$$

Note that eqs. (52) have the same form as the single-phase transport equation. However, the coefficients represent pooled effects of transport in all phases. Note also that interphase mass transfer terms occur in the phase-summed equation only as a sum over all components which will produce mild coupling between the equations, facilitating their solution. The transport equations are strongly coupled to the flow equations because of the dependence on Darcy velocities. The flow equations are themselves coupled to the transport equations only through interphase transfer terms and via concentration dependence of phase densities. As this coupling is often mild, serial solution methods for flow and transport equations with suitable corrections to account for interphase mass transfer and phase density changes provide feasible and efficient computational procedures.

4 OVERVIEW OF AREAS FOR FUTURE RESEARCH

While much progress has been made in the development of theoretical models and numerical methodology for the solution of problems involving multiphase flow and transport in geologic media, many unresolved issues remain. We will review some of the difficult and pressing issues here briefly.

In spite of the fact that three-phase permeability–saturation–capillary pressure relations have been a subject of research for over 50 years, experimental studies have been few in number, and theoretical models, even for non-hysteretic saturation paths, have not been rigorously validated. While models for hysteresis and nonwetting fluid entrapment effects have been developed, they remain fairly speculative. Most saturation–capillary pressure models are based on the Leverett scaling concept and the implicit assumption that wetting phase residual saturation is independent of fluid pair, which has been subject to limited direct testing. Scaling coefficients are often estimated from fluid interfacial tension values, assuming interfacial tensions are constant and variations in contact angles for different fluid pairs are small. However, situations may arise in which contact angles are fluid dependent or in which interfacial tension and/or contact angles change with time due to changes in fluid composition (e.g. McBride, Simmons & Carey, 1992). Such changes may be natural, due to gradual mass transfer between phases, or they may be engineered, e.g. via introduction of surfactants. Surfactants may affect three-phase saturation–capillary pressure relations by changing surface or interfacial tension, contact angles, or residual saturations. Most capillary pressure models are also predicated on the supposition that oil is the intermediate wettability phase. The presence of hydrophobic solids may result in fully or partially oil-wet systems, which will exhibit markedly different behavior than water-wet systems.

An equally vexing problem for field-scale modeling of multiphase flow and transport involves how to accommodate porous media heterogeneity at various scales. Geologic media exhibit variations in permeability and capillary pressure relations over various scales, reflecting variations in grain size distribution, stratigraphic features, local fracture patterns, etc. Since continuum models necessarily involve averaging at the subcontinuum scale, details of smaller-scale heterogeneity are implicitly accommodated in the continuum parameters. However, the critical problem is how much detail can be safely integrated out in this fashion, and how much can be dealt with explicitly. In other words, what scale of averaging is permissible? A simple answer to this question is, unfortunately, not possible. In fractured porous media, an approach that is sometimes employed is to distinguish two averaging scales, one for fractures and one for interfracture rock material, and to model the system as a bicontinuum with inter-region mass transfer defined by various empirical or physically based relations. When observation scales do not coincide with model scales, or if model scales are improperly selected, apparent anomalies in porous medium behavior may be observed. For example, small-scale heterogeneity in a porous medium may give rise to unstable flow and fingering, which cannot be described by a coarse-scale continuum model. Vertically integrated models, predicated on the assumption of vertical equilibrium, are often used to simplify numerical treatment of single-phase and multiphase flow problems. Such models implicitly introduce vertically averaged permeability–saturation–capillary pressure relations, which must be determined to model accurately the field-scale system.

Effects of heterogeneity on the apparent scale dependence of dispersion have already been mentioned. Another problem involving transport is the notion of equilibrium phase partitioning. If differential phase velocities become large enough, kinetics of mass transfer between phases must be considered. Even if equilibrium partitioning occurs at the pore scale, it may not apply at the implied scale of the model. If the model scale is large relative to the scale of heterogeneity (e.g. correlation scale for random heterogeneity or fracture spacing for fractured rock), as is often the case for field-scale simulations with coarse grids, sub-grid scale heterogeneity may inhibit attainment of local equilibrium.

Other problems which are not well understood, but which may have significant effects on contaminant transport and site remediation, include nonisothermal flow, transport of dispersed colloidal material (e.g. 'blobs' of organic liquid in aqueous suspension (Wan & Wilson, 1994)), multispecies chemical interactions, and biochemical transformations. Site remediation based on steam flooding considering nonisothermal flow and thermal induced phase changes has been investigated numerically by Falta *et al.* (1992). Brown *et al.* (1994) have investigated surfactant enhanced remediation, considering effects of surfactants on permeability–saturation–capillary pressure relations, effects of surfactants on phase partitioning of contaminants from a nonaqueous phase liquid to a dispersed colloidal phase, and the transport and chemical reactions of the surfactant itself. Chemical and biochemical reactions of contaminants in the subsurface can introduce significant complexity in models to accommodate the many possible reaction sequences, reaction kinetics, and to characterize the fate and transport of co-reactants. For example, aliphatic and aromatic hydrocarbons generally undergo aerobic bio-oxidation, the rate of which may be limited by oxygen, nutrients, or co-metabolites under various conditions. Chlorinated solvents undergo anaerobic dechlorination, which may be mediated by the presence of various reducing agents and co-metabolites. Sophisticated models for contaminant biotransformation have been developed and compared with laboratory studies (e.g. Borden & Bedient, 1986; Chen *et al.*, 1992). However, uncertainty in model parameters limits their practical utility in the field. Many challenges also remain related to the development of efficient, accurate, and robust numerical simulators for multiphase flow and transport problems. Because of their strong coupling and highly nonlinear behavior, multiphase flow equations provide a challenging numerical problem. Adaptive procedures for handling variables with implicit and explicit numerical formulations and for accommodating zones in which certain phase equations are inactive hold promise for improving numerical efficiency (Katyal & Parker, 1992; Katyal, Kaluarachchi & Parker, 1992). Numerical procedures to accommodate varying time scales at which flow and transport processes occur may also enhance efficiency. Likely increases in the use of parallel architecture computers will require the development of new solution algorithms to exploit their potential fully. Although the heavy computational burden of performing field-scale simulations of three-dimensional multiphase flow and multicomponent transport has been prohibitive, advances in computer technology seem destined to minimize this limitation in the future and to make the accurate characterization of the physical processes of paramount importance.

5 A PRACTICAL FIELD-SCALE MODELING APPROACH FOR LNAPL SPILLS

A practical solution to the trade-off between computational cost versus accurate representation of the field-scale system is to employ vertical integration to reduce the numerical dimensionality and degree of nonlinearity. Two-dimensional areal multiphase flow models, based on the assumption of three-phase vertical equilibrium, have been used for many years to model petroleum reservoirs (Martin, 1968), and have more recently been applied to model LNAPL spills in groundwater (Hockmuth & Sunada, 1985; Kaluarachchi, Parker & Lenhard, 1990, 1992). In this final section, we will discuss a practical modeling approach, based on the vertical equilibrium method, for NAPL, water, and air flow, and for dissolved phase transport of a multicomponent mixture with oxygen-limited biodecay.

5.1 Vertical integration of flow equations

Our interest is in a flow domain bounded at the top by the atmosphere and at the bottom by the lower boundary of an unconfined aquifer. We designate the elevations of these physical boundaries as z_u and z_l, respectively. As discussed by Parker & Lenhard (1989), vertical integration of Darcy's equations (1) for water and oil, subject to the assumption of local vertical equilibrium discussed in Section 2.5, yields

$$Q_{w_i} = -T_{w_{ij}} \frac{\partial \Psi_w}{\partial x_j} \tag{53a}$$

$$Q_{o_i} = -T_{o_{ij}} \frac{\partial \Psi_o}{\partial x_j} \tag{53b}$$

where Q_{w_i} and Q_{o_i} are vertically integrated fluxes [L^2T^{-1}] of water and oil in the i direction ($x_i=x,y$) of the form

$$Q_{p_i}=\int_{z_l}^{z_u} q_{p_i} dz \tag{54}$$

and transmissivities are defined by

$$T_{w_{ij}}=\int_{z_l}^{z_u} K_{w_{ij}} dz \tag{55a}$$

$$T_{o_{ij}}=\int_{z_l}^{z_u} K_{o_{ij}} dz \tag{55b}$$

where conductivities refer to horizontal values, since Q_{p_i} describes flow in the horizontal direction. To interpret Q_{p_i}, note that it corresponds to the volume of flow in the horizontal i direction per unit time per unit length horizontally perpendicular to the i direction (for example, volume of flow through a boundary per length of boundary perimeter). Integrating the liquid continuity equations (3) over the vertical domain yields

$$\frac{\partial V_w}{\partial t}=-\frac{\partial Q_{w_i}}{\partial x_i}+J_w \tag{56a}$$

$$\frac{\partial V_o}{\partial t}=-\frac{\partial Q_{o_i}}{\partial x_i}+J_o \tag{56b}$$

where J_w and J_o are vertically integrated source–sink terms [LT^{-1}], and V_w and V_o are total water and oil volumes per horizontal area [L] at a point in the x–y plane (*fluid specific volumes*) defined by

$$V_w=\int_{z_l}^{z_u} \phi S_w dz \tag{57a}$$

$$V_o=\int_{z_l}^{z_u} \phi S_o dz \tag{57b}$$

Combining (53) and (56) yields the governing equations for areal flow of water and oil as

$$\frac{\partial V_w}{\partial t}=\frac{\partial}{\partial x_i}\left(T_{w_{ij}}\frac{\partial \Psi_w}{\partial x_j}\right)+J_w \tag{58a}$$

$$\frac{\partial V_o}{\partial t}=\frac{\partial}{\partial x_i}\left(T_{o_{ij}}\frac{\partial \Psi_o}{\partial x_j}\right)+J_o \tag{58b}$$

In general, V_w and V_o are functions of Ψ_w, Ψ_o, and h_a, and the left-hand sides of (58) may be expanded as

$$\gamma_{ww}\frac{\partial \Psi_w}{\partial t}+\gamma_{wo}\frac{\partial \Psi_o}{\partial t}+\gamma_{wa}\frac{\partial h_a}{\partial t}=\frac{\partial}{\partial x_i}\left(T_{w_{ij}}\frac{\partial \Psi_w}{\partial x_j}\right)+J_w \tag{59a}$$

$$\gamma_{ow}\frac{\partial \Psi_w}{\partial t}+\gamma_{oo}\frac{\partial \Psi_o}{\partial t}+\gamma_{oa}\frac{\partial h_a}{\partial t}=\frac{\partial}{\partial x_i}\left(T_{o_{ij}}\frac{\partial \Psi_o}{\partial x_j}\right)+J_o \tag{59b}$$

with capacity coefficients defined by

$$\gamma_{pq}=\partial V_p/\partial \Psi_q \tag{60}$$

in which $p,q=a,o,w$ are phase indices and $\Psi_a=h_a$. While eqs. (58) represent the general form of the vertically integrated oil–water flow equations, it is instructive (and reassuring) to note that, when no oil is present and air pressure is constant, (58a) reduces to the conventional areal groundwater flow equation

$$\phi_e\frac{\partial Z_{aw}}{\partial t}=\frac{\partial}{\partial x_i}\left(T_{w_{ij}}\frac{\partial Z_{aw}}{\partial x_j}\right)+J_w \tag{61}$$

where ϕ_e is the specific yield of the aquifer. This is because when no oil is present, $\Psi_w=Z_{ao}=Z_{aw}$ and $\gamma_{ww}+\gamma_{wo}+\gamma_{wa}=\phi_e$.

Air flow may be modeled in an analogous manner, except that phase compressibility and vertical leakage between the unsaturated zone and the atmosphere may need to be considered. The vertically integrated Darcy equation for air flow is written as

$$Q_{a_i}=-T_{a_{ij}}\frac{\partial h_a}{\partial x_j} \tag{62}$$

where Q_{a_i} is the vertically integrated air flow rate [L^2T^{-1}], h_a is the air pressure expressed in equivalent water height, and $T_{a_{ij}}$ is the air transmissivity defined by

$$Q_{a_i}=\int_{z_{ao}}^{z_{ua}} q_{a_i} dz \tag{63a}$$

$$T_{a_{ij}}=\int_{z_{ao}}^{z_{ua}} K_{a_{ij}} dz \tag{63b}$$

where q_{a_i} is the Darcy velocity of air, $K_{a_{ij}}$ is the air conductivity, and the limits of integration are from the air–oil table, Z_{ao} (or air–water table in the absence of oil), to an effective upper elevation for horizontal gas flow, Z_{ua}. The latter may be operationally defined as the bottom of a layer of low air permeability (if one exists), as the ground surface if it is sealed, or as the elevation interpolated through upper screen elevations for vacuum wells if leakage from the ground surface is considered. Air flow between Z_{ao} and Z_{ua} is modeled as horizontal and air flow between Z_{ua} and the ground surface is treated as vertical.

Integrating the gas phase continuity equation (3) over the vertical domain yields

$$\frac{1}{\rho_a}\frac{\partial \rho_a V_a}{\partial t}=-\frac{\partial Q_{a_i}}{\partial x_i}+J_a \tag{64a}$$

where V_a is the volume of air per horizontal area [L] at a point in the x–y plane defined by

$$V_a=\int_{z_{ao}}^{z_{ua}} \phi S_a dz \tag{64b}$$

and J_a is a vertically integrated source–sink term [LT^{-1}].

Combining (62) and (64) and expanding the time derivative term yields

$$\gamma_{aa}\frac{\partial h_a}{\partial t}+\gamma_{ao}\frac{\partial \Psi_o}{\partial t}=\frac{\partial}{\partial x_i}\left(T_{a_{ij}}\frac{\partial h_a}{\partial x_j}\right)+J_a \tag{65a}$$

where the air capacity coefficients are defined by

$$\gamma_{aa}=\frac{V_a}{\rho_a}\frac{\partial \rho_a}{\partial h_a}+\frac{\partial V_a}{\partial Z_{ao}} \tag{65b}$$

$$\gamma_{ao}=\frac{\partial V_a}{\partial \Psi_o} \tag{65c}$$

Note that the first term reflects the gas phase compressibility and the second term reflects changes in air volume associated with liquid level changes. In the absence of free oil, $Z_{ao}=Z_{aw}$ and $\Psi_o=\Psi_w$.

Air leakage between the soil and the atmosphere will occur unless an impermeable layer intervenes. Leakage may be described in the manner of Benson, Huntley & Johnson (1993) as

$$J_a=-\frac{K_{sw_v}h_a}{\eta_{ra}Z_a} \tag{66}$$

where K_{sw_v} is the vertical hydraulic conductivity, h_a is the gauge air pressure in the soil (i.e. measured relative to atmospheric pressure), η_{ra} is the ratio of air to water viscosity (0.018), and Z_a is the average distance from the ground surface to the region of air flow. The latter may be approximated as $Z_{gs}-(Z_{ua}+Z_{ao})/2$, where Z_{gs} is the ground surface elevation, Z_{ua} is the upper elevation of horizontal gas flow region, and Z_{ao} is the air–oil table elevation (or air–water table elevation in the absence of oil).

5.2 Vertically integrated constitutive relations

The relationships of phase specific volumes and transmissivities with table elevations are necessary to solve the vertically integrated flow equations. For the wetting and nonwetting phases (i.e. water and air), a sharp interface model can be used to relate air and water specific volumes and transmissivities to table elevations in a manner analogous to that used in conventional vertically integrated groundwater flow models. The relevant table elevations for air and water will be the air–oil and oil–water table elevations, respectively, if oil is present, or the air–water table in the absence of oil.

For the intermediate wettability phase, i.e. NAPL, sharp interface approximations will prove inaccurate in most practical circumstances. Thus, (55b) and (57b) must be integrated considering variations in oil saturation and relative permeability with elevation. This may be performed using a suitable model for three-phase permeability–saturation–capillary pressure relations, as discussed in Section 2.

Hysteresis in the vertically integrated relations is of great importance, particularly in defining the oil specific volume. It is useful to consider the total oil specific volume to be accounted for as follows:

$$V_o=V_{of}+V_{og}+V_{or} \tag{67}$$

where V_{of} is the free or hydraulically mobile oil specific volume, V_{og} is the residual oil specific volume in the unsaturated zone, and V_{or} is the residual oil specific volume in the saturated zone. Free oil specific volume is a function of the apparent product thickness, $H_o=Z_{ao}-Z_{ow}$, and of the soil and fluid properties. A typical relationship for a sandy soil is illustrated in Fig. 7. Note that a certain degree of hysteresis in free oil specific volume versus product thickness may occur.

Residual oil specific volume is strictly a function of the wetting history at a given location. Residual oil in the saturated zone will be induced during periods of water imbibition, due to incomplete displacement of oil by water from soil pores. Water imbibition occurs when oil–water capillary pressure decreases, which in turn occurs during periods of rising oil–water table elevations. Therefore, the residual oil specific volume in the saturated zone is controlled by the difference between the current oil–water elevation and the historical minimum oil–water table elevation. Similarly, residual oil in the unsaturated zone occurs when the air–oil capillary elevation pressure increases, which in turn occurs during periods of falling air–oil table elevations. Therefore, the residual oil specific volume in the unsaturated zone is controlled by the difference between the current air-oil table elevation and the historical maximum air–oil table elevation.

Oil transmissivity is controlled by the vertical distribution of free oil, which is related to the current apparent product thickness. Assuming vertical equilibrium pressure distributions, the vertical distribution of fluid saturation, and thus relative permeability, may be computed and the vertically integrated behavior determined. A typical relationship between oil transmissivity and free oil specific volume is illustrated in Fig. 10.

5.3 Mobile zone mass transport

We turn now to the consideration of dissolved phase transport for species originating from a NAPL plume in porous or fractured media. The NAPL plume is assumed to be immobile (e.g. at residual saturation or hydraulically contained) and soluble contaminants are gradually released to the groundwater. The aquifer is conceptualized as a bimodal porous medium with mobile water occupying a fraction of the pore space (e.g. fractures or foliation planes) and immobile water in the remainder (e.g. relatively impermeable matrix). Transport occurs by convection and dispersion in

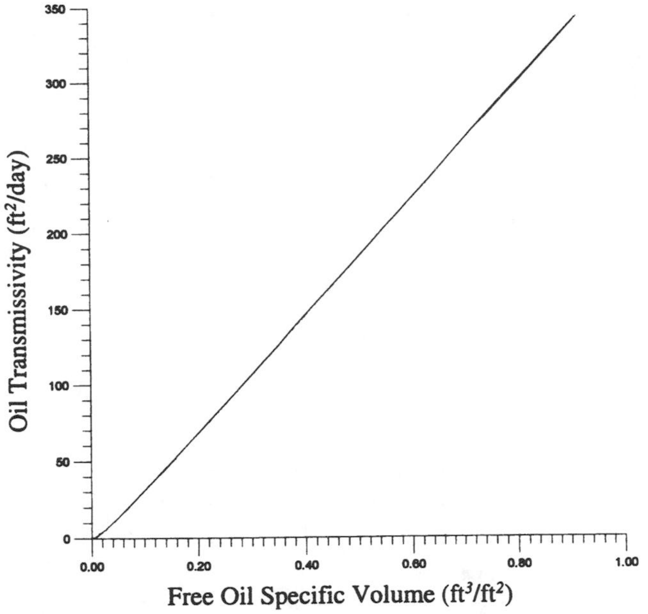

Fig. 10 Oil transmissivity versus free oil specific volume for gasoline in a sandy soil.

mobile zones with diffusive mass transfer between mobile and matrix zones. Equilibrium adsorption, oxygen-limited biodecay or first-order biodecay may be considered within both the mobile and matrix zones.

The vertically integrated governing equation for transport of dissolved species in the mobile regions may be written as

$$(\phi_m+\rho_b f_m k_\alpha)\frac{\partial C_{m\alpha}}{\partial t}=\frac{\partial}{\partial x_i}\left(\phi_m D_{ij}\frac{\partial C_{m\alpha}}{\partial x_j}\right)-\frac{\partial(q_i C_{m\alpha})}{\partial x_i}$$
$$-\gamma_{m\alpha}-Q_v C^{w}_{\alpha}+\frac{S_\alpha}{L}-M_\alpha \qquad (68)$$

where $C_{m\alpha}$ is the aqueous phase concentration of species α (contaminant or oxygen) in the mobile regions [ML^{-3}], t is time [T], x_i and x_j are horizontal spatial coordinates ($i,j=1,2$) [L], ϕ_m is the mobile zone porosity [L^3L^{-3}], ρ_b is the aquifer dry bulk density [ML^{-3}], f_m is the fraction of the aquifer solids that are in the mobile zone, k_α is the solid–water distribution coefficient [L^3M^{-3}], D_{ij} is a dispersion tensor in the mobile region [L^2T^{-1}], q_i is the Darcy velocity for water in the i direction [LT^{-1}], $\gamma_{m\alpha}$ is the decay rate for species α in the mobile region [$ML^{-3}T^{-1}$], Q_v is the volumetric withdrawal (+) or injection (−) rate of water per unit volume of aquifer [T^{-1}], C^w_α is the concentration of α in water being withdrawn or injected, S_α is a source term due to dissolution of contaminant from the separate phase plume [$ML^{-3}T^{-1}$], L is the aquifer thickness [L], and M_α is the net rate of mass transfer between fracture and matrix zones (positive for loss to mobile zone). For steady-state flow, mass conservation for the water phase requires

$$\frac{\partial q_i}{\partial x_i}=-Q_v \qquad (69)$$

If Q_v represents distributed recharge, it is equal to the recharge rate [LT^{-1}] divided by aquifer thickness. For point sources or sinks, it is the volumetric rate divided by aquifer thickness and the nodal area. Expanding the second term on the right-hand side of (68), using (69) after multiplying by $C_{m\alpha}$, yields

$$(\phi_m+\rho_b f_m k_d)\frac{\partial C_{m\alpha}}{\partial t}=\frac{\partial}{\partial x_i}\left(\phi_m D_{ij}\frac{\partial C_{m\alpha}}{\partial x_j}\right)-q_i\frac{\partial C_{m\alpha}}{\partial x_i}-\gamma_{m\alpha}+\frac{S_\alpha}{L}$$
$$-M_\alpha-(C^w_\alpha-C_{m\alpha})Q_v \qquad (70)$$

For pumping wells, we assume $C^w_\alpha=C_{m\alpha}$, which drops the last term in (70). For injection wells, the concentration of injected water, C^w_α, must be specified.

The dispersion tensor in the mobile pore region is characterized by (Bear, 1972)

$$D_{ij}=\frac{1}{\phi}\left(A_T\bar{q}\delta_{ij}+(A_L-A_T)\frac{|q_iq_j|}{\bar{q}}\right) \qquad (71)$$

where A_L and A_T are longitudinal and transverse dispersivities [L], q_i and q_j are Darcy velocities in the i and j directions, $\bar{q}=(\Sigma q_i^2)^{1/2}$ is the magnitude of the resultant water velocity, and δ_{ij} is Kronecker's delta.

5.4 Treatment of fracture–matrix interactions

For problems involving nonequilibrium mobile–immobile region mass transfer, the movement of dissolved species between mobile and immobile regions (e.g. fractures and soil matrix) is characterized by a first-order mass transfer function of the form

$$M_\alpha=\lambda_\alpha(C_{m\alpha}-C_{im\alpha}) \qquad (72)$$

where $C_{im\alpha}$ is the aqueous concentration of species α in the immobile pore region and λ_α is the fracture–matrix mass transfer coefficient for species α [T^{-1}]. A mass balance for the immobile pore water, considering adsorption and decay, may be written as

$$(\phi_{im}+\rho_b f_{im} k_\alpha)\frac{\partial C_{im\alpha}}{\partial t}=\lambda_\alpha(C_{m\alpha}-C_{im\alpha})-\gamma_{im\alpha} \qquad (73)$$

where f_{im} is the fraction of solid mass in the immobile region, $\phi_{im}=\phi-\phi_m$ is the immobile zone porosity, where ϕ is total porosity, and $\gamma_{im\alpha}$ is the decay rate for species α in the matrix. Mass transfer between mobile and immobile pore water is controlled by diffusion and depends on the geometry of the diffusion path. Following the approaches of van Genuchten (1985) and Parker & Valocchi (1986), and using the Millington and Quirk model to estimate the effective diffusion coefficient, yields

$$\lambda_\alpha = \frac{a\phi_{im}^{7/3} D_{o\alpha}}{L_{im}^2} \tag{74}$$

where ϕ_{im} is the immobile zone porosity, $D_{o\alpha}$ is the diffusion coefficient of α in bulk water [L^2T^{-1}], L_{im} is a diffusion length, and a is a geometry factor. For a system of spherical aggregates, L_{im} is the average aggregate diameter and $a=60$ (Parker & Valocchi, 1986). For a system of parallel fractures, L_{im} is the average distance between fractures (or fracture zones) and $a=60/2\pi \approx 10$. Assuming the bulk density in the mobile and immobile zones are approximately equal, then

$$f_m = \frac{\phi_m}{\phi} \tag{75a}$$

$$f_m = 1 - \frac{\phi_m}{\phi} \tag{75b}$$

Note that if $\phi_m=1$, then $f_m=1$ and $f_{im}=0$, and the transport model reduces to the conventional convection-dispersion model.

5.5 Source due to hydrocarbon dissolution

Contaminant dissolution from NAPL is considered as a spatially and temporally distributed source in the transport equation. We represent the oil–water mass transfer rate as the sum of two mechanisms: (1) source due to leaching of contaminant species by water infiltrating vertically through the NAPL, and (2) removal by groundwater flowing transversely past NAPL in the vicinity of the water table. The first mechanism is handled through the pumping/recharge term $Q_v C_\alpha^w$ in the transport equation (68) assuming equilibrium between infiltrating water and the NAPL as

$$Q_v C_\alpha^w = \frac{-q_u C_\alpha^{eq}}{L} \tag{76}$$

where q_u is the recharge rate, C_α^{eq} is the aqueous concentration of species α [ML^{-3}] in equilibrium with the current oil phase composition, and L is the effective aquifer thickness. The relationship between C_α^{eq} and the liquid hydrocarbon plume composition will be discussed later.

The second mechanism of oil–water mass transfer is treated using the first-order mass transfer model of Pfannkuch (1984)

$$S_\alpha = k_o \bar{q}(C_\alpha^{eq} - C_{m\alpha}) \tag{77}$$

where k_o is a dimensionless oil–water mass transfer coefficient, $\bar{q}$ is the magnitude of the groundwater velocity, C_α^{eq} is the equilibrium dissolved concentration, and $C_{m\alpha}$ is the actual concentration in the mobile region at the end of the previous time-step. Since (77) only applies if liquid hydrocarbon is present at a location, k_o is defined to be zero at nodes where no separate phase hydrocarbon is present. Since the composition of the hydrocarbon changes slowly, C_α^{eq} is lagged by one time-step. Note that $\bar{q}$, C_α^{eq}, and $C_{m\alpha}$ will vary spatially within the solution domain. The oil–water mass transfer coefficient may be approximated by

$$k_o = \frac{A_V}{L} \tag{78}$$

where A_V is the vertical dispersivity of the aquifer [L] and L is the effective aquifer thickness.

The liquid hydrocarbon plume is initially characterized by an areal distribution of liquid phase hydrocarbon mass per area and by the mass fraction of contaminant species α in the hydrocarbon phase. At each time-step in the numerical solution, the mass remaining in the hydrocarbon plume is updated at each spatial location. Since the hydrocarbon is assumed to be immobile, a mass balance for each species may be written in time-integrated form as

$$\Delta m_\alpha = m_\alpha^t - m_\alpha^{t+1} = (q_u C_\alpha^{eq} + S_{\alpha+} B_\alpha + E_\alpha)dt \tag{79}$$

where m_α^{t+1} and m_α^t are the mass per area of species α in the oil phase at the current and previous time-step, q_u is the water recharge rate through the NAPL [LT^{-1}], S_α is the non-equilibrium oil–water mass transfer rate from (77), B_α is the biodecay rate of α in the NAPL plume, E_α is the volatilization rate [ML^{-2}], and dt is the time-step duration. Biodecay is assumed to be oxygen-limited, and oxygen in infiltrating water is assumed to react with species proportional to their mole fraction in the NAPL as

$$B_\alpha = \frac{q_u C_{xu} F_\alpha}{\Re_\alpha} \tag{80}$$

where C_{xu} is the oxygen concentration in infiltrating water, F_α is the mole fraction of species α in NAPL, and $\Re_\alpha$ is a stoichiometric factor representing the mass of oxygen consumed per mass of contaminant species α. The rate of volatilization from the product may be estimated assuming equilibrium between the hydrocarbon and vapor at the air–product interface using the relation

$$E_\alpha = k_{a\alpha} C_\alpha^{eq} \tag{81}$$

where $k_{a\alpha}$ is a diffusive mass transfer coefficient [LT^{-1}], which may be approximated by the linear diffusion relation using the Millington–Quirk model to estimate the effective diffusion coefficient as

$$k_{\alpha a} = \frac{\phi_a^{4/3} D_{a\alpha} H_\alpha}{L_a} \tag{82}$$

where $D_{a\alpha}$ is the diffusion coefficient for α in bulk air, H_α is a dimensionless Henry's coefficient, and L_a is an effective air diffusion path length, e.g. the distance from a residual NAPL zone to the ground surface, and $\phi_a = \phi(1-S_m)$ is the air-filled porosity.

A mass balance for the phase-separated hydrocarbon may be written as

$$m^{t+1} = m^t - \sum_\alpha \Delta m_\alpha \tag{83}$$

where m^{t+1} and m^t are the total hydrocarbon mass per area in the oil phase at the current and previous time-step. Several species or pseudo-species may be considered, e.g. BTEX, C10–C20 hydrocarbons, etc. The oil phase mole fraction of species α, F_α, is updated from the current species mass per area as

$$F_\alpha = \frac{\dfrac{m_\alpha}{W_\alpha}}{\sum_\alpha \dfrac{m_\alpha}{W_\alpha}} \tag{84}$$

where W_α is the molecular weight of species α, and the sum in the denominator is over all species (or pseudo-species) in the hydrocarbon phase, such that the sum of F_α for all species is equal to unity. The equilibrium aqueous concentration corresponding to this oil phase composition may be computed as

$$C_\alpha^{eq} = F_\alpha C_\alpha^* \tag{85}$$

where C_α^* is the solubility of pure α in water. Values of F_α and C_α^* are updated at each node in the numerical mesh at the end of each time-step. Up to five soluble contaminant species may be modeled. If the sum of the initial mole fractions is less than unity, the remaining hydrocarbon mass is assumed to be 'inert'.

5.6 Oxygen transport and biodecay

Oxygen transport is modeled by (68) in the same manner as contaminant species, except that source terms are treated differently. Oxygen inflow on upstream boundaries is controlled by (73) by specifying the dissolved oxygen concentration in groundwater upstream from the model domain. Oxygen additions to groundwater occur due to distributed recharge and to groundwater injection via the $Q_v C_\alpha^w$ term in (68), except at nodes where NAPL occurs. For the latter, oxygen in recharge water flowing through the NAPL is assumed to be depleted due to microbial activity associated with degrading NAPL (eq. (80)). Oxygen concentrations in natural recharge water and in injection wells are specified by the user. Oxygen additions to groundwater may also be modeled using a mixing function analogous to (77) as

$$S_x = k_x \bar{q}(C_x^{eq} - C_{mx}) \tag{86}$$

where S_x is the mass loading rate of oxygen per area, k_x is a mixing coefficient, $\bar{q}$ is the resultant groundwater velocity, C_x^{eq} is the dissolved concentration of oxygen in the unsaturated zone that can be maintained by diffusion from the atmosphere, and C_{mx} is the dissolved oxygen concentration in the mobile region. It may be more appropriate to employ (86) when recharge rates are very low. With higher recharge rates, (86) may be disregarded since most oxygen addition will occur due to recharge.

Two options are considered for handling biodecay. One option is to assume apparent first-order decay for contaminant species, with decay rates given by

$$\gamma_{m\alpha} = \phi_m \mu_{m\alpha} C_{m\alpha} \tag{87a}$$

$$\gamma_{im\alpha} = \phi_{im} \mu_{im\alpha} C_{im\alpha} \tag{87b}$$

where $\mu_{m\alpha}$ and $\mu_{im\alpha}$ are apparent first-order decay coefficients. If this approach is used, oxygen transport does not need to be modeled, but estimation of first-order decay coefficients is subject to a great deal of uncertainty.

A more rigorous approach to computing decay rates assumes oxygen to be a major factor limiting microbial activity in groundwater. An approach is employed which is similar to that of Borden & Bedient (1986), as implemented numerically by Rifai *et al.* (1988) for single contaminant species transport. Here, we extend the method to consider transport and decay of multiple contaminant species in groundwater that may have mobile and immobile pore regions, and to consider a maximum decay rate associated with limiting factors other than oxygen. The method is based on the assumption that oxygen consumption is essentially instantaneous relative to groundwater velocities. This assumption enables rate expressions to be solved sequentially with the transport equations for oxygen and contaminants.

Following this approach, the solution to the transport equations is obtained in two steps. First, the transport equations for oxygen and for all contaminant species are solved independently at each time-step with decay rates $\gamma_{m\alpha}$ and $\gamma_{im\alpha}$ set to zero. Secondly, species concentrations are corrected for decay before proceeding to the next time-step. The correction step involves the following calculations for each node in the numerical model:

$$C'_{m\alpha} = max\left(0, C_{m\alpha} - \frac{C_{mx} C_{m\alpha}}{\Re_\alpha \Gamma_{m\alpha} \Sigma C_{m\alpha}}, C_{m\alpha} - \frac{R_{max} \Delta t \Gamma_{m\alpha} C_{m\alpha}}{\Sigma \Gamma_{m\alpha} C_{m\alpha}}\right) \tag{88a}$$

$$C'_{mx} = max\left(0, C_{mx} - \Sigma \Re_\alpha \Gamma_{m\alpha}(C_{m\alpha} - C'_{m\alpha})\right) \tag{88b}$$

$$C'_{im\alpha} = max\left(0, C_{im\alpha} - \frac{C_{mx} C_{im\alpha}}{\Re_\alpha \Gamma_{im\alpha} \Sigma C_{im\alpha}}, C_{im\alpha} - \frac{R_{max} \Delta t \Gamma_{im\alpha} C_{im\alpha}}{\Sigma \Gamma_{im\alpha} C_{im\alpha}}\right) \tag{88c}$$

$$C'_{imx} = max\left(0, C_{imx} - \Sigma \Re_\alpha \Gamma_{m\alpha}(C_{im\alpha} - C'_{im\alpha})\right) \tag{88d}$$

where C_{mx} and C_{imx} refer to oxygen concentrations in mobile and immobile regions, $C_{m\alpha}$ and $C_{im\alpha}$ refer to individual contaminant species concentrations, Σ indicates summation over

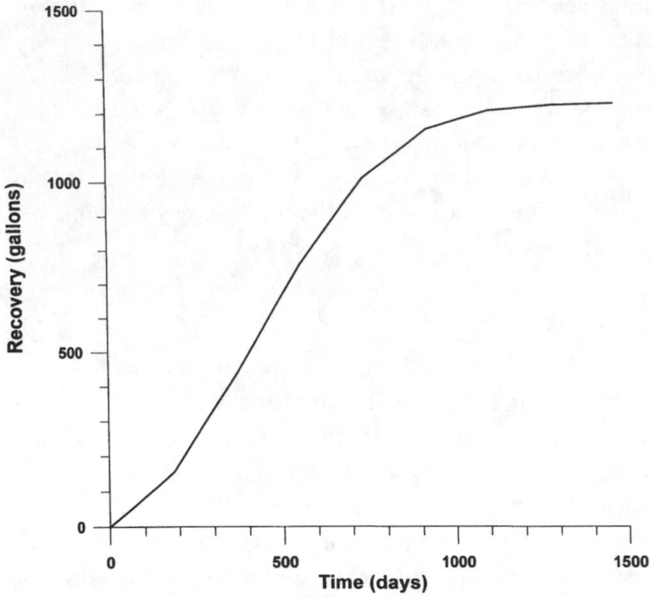

Fig. 11 Free product recovery versus time predicted for the example given in the text using three recovery wells.

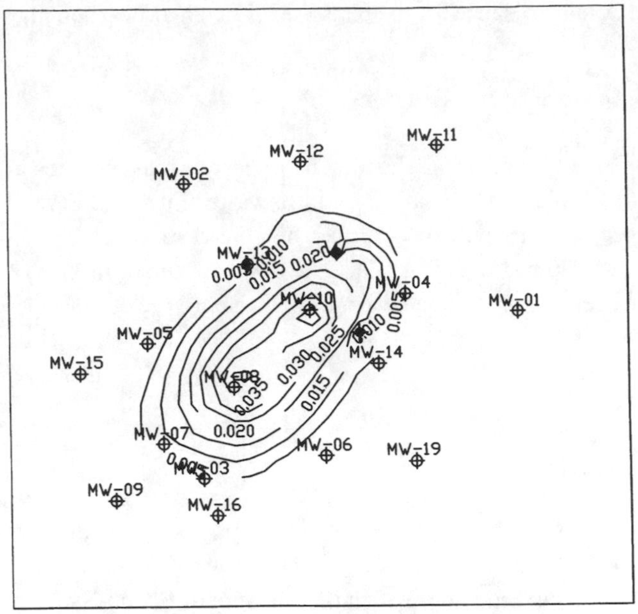

Fig. 12 Predicted spatial distribution of residual oil specific volume ft³/ft³ after reaching asymptotic product recovery and locations of monitoring wells used to estimate initial oil distribution for the example given in the text.

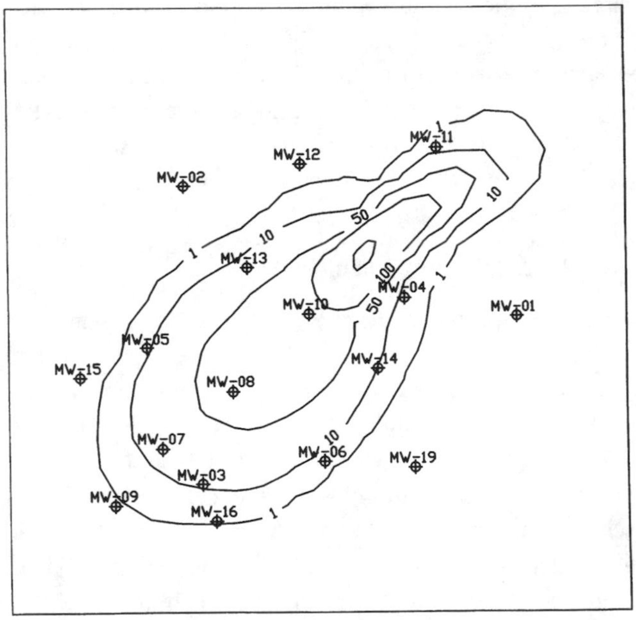

Fig. 13 Predicted dissolved benzene in groundwater (μg/ℓ) after 5 years of pump-and-treat system operation for the example given in the text.

all contaminant species α, $\Re_\alpha$ is the oxygen stoichiometric coefficient, R_{max} is the maximum total contaminant decay rate [MT^{-1}] due to limitations other than oxygen, and $\Gamma_{m\alpha}$ and $\Gamma_{im\alpha}$ are retardation coefficients defined by

$$\Gamma_{m\alpha} = 1 + \frac{\rho_b f_m k_\alpha}{\phi_m} \tag{89a}$$

$$\Gamma_{im\alpha} = 1 + \frac{\rho_b f_m k_\alpha}{\phi_{im}} \tag{89b}$$

Concentrations without a prime in (88) represent values computed by the numerical solution prior to correction, and concentrations with a prime are corrected concentrations. The algorithm assumes that limited oxygen is divided among competing contaminants proportional to their aqueous phase concentrations.

5.7 Example application

The mathematical models for multiphase flow and multicomponent transport have been implemented numerically in the programs ARMOS and BIOTRANS (Environmental Systems & Technologies, Inc., 1994a,b). We will consider here an application to a problem involving a gasoline spill in a sandy unconfined aquifer. From monitoring well fluid level data, the water and oil gradients and the distribution of free oil were determined. The estimated initial free oil volume was 6500 gallons. A recovery system was designed to control free product migration and to recover liquid product. Optimal well placement and pumping rates were determined by trial and error using the numerical model to achieve maximum product recovery with three recovery wells. With a total water pumping rate of 4.5 gallons min^{-1}, the model predicts an asymptotic product recovery of 1250 gallons, or about 20 per cent of the initial free product volume, in about 3.5 years (Fig. 11). The predicted distribution of residual hydrocarbon after reaching asymptotic recovery is shown in Fig. 12.

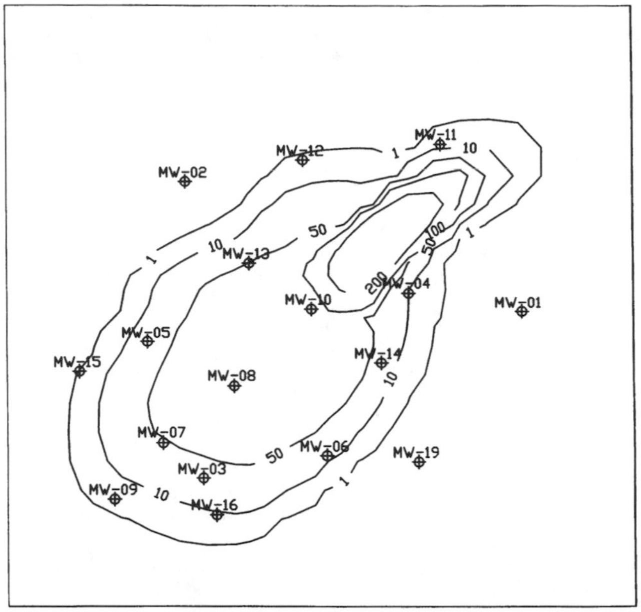

Fig. 14 Predicted dissolved TEX in groundwater (μg/ℓ) after 5 years of pump-and-treat system operation for the example given in the text.

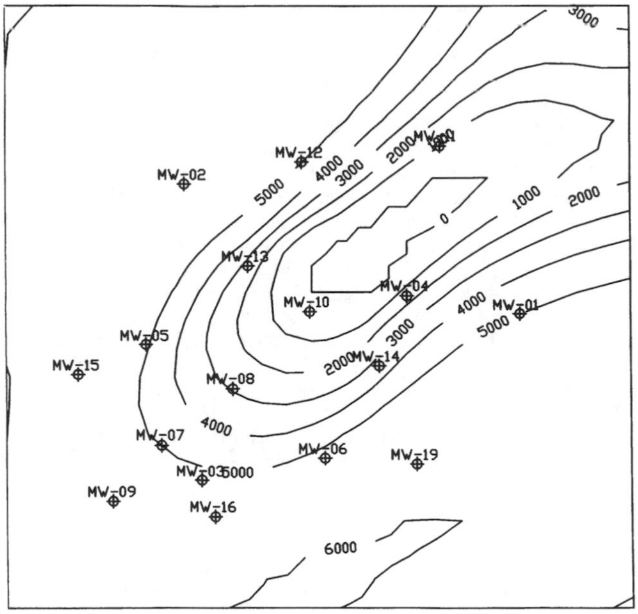

Fig. 15 Predicted dissolved oxygen in groundwater (μg/ℓ) after 5 years of pump-and-treat system operation for the example given in the text.

Dissolved phase transport of benzene, TEX (sum of toluene, ethylbenzene, and xylenes), and oxygen was simulated for a 5 year period, assuming continued operation of the three wells used for product recovery. Predicted dissolved benzene, TEX, and oxygen concentrations in groundwater after 5 years of pump-and-treat system operation are shown in Figs. 13–15, respectively. The maximum benzene concentration is slightly higher than the maximum TEX concentration, reflecting the higher mole fraction of TEX in gasoline compared with benzene. The zones of highest contaminant concentrations correspond to the area of depleted oxygen downgradient of the residual hydrocarbon plume, which acts as a source of soluble hydrocarbons over time.

REFERENCES

Acar, Y. B., Hamidon, A. B., Field, S. D. & Scott, L. (1985). The effect of organic fluids on hydraulic conductivity of compacted kaolinite. *Hydraulic Barriers In Soil and Rock*, ASTM STP, 874, 171–187.

Aziz, K. & Setari, A. (1979). *Petroleum Reservoir Simulation*. London: Applied Science Publications.

Baker, L. E. (1988). Three-phase relative permeability correlations. *Society of Petroleum Engineering*, SPE/DOE 17369, pp. 539–554.

Bear, J. (1972). *Dynamics of Fluids in Porous Media*. New York: Dover Publications.

Bear, J., Braester C. & Menier, P. C. (1987). Effective and relative permeabilities of anisotropic porous media. *Transport in Porous Media*, 2, 301–316.

Benson, D. A., Huntley, D. & Johnson P. C. (1993). Modeling vapor extraction and general transport in the presence of NAPL mixtures and nonideal conditions. *Groundwater*, 31, 437–445.

Borden, R. C. & Bedient, P. B. (1986). Transport of dissolved hydrocarbons influenced by oxygen-limited biodegradation. *Water Resources Research*, 22, 1973–1982.

Brown, C. L., Pope, G. L., Abriola, L. M. and Sepehroori, K. (1994). Simulation of surfactant-enhanced aquifer remediation. *Water Resources Research*, 30, 2959–2977.

Chen, Y.-M., Abriola, L. M., Alvarez, P. J. J., Anid, P. J. & Vogel, T. M. (1992). Modeling transport and biodegradation of benzene and toluene in sandy aquifer material: comparisons with experimental measurements. *Water Resources Research*, 28, 1833–1847.

Corey, A. T. (1986) *Mechanics of Immiscible Fluids in Porous Media*. Littleton, Colorado: Water Resources Publications.

Dullien, F. A. I. (1979). *Porous Media: Fluid Transport and Pore Structure*. San Diego, California: Academic.

Environmental Systems & Technologies, Inc. (1994a). *ARMOS, Areal Multiphase Organic Simulator for Free Phase Hydrocarbon Migration and Recovery*. Version 5, Program User and Technical Guide.

Environmental Systems & Technologies, Inc. (1994b). *BIOTRANS, Areal Multicomponent Transport and Biodegradation of Dissolved Contaminants from NAPLs in Groundwater*. Program User and Technical Guide.

Falta, R. W., Pruess, K., Javandel, I. & Witherspoon, P. A. (1992). Numerical modeling of steam injection for the removal of nonaqueous phase liquids for the subsurface 1. Numerical formulation. *Water Resources Research*, 28, 433–449.

Gelhar, L. W. (1986). Stochastic subsurface hydrology from theory to applications. *Water Resources Research*, 22, 135S–145S.

Greenkorn, R. A. (1983). *Flow Phenomena in Porous Media*. 550 pp. New York: Marcel Dekker.

Havercamp, R. & Parlange, J.-Y. (1983). Prediction of water retention curve from particle size distribution, 1, Sandy soils without organic matter. *Soil Science*, 142, 325–339.

Hockmuth, D. P. & Sunada. D. K. (1985). Ground water model of two phase immiscible flow in coarse material. *Groundwater*, 23, 617–626.

Huntley, D. & Hawk, R. N. (1992). Non-aqueous phase hydrocarbon saturations and mobility in a fine-grained, poorly consolidated sandstone. *Proceedings of Petroleum Hydrocarbons and Organic Chemicals in Groundwater*. Houston: API/NGWA.

Kaluarachchi, J. J. & Parker, J. C. (1987). Effects of hysteretics of water flow in the unsaturated zone. *Water Resources Research*, 23, 1967–1976.

Kaluarachchi, J. J. & Parker, J. C. (1989a). An efficient finite element method for modeling multiphase flow. *Water Resources Research*, 25, 43–45.

Kaluarachchi, J. J. & Parker, J. C. (1989b). Multiphase flow in porous media with a simplified model for oil entrapment. *Transport in Porous Media*, 7, 1–14.

Kaluarachchi, J. J., Parker, J. C. & Lenhard, R. J. (1990). A numerical model for a real migration of water and light hydrocarbon in unconfined aquifers. *Advances in Water Resources*, 13, 29–40.

Kaluarachchi, J. J., Parker, J. C. & Lenhard, R. J. (1992). Modeling flow in three fluid phase porous media with nonwetting fluid entrapment. In *Proceedings of the Conference on Subsurface Contamination by Immiscible Fluids*, ed. K. U. Weyes. Rotterdam: A. A. Balkema, pp. 203–209.

Katyal, A. K., Kaluarachchi, J. J. & Parker, J. C. (1992). Evaluation of methods for improving the efficiency and robustness of multiphase flow equations. In *Subsurface Contamination by Immiscible Fluids*, ed. K. U. Weyer, Rotterdam: Balkema, pp. 203–209.

Katyal, A. K. & Parker, J. C. (1992). An adaptive solution domain algorithm for solving multiphase flow equations. *Computers in Geosciences*, 18(1), 1–9.

Kemblowski, M. W. & Chiang, C. Y. (1990). Hydrocarbon thickness fluctuations in monitoring wells. *Groundwater*, 28, 244–252.

Kool, J. B. & Parker, J. C. (1987). Development and evaluation of closed-form expressions for hysteretic soil hydraulic properties. *Water Resources Research*, 23, 105–114.

Kool, J. B. & Parker, J. C. (1988). Analysis of the inverse problem for unsaturated transient flow. *Water Resources Research*, 24, 817–830.

Land, C. S. (1968). Calculation of imbibition relative permeability for two- and three-phase flow from rock properties. *Transactions of the American Institute of Mining, Metallurgy and Petroleum Engineering*, 243, 149–156.

Lenhard, R. J. & Parker, J. C. (1987a). Measurement and prediction of saturation-pressure relationships in three phase porous media systems. *Journal of Contaminant Hydrology*, 1, 407–424.

Lenhard, R. J. & Parker, J. C. (1987b). A model for hysteretic constitutive relations governing multiphase flow, 2, Permeability-saturation relations. *Water Resources Research*, 23, 2197–2206.

Lenhard, R. J. & Parker, J. C. (1988). Experimental validation of the theory of extending two phase saturation-pressure relations to three phase systems for monotonic saturation paths. *Water Resources Research*, 24, 817–830.

Lenhard, R. J., Parker, J. C. & Kaluarachchi, J. J. (1989). A model for hysteretic constitutive relations governing multiphase flow, 3, Refinements and numerical simulations. *Water Resources Research*, 25, 1727–1736.

Lenhard, R. J., Parker, J. C. & Kaluarachchi, J. J. (1991). Computing simulated and experimental hysteretic two-phase fluid flow phenomena. *Water Resources Research*, 27, 2113–2124.

Leverett, M. C. (1941). Capillary behavior in porous solids. *Transactions of the American Institution of Mining Metallurgy and Petroleum Engineering*, 142, 152–169.

Lyman, W. J., Reehl, W. F. & Rosenblatt D. H. (1982). *Handbook of Chemical Property Estimation Methods*. New York: McGraw-Hill.

McBride, J. F., Simmons, C. S. & Carey, J. W. (1992). Interfacial spreading effects on one-dimensional organic liquid imbibition in water-wetted porous media. *Journal of Contaminant Hydrology*, 11, 1–25.

Mantoglou, A. & Gelhar, L. W. (1987). Effective hydraulic conductivities of transient unsaturated flow in stratified soils. *Water Resources Research*, 23, 57–67.

Martin, J. C. (1968). Partial integration of equations of multiphase flow. *Trans. SPE of AIME*, 243, 370–380.

Millington, R. J. & Quirk, J. P. (1959). Permeability of porous media. *Nature (London)*, 183, 387–388.

Mualem, Y. (1976). A new model for predicting the hydraulic conductivity of unsaturated media. *Water Resources Research*, 12, 513–522.

Parker, J. C. & Lenhard, R. J. (1987). A model for hysteretic constitutive relations governing multiphase flow, 1. Saturation-pressure relations. *Water Resources Research*, 23, 2187–2196.

Parker, J. C. & Lenhard, R. J. (1989). Vertical integration of three-phase flow equations for analysis of light hydrocarbon plume movement. *Transport in Porous Media*, 5, 187–206.

Parker, J. C., Lenhard, R. J. & Kuppasamy, T. (1987). A parametric model for constitutive properties governing multiphase flow in porous media. *Water Resources Research*, 23, 618–624.

Parker, J. C. & Valocchi, A. J. (1986). Constraints on the validity of equilibrium and first-order kinetic transport models in structured soils. *Water Resources Research*, 22, 399–407.

Parker, J. C., Zhu, J. L., Johnson, T. G., Kremesec, V. J. & Hockman, E. L. (1994). Modeling free product migration and recovery at hydrocarbon spill sites. *Groundwater*, 32, 119–128.

Pfannkuch, H. (1984). Determination of the contaminant source strength from mass exchange processes at the petroleum groundwater interface in shallow aquifer systems. In *Proceedings of the International Conference on Petroleum Hydrocarbons and Organic Chemicals in Groundwater*. Houston: NWWA/API, pp. 111–129.

Rifai, H. S., Bedient, P. B., Borden, R. C. & Haasbeek, J. F. (1988). *Bioplume II: A Computer Model of Two-Dimensional Contaminant Transport Under the Influence of Oxygen Limited Biodegradation in Ground Water*. U.S. Environmental Protection Agency Report, EPA/600/8–88/093a.

Schiegg, H. O. (1983). Considerations on water, oil, and air in porous media. *Water Science Technology*, 17, 467–476.

Stone, H. L. (1970). Probability model for estimating three phase relative permeability. *Journal of Petroleum Technology*, 22, 214–218.

Stone, H. L. (1973). Estimation of three phase relative permeability and residual oil saturation. *Journal of Petroleum Technology*, 12, 53–61.

van Genuchten, M. T. (1980). A closed form equation for predicting the hydraulic conductivity of unsaturated soils. *Soil Science Society of America Journal*, 44, 892–898.

van Genuchten, M. T. (1985). *A General Approach for Modeling Solute Transport in Structured Soils*. Proceedings of the 17th International Congress of Hydrogeology of Rocks of Low Permeability, IAH, Jan. 7–12.

Wan, J. & Wilson, J. L. (1994). Colloid transport in unsaturated porous media. *Water Resources Research*, 30, 857–864.

VI

A view to the future

1 Stochastic approach to subsurface flow and transport: a view to the future

SHLOMO P. NEUMAN
University of Arizona

1 STATISTICAL CHARACTERIZATION OF GEOLOGIC COMPLEXITY

Subsurface flow and transport parameters such as permeability, porosity and dispersivity have been traditionally viewed as well-defined local quantities that can be assigned unique values at each point in space. Yet subsurface flow and transport take place in a complex geologic environment whose lithologic, petrophysical and structural makeups vary in ways that cannot be predicted deterministically in all of their relevant details. These makeups tend to exhibit discrete and continuous variations on a multiplicity of scales, causing flow and transport parameters to do likewise. In practice, such parameters can at best be measured at selected well locations and depth intervals, where their values depend on the scale (support volume) and mode (instrumentation and procedure) of measurement. Estimating the parameters at points where measurements are not available entails a random error. Quite often, the support of measurement is uncertain and the data are corrupted by experimental and interpretive errors. These errors and uncertainties render the parameters random and the corresponding flow and transport equations stochastic.

Though the uncertain nature of flow and transport parameters is now widely recognized, there does not yet appear to be a consensus about the best way to deal with it mathematically. The most prevalent approach has been that represented by the geostatistical school of thought. According to this philosophy, parameter values determined at various points within a more-or-less distinct hydrogeologic unit can be viewed as a sample from a random field defined over a continuum. This random field is characterized by a joint (multivariate) probability density function or, equivalently, its joint ensemble moments. Thus, a parameter such as (natural) log hydraulic conductivity $Y(\mathbf{x})=\ln K(\mathbf{x})$ varies not only across the real space coordinates $\mathbf{x}$ within the unit, but also in probability space (this variation may be represented by another 'coordinate' ξ which, for simplicity, we shall suppress). Whereas spatial moments are obtained by sampling $Y(\mathbf{x})$ in real space (across $\mathbf{x}$), ensemble moments are defined in terms of samples collected in probability space (across ξ).

Stochastic theories of flow and transport typically assume that all requisite statistical properties of the random parameters (including initial conditions and forcing terms) can be inferred from measurements. To render such inferences statistically meaningful, there is a need for relatively large samples of measured parameter values. Yet the number of flow and transport parameters that can be measured at any given site is usually quite limited. It is therefore important to augment such data with other hard or soft information which is related to it quantitatively or qualitatively. This is why there have been in recent years so many attempts to discover relationships between geological, geophysical and, to a lesser extent, geochemical signatures of the subsurface environment and parameters such as permeability and porosity. Much of this effort has centered on petroleum reservoirs and on low-permeability fractured rocks considered potentially suitable for underground (primarily nuclear) waste storage. Among the most promising developments in this area are geophysical methods (such as seismic and electromagnetic geotomography) which remotely image the subsurface in a way that correlates with permeability (Ramirez, 1986), porosity, and fluid or salt content (SKB, 1993). The further development of such methods of field investigation, and of computational algorithms which take advantage of such correlations to help reduce uncertainty about subsurface flow and transport conditions, constitutes important challenges for future research. Recent examples of relevant algorithms include the procedures to jointly invert hydrologic and seismic data due to Rubin, Mavko & Harris (1992) and Hyndman, Harris & Gorelick (1994).

The recognition that geology is complex and uncertain has prompted the development of geostatistical methods to help reconstruct it on the basis of limited data. The state of the art is most advanced among petroleum specialists whose motivation is the development of optimum management strategies for oil reservoirs. An example is the recent case study described by Rossini *et al.* (1994). The reservoir shows extreme variability, comprising all the transitional lithologies from sand to dolomite. Due to the limited continuity of its lithologic facies, their spatial distribution is difficult to identify. The authors started by identifying deterministically three hydraulically separated layers in a vertical cross-section, distinguished in part by their diverse original gas/oil and oil/water contacts. After developing a petrophysical criterion to distinguish between sandy and dolomitic facies, they established a frequency distribution of porosities, and a correlation between log permeabilities and porosities, for each facies on the basis of core data. Likewise, they developed horizontal and vertical indicator semivariograms of porosity for each facies. Next, the authors divided the three layers into 1 645 000 cells measuring 50 m horizontally and 0.5 m vertically. They then generated ten equally likely random images of the facies across this grid by conditional stochastic indicator simulation, and assigned random porosities and permeabilities to grid blocks within each facies. To simulate flow through each of the ten generated reservoirs, the authors superimposed a coarse grid over the original fine grid and assigned an 'upscaled' porosity and permeability to each coarse grid block. More will be said about upscaling later. The case study underscores the need to continue developing advanced techniques for the reconstruction of formation architecture under uncertainty so as to enhance the simulation of flow and transport. A more formidable challenge is to base such a reconstruction not only on the geostatistical interpolation of spatial data but also on well-founded geodynamic models of basin evolution. An example of such a model has recently been described by Koltermann & Gorelick (1992).

The extent to which continuum geostatistical and stochastic concepts may or may not apply to fractured rocks has been the subject of intense research and debate for well over a decade. In such rocks, flow and transport often take place preferentially through discrete fractures and channels. Usually, some of these discontinuities can be identified and mapped in surface outcrops, boreholes and subsurface openings. This has led to the widely held belief that it should be possible to delineate the geometry of the subsurface 'plumbing system' through which most flow and advective transport must take place. Many consider it especially feasible to construct realistic models of fracture networks deterministically or stochastically. Typically, such networks consist of discrete polygonal or oval-shaped planes of finite size, embedded in an impermeable, or at times permeable, rock matrix. Each plane is assigned effective flow and transport properties, usually at random; in some single-fracture studies, these properties are further treated as random fields defined at each point in the fracture plane. Fracture network models containing thousands of planes have been used to simulate flow and tracer migration at several experimental sites, most notably in crystalline rocks of the Site Characterization and Validation (SCV) complex at the Stripa mine in Sweden and the Fanay-Augeres mine in France. Cacas *et al.* (1990a,b) claim to have validated the fracture network modelling approach at the Fanay-Augeres mine.

The conceptual framework behind the discrete fracture modeling approach has been criticized by Neuman (1987, 1988) as being contrary to experimental evidence. Neuman also questioned the practicality of the approach on the grounds that existing field techniques make it extremely difficult, if not impossible, to reconstruct with any reasonable degree of fidelity either the geometry of the subsurface plumbing system or the flow and transport properties of its individual components (fractures and channels). Indeed, a similar conclusion was recently reached by Tsang & Neuman (1997) based on extensive experience gained during the six-year international INTRAVAL project. The authors pointed out that several INTRAVAL field hydraulic and tracer experiments have proven equally amenable to analysis by discrete and continuum models, rendering the validation of either approach difficult. The best models appeared to be those that were neither too simplistic nor too complex. A recent summary of the international Stripa project (SKB, 1993) has concluded that, while it has been possible to construct working fracture network models with thousands of discrete planes for the SCV site by calibrating them against observed hydraulic and tracer data, these models have generally not performed better than much simpler and more parsimonious continuum models, and have sometimes been worse.

This and other recent work provide support for an earlier premise by Neuman (1987, 1988) that flow and transport in many fractured rock environments should be amenable to analysis by continuum models which account adequately for medium heterogeneity. In many cases, it may be possible to distinguish deterministically between distinct zones of low and high permeability on scales not much smaller than the domain of interest (as has been done for a block of crystalline rock at Chalk River in Ontario, Canada, by Carrera *et al.* (1990)), and then to treat the internal properties of each such discrete unit as random fields. The most recent evidence that the latter idea often works can be found in the theses of Kostner (1993), Ando (1995) and Guzman (1997). The first two theses demonstrate that hydraulic and tracer tests at the Fanay-Augeres mine can be reproduced by means of continuum indicator geostatistics, and continuum stochastic flow and transport models, with greater fidelity than has been done previously with discrete fracture network models. The last thesis demonstrates that air-permeability data from unsaturated fractured tuffs near Superior, Arizona, are likewise amenable to continuum geostatistical analysis, exhibiting both anisotropic and random fractal behaviors (more about fractals later). It is important to conduct additional field studies in various fractured rock environments, on various scales and under varied conditions of flow and transport, to continue improving our understanding of these phenomena and the corresponding analytical tools. Another remaining challenge is to develop analytical tools for special situations such as those found within the igneous-metamorphic complex at Mirror Lake in New Hampshire (Hsieh and Shapiro, 1993), where numerous discrete zones of extremely high permeability but limited extent (which makes them difficult to identify and map out) are embedded within a much less permeable fractured country rock.

The theory of geostatistics, as well as most current stochastic theories of flow and transport, deal, to a large extent, in relationships between ensemble moments. Yet these moments are theoretical artifacts which cannot be evaluated directly in any unique hydrogeological setting because there is no ensemble of statistically similar settings that could be sampled in probability space. Reliance on such artifacts is therefore motivated not by objective reality but by mathematical convenience; geostatistical and stochastic theories provide a convenient, albeit artificial, framework within which spatial (as well as temporal) variability and uncertainty can be accommodated with relative ease. To accept this framework requires a leap of faith which some find disturbing. There is, in my view, little chance of resolving this philosophical dilemma; those who adhere to the existing framework must challenge themselves to demonstrate more convincingly that it often leads to results which are consistent with observations. A formidable challenge for those who reject the current framework is to search for an equally powerful alternative that would be philosophically more satisfying.

The parameter that is most commonly taken to vary randomly in space is the log permeability $Y(\mathbf{x})$. The literature indicates that when $Y(\mathbf{x})$ is sampled at various points $\mathbf{x}$ in a given hydrogeologic unit, the data often fall close to a straight line on normal probability paper. This means that the data can be considered to form a sample from a univariate Gaussian probability distribution. Though this does not necessarily imply that $Y(\mathbf{x})$ is a multivariate Gaussian field, it is nevertheless common to assume so. The latter is a far-reaching and controversial assumption, which implies that if the first two multivariate ensemble moments of $Y(\mathbf{x})$ are specified, so is the entire multivariate distribution of this field, including all its higher moments. Hence only the first two moments of $Y(\mathbf{x})$ need to be inferred from the data. There are those who feel that the multivariate Gaussian assumption renders $Y(\mathbf{x})$ too smooth to allow the reproduction of elongated high-permeability channels and low-permeability barriers that might otherwise be present in the field (Gomez-Hernandez & Wen, 1994). Such channels and barriers may however be generated without sacrificing the simplicity of the multi-Gaussian hypothesis by allowing $Y(\mathbf{x})$ to be autocorrelated over large distances, as is typical of random fractals (Grindrod & Impey, 1992). Many feel that, in the absence of direct evidence to the contrary, there is no justification to reject the parsimonious multi-Gaussian hypothesis. Checking the importance and testing the validity of this hypothesis remains a challenge for future research.

Regardless of whether or not the multi-Gaussian hypothesis is valid for $Y(\mathbf{x})$, one seldom has sufficient data to infer moments of order higher than two for this or any other subsurface flow and transport parameter. To infer the first two moments one usually starts by postulating a parametric model for the spatial mean of the data in the form of a (typically polynomial) drift function, and a parametric model for their spatial variance–covariance structure in the form of a theoretically admissible (positive-semidefinite) spatial covariance function. One then estimates the parameters of these models by fitting the latter in some optimal way to empirically determined values of the corresponding functions, e.g. Samper & Neuman (1989). It is often possible to fit different models to the same data, rendering the estimation process nonunique. Several formal maximum likelihood criteria to discriminate between such alternative models have been explored by the above authors.

An alternative to identifying the spatial covariance of $Y(\mathbf{x})$ is to identify the semivariogram of its spatial increments. If the value of this semivariogram stabilizes at a fixed 'sill' value, or tends asymptotically to a fixed sill, as the distance between such pairs of data increases, one usually takes this to imply that fluctuations in $Y(\mathbf{x})$ about the spatial drift function are statistically homogenous across the field (strictly so if $Y(\mathbf{x})$ is taken to be multi-Gaussian, weakly so or to second-order otherwise). If the spatial drift is additionally constant, one takes this to imply the same about $Y(\mathbf{x})$. One must then invoke ergodicity to justify viewing the spatial mean and covariance (or semivariogram) of such a homogeneous random field as a representation of its ensemble mean and covariance (or semivariogram). Since ergodicity cannot be proven in the absence of an ensemble, this last step constitutes not a testable but merely a working hypothesis on which the entire edifice of stochastic data analysis must ultimately rest. As already mentioned, it poses a philosophical dilemma which to date has not been fully resolved to everyone's satisfaction.

When the semivariogram of $Y(\mathbf{x})$ continues to climb without apparent limit as the distance between data points increases, $Y(\mathbf{x})$ cannot be considered statistically homogeneous. At times, this climb can be arrested by postulating a drift. There is, however, growing evidence (Gelhar, 1993, table 6.1) that the integral scale of $Y(\mathbf{x})$, a measure of the distance beyond which the spatial correlation of this variable appears insignificant and its semivariogram appears to stabilize when one employs standard methods of geostatistical inference, increases consistently with the size of the domain under investigation. This suggests that stationarity may be an artifact of the scale of observation and method of inference rather than a true property of $Y(\mathbf{x})$.

Indeed, when sample semivariograms are plotted on logarithmic paper, rather than on arithmetic paper as has been the standard procedure, the data often lie close to a straight line, implying that the semivariogram grows approximately as a power 2ω of the distance without ever reaching a stable value. Such behavior is being observed at an increasing number of sites on distance scales ranging from a few meters to 100 km (Neuman, 1995). The value of ω is not the same at each site, though it has been found to lie between 0.2 and 0.3 in several recent studies, including that of Guzman (1997). Nevertheless, when one juxtaposes the apparent sills and integral scales of semivariograms from many different sites, one finds that they fit a generalized power-law model with $\omega \approx 0.25$ (Neuman, 1994). This generalized behavior of $Y(\mathbf{x})$ had been deduced earlier by Neuman (1990) from the observed scale-dependence of juxtaposed apparent dispersivities reported for a large number of tracer studies worldwide. The validity of this deduction has been the subject of intense debate (Neuman, 1991, 1993a), which may not close until either confirmation or refutation comes to rest on firmer ground. To the extent that further research will continue confirming this or other generalized patterns of behavior associated with $Y(\mathbf{x})$ or other parameters of relevance to flow and transport, a much more profound challenge will be to explain how the natural subsurface environment evolves toward such a highly organized structure. An answer will most probably have to be sought within the joint framework of traditional earth sciences and newly emerging theories of self-organized complex systems.

A power-law semivariogram is indicative of a nonhomogeneous $Y(\mathbf{x})$ field which has neither a well-defined mean nor finite variance or integral scales, rendering it spatially correlated on all distance scales. Instead, the field possesses homogeneous spatial increments (such fields are called 'intrinsic' in the traditional language of geostatistics). Power-law behavior implies that $Y(\mathbf{x})$ is a random fractal with dimension $D=E+1-\omega$, where E is the topological dimension of the domain under investigation and ω is the so-called Hurst coefficient. Theoretically, ω is restricted to values between 0 and 1. When $\omega=0.5$, the increments are uncorrelated in E-dimensional space and thus resemble Brownian motion. When $\omega>0.5$, the increments are positively correlated and show relatively smooth variations characterized by long-range persistence of positive and negative values (this is the well-known Hurst phenomenon exhibited by time series of flows in major rivers such as the Nile). When $\omega<0.5$, the increments are negatively correlated and $Y(\mathbf{x})$ shows noisy behavior called 'antipersistence'. Regardless of what the magnitude of ω may be, a $Y(\mathbf{x})$ field correlated on a multiplicity of scales gives rise to flow and transport behavior that is fundamentally different from that which may develop in a homogeneous $Y(\mathbf{x})$ field.

As we shall soon see, most existing stochastic theories of flow and transport treat either $Y(\mathbf{x})$ or its fluctuations about the mean as homogeneous random fields. Theories which account explicitly for the multiscale nature of subsurface flow and transport have made their debut in the literature only quite recently, most notably in a book edited by Cushman (1990). Their further development and validation pose a grand challenge to theoreticians, computational analysts and experimentalists well into the future.

2 COMPUTATIONAL STOCHASTIC MODELS OF FLOW AND TRANSPORT

If the statistical properties of relevant random parameters can be inferred from measurements, the stochastic flow and transport equations can be solved numerically by (conditional) Monte Carlo simulation and the results analyzed statistically. The statistics most commonly computed from such simulations include (sample conditional) mean hydraulic heads and gradients, volumetric water fluxes and seepage velocities, solute concentrations and mass fluxes, and plume spatial as well as temporal moments. The true counterparts of these sample means constitute optimum (unbiased) predictors of system behavior under uncertainty. Another statistic commonly computed from Monte Carlo simulations is the (sample conditional) variance of the associated prediction errors.

The Monte Carlo approach is conceptually straightforward and has the advantage of applying to a very broad range of both linear and nonlinear flow and transport problems. However, it has a number of potential drawbacks. To properly resolve high-frequency space-time fluctuations in the random parameters (including random initial and forcing terms), it is necessary to employ fine numerical grids in space-time. To avoid artificial boundary effects, these grids must span large space-time domains. Each sample calculation may therefore place a heavy demand on computer time and storage, especially when one deals with transient, two- and three-dimensional nonlinear flow and transport in strongly heterogeneous media (where the discretized governing equations may become stiff). To insure that the sample output moments converge to their (generally unknown) theoretical ensemble values, a very large number of Monte Carlo runs are often required. Even if some sample moments appear to stabilize after a sufficiently large number of runs, there is generally no guarantee that they have in fact converged.

Part of the problem may stem from the manner in which random fields are generated numerically, and part from the way in which the flow and transport equations are solved; there are multiple sources of potential error which are difficult to either ascertain or control. Hence, the widely accepted practice of viewing Monte Carlo studies as a standard against which one judges the accuracy of other approximate solutions (analytical or otherwise) entails a tangible risk of misjudgment. Indeed, Hsu, Zhang & Neuman (1996) have recently raised a question about the accuracy of some published Monte Carlo studies and their ability to represent correctly higher-order effects on transport. To conduct a very large number of high-resolution Monte Carlo simulations is often not feasible with present day hardware and software within reasonable time and budget constraints. Hence, major challenges for the future lie in the development of accurate and efficient computational tools to conduct such simulations, and in the derivation of reliable criteria by which to ascertain the quality of their results.

Until, and unless, these challenges are met, there remains a strong incentive to pursue alternative computational approaches capable of predicting as accurately and efficiently as possible flow and transport in randomly heterogeneous media. Included among these alternatives are traditional deterministic models. As system outputs are generally nonlinear in the controlling parameters, the conditional mean outputs of Monte Carlo simulations are generally different from outputs one would obtain upon simply replacing the parameters in standard deterministic models by their (conditional) mean values. Such deterministic outputs would generally be biased and therefore less than optimal.

To render deterministic models less biased, there has been an intensive search in the literature for 'equivalent' parameters that could be used to replace their suboptimal counterparts. The search has focused in large part on methods of 'upscaling' which ascribe equivalent parameters to the grid blocks of numerical flow and transport models on the basis of smaller-scale random (or nonrandom) parameter values. Traditionally, upscaling has been conducted numerically based on more-or-less *ad hoc* criteria of equivalence. More rigorous theoretical criteria of equivalence have recently been proposed for hydraulic conductivity by Indelman and Dagan (1993), but these are not easy to implement in practice. A method of upscaling dispersivities has also been proposed by Dagan (1994a).

A major conceptual difficulty with upscaling in real space-time (as opposed to Fourier–Laplace space, which offers a viable but relatively unexplored alternative for the unconditional case) is that it postulates local relationships between (conditional) mean driving forces and fluxes (Darcy's and Fick's laws) when in fact these relationships are generally nonlocal (Cushman & Ginn, 1993; Neuman, 1993b; Neuman & Orr, 1993). Another conceptual difficulty with traditional upscaling is that it requires the *a priori* definition of a numerical grid in the absence of firm theoretical guidelines for its selection. Hence, a major challenge for the future is to either place upscaling on a firmer theoretical footing than has been possible thus far, or to continue developing alternative ways of predicting flow and transport deterministically in a manner consistent with (conditional) stochastic theory.

Deterministic alternatives to (conditional) Monte Carlo simulation seek to predict flow and transport under uncertainty without having to generate random fields or variables. One approach is to write a system of partial differential equations satisfied approximately by the first two (conditional) ensemble moments of a quantity such as hydraulic head or solute concentration, then to solve these equations numerically on a grid in space-time. For example, Graham & McLaughlin (1989) used a distributed parameter Kalman filter coupled with discrete, low-order Eulerian transport equations to predict the first two moments of concentration by finite differences, conditioned on measurements of log transmissivity, hydraulic head and concentration. Contrary to upscaling, in which a grid is defined *a priori* on the basis of more-or-less *ad hoc* criteria, here the grid is defined *a posteriori* based on the degree of smoothness one expects the moment functions to exhibit. This degree of smoothness is in turn controlled to a large extent by the distribution of conditioning points in space-time. In most cases such points are sparse enough to insure that the conditional moment functions fluctuate at much lower spatial and temporal frequencies than do their random counterparts. Hence the grid required to resolve the former is generally much coarser than that required to resolve the latter.

Based on the latter, one might expect numerical schemes based on moment equations to require much less computer time and storage than that necessary for high-resolution Monte Carlo simulations. Yet the particular numerical scheme just described is computationally demanding because the coefficient matrix of the finite difference equations includes covariance terms, which render most of its entries nonzero. In contrast, finite difference matrices which arise from the original transport equation are usually sparse and therefore much easier to store and invert. The governing moment equations are themselves based on a nonasymptotic approximation in which terms neglected are of the same order as terms retained (Dagan & Neuman, 1991). This leaves us with the challenge to develop efficient computational methods for conditional moment equations that rest on a firm theoretical foundation.

I shall refer to numerical methods based on conditional

moment equations as 'smoothing,' to distinguish them from methods based on 'upscaling'. I further wish to propose that such methods of smoothing be in the future based on rigorous conditional moment equations of the kind developed for steady state saturated flow by Neuman & Orr (1993), and for advective transport by Neuman (1993b). Their equations are integrodifferential and therefore nonlocal, the first in space and the second in space-time. The kernels under the corresponding integrals constitute nonlocal parameter functions which, together with local parameters outside the integrals, are conditional on data and therefore nonunique. Though the conditional kernels are theoretically well-defined, they cannot be evaluated directly without either high-resolution Monte Carlo simulation or approximation; a third option is to estimate them indirectly by means of inverse methods. Hence one major challenge is to find suitable approximations for these kernels and/or appropriate inverse methods to estimate their values.

Another major challenge is to develop efficient computational schemes for the solution of nonlocal mean equations. The special case of unconditional nonlocal equations lends itself to efficient numerical analysis in Fourier–Laplace space where such equations take on a local form; one example is the spectral approach of Deng, Cushman & Delleur (1993), which takes advantage of the fast Fourier transform. The same does not generally apply to conditional equations in which the integrals are no longer standard convolutions in space-time. An example of how such equations may be solved to first-order has been described recently in connection with nonreactive advective transport by Zhang & Neuman (1995). They solved the conditional mean concentration equation analytically at early time and expressed it in pseudo-Fickian form at later time. The early time solution, modified after Batchelor (1952), is local in space-time. The pseudo-Fickian equation is nonlocal in time, involving a conditional space-time dependent dispersion tensor which the authors evaluate numerically along mean 'particle' trajectories. The equation lends itself to accurate solution by standard Galerkin finite elements on a relatively coarse grid. It also allows explicit evaluation of the conditional variance–covariance of concentrations together with various spatial and temporal moments of relevance. The method derives much of its strength from reliance on partial localization of the otherwise nonlocal conditional mean transport equation. Such localization may not always be possible; Cushman, Hu & Deng (1995) have shown that, in the case of reactive transport, it is mandatory to account for nonlocal effects if accuracy is desired for moments beyond the first. This underscores the need for more general and powerful computational tools to deal with nonlocal conditional moment equations of flow and transport. It is presently unclear to what extent nonstationary spectral methods such as the one proposed by Li & McLaughlin (1991) may prove useful in this context.

Conditional mean flow and transport equations of the above genre involve conditional local and nonlocal parameters which depend not only on medium properties but also on the information one has about these properties (scale, location, quantity and quality of data). As such, the parameters are nonunique. Darcy's law and Fick's analogy are generally not obeyed by the flow and transport predictors except in special cases or as approximations. Such approximations entail localization which may yield familiar-looking differential equations in which, however, the hydraulic conductivity, seepage velocity and dispersivity are nonunique, information-dependent parameters. This helps to explain why the results of traditional deterministic model calibration exercises tend to vary continuously as more and more data are incorporated in the model. A major educational challenge is to have the profession recognize this intimate relationship between deterministic and stochastic models and its implications concerning the inherently nonunique nature of traditional deterministic flow and transport parameters.

An equally important challenge is to develop computational tools that are able to handle not only the forward problem of predicting flow and transport under uncertainty but also the inverse problem of estimating conditional mean local and nonlocal parameters of the corresponding moment equations from observed system behavior. These tools must consider explicitly that the very act of parameter estimation, by virtue of its reliance on an expanded data base, modifies the (data-dependent) parameters which it is designed to estimate. Some inverse methods developed for traditional deterministic numerical models are able to estimate not only flow and transport parameters, including initial and forcing terms, but also geostatistical parameters such as variance and correlation scale (Carrera & Neuman, 1986; Samper & Neuman, 1986). It would be desirable for the proposed new generation of inverse methods, designed for conditional moment equations, to possess similar capabilities.

The aforementioned conditional moment theories due to Neuman & Orr (1993) and Neuman (1993b) are based on the assumption that all quantities are defined and measured on one consistent support scale in space-time. In reality, measurements and observations are often made on a variety of spatial and temporal scales at any given site. While the available theories allow integration of predicted quantities (outputs) in space-time so as to compare them with observations made on scales larger than the nominal support, they do not accommodate parameters and forcing terms (inputs) on any but the latter scale. Hence an important challenge for future research is to develop conditional moment equations which can account in a more complete way for data collected on multiple supports.

3 ANALYTICAL AND QUASIANALYTICAL STOCHASTIC MODELS OF FLOW AND TRANSPORT

Regardless of the extent to which the above challenges may eventually be met, it will always remain difficult to generalize the results of numerical studies, Monte Carlo or otherwise, in the way that is often possible with analytical theories. Hence among major challenges for the foreseeable future will remain the further development of analytical and quasianalytical solutions to various theoretical and real-world problems. By far the majority of such theories have so far focused on steady state water flow in unbounded, saturated porous media under a uniform mean hydraulic gradient, and on nonreactive advective–diffusive tracer transport superimposed on such flow. Randomness has been traditionally associated with a scalar hydraulic conductivity field $K(\mathbf{x})$, defined at each point $\mathbf{x}$ in a fictitious porous continuum, such that $Y(\mathbf{x})=\ln K(\mathbf{x})$ is multivariate Gaussian with constant ensemble mean $\langle Y \rangle$, variance σ_Y^2 and finite principal integral scales $\lambda_1, \lambda_2, \lambda_3$. This renders $Y(\mathbf{x})$ statistically homogenous and anisotropic with well-defined moments of all order (in σ_Y). In many, though not all, cases, $Y(\mathbf{x})$ has been taken to have an exponential (spatial or auto-) correlation structure.

Many interesting closed-form solutions have been obtained by linearization for unconditional flow and transport. Closure by linearization (to first-order in σ_Y^2) nominally restricts these results to mildly heterogeneous porous media in which $\sigma_Y^2<1$. Comparisons with Monte Carlo simulations suggest that some of these results may in fact constitute good approximations for cases where σ_Y^2 exceeds 2 by a small margin. Many of these results are summarized in two recent books by Dagan (1989) and Gelhar (1993); others have been published more recently. They include explicit expressions for effective hydraulic conductivity under statistical anisotropy; autocovariances or semivariograms of hydraulic head; auto- and cross-covariances of velocities; pre-asymptotic effective dispersivities as functions of mean plume travel distance (or residence time) and their asymptotic (Fickian) limits; the variance of concentration; and spatial as well as temporal plume moments.

Several large-scale, long-term natural gradient tracer experiments have been conducted in recent years, some of which seem to support linearized predictions about the way a plume spreads longitudinally under the above conditions. The question of whether transverse spread is correctly predicted by linearized theories is more difficult to ascertain on the basis of these experiments (Zhang & Neuman, 1990). Another unresolved issue which is presently debated in the literature concerns the role that local dispersion plays in the mixing or dilution of solute on the field-scale, and its impact on predictive uncertainty. Resolution of this issue is important because if Kapoor & Gelhar (1994) are correct that one must always consider local dispersivity no matter how small, then one would need to invest in much more detailed field sampling and higher-resolution modeling than might otherwise be required. Our own analysis suggests that local dispersion can be safely disregarded for many but local purposes when the corresponding dispersivity is small.

The range of linearized analytical solutions is being continuously expanded; to date, solutions have been published which account to various degrees for transient flow, boundary effects, nonuniform mean flows (primarily toward point and line sources), a drift in $Y(\mathbf{x})$, multiscale (including fractal) behavior of $Y(\mathbf{x})$, reactive transport and unsaturated flow. Such simplified solutions are prized for their relative mathematical simplicity and the theoretical insight they provide into flow and transport under uncertainty. The price of this simplicity and insight is a restriction on the range of real-world conditions to which the solutions apply.

A major challenge facing those who work on reactive transport is to insure that its mathematical description reflects correctly the physics and chemistry of the phenomenon, and that the spatial and temporal variation of the relevant parameters can be measured under realistic field conditions.

Though there have been numerous field tracer tests involving reactive chemicals, there have so far been very few opportunities to sample reliably the spatial and temporal distributions of parameters which control such reactions *in situ*. On the other hand, excellent and extensive field data have been generated in recent years which should prove useful in the study of unsaturated flow and transport in heterogeneous soils. There is a need to carry such experiments deeper underground and into fractured rock environments.

One way to extend the range of problems to which analytical stochastic solutions can be applied is to condition them on measured data. Rubin (1991a,b) and Rubin & Dagan (1992) condition first-order analytical velocity moments on hydraulic data via cokriging, generate corresponding Gaussian velocity fields by conditional Monte Carlo simulation, and track one or two particles through each simulated field. They then convert conditional one- and/or two-particle spatial moments into conditional ensemble concentration moments by assuming that the corresponding mean concentration is spatially Gaussian. Bellin, Rubin & Rinaldo (1994) likewise generate conditional velocity fields geostatistically by means of a first-order quasianalytic approach, but then track a cloud of particles across each field. The idea of conditioning first-order analytical velocity moments on hydraulic data via cokriging has also been employed by Zhang & Neuman (1995) in their analytical–numerical solution of the conditional transport equations described earlier. Quasianalytical methods of conditioning have been used earlier to solve the steady state inverse flow problem via conditional probability by Dagan (1985) and Rubin & Dagan (1987), and via cokriging by Hoeksema & Kitanidis (1984) and Kuiper (1986). There should be challenging opportunities to develop additional and more advanced methods of combining analytical and numerical methods to solve forward and inverse problems of flow and transport in heterogeneous media.

Another way to extend the range of problems to which analytical stochastic solutions can be applied is to include in them terms of order higher than σ_Y^2. Linearized expressions for effective hydraulic conductivity under uniform mean flow in bounded (Paleologos, Neuman & Tartakovsky, 1996) and unbounded (Gelhar & Axness, 1983) media have been extrapolated into the domain of large σ_Y^2 values based on the so-called Landau–Lifshitz conjecture. For statistically isotropic media, this conjecture has been validated by Monte Carlo studies up to at least $\sigma_Y^2=7$ (Neuman & Orr, 1993), and theoretically to second order (Dagan, 1993). A second-order analysis under conditions of statistical anisotropy suggests that the effective conductivity is mildly dependent on the shape of the $Y(\mathbf{x})$ correlation function, a fact not accounted for by the Landau–Lifshitz conjecture (Indelman & Abramovich, 1994). This notwithstanding, the extrapolated form appears to have independent field support from cross-hole hydraulic test data in fractured granites at Oracle, Arizona, where σ_Y^2 exceeds 7 (Neuman & Depner, 1988).

Second-order corrections to flow have recently been developed by Deng & Cushman (1995) and Hsu *et al.* (1996). Dagan (1994b) developed a second-order correction for pre-asymptotic and asymptotic transverse dispersivity which relies on a first-order solution of the flow problem; he found this correction to have virtually no impact on the dispersivity. Hsu *et al.* (1996) extended this analysis by accounting for second-order flow, and found it to have a significant impact on transverse, and a lesser impact on longitudinal, dispersion. Monte Carlo studies have so far been unable to reproduce this second-order effect. It is clear that major challenges lie ahead in extending stochastic analytical solutions to the domain of strongly heterogeneous media and in validating them numerically and experimentally.

Special methods of analysis need to be developed for nonlinear equations such as those which govern multiphase flow and transport. The local conditional moment equations developed by Neuman & Orr (1993) for steady state saturated flow should provide the basis for solving such problems deterministically without linearization. Chen & Neuman (1996) were successful in applying reliability theory to the establishment of criteria for the onset of instability and fingering during wetting front infiltration into a strongly heterogeneous soil. These and other nonlinear problems constitute a virtually unexplored frontier which should provide ample excitement for adventurous researchers well into the 21st century.

4 SUMMARY OF KEY POINTS

It has become common to treat the spatial variability of subsurface flow and transport parameters geostatistically. The geostatistical approach treats parameters determined at various points within a more-or-less distinct hydrogeologic unit as a sample from a random field, defined over a continuum. Stochastic theories assume that all requisite statistical properties of such random fields can be inferred from such data.

The number of flow and transport parameters that can be measured at any given site is usually limited. It is therefore important to augment them with other relevant hard and soft information. In recent years, there has been impressive progress in the development of geophysical imaging techniques which yield indirect information about the subsurface distributions of permeability, porosity and fluid or salt content. It is important to continue the development of such techniques and of quantitative methods which use geophysical images to help reduce hydrologic uncertainty.

Another promising approach which requires further development is the geostatistical reconstruction of lithology from limited data and its correlation with porosity and permeability. A more formidable challenge is to aid such reconstructions with well-founded geodynamic models of basin evolution.

It is generally not feasible to delineate the subsurface 'plumbing system' of fractured rocks by considering thousands of discontinuities and their individual flow and transport properties. A more feasible approach is to subdivide such rocks into relatively few distinct zones of low and high permeability and to treat each such zone either as a random continuum or as the superposition of several such continua. There is a need to conduct additional field studies in various fractured rock environments, on various scales and under diverse conditions of flow and transport, to further identify situations in which this premise does and does not hold. A major challenge is to develop well-founded methods of analysis for rocks which can be represented neither as juxtaposed nor as superimposed continua.

Geostatistical and stochastic theories deal with ensemble statistics, which cannot be inferred from data without invoking ergodicity. Our inability to test the ergodic hypothesis constitutes a philosophical dilemma which some find disturbing. Those who adhere to the existing conceptual framework face the challenge of demonstrating that it is consistent with observations. Those who reject it face the formidable challenge of supplanting the existing framework with an equally powerful alternative that is philosophically more satisfying.

Though hydraulic conductivity data often appear to fit a univariate lognormal probability distribution, there is no direct evidence that the log hydraulic conductivity $Y(\mathbf{x})$ is multivariate Gaussian. It is important to continue examining the consequences of this common multi-Gaussian hypothesis and to validate it.

There is growing evidence that the integral scale of $Y(\mathbf{x})$ increases consistently with the size of the domain under investigation. This suggests that stationarity may be an artifact of the scale of observation and method of inference rather than a true property of $Y(\mathbf{x})$.

New methods of inference have recently yielded power-law semivariograms of $Y(\mathbf{x})$ which suggest that it behaves locally as a random fractal on distance scales of at least up to 100 km. Though the power varies from site to site, the juxtaposition of data from many sites supports a generalized model with power $2\omega \approx 0.5$. To the extent that further research will continue revealing this or other generalized patterns of behavior, we will need to address the profound question of how the natural subsurface environment evolves toward such a high level of self-organization.

One may expect flow and transport in multiscale geologic media to show very different behavior than in statistically homogeneous media. There is a need to continue developing analytical and computational stochastic models that account for such behavior, and to validate them by field observation.

High-resolution (conditional) Monte Carlo simulation can be used to solve a broad range of linear and nonlinear stochastic flow and transport problems. It is, however, computationally demanding and prone to errors which are difficult to identify and control. Hence, the widely accepted practice of viewing Monte Carlo results as a standard against which to judge the accuracy of other approximate solutions (analytical or otherwise) entails a risk of misjudgment. Major challenges for the future lie in the development of accurate and efficient computational tools to conduct such simulations, and in the derivation of reliable criteria by which to ascertain the quality of their results.

A popular way to simulate (conditional) mean behavior efficiently is to use standard numerical models with equivalent or upscaled flow and transport parameters. There are at least two major conceptual difficulties with this approach. Hence, a major challenge for the future is either to place upscaling on a firmer theoretical footing or to continue developing alternative numerical methods directed toward the same goal.

A promising alternative to upscaling are numerical methods based on formal equations satisfied by the (conditional) moments of state variables such as head and concentration. The latter are often integro-differential equations with local and nonlocal parameters that are conditional on data and therefore nonunique. A major challenge is to estimate these space- and time-dependent parameters directly by approximation and/or indirectly by inverse methods. Another major challenge is to develop efficient computational schemes for the solution of nonlocal conditional moment equations.

Existing conditional moment theories are based on the assumption that all quantities are defined and measured on one consistent support scale in space-time. An important challenge for future research is to develop conditional moment equations that account for data collected on multiple supports.

Among major challenges for the foreseeable future will remain the further development, and experimental validation, of analytical and quasianalytical solutions to various theoretical and real-world problems. Most existing solutions are based on a first-order approximation which nominally restricts them to mildly heterogeneous media. While some predictions based on these solutions appear to have partial support in experiment, other remain unconfirmed and controversial. One issue which has been the subject of debate concerns the role that local dispersion plays in the mixing or dilution of solute on the field-scale, and its impact on predictive uncertainty.

A major challenge facing those who work on reactive transport is to insure that its mathematical description reflects correctly the physics and chemistry of the phenomenon, and that the spatial and temporal variation of the relevant parameters can be measured under realistic field conditions. Though there have been numerous field tracer tests involving reactive chemicals, there have so far been very few opportunities to reliably sample the spatial and temporal distributions of parameters which control such reactions *in situ*.

Excellent and extensive field data have been generated in recent years which should prove useful in the study of unsaturated flow and transport in heterogeneous soils. There is a need to carry such experiments deeper underground and into fractured rock environments.

One way to extend the range of problems to which analytical stochastic solutions can be applied is to condition them on measured data. There seem to be many opportunities for the further application of such quasianalytical models to a variety of forward and inverse flow and transport problems.

Second-order corrections to flow and transport are

presently making their debut in the literature. It appears that second-order flow has a significant impact on transverse dispersivity. Major challenges lie ahead in extending stochastic analytical solutions to the domain of strongly heterogeneous media and in validating them numerically and experimentally.

Special methods of analysis need to be developed for nonlinear equations such as those which govern multiphase-phase flow and transport. Conditional moment equations developed recently for steady state saturated flow should provide a basis for solving such problems deterministically without linearization. There has been recent success in applying reliability theory to the establishment of criteria for the onset of instability and fingering during wetting front infiltration into a strongly heterogeneous soil. These and other nonlinear problems constitute a virtually unexplored frontier which should provide ample excitement for adventurous researchers well into the 21st century.

ACKNOWLEDGMENT

This work was supported in part by the US Nuclear Regulatory Commission under contract number NRC-04-90-51.

REFERENCES

Ando, K. (1995). Continuum stochastic modeling of flow and transport in a crystalline rock mass. M.S. thesis, The University of Arizona, Tucson.

Batchelor, G. K. (1952). Diffusion in a field of homogeneous turbulence, II, The relative motion of particles. *Proceedings of the Cambridge Philosophical Society*, 48, 345–363.

Bellin, A., Rubin, Y. & Rinaldo, A. (1994). Eulerian-Lagrangian approach for modeling of flow and transport in heterogeneous geological formations. *Water Resources Research*, 30(11), 2913–2924.

Cacas, M. C., Ledoux, E., Marsily, G. de, Barbreau, A., Calmels, P., Gaillard, B. & Margritta, R. (1990a). Modelling fracture flow with a stochastic discrete fracture network: Calibration and validation: 1. The flow model. *Water Resources Research*, 26(3), 479–489.

Cacas, M. C., Ledoux, E., Marsily, G. de., Barbreau, A., Calmels, P., Gaillard, B. & Margritta, R. (1990b). Modelling fracture flow with a stochastic discrete fracture network: Calibration and validation: 2. The transport model. *Water Resources Research*, 26(3), 491–500.

Carrera, J. & Neuman, S. P. (1986). Estimation of aquifer parameters under transient and steady state conditions: 1. Maximum likelihood method incorporating prior information. *Water Resources Research*, 22(2), 199–210.

Carrera, J., Heredia, J., Vomvoris, S. & Hufschmied, P. (1990). Fracture flow modelling: Application of automatic calibration techniques to a small fractured monzonitic gneiss block. In *Hydrogeology of Low-Permeability Environments vol. 2*, eds. S. P. Neuman & I. Neretnieks. Hydrogeology Selected Papers. IAH, Hannover: Verlag Heinz Heise, pp. 115–167.

Chen, G. & Neuman, S. P. (1996). Wetting front instability in randomly stratified soils. *Physics of Fluids*, 8(2), 353–369.

Cushman, J. H., ed. (1990). *The Dynamics of Fluids in Hierarchical Porous Media*. San Diego, California: Academic Press.

Cushman, J. H. & Ginn, T. R. (1993). Non-local dispersion in media with continuously evolving scales of heterogeneity. *Transport in Porous Media*, 13(1), 123–138.

Cushman, J. H., Hu, B. X. & Deng, F.-W. (1995). Nonlocal reactive transport with physical and chemical heterogeneity: Localization errors. *Water Resources Research*, 31(9), 2219–2237.

Dagan, G. (1985). Stochastic modeling of groundwater flow by unconditional and conditional probabilities: The inverse problem. *Water Resources Research*, 21(1), 65–72.

Dagan, G. (1989). *Flow and Transport in Porous Formations*. New York: Springer-Verlag.

Dagan, G. (1993). Higher-order correction of effective conductivity of heterogeneous formations of lognormal conductivity distribution. *Transport in Porous Media*, 12, 279–290.

Dagan, G. (1994a). Upscaling of dispersion coefficients in transport through heterogeneous formations. In *Computational Methods in Water Resources X*, eds. A. Peters, G. Wittum, B. Herling, U. Meissner, C. A. Brebbia, W. G. Gray & G.F. Pinder. Dordrecht: Kluwer Academic Publishers, pp. 431–439.

Dagan, G. (1994b). An exact nonlinear correction to transverse macrodispersivity for transport in heterogeneous formations. *Water Resources Research*, 30(10), 2699–2705.

Dagan, G. & Neuman, S. P. (1991). Nonasymptotic behavior of a common Eulerian approximation for transport in random velocity fields. *Water Resources Research*, 27(12), 3249–3256.

Deng, F.-W. & Cushman, J. H. (1995). On higher-order corrections to the flow-velocity covariance tensor. *Water Resources Research*, 31(7), 1659–1672.

Deng, F.-W., Cushman, J. H. & Delleur, J. W. (1993). A fast Fourier transform stochastic analysis of the contaminant transport problem. *Water Resources Research*, 29(9), 3241–3247.

Gelhar, L. W. (1993). *Stochastic Subsurface Hydrology*. Englewood Cliffs, New Jersey: Prentice Hall.

Gelhar, L. W. & Axness, C. L. (1983). Three-dimensional stochastic analysis of macrodispersion in aquifers. *Water Resources Research*, 19(1), 161–180.

Gomez-Hernandez, J. J. & Wen, X.-H. (1994). To be or not to be MultiGaussian. Stanford Center for Reservoir Forecasting, Rep. 7, School of Earth Sciences, Stanford University.

Graham, W. & McLaughlin, D. (1989). Stochastic analysis of nonstationary subsurface solute transport: 2. Conditional moments. *Water Resources Research*, 25(11), 2331–2355.

Grindrod, P. & Impey, M. P. (1992). Fractal field simulations of tracer migration within the WIPP Culebra Dolomite. Rep. IM2856–1, Version 2, Intera Information Technologies, Denver, Colorado, March, 1992.

Guzman, A. G. (1997). Air permeability tests in unsaturated fractured tuffs at the Apache Leap Tuff Site near superior, Arizona. Ph.D. dissertation, The University of Arizona, Tucson.

Hoeksema, R. J. & Kitanidis, P. K. (1984). An application of the geostatistical approach to the inverse problem in two-dimensional groundwater modeling. *Water Resources Research*, 20(7), 1009–1020.

Hsieh, P. A. & Shapiro, A. M. (1993). Hydraulic characteristics of fractured bedrock underlying the FSE well field at the Mirror Lake site, Grafton County, New Hampshire. In *Field Trip Guidebook for Northeastern United States*, eds. J. T. Chaney & J. C. Hepburn. GSA, Annual Meeting, Boston, Massachusetts, October 25–28.

Hsu, K.-C., Zhang, D. & Neuman, S. P. (1996). Higher-order effects on flow and transport in randomly heterogeneous porous media. *Water Resources Research*, 32(3), 571–582.

Hyndman, D. W., Harris, J. M. & Gorelick, S. M. (1994). Coupled seismic and tracer test inversion for aquifer property characterization. *Water Resources Research*, 30(7), 1965–1977.

Indelman, P. & Abramovich, B. (1994). A higher-order approximation to effective conductivity in media of anisotropic random structure. *Water Resources Research*, 30(6), 1857–1864.

Indelman, P. & Dagan, G. (1993). Upscaling of permeability of anisotropic heterogeneous formations: 1. The general framework. *Water Resources Research*, 29(4), 917–923.

Kapoor, V. & Gelhar, L. W. (1994). Transport in three-dimensionally heterogeneous aquifers: 2. Predictions and observations of concentration fluctuations. *Water Resources Research*, 30(6), 1789–1801.

Koltermann, C. E. & Gorelick, S. M. (1992). Paleoclimatic signature in terrestrial flood deposits. *Science*, 256, 1775–1782.

Kostner, A. (1993). Geostatistical and numerical analysis of flow in a crystalline rock mass. M.S. thesis, The University of Arizona, Tucson.

Kuiper, L. K. (1986). A comparison of several methods for the solution of the inverse problem in two-dimensional steady state groundwater flow modeling. *Water Resources Research*, 22(5), 705–714.

Li, S.-G. & McLaughlin, D. A. (1991). Nonstationary spectral method for solving stochastic groundwater problems: Unconditional analysis. *Water Resources Research*, 27(7), 1589–1605.

Neuman, S. P. (1987). Stochastic continuum representation of fractured rock permeability as an alternative to the REV and fracture network concepts. In *Rock Mechanics: Proceedings of 28th U.S. Symposium*, eds. I. W. Farmer, J. J. K. Daemen, C. S. Desai, C. E. Glass & S. P. Neuman. Rotterdam: A. A. Balkema, pp. 533–561.

Neuman, S. P. (1988). A proposed conceptual framework and methodology for investigating flow and transport in Swedish crystalline rocks, Arbetsrapport 88–37. SKB Swedish Nuclear Fuel and Waste Management Co., Stockholm, September 1988.

Neuman, S. P. (1990). Universal scaling of hydraulic conductivities and dispersivities in geologic media. *Water Resources Research*, 26(8), 1749–1758.

Neuman, S. P. (1991). Reply to Comment by M. P. Anderson. *Water Resources Research*, 27(6), 1381–1382.

Neuman, S. P. (1993a). Eulerian-Lagrangian theory of transport in space-time nonstationary velocity fields: Exact nonlocal formalism by conditional moments and weak approximation. *Water Resources Research*, 29(3), 633–645.

Neuman, S. P. (1993b). Comment on 'A Critical Review of Data on Field-Scale Dispersion in Aquifers' by L. W. Gelhar, C. Welty & K. R. Rehfeldt. *Water Resources Research*, 29(6), 1863–1865.

Neuman, S. P. (1994). Generalized scaling of permeabilities: Validation and effect of support scale. *Geophysical Research Letters*, 21(5), 349–352.

Neuman, S. P. (1995). On advective dispersion in fractal velocity and permeability fields. *Water Resources Research*, 31(6), 1455–1460.

Neuman, S. P. & Depner, J. S. (1988). Use of variable-scale pressure test data to estimate the log hydraulic conductivity covariance and dispersivity of fractured granites near Oracle, Arizona. *Journal of Hydrology*, 102(1–4), 475–501.

Neuman, S. P. & Orr, S. (1993). Prediction of steady state flow in nonuniform geologic media by conditional moments: Exact nonlocal formalism, effective conductivities and weak approximation. *Water Resources Research*, 29(2), 341–364.

Paleologos, E. K., Neuman, S. P. & Tartakovsky, D. (1996). Effective hydraulic conductivity of bounded, strongly heterogeneous porous media. *Water Resources Research*, 32(5), 1333–1341.

Ramirez, A. (1986). Recent experiments using geophysical tomography in fractured granite. *Proceedings of the IEEE*, 74(2), 347–352.

Rossini, C., Brega, F., Piro, L., Rovellini, M. & Spotti, G. (1994). Combined geostatistical and dynamic simulations for developing a reservoir management strategy: A case study. *Journal of Petroleum Technology*, 46(3), 979–985.

Rubin, Y. (1991a). Prediction of tracer plume migration in disordered porous media by the method of conditional probabilities. *Water Resources Research*, 27(6), 1291–1308.

Rubin, Y. (1991b). The spatial and temporal moments of tracer concentration in disordered porous media. *Water Resources Research*, 27(11), 2845–2854.

Rubin, Y. & Dagan, G. (1987). Stochastic identification of transmissivity and effective recharge in steady groundwater flow: 1. Theory. *Water Resources Research*, 23(7), 1185–1192.

Rubin, Y. & Dagan, G. (1992). Conditional estimation of solute travel time in heterogeneous formations: Impact of transmissivity measurements. *Water Resources Research*, 28(4), 1033–1040.

Rubin, Y., Mavko, G. & Harris, J. (1992). Mapping permeability in heterogeneous aquifers using hydrologic and seismic data. *Water Resources Research*, 28(7), 1809–1816.

Samper, F. J. & Neuman, S. P. (1986). Adjoint state equations for advective-dispersive transport. In *Finite Elements in Water Resources, Proceedings of the 6th International Conference*, eds. A. Sa da Costa, A.M. Baptista, W.G. Gray, C.A. Brebbia & G.F. Pinder. Southampton/Berlin: Computational Mechanics Publications/Springer Verlag, pp. 423–438.

Samper, F. J. & Neuman, S. P. (1989). Estimation of spatial covariance structures by adjoint state maximum likelihood cross-validation. *Water Resources Research*, 25(3), 351–362.

SKB (1993). Annual Report 1992, SKB Tech. Rep. 92–46, Swedish Nuclear Fuel and Waste Management Co., Stockholm, Sweden, May 1993.

Tsang, C.-F. & Neuman, S. P. (1997). Introduction and general comments on INTRAVAL Phase 2, Working Group 2, Test Cases. In *The International Intraval Project, Phase 2*, Paris: NEA/OECD, in press.

Zhang, Y.-K. & Neuman, S. P. (1990). A quasilinear theory of non-Fickian and Fickian subsurface dispersion: 2. Application to anisotropic media and the Borden site. *Water Resources Research*, 26(5), 903–913.

Zhang, D. & Neuman, S. P. (1995). Eulerian-Lagrangian analysis of transport conditioned on hydraulic data: 1. Analytical-numerical approach. *Water Resources Research*, 31(1), 39–51.